AF352682

The Waning of the Mediterranean, 1550–1870

The Waning of the Mediterranean,
1550–1870

A Geohistorical Approach

FARUK TABAK

The Johns Hopkins University Press
Baltimore

The Johns Hopkins University Press
2715 North Charles Street
Baltimore, Maryland 21218-4363
www.press.jhu.edu

Library of Congress Cataloging-in-Publication Data

Tabak, Faruk.
The waning of the Mediterranean, 1550–1870 : a geohistorical approach / Faruk Tabak.
p. cm.
Includes bibliographical references and index.
ISBN-13: 978-0-8018-8720-8 (hardcover : alk. paper)
ISBN-10: 0-8018-8720-8 (hardcover : alk. paper)
1. Human geography—Mediterranean Region—History. 2. Human ecology—
Mediterranean Region—History. 3. Mediterranean Region—History. 4. Mediterranean
Region—Environmental conditions. I. Title.
GF541.T23 2007
909'.0982205—dc22   2007020364

A catalog record for this book is available from the British Library.

CONTENTS

Acknowledgments   *vii*

Introduction: *Unrelieved Weight of Wealth in the Inner Sea*   1

PART I   OF CITIES OF SAINTS AND RICH TRADES

1 Empires and Empire-Building City-States   33

2 City-States and the Inner Sea   84

3 Eclipse of the City-States and the Resurfacing of the Mediterranean   134

PART II   OF MALARIAL PLAINS AND ARBOREAL HILLS

4 Reversal in the Fortunes of the Plains   189

5 New World of the Hills   242

Conclusion: *The Mediterranean between the Leek-Green Sea and the Green Sea*   299

Notes   *309*

Bibliography   *369*

Index   *417*

"All that is finished, let it fade," said Yates, but not before the author acknowledges the immense debt he incurred in completing this book. Immanuel Wallerstein, Terence K. Hopkins, and Çağlar Keyder read and commented on an earlier version. Their support and guidance has been central not only for this book, but more generally in my intellectual formation. I have benefited greatly from Ravi Palat's comments, who read and reread the manuscript more than he would have liked. At Georgetown I had access to John R. McNeill's intimate and immense knowledge of Mediterranean and environmental history. The work has gained immensely from his thorough critique. Scott Redford's comments were of great value in revising the manuscript for the Johns Hopkins University Press. Tim Beach provided perceptive suggestions for its improvement. All three generously devoted their time to read a colleague's work in entirety. I thank them for their generosity and kindness. Special thanks are due to Henry Tom, of the Press, who has been unfailingly supportive of the project from its beginning.

The Waning of the Mediterranean, 1550–1870

Introduction

Unrelieved Weight of Wealth in the Inner Sea

From the fourteenth century to the turn of the sixteenth, the Mediterranean was a world unto itself, a world-economy. The economic fabric of this world was initially woven by city-states strung along the northern shores of the Italian peninsula, and it was under their aegis that "the whole sea shared a common destiny, with identical problems and general trends if not identical consequences."[1] Thus was the history of the Mediterranean in its heyday, from the sugarcane fields in Madeira to the spice emporium in Alexandria and from the silver mines of Bohemia to the gold mines of Takrūr. In the historiography of the region, this shared unitary history rapidly fades into obscurity past the region's prime—past the early seventeenth century, to be precise. For most scholars, it was then that the steady shift in the terminus of the spice trade from the Levant to the ports on the Atlantic and the North Sea was finally sealed to the benefit of the latter.[2] The conventional designation of the close of the sixteenth century as the onset of the erosion of the Mediterranean's gilded age has not been without its ramifications, however. Rare if not outright extinct, as a result, have become ecumenical histories of the Inner Sea in studies chronicling developments from the turn of the seventeenth century. In a perverse fashion, the very paucity of holistic accounts of the basin has eventually turned into one of the distinguishing features of its age of twilight—in stark contrast, of course, with the depictions of its heyday, sketched almost habitually with broad idiographic, colorful strokes. This volte-face in the historiography of the Inner Sea away from holistic accounts to singular and sectorial histories can therefore be attributed to the underlying assumptions that the forces that had fostered unity along the Mediterranean had, in its autumn, lost their coherence and that the destiny shared by polities in the basin at the zenith of its power had consequently ceased to be common. Hence the divergent and solitary destinies of its denizens from the first half of the seventeenth century, if not earlier.[3]

Definitely, the Mediterranean had, by the opening decades of the sixteenth century, started to lose its centrality in world-economic flows when the union of the kingdoms of Aragón and Castile had led to a gradual but steady re-anchoring of the Habsburg empire in the Atlantic, under Charles V in particular. Especially after 1563, when the Manila galleons started to traverse the Pacific on a regular basis, Philip II's empire became veritably globe spanning.[4] In the process, Venice's reign was gravely weakened by the shift in the center of gravity of world-economic flows away from the Mediterranean, the *Mare Nostrum* of the city of St. Mark. The Venetian, and to a lesser degree the Genoese, colonial empires suffered dismemberment contemporaneously with, but not necessarily as a result of, the relentless westerly march of the house of Osman. For one, the port-cities of the western Mediterranean, from Tripoli to Oran, fell to the Ottomans only after the region was, for all practical purposes, vacated by the Genoese merchants, first, in favor of the Atlantic and, later, in favor of royal finance at the employ of the Castilian throne. By the closing decades of the sixteenth century, the Sublime Porte itself had already turned east to contain Shah Abbas and secure the flow of the silk trade through its imperial dominions.[5] By and by, the Mediterranean was fast turning into a millpond as the Indian Ocean was penetrated by Portuguese galleys and the Atlantic connected by trunk lines. Correlatively, the unity of the Mediterranean basin was being inescapably and irreparably fractured much to the chagrin of its august custodian, the Serene Republic. The *Mare Internum* was thus falling prey to subversive forces operating from within and without, and mostly beyond the control of the Serenissima.

With the change in oceans, mapping the spectacular geographical expansion of the world-economy in the long sixteenth century along the shores of the Atlantic—the all-encompassing ocean, *al-Bahr al-Muhit*—has understandably become more appealing a historiographical task than charting the unmaking of the Mediterranean. After all, the seventeenth century in the Mediterranean was a period of "dismal depression and sordid melancholy."[6] And the light that lit its skies in the first half of the century got even dimmer after 1657, when the age of the Genoese, *el siglo de los Genoveses,* which had been inaugurated in the 1550s, came to a terminal end. The control the bankers from the city of St. George had established over the world-economy's precious metal flows started to erode at a relatively rapid pace. This was because the shift in the terminus of silver arriving from the American mines, from Seville to Amsterdam, undermined the reign of Genoese banker-financiers, who were promptly sidelined by the Portuguese New Christians, *marranos*, "the front men for the merchants of Amsterdam."[7]

Ironically, the Genoese bankers, even before they were outmaneuvered by the Dutch or their front men, actively contributed to the shift in direction of world-economic flows from the shores of the Italian peninsula to those of the North Sea. For what abetted this northward shift was the alacrity with which the bankers from the city of St. George assumed the responsibility in 1557 of administering monetary flows radiating out of Seville, the commercial nerve center of Habsburg Spain, into Milan and Antwerp, the empire's military outposts watching over the Mediterranean and the North Sea, respectively. The privilege the Catholic king bestowed on the Genoese bankers provided these men of money with access to the incoming American silver, a privilege given with the expectation that the bankers convert the Castilian crown's vast yet irregular sources of income into regular streams of revenue.

In the process, Genoese bankers' masterful orchestration of bullion and financial flows across and beyond the Habsburg realm momentarily reincarnated the splendors of the Mediterranean by reviving the spice trade in the latter half of the sixteenth century. In no small degree, this revival was owed to the Genoese banker-financiers' ability to make payments in gold, on account of their sizeable reserves, in Philip II's name in Antwerp and Milan. This invaluable service rendered to the Prudent King and the Castilian crown allowed Genoese bankers to lay claim over the fabled cargoes of American silver.[8] They, in turn, unloaded this relatively bounteous and hence inferior metal in the ports of the Mediterranean.[9] The influx of American silver into the Mediterranean via Genoese Seville from the mid-sixteenth century prompted the return of the spice trade back to its centuries-old Levantine route. The abundance of the white metal in the Mediterranean served to restore the spice trade, given the soaring demand for silver in the Indian Ocean and Ming China, as attested by the introduction of a single-whip tax that had to be paid in silver.[10] The establishment of Ottoman naval presence in the Red Sea and the Indian Ocean, however feeble, reflected and abetted the revival of the spice trade.[11]

This latter-day prosperity notwithstanding, the apportionment of the relatively scarce and precious gold and the abundant and lowly silver, respectively, between the ports of the North Sea and the Mediterranean subtly testifies to a change in the fortunes of the two seas long before the turn of the seventeenth century. And the stability generated by adherence to gold in the minting of money in the north stood in striking contrast to the uncertainties precipitated by the violent price fluctuations along the banks of the Mediterranean, caused by the ceaseless inundation of silver monies. Certainly, return the spice trade did back to its time-honored Levantine route in the latter half of the sixteenth century following the arrival of American silver.[12] But the economic boost generated by the revival of the Levantine route proved short-lived at best, and failed to fully restore the status quo of the fifteenth century that had endowed the Inner Sea exclusively with the prosperity of the rich spice trade. On the contrary, it tilted the balance farther away from the Mediterranean and toward the North and the Baltic seas.[13]

The solidity and longevity of this northward shift is best captured by the diverging trajectories of northerly Antwerp and southerly Lyon during the seventeenth century. In the first half of the sixteenth century, Antwerp was the most prominent financial center on the North Sea, and Lyon on the Mediterranean. Both cities, along with Medina del Campo of Castile, suffered a drastic reversal in their fortunes from the late 1550s.[14] All three were dethroned, almost simultaneously, when the Genoese bankers captured the envied role of bankers to the Spanish throne from the Fuggers, and became the exclusive handlers of the Potosí silver—and thereby of the world-economy's wealth. Following the fall of the house of Fugger, Antwerp had the most to lose from the assumption by the Genoese *nobili* of the position of bankers to the Catholic king. Philip II's grant of Spanish contracts, *asientos*, to the men of money from the city of St. George furnished them from 1557 with exclusive command over Seville's silver treasure and diminished Antwerp's control over the world-economy's precious metal flows. This, expectedly, deprived Antwerp's financiers of the pivotal position they had occupied in the first half of the sixteenth century. All the same, the city resurfaced from the 1560s, this time as an important industrial center and a financial intermediary, and remained so for a considerable length of time.[15]

Lyon, by contrast, hit equally hard by the departure from its fairs of the Genoese bankers, failed to exhibit a similar resilience. As the Genoese *nobili* acquired the power to single-handedly determine the location of money markets north and south of the Alps, the Lyon fairs were supplanted by those of Lons-le-Saulnier, Montluel, Chambéry, and finally Piacenza in 1579.[16] The rise of competing money markets under the Genoese bankers' long shadow paved the way for Lyon's imminent and relatively prolonged eclipse—and that of the Midi along with it.[17] Given its geographical proximity to Genoa, the city no doubt remained an economic center, yet it had to resign itself to perform roles of lesser significance than previously. In the seventeenth century it survived merely as a regional and supplemental money market: the radius of its bankers' financial transactions shrank precipitously to barely blanket the city's immediate hinterland.[18] In the final analysis, Lyon was a financial center subject to the fluctuations in the fortunes of the Italian city-states and not, like Antwerp, to that of the vibrant Amsterdam-centered world-economy.

The transformations set in motion during the age of the Genoese were not merely confined to the realm of finance, however. They were more comprehensive than that. For one, the growing sway of the Genoese haute finance after it was granted the Spanish *asientos* bestowed upon the city's men of money the opportunity to withdraw from commerce at the zenith of their economic might.[19] The diminishing presence of the merchants of St. George in the trading worlds of the Atlantic and the western Mediterranean rendered its world-economic presence primarily financial, and thereby ethereal. The metamorphosis in the nature of Genoese rule, in turn, refashioned the commercial networks of the basin in two ways. On the one hand, Genoese bankers' dexterity in handling the American treasure served to kindle a revival of the spice trade. As long as the rich trades remained buoyant, thanks to the ethereal control the Genoese bankers exercised over bullion flows, the destinies of the corporeal denizens of the sea, that of the Signoria and the "Grand Turk," remained welded to one another at least until the opening decades of the seventeenth century.[20] For they both benefited from the resumption of the spice trade, the Ottoman dominions as its main conduit and Venice as its central point of disembarkation and redistribution.[21] In other words, the Genoese *nobili*'s specialization in finance under the leadership of the Grimaldi, Pinelli, Lomellini, Spinola, and Doria families proved, in the short run, to be a boon for the Levant and, by association, for the Venetian trade.

On the other hand, the lacunae left behind by the egress from commerce of the merchants of St. George facilitated the arrival of northern ships, first, into the western Mediterranean, the Genoese merchants' principal locus operandi, and later into the Levant. In the eastern Mediterranean, Ragusan ships took over the task of carrying bulky commodities—grain, salt, and wool—that the Genoese traders controlled prior to their withdrawal from trade.[22] These twin processes gravely undermined the Venetian merchants' dominion in the Levant from the late sixteenth century.[23] The reverberations of the establishment of Genoese financiers' discreet rule over the Atlantic Mediterranean were thus inevitably felt by the basin's empires, city-states, and territorial states alike—as were the aftershocks of its fall. As Genoese paramountcy started exhibiting signs of enfeeblement and erosion from the first half of the seventeenth century, so did not only, if not unexpectedly, the house of Habsburg under Philip IV and

Charles II—the royal dispensers of *asientos*. The house of Osman and the Signoria, whose fortunes were closely interwoven on account of the restoration of the spice trade, suffered concordantly.[24] It was then with the onset of the Cretan crisis in 1654 that the long period of peace between the Ottoman empire and the Venetian Republic that had started in the 1570s came to an end, and the island became an Ottoman possession in 1669.[25] The tremors propagated by the passing of the age of the Genoese and the attendant contraction in economic activity were therefore experienced in all corners of the Mediterranean, notwithstanding the arrival and growing presence of northern ships in the basin's fold.

True, the rise and consolidation of Dutch hegemony strengthened and sealed the North Sea—and the Baltic Sea—bound shift in the center of accumulation of the capitalist world-economy as confirmed by the fact that the Genoese *nobili vecchi* presided principally over the affairs of the Atlantic and the North Sea, and only secondarily over that of the Mediterranean. And the revision in the world-economy's division of labor following the great expansion of the long sixteenth century, which spanned from 1450 to 1650, inescapably undermined the basin's relative position vis-à-vis the north. But the invasion of the Mediterranean by the ships of the Atlantic in the 1590s failed to turn it into a northern lake all at once. Fortuitously for the Republic of St. Mark, the commercial nerve center of the *Mare Internum*, the trading networks put in place by its competitors in the twilight of its power lacked longevity. For example, both the Armenian merchants who presided over the Safavid silk trade and established their presence in all principal commercial centers of the time, and the Jewish merchants who became important players in the Levant trade and the continental fairs, presided over the trading world of the region only for relatively short periods of time.[26] The French, and the British merchants in particular, ruled the waves of the Mediterranean episodically: the sudden rise of the port-city of Livorno during the late sixteenth century as the warehouse of northern merchants did not culminate in its commercial supremacy in the following centuries.[27] Nor did the Levant Company manage to establish the kind of imposing presence the East India Companies did in the Indian Ocean.

The relatively narrow compass of the operations of emerging or rival merchant fleets and the brevity of their presence in the Mediterranean ports of call almost guaranteed that the Serenissima would not be unceremoniously dislodged from its millpond. The Venetian commercial fleet, still sizeable in the seventeenth and eighteenth centuries, continued to ply the waters of the Mediterranean. The Genoese men of money, former *hombres de negocios,* were forced to take up residence in the Mediterranean as their presence in the Castilian court, especially under Philip IV and the count-duke of Olivares, failed to generate the fabled riches of the sixteenth century.[28] At times, they even served as traders, a vocation they had largely disdained from the mid-sixteenth century.[29] And as to be expected, they went on to function as financiers to the states in the region, giving the Mediterranean economy a big stimulus through their subscription to state loans, from Austria to France. Equally significant, both city-states played key roles in the industrial expansion of the seventeenth and eighteenth centuries. Following in the footsteps of Venetian industrialization, which had its heyday in the latter half of the sixteenth and the opening decades of the seventeenth centuries, the Genoese bankers, who were forced to resume their affairs in the Mediterranean following the reversal in the fortunes of the Spanish throne, started investing, in part, in manufactures. They crafted their wares according to the demands of their former headquarters, Cádiz and Lisbon, indisputably to obtain gold from the latter and silver from the former.[30]

In fact, the contrasting yet complementary fortunes of the twin republics of Venice and Genoa during the second sixteenth century, which stretched from the 1550s to the 1650s, invested the Mediterranean with remarkable resilience at the twilight of its power. The growing role of manufacturing in Venetian economic life during the hiatus in the Levant spice trade in the first half of the sixteenth century on the one hand, and the withdrawal of Genoese merchants from commerce from 1557 and their subsequent specialization in handling the wealth of the world-economy on the other, provided a double stimuli to the Mediterranean trade. The first stimulus was given by an acceleration in the tempo of economic activity that flowed from the industrialization drive of the second sixteenth century which, in the Mediterranean, was spearheaded by Venice, only to be later joined by Genoa, Antwerp, Geneva, and Lyon.[31] The second stimulus was furnished by the easterly flow of Seville-bound Potosí silver, safeguarded by Genoese bankers. In concert, the steady infusion of silver and the increasing salience of manufacturing in Venetia and Genoa undergirded and bolstered general economic activity in the basin. They served to diversify the range of goods that crisscrossed, and created new multilateral dependencies across the shores of the *Mare Internum*.

Concurrently, these two developments set the tune and pace of economic change in the Inner Sea, granting it the opportunity to make the transition to its new, albeit diminished, status in a relatively smooth and leisurely manner. To a great extent, then, the afterglow of the Mediterranean was underwritten by Genoese haute finance, and lasted well into the late seventeenth century as these bankers went on to participate in Spanish contracts long after the Castilian throne's bankruptcy in 1627.[32] However dim the lights of this "afterglow" were and however sparse its "autumnal fruits,"[33] the age of the Genoese reshaped the basin's common destiny rather than undid it. This common destiny did not come immediately unglued when the spice trade changed its itinerary irreversibly from the turn of the seventeenth century and reached the North Sea by way of Lisbon, Antwerp, and later Amsterdam, and even when subsequently the flow of American silver was diverted away from the Mediterranean.[34] Yet, the reflexive assumption in the region's historiography that, with the erosion of the power of the city-states and the loss of the spice trade, the Mediterranean unavoidably lost its economic coherence tends to confine holistic analyses of the basin exclusively to its heyday. The premise of lost, fractured economic unity, assumed to have set in past its prime, sanctions in turn the singular and individual histories of the polities that were quartered along the basin, be they city-states or empires.

This is why the histories neither of the city-states that towered over the Mediterranean, the cities of St. Mark and St. George, nor that of the empires that boxed it in, the Habsburg and Ottoman empires, have been immune to this particularistic reading. In historical studies that investigate the waning of the Mediterranean, the ecumenical setting of the golden age of the basin fades into the background, only to be supplanted by differential and singular settings from the seventeenth century.[35] Paradoxically, the common thread running through these disparate accounts is that they all shoulder, singly and severally, the task of chronicling the passing of the region's golden age, albeit from different vantage points, as the specter of the Mediterranean's resplendent past casts a long shadow over its history in the ages of baroque and enlightenment. That the Inner Sea ceased from the early decades of the seventeenth century to be the integrative force it had once been remains a widely shared assumption in the historiography of the region, with the notable exception of that of France, which, not coincidentally, was as much an inhabitant of the North Sea as it was of the Mediterranean, emblematic of which was the long-running rivalry between Lyon and Paris. Thanks to the Rhône corridor, the Midi was, and still is, not segregated from temperate France.[36]

The portrayal of the economic fabric of the Mediterranean as lacerated and torn after it lost its luster thus does not fit into the mold cast by Braudel in his depiction of the basin at its height. In all, his analysis lays down in detail the foundations on which the historical unity, physical and human, of the lands enveloping it were built. Since this unity is considered to have collapsed with the rise of the Baltic and the Atlantic, his line of inquiry has therefore not been fully embraced by the students of the region in their analyses of the developments that reshaped it in the seventeenth and eighteenth centuries. Without doubt, the long and drawn-out wars fought at the turn of the sixteenth century signaled with certainty that the empires enveloping the Mediterranean had all turned their backs on it. Excepting the Twelve Years' Truce between 1609 and 1621, the Castile-based Habsburg empire fought its wars on the shores of the North Sea. The theater of action for the Sublime Porte was not in the Mediterranean but on the empire's eastern front: it was against the Safavid empire that the military campaigns were organized in 1578–90 and 1603–39 to secure the northern-bound flow of the silk trade.[37] In France, the wars of religion were concentrated for the most part in its northern provinces.

Then again, even the age of the Genoese that prompted the revival of the Mediterranean after the 1550s owed its prosperity, as emphasized above, to the ability of its merchants and bankers to preside over the northerly and easterly flow of precious metals bound for Seville from the Americas. The regular dispatch of gold to the Spanish Netherlands by the Genoese bankers rendered possible the easterly flow of Potosí silver to the Levant. Stated differently, it was the vibrancy of the Baltic and the North seas that stimulated economic activity along the banks of the Mediterranean even during the latter half of the sixteenth century. Whatever its provenance, this generalized economic activity, relayed to the Inner Sea by the Genoese bankers, did not fail to provide a considerable degree of coherence to it.[38]

Since this coherence, revamped though it was after the sixteenth century, is overlooked in many a historical account, the analytical terrain charted by Braudel has not served as an especially fertile ground for studies mapping the "collective destinies" and "general trends" of the Mediterranean during the ages of baroque and reason. "Sectorial" histories that have come to supplant holistic accounts of the basin underplay, by design, the general trends of the seventeenth and eighteenth centuries, and portray the basin as subject to the ebb and flow of local, singular forces. The Spanish empire's silver and copper ages that followed its *Siglo de Oro*, the Italian city-states' dwindling fortunes, or the withering away of the Ottoman empire's classical age are all accounted for with reference to a set of sui generis factors, not necessarily shaped by forces spanning the Mediterranean as a whole. It is as if the Mediterranean was torn asunder by the tempestuous gales of the Atlantic and what was left behind was merely the debris of the Inner Sea. In this frame of reference, the insularity of "regional" histories is considered to have been brought to an end by the arrival of the steamship in the Mediterranean in the mid-nineteenth century, which is credited with unleashing forces of change that brought economic dynamism back to the basin.[39] Furthermore, the loss by Bourbon Spain of its American empire and the changes wrought by the Napoleonic era in the Iberian and Italian peninsulas neatly separate the histories of these two peninsulas in the nineteenth century from their histories in the preceding two centuries.[40]

Yet, reasons for the fractured state of Mediterranean history do not necessarily stem from the insular nature of the individual historiographies that constitute it. Rather, they lie in the field of Mediterranean studies at large, however defined. If the Mediterranean "lived and breathed with the same rhythms," then the relatively marginal role attached to these rhythms by the historians of the region in their analyses of its eclipse should be brought under scrutiny so that the path "individual" historiographies should or could take can be determined correspondingly.[41] In other words, the task of recovering these centuries from obscurity should start by restoring the unity of the Mediterranean and the role Genoa and Venice played in unison in preserving that unity and in plotting the "general trends" and "collective destinies" that this unity implied rather than assume that the basin's economic structures became subject to inexorable disintegration.

So, if the buoyancy of the Baltic and the North seas revitalized economic activity on the shores of the Mediterranean, and if the Genoese bankers' ability to preside over the bullion flows rerouted via Seville generated a synergy between these two aquatic worlds, this was because, despite the afterglow of the Mediterranean, the latter half of the sixteenth century was shaped first and foremost by the rise of *pax Neerlandica*. The northerly shift in the center of gravity of the world-economy was neither speedy nor sudden.[42] But the emphasis on the revival of the Levant spice trade, by keeping the spotlight on the Mediterranean from the 1550s, tends to underplay this major sea change and, more significant, the revision this shift brought about in the division of labor in the Greater Mediterranean. Navigating in the dimming lights of the Mediterranean when the lighthouse on the Baltic was in full glow has its drawbacks.

General Trends and Common Destinies

The septentrional shift in the center of the capitalist world-economy away from the northern coasts of the Italian peninsula cannot simply be reduced to the two reigning merchant republics' *diminutio capitis*. Indeed, it was not by accident or chance that the three principal processes, which spanned the Mediterranean as a whole and provided it with a common destiny in its autumnal age, were all initiated by the Venetian and Genoese merchants and financiers to strengthen and perpetuate their dominion over the Inner Sea. The first of these processes was the steady westerly relocation, owing to the labors of the Venetian and Genoese merchants, of the lucrative oriental crops, such as sugar and cotton.

The economic devolution of the Inner Sea was later compounded by the northerly relocation of commercial grain production, this time owing to the labors of the Dutch merchants.[43] This double movement resulted in a concomitant contraction in the cultivation of tropical and bread crops, first, on the eastern and, later, on the western flanks of the Mediterranean.[44] The transformation of Baltic grain trade into the mother trade, *Moedernegotie,* of Dutch merchants and the preeminent role assumed by these merchants in the production and circulation of tropical crops (and spices) gave this double movement a long lease on life.[45]

The departure of sugar and the transformation of cotton from plantation crop to cash crop were followed by the growing popularity of the Mediterranean tree crops. Surely, thriving olive oil (and grain) trade was captured by Marseille, yet Genoa not only was as central a port as was Marseille in these trades in the seventeenth and eighteenth centuries. But also the financiers of the city of St. George had a considerable stake in Marseille's fortunes. Wine trade continued to thrive on the western flanks of the Mediterranean. Gascony, southern Iberia, Duoro valley, Languedoc, along with Tenerife and Madeira, dispatched their wines to the British isles. Yet there was also a flourishing trade in the east, via Burgundy, catering not only to the Dutch but to the nouveaux riches of the region that stretched from the Baltic to the Black Sea.[46] In this trade, Venetian merchants played a significant role.[47] The commercial crops of the Mediterranean's heyday may have migrated elsewhere, but the resuscitation of the region's indigenous crops opened up new avenues of commercial activity for the twin city-states, at least initially.

If the first general trend pertained to the realm of commercial agriculture and, more specifically, to its contraction and the subsequent resurgence of tree crops, the second process that underlay the unity of the Mediterranean involved changes within the field of manufacturing. This involved the percolation of manufacturing, primarily of textile industries, into the countryside from the turn of the seventeenth century due to the fierce competition the Venetian woolen industries encountered from the north. Yet, unlike the late fourteenth- and fifteenth-century relocation of fustian manufacturing across the Alps in southern Germany, which depended on the availability of raw cotton from the Levant, this wave of ruralization possessed a longevity that the former wave sorely lacked. This was largely because the principal raw material, wool, unlike cotton and silk in the earlier epoch, was easily available to producers even in the most far-flung corners in north of the Brenner Pass. That the textile industries in the north, too, resorted to a similar strategy of putting-out most, if not all, stages of production to rural regions or alternatively of placing new textile industries outside existing corporate structures—even in urban settings—extended the historical longevity of the process into the late eighteenth century.[48]

Indeed, it was the dissemination of manufacturing north of the Alps in the seventeenth and eighteenth centuries that demanded the services, mercantile and financial, of the imperial cupola of the basin's golden age: Venice and Genoa. Venetian merchants were to be found in the region's fairs and cities, delivering fabric to Leipzig and Cracow, from Florence, Milan, and "even Venice."[49] The men of money from the city of St. George, as well, placed their accumulated wealth in the service of the city's manufacturing industries, located mostly in Genoa's immediate vicinity—Gavi, Serravalle, Busalla, and Voltaggio—especially during the seventeenth century.[50] Moreover, the growing financial requirements of the states north and east of the Brenner Pass and Pontebba pursuant to the change in the direction and nature of economic flows were partly met by the Genoese men of money. Austrian Lombardy profited from the return of the Genoese capital back to the Mediterranean, not to mention of course France, Lombardy, and Venice.

The third and final process was the reorganization of the locus operandi of the Venetian merchants and Genoese bankers from the seventeenth century in accordance with the diminishing significance of maritime trade in the Mediterranean for the former and with the loss of the latter's position as bankers to the Spanish throne. In the eastern Mediterranean where Venetian merchants enjoyed unrivaled maritime supremacy—as the masters of "the empire of the waves"[51]—this was a change that tipped the balance in favor of overland trade. As routes connecting the Ottoman dominions and the Adriatic Sea to Amsterdam and the Baltic came to be traversed more heavily than before, the new heart of the *Mare Internum* was no longer exclusively maritime but terrestrial as well.[52] Fortuitously, the return of the Genoese capital to the region in the form of state loans—extended to the house of Habsburg, among others—helped bolster the trend that affixed Venetian (and Florentine) merchants' dealings and fortunes in a region stretching from Cracow to Nuremberg and Leipzig.[53] This process, too, like the first two, had a long historical lifespan, since so long as Amsterdam remained the center of gravity of the world-economy, the strong magnetic field its economic presence generated subjugated the Mediterranean to its demands. Therefore, the restructuring of commercial networks to the south and east of Amsterdam and to the Dutch merchants' content was anything but short-lived.[54] Stated briefly, these triple processes did not lose momentum with the eclipse of the twin city-states and the end of the century of the Genoese. Quite the contrary: they gained strength under the aegis of Dutch rule, and lasted well into the closing decades of the eighteenth century. In other words, the three processes outlined above were vested with a gift of time, and constituted the general trends that came to reshape the economic life of the Mediterranean largely because they were in consonance with the systemic needs of *pax Neerlandica*, and not solely because of the lingering might or the mastery of the twin city-states.

Subject thus to the rhythms of the Amsterdam-centered world-economy, the basin followed a trajectory which, in many respects, was more analogous to that of its northern counterpart than not. To start with, the Italian city-states' industrial renaissance in the mid-sixteenth century was cut short at the turn of the seventeenth. But so was Antwerp's. The basin's manufacturing activities migrated from their protected, urban enclaves back to the city-states' *contadi* or, in the case of the empires, into lesser cities and the deep countryside. But so did those in Amsterdam. As a result of this egress from the urban centers, industrial activity, which in the sixteenth century was concentrated in the Netherlands and northern Italy, dissipated therefrom to both sides of the Flanders-Florence axis in the seventeenth century.[55] The empires in the basin, too, followed these trends closely. In Spain, the industry spilled out of the towns and into the countryside, a trend from which neither the Ottoman nor the French industries were exempt.[56] Venetian merchants tirelessly canvassed southern Germany, expanding the territorial compass of their putting-out, *verlag*, operations. But so did the Dutch merchants, expanding into the hinterland of Liège and Ardennes, just to name two locations.[57]

If, broadly speaking, the basin's manufacturing and mercantile activities did not set it drastically apart from its northern counterpart, the same, however, cannot be said of its agricultural performance. Its agriculture and "antiquated" rural social structures are usually singled out as the main culprits for its poor economic performance prior to the nineteenth century.[58] Eighteenth- and nineteenth-century reformers and savants attributed the deplorable state of the region's agriculture to the draconian rule and political weight of landowners, tax-farmers, and large flock owners as well as to the complexity of patterns of ownership and tenure.[59] From its family structures and crop-rotation patterns to its systems of land tenure and migration, Mediterranean agriculture—as a whole, including that of its leading city-states—is said to have been innocent of the agronomic advances registered on both sides of the English Channel, at least until the mid-nineteenth century.[60] Unavailable though holistic accounts of the evolution of the basin's agriculture after the sixteenth century may be, the Mediterranean countryside is habitually portrayed as mirroring all the symptoms of the region's long-term economic malaise.[61] The relative deterioration in rural producers' position due to the increasing sway of sharecropping and tax-farming arrangements; the appearance of *fattoria*- and *çiftlik*-type large estates in the Italian peninsula and along the shores of the Black, Aegean, and Adriatic seas, worlds apart from both the manorial estates east of the Elbe and the capitalist farms of the North; concentration of land under the hands of ecclesiastical bodies and lay nobles; contraction in the arable to the benefit of livestock agriculture; and poor agricultural yields are all seen as symptoms of the Mediterranean's lagging agricultural performance.[62]

Yet, even if the aforementioned traits can be said to have colored the Mediterranean countryside, they were the properties not of the basin per se, but of its low-lying landscapes, not only limited in extent and size but also either inhabited sporadically or mostly deserted starting from the 1620s. Unlike the vast Atlantic and Vistula plains north of the Alps and the Pyrenees, it was the sheltering highlands of the Mediterranean where most of the region's vital economic activities and populations were located in the seventeenth through the mid-nineteenth centuries.[63] This was scarcely the case during the expansionary long sixteenth century, from 1450 to 1650, as well as the previous period of prolonged expansion, from 1000 to 1250/1300, when the low landscapes of the region were vigilantly brought under the plow, tilled, and inhabited on a regular basis.[64] The shift in the epicenter of the basin's economic life from its lower altitudes to its higher elevations was therefore a momentous change, and represented a dramatic departure from the landscape the region presented to the traveler in the fifteenth and sixteenth centuries. Bread crops, for one, and wheat in particular, which invaded the plains and the newly reclaimed lowlands during the long sixteenth century were eventually dethroned, and exiled in the seventeenth and eighteenth centuries in the *cultura promiscua* of the hilly regions of the Inner Sea.

Ironically enough, it was during its autumn, and not heyday, that the Inner Sea assumed, or reassumed, if you will, what we today recognize as its quintessential properties. Vine and olives—the "civilizational crops" of the Mediterranean—which, in its golden age, were commercially produced in select locales, extended their dominion mostly from the seventeenth century throughout the width and breadth of the basin. Again, livestock agriculture which, in the region's high age, was overseen by corporate bodies, such as the Mesta in Iberia, the Dogano in Naples, or tribal confederations in Anatolia, witnessed the role of stationary sheep grow at the expense of transhumant sheep.[65] Meanwhile, the relatively lucrative crops of the previous era took up residence in the plantations in the Atlantic. In fine, the impressive range of crops that crisscrossed the Mediterranean in its age of splendor diminished in kind and volume as the basin's olden, "traditional" crops assumed a higher profile in the region's economic livelihood.

It was, therefore, during the seventeenth and eighteenth centuries that the hold in the basin of the trinity of the Mediterranean—wheat, tree crops (olives, vines), and small livestock (sheep and goats)—was fully reestablished.[66] The growing significance of the region's trademark triad of crops mirrored more than the region's diminishing economic stature, however. Changes in flora naturally flowed from and accompanied changes in the region's ecology: the coastal plains of the Mediterranean, home formerly to mono-crop and commercial cultivation, were largely deserted and taken over by swamps, wetlands, and reeds—not to mention the fauna that thrived in such environments: the mosquito, snakes, storks, and lizards.[67] These lands, if not tilled intermittently, were put to work mainly as winter grazing pastures.[68] With the growing dissemination of tree crops throughout the region, the hillsides became more densely settled than before. Correlatively, there was a shift in the basin's center of gravity from its low-lying lands to its hillsides and mountains.

The villages previously located among the crops dotting the plains were now affixed between the forests of the mountainsides and the crops of the plains and valleys. Villages that had "slid downhill" in the expansionary period of the fifteenth and sixteenth centuries to take advantage of the opportunities offered by the level low-lying plains witnessed, first, a slowdown in the pace of, and later a halt in, the colonization of low landscapes.[69] This spatial rearrangement was in consonance with the passing of the age of the twin city-states, for, as the empires of the waves, the thalassic order these merchant republics fashioned was built on the primacy of port-cities, coastal plains, and oriental crops. Large-scale cultivation of oriental crops along the banks and in the islands of the Atlantic undermined the foundations of the Venetian *stato do mar.* The Mediterranean landscape changed from the late sixteenth century to the advantage of its higher altitudes in tandem with the region's gradual loss of economic prominence and the transformations that this demotion entailed. Contributing to these ecological transformations that favored the hills was the advent of the crops of the Columbian exchange. Whereas the newly arriving oriental crops blessed the maritime plains from the eighth or ninth century to the sixteenth, it was the arrival of American food crops that fittingly helped some of the inhabitants, mostly at the higher altitudes of the region, to escape from the scourges of crop shortages and famine. In perfect concordance with the declining significance of wheat and the upward ascent of rural settlements, these crops, as substitute cereals, helped buttress the transfer of rural settlements away from pestiferous lowlands.

If the oriental crops undergirded and symbolized the heyday of city-states, the resurgence of tree crops and the popularity of American food crops reflected the enfeeblement of the maritime republics—and the economic devolution that accompanied it. All in all, it was the hollowing out of the Mediterranean lowlands in tandem with the rise of the Baltic as the provider of commercial grain that caused the retreat in the arable.[70] This was followed by the growing dissemination and popularity of tree crops on the hillsides and over the versants of the mountains enveloping the Mediterranean. In a similar fashion, the spread of manufacturing into the countryside and the competition in woolen textiles that pitted all against all facilitated the conversion of considerable segments of the arable into grazing fields. Moreover, the changing balance in favor of overland trade increased demand for haulage animals and their staple, oats, and lesser cereals, most of which, unlike wheat, did not require the expansive fields of the lowlands: this development promoted fodder cultivation.[71]

The general trends of the mid-sixteenth century through the turn of the nineteenth, in sum, were instrumental in the relocation of the center of gravity of the Mediterranean from its plains to its hillsides and mountains. A shift of this nature and magnitude could not have occurred, and did not occur, in a brief span of time. Instead, this was a prolonged and slow-moving process with a lifespan of over three centuries, lasting from the 1550s to the 1870s, if not later. And this sea change was precipitated, consolidated, and furthered by a wide array of forces ranging, among others, from climatic to economic, from ecological to pathological, and from cycles of hegemony to the secular trends of the world-economy. The great agrarian cycle à la Le Roy Ladurie that spanned for more than three centuries was comprised of different temporalities that sequentially and cumulatively reinforced the shift away from the plains and patterned the common destiny of the Mediterranean.[72] These temporalities were threefold, overlapping, complementary, and of differing durations. They were of differing lineage, too: geohistorical, world-hegemonic, and world-economic.

Framing the great cycle was the return of the Little Ice Age, which stretched roughly from the 1550s to the 1870s.[73] Even though the Little Ice Age started, *grosso modo,* at the turn of the fourteenth century, the decisive ecological shift set in with the onset of Braudel's "second sixteenth century" in the 1550s. Of the wide array of climatic and environmental changes wrought by the Little Ice Age over the course of the following three centuries, it was the prevalence of wetter (and colder) conditions as well as increased climatic variability that are of significance for our analysis. The dawn of the Little Ice Age may have heralded the onset of cool summers and snowy winters in the north, but wetter conditions were manifested in the Mediterranean climate in the form of increased precipitation as torrential rain and, at times, as increased snowfall on higher altitudes. Witness the emergence of the snow trade from the seventeenth century onward as a popular economic activity from the Sierra Espuña and Ventoux to Mount Lebanon, the Taurus Mountains, and Crete.[74] Overall, however, with the notable exceptions of the Alps and the Pyrenees, the melting of snow had, and continues to have, limited significance.[75]

Increased fluvial activity and incidences of flooding, on the other hand, when coupled with the climatic variability of the Little Ice Age, turned the cultivation of lowlands into a toilsome and worse, unpredictable, undertaking.[76] Given the rugged relief of the Mediterranean, increased hydrological activity caused the inundation of river beds and wadis with raging waters.[77] Streams and rivers, usually dry during the long, arid summer months, frequently flooded in the rainy season.[78] Crucially, since deforestation prompted by the reclamation and settlement efforts of the 1000–1550 period had already diminished the absorptive capacity of soil on mountain and hillsides, valley bottoms and plains on the skirts of the mountainous range found themselves exposed to the vicissitudes of intermittent flooding.[79] And as a result of increased fluvial discharge, the marshlands, generated and nourished by this aquatic activity, extended their dominion in, and at the expense of, the region's lowlands, which happened to be its premium croplands. The deltas expanded from the Ebro valley to the Rhône, from the river Arno to the river Meander, but the fertile soil that accrued as a result along their banks was not easy to bring, much less keep, under cultivation.[80]

The disastrous consequences of the contraction in bread lands was more pronounced at the edges and valley floors of the mountains of the Mediterranean. Certainly, the intensity of these fluvial activities waxed and waned over time, especially given the long lifespan of the Little Ice Age. Periods of flooding, as discussed in the following chapters, were more concentrated in time than not, with stretches of thirty to forty years of increased and recurrent fluvial activity marking each century. Nonetheless, the persistence, time and again, of increased precipitation and recurring inundations turned the cultivation of the basin's lowlands into an exacting and unpredictable task. The scenario that unfolded throughout the length and breadth of the Mediterranean, in the environs of Cádiz, in Malàga, Roman Campagna, Sardinia, Corsica, Cilicia, and Palestine, among others, was perhaps not surprisingly similar.[81] In Edirne, for instance, after the flood of 1688–89, horticulture and flax cultivation supplanted cereal production, the output of which proved to be vexingly erratic.[82]

What the wetter conditions of the Little Ice Age and the heightened hydrological activity that resulted from it prompted was the relocation of rural settlements from the marshy environments of the new ecological setting to the ambient hills that were shielded from the unhealthy vapors of the swampy and fever-ridden lowlands. The first temporality that neatly framed the great agrarian cycle was thus set by the ecological pulse of the natural world, a deep-flowing current in the history of the Mediterranean from the 1560s.

The second temporality was shorter in span than that of the Little Ice Age. It stretched from c. 1590 to c. 1815, and bracketed a transformation that took place in the world-economy's division of labor as the ascending hegemon, first, the United Provinces and, later, Great Britain embarked on reshaping the global economic space according to their needs and means. In the process, the order fashioned under *pax Neerlandica* turned the Baltic grain trade into its "mother trade."[83] The shift in the center of commercial agriculture became manifest by the invasion of the Inner Sea by Baltic grain in the 1590s. Not only the breadbaskets of the Mediterranean (i.e., Sicily, Apulia, Sardinia, Genoese Romania, and occasionally the Ottoman lands), but the lowlands that were reclaimed and put under the plow from the eleventh century to the 1550s, were profoundly affected by the relocation of commercial agriculture away from the coastal plains of the Inner Sea. This proved to be a long-lasting process. Since even when the east of the Elbe, the granary of *pax Neerlandica,* was supplanted over time by new sites of primary production that came to provision the urban centers of the core regions of the world-economy, these sites of production were all situated, without exception, even at more northerly latitudes than Poland, Silesia, and Danzig. When seen from the vantage point of the Mediterranean, the basin played no part in the new geographical distribution of commercial cereal production, at least until after the Napoleonic era.[84] Put differently, the area devoted to cereals in the basin consequently went on to contract from the advent of *pax Neerlandica* to the advent of *pax Britannica.* In fine, the second temporality was constituted by the cycles of hegemony of the world-system.

The third temporality was framed by the world-economy's secular trend, that of the seventeenth-century crisis (c. 1620/1650–1750).[85] Following on the footsteps of the demographic slowdown of the seventeenth century that reduced hands available for work and the demand for grain, the crisis facilitated the conversion in most parts of the Mediterranean of the arable into, among others, pastureland and orchards, prompting a contraction in corn land.[86] The contraction became more acute in the light of the fact that, with the relatively steep decline in the prices of agricultural goods, lesser grains (and legumes) grown for subsistence staged a comeback at the expense of wheat.[87] The wheat-gold fields of the Mediterranean lowlands became a distant memory of its golden age in the sixteenth century. In their stead, the light, dry, and stony soils of its hillsides were put to use, and the ecology of the basin underwent a remarkable change. The secular rhythms of the world-economy therefore constituted the third and final temporality of the agrarian cycle. These three developments cursorily sketched above led, singly and collectively, to a contraction in the arable in the low-lying plains of the Mediterranean. At the height of the great agrarian cycle, arable production was faced by a triple threat, a formidable threat indeed, which turned continuous cultivation into a most daunting task. For the most part during the great agrarian cycle, the plain of Roussillon, the Ebro, Po, and Arno valleys, the maritime plains of Asia Minor, the countryside around Salonika and the Algerian Mitidja, to name a few, were continually threatened by waves of sudden floods and the insalubrious environment bred by relentless flooding in the flatlands of the basin—littoral and inland.[88] What was striking about the Tavoliere of Fuggia, Puglia, or the Aegean coast until the nineteenth century was not how well tilled the fields were but how vacuous the coastline was.[89]

To recapitulate, then, the great agrarian cycle and the autumn of the Mediterranean were intimately intertwined. The temporalities that constituted the great agrarian cycle were threefold and led to the abandonment of the basin's plains and low landscapes. The processes that constituted the waning of the Mediterranean were also threefold and crafted a new balance between the lower and higher altitudes of the basin to the injury of the former, and compelled rural producers to diversify their crop-mix to the detriment of low-yielding cereal crops. The two worked in concert to bring about the withdrawal of permanent cultivation from the low landscapes, encouraged crop diversification by closely incorporating the region's higher altitudes, and brought the diverse ecological setting this shift offered within the compass of rural households' agro-sylvan operations.

Curiously enough, the onset of all three processes had their origins somewhere between the 1550s and the 1650s—at the height of the age of the Genoese, that is. The concurrence may have been coincidental at the beginning, but the Genoese rule over the Mediterranean provided added incentive to the desertion of the lowlands for two reasons. First, the return of the spice trade back to its old Levant route and the influx of fresh silver brought a degree of prosperity that boosted demand for a wide range of goods, agricultural as well as industrial.[90] The resultant shift to more lucrative crops or occupations undermined the unchallenged position of basic bread grains, a status acquired during the 1450–1550 period, and their primary loci of production, the lowlands and the plains. Second, the inflationary wave generated by the influx of silver pushed prices higher in the Mediterranean than in the other breadbaskets of the world-economy, east of the Elbe in particular.[91] By the time these inflationary pressures started to subside in the 1620s, the shift in the granary of the world-economy was already completed at the expense of the Mediterranean.[92] In brief, Genoa's growing ethereal presence was of signal import in setting the great cycle in motion.

Also, the age of the Genoese was poised to play the ideal host to the start of the great agrarian cycle, because at its commencement, the ecological levy exacted by centuries-long processes of assarting and deforestation was about to turn menacing.[93] The debilitating pace of human and economic activity, too, had left its deep marks on the Mediterranean landscape. Broadly speaking, the pace of expansion from the eleventh to the sixteenth centuries—the 1250/1300–1450 depression excepted—was frenetic. The assart of forests in the north and the reclamation of plains (swamp, marsh, moor, and fen) in the south had radically altered the region's landscape long before the long sixteenth century.[94] At the same time, the accelerating rhythm of urbanization hastened the tempo of colonization of the higher altitudes of heavily forested mountains for industrial exploitation, such as shipbuilding, mining, and fuel.

The peril of erosion was hence looming large on the horizon when, roughly around 1300, the Little Ice Age with its colder and wetter winters, poised to expose the ecological strains occasioned by the colonization of higher altitudes and deforestation, set in.[95] Fortuitously or not, the 1250/1300–1450 downswing, amplified by the demographic toll taken by the Black Death, decelerated the vertiginous pace of urbanization and hence of the colonization of higher elevations. All the same, the momentum of economic activity picked up with the onset of the long sixteenth century in 1450 or so. Thanks to the momentary return in the 1450–1560 period of global warming in the form of the Medieval Optimum, which temporarily reversed the wetter conditions of the c. 1300–1450 era, the basin was given a blissful reprieve by the momentary cessation of erosion despite the resumption of urbanization and deforestation.[96] The respite offered by the Little Optimum was cut short by the return of the Little Ice Age in the 1550s, and the toll of the frantic pace of urbanization and deforestation on the basin's ecology became promptly and painfully manifest.[97] By the end of the sixteenth century, for instance, the scarcity of timber had become widespread throughout the Mediterranean.[98] As increasing precipitation and wetter conditions returned from the 1550s, the threat of erosion that had remained dormant during the Medieval Optimum for a century or so resurfaced in earnest.

The process may have picked up speed during the age of the Genoese, but its historical lifespan was much longer than that. For the contraction in the plowed fields of the Mediterranean and the desertion of its lowlands were, as argued above, long-term and deep-seated movements. That is why the completion, sequentially, of the three processes, the beginnings of which were summarily related above, brought the movement to an end from the mid-eighteenth century. Though the region-wide contraction in the arable lasted until the 1870s, the process started to turn around, timidly at first, with the conclusion of the shortest of the three processes: the seventeenth-century crisis. After 1750, in concomitance with the onset of the great expansion of the world-economy, demand for cotton and, to a lesser extent, wheat picked up. The inexorable march of cotton, a crop in need of irrigation and of level and extensive land for large-scale cultivation, flagged the first signs of a turnabout in the exploitation of lowlands. Low landscapes that met both conditions experienced a reversal in their fortunes, from neglect to mild interest.[99]

Since draining lowlands and ameliorating conditions of work in the plains after an extended period of inattention proved to be taxing, financially and otherwise, recourse to coerced labor in the bonification of marshy lands that had become home to fever-inducing maladies remained as the prevalent option, as was the case in the large estates, *çiftlik*s, in the Balkans and elsewhere.[100] All in all, however, the marshes and swamps went on to populate most of the basin's lower altitudes, though efforts or plans to turn the wilderness into arable had appeared on the horizon. Appropriately enough, the reversal of a process as long-term and large-scale as the upward movement described here proved to be equally gradual and drawn-out: it may have commenced with the halfhearted and partial recolonization of the basin's lower altitudes and coastal plains. Evident though the first signs of growing interest in bringing the plains back under the plow may have become in the 1750s, it was not until after the Napoleonic Wars and the mid-Victorian economic boom that the movement of reclamation and (re)settlement gained sustained momentum, turning the lowlands into respectably populated stretches of land.

That is, if the turnaround in the fortunes of the low landscapes became palpable after the seventeenth-century crisis came to a close in the 1750s, the second stimulus that served to restore the economic health of the low-lying plains was the onset of the new cycle of hegemony that refashioned the world-economy's division of labor concomitant with Britain's ascent to position of paramountcy after 1815. In agriculture, the new order found embodiment in the dismantlement of the Corn Laws "at home" in the 1840s and a corresponding explosion in grain production "overseas."[101] The mid-Victorian agricultural boom that ensued stimulated cereal cultivation not only in the vast expanses abutting into the Black Sea, but in the temperate world.[102] The Mediterranean proved no exception: the return to, and the recolonization of, the lower-lying lands gained velocity following the passing in 1858 of the Ottoman Land Code, which helped establish property rights over unreclaimed land and wasteland, thereby opening the plains to incursions by would-be landowners; in Spain, the abolition of the powerful Mesta in the 1830s lifted restrictions on converting the pastures, previously earmarked for grazing, into arable. The kingdoms of Naples, Sicily, and Sardinia, the former breadbaskets of the Mediterranean, all witnessed similar transformations from the turn of the nineteenth century, beginning with the Napoleonic era.[103]

What sealed the solidity and fixity of the process of colonization of the low landscapes was, of course, the end of the longest of the three processes (or cycles, if you will) enumerated above. The great cycle that was ushered in by the advent of the Little Ice Age in the 1550s was completed, fittingly, by its end in the 1870s, for the simple reason that the return of warmer conditions and decreased climatic variability rendered the reclamation of swamps incomparably easier.[104] The arrival of quinine—which later came to be known as the "Jesuit's bark"—from Spanish Peru and its diffusion by the Genoese, at least in some parts of the Mediterranean and not incidentally from the 1550s, had provided temporary and fleeting relief against malaria, the most forbidding affliction disseminated widely by the spread of swamps and moors.[105] As can be surmised, the reprieve granted by the administration of quinine could not be turned into a lasting remedy in that how and through which channels the ailment was broadcast still remained a mystery, and malaria went on to vex the inhabitants of the plains. Only from the 1840s, with the greater employment of quinine, did malaria cease to be a deadly disease. Progressively, quinine was transformed into a tool of empire-building, particularly in tropical Africa, in the hands of colonial powers after cinchona plantations were built and a steady supply of quinine was ensured.[106] In the closing decades of the nineteenth century, at the concurrent completion of the Little Ice Age and the great agrarian cycle, when the mosquito was singled out as the carrier and purveyor of the malady, the spread of malaria was at last successfully controlled. The recognition that the debilitating disease was transmitted by mosquitoes and not by what at the time was called "malarious air" was pivotal, for the eradication of malaria was contingent upon the elimination of the breeding places of mosquitoes, the swamps.[107]

Fundamentally, it was the systematic draining of swamps throughout the basin that sealed the fate of a movement which, with its ups and downs, lasted for over three centuries.[108] Turning marshlands and swamps into arable was an equally lengthy process, at times requiring the skills of a *sluismeester*, manager of the drainage works. Although projects of drainage were launched from early on, under the able command of legendary Humfroy Bradley, "maître des digues de France," or the Caymans and the Hoeffts in Provence and Languedoc, it was not until well into the mid-nineteenth century that the movement gained steady momentum.[109] It was during the third quarter of the twentieth century that the process of thoroughly reclaiming the basin's swamp-ridden lowlands was largely brought to completion.[110] Most of the undertakings embarked upon during *pax Britannica* placed the land retrieved from the wild at the disposal of producers, preferably but not necessarily of landlords, in order to fortify the foundations of the world agrarian order that provisioned the world-economy's core zones.

Given the aforementioned transformations in the Mediterranean landscape and the resultant relocation of economic activity away from its plains during the region's autumn, it becomes clear that depicting agrarian change in the new ecological setting of the basin by simply employing indexes devised with the vast and level plains north of the Alps in mind does not do justice to the complexity of the circumstance at hand.[111] On account of the narrowness of the low-lying plains of the basin and the limited size of the fields on its hillsides, gauging only yield ratios and gross cereal output fails to capture the variegated nature of its output. So does tracing the extent of mechanization, given the non-level character of most of its surfaces. The changed landscape demands a line of analysis that reflects the fact that the center stage of the Mediterranean in its waning years was not the plains or cereal agriculture, but its highlands and mountains and its polyculture.[112] Plotting the great agrarian cycle does just that, as we elaborate below. To reiterate, the temporalities that framed the waning of the Mediterranean and laid out its common destiny were threefold: the secular rhythms of the world-economy, the cycles of hegemony of the world-system, and the ecological pulse of the natural world. Accordingly, the center of gravity of economic life along the shores of the Mediterranean shifted from its low-lying lands to its hillsides and mountains—and back—during the great agrarian cycle. Naturally, the relocation of the basin's economic nuclei from its coastal plains to its hillsides, mountains, and high valleys transformed the Mediterranean landscape, including types of settlement and ways of life that populated it as well as the composition of the flora and fauna that inhabited it.[113]

As can be surmised, the factors underlying a geohistorical movement of this scope, a movement formidable enough to radically transform the basin's landscape, were manifold. These factors, despite deep-seated differences in their provenance and constitution, are analyzed in this study in a double register.

Changes that accompanied the shift in the world-economy's center of gravity from the Mediterranean to the North Sea and the Atlantic constitute the first register. The dynamics unleashed by the labors of the Genoese and Venetian merchants and financiers to revive, in some form, their eroding hegemony in the Mediterranean from the sixteenth century are investigated under this heading, for this process framed the "general trends" the whole sea shared under the watchful eyes of the Italian city-states. It is in this register that the three temporalities that successively disadvantaged the plains and the lowlands are chronicled.

Second, there is a detailed examination of the changes that accompanied the shift in the basin's center of gravity from the lowlands to the hillsides, and the great agrarian cycle, from the 1550s to the 1870s, that temporally framed this geohistorical movement. And the Mediterranean's "common destiny" is charted within the parameters set by this cycle. A cycle of this historical breadth, spanning over three centuries, was undoubtedly governed by multiple temporalities that stretched from that of the natural world to that of the world-system. Emphasis is placed on the reemergence of the Mediterranean trinity— small livestock, tree crops, and wheat—and on how the passing of the primacy of cereal husbandry, by facilitating the transition to a diversified crop-mix, served as an underlying factor in the recasting of the basin's landscape. It is in this register that the triple dynamics that paved the way for the hills and mountains to supplant the plains as the center stage of Mediterranean agriculture are examined.

Three temporalities and three processes, then, account for the changes in the basin's agrarian structures, settlement patterns, and landscape in its autumnal times. It is in this double register that the transformations that reshaped the Mediterranean are framed and investigated. For these registers keep track of two sets of movements, of different pedigree, duration, and nature, that complemented and reinforced each other in a symbiotic fashion.

The Structure of the Book

It was not by chance that the waning of the Mediterranean proved to be as lengthy as the great agrarian cycle. It had, after all, been a world-theater for most of the period from the fourteenth through the sixteenth centuries. Its waning, befitting its majestic stature, proved to be long and drawn-out as well: its era of "afterglow" from the mid-sixteenth to the mid-seventeenth centuries, largely coterminous with the age of the Genoese, followed by an era of "autumn" until the mid-nineteenth, lasted, or lingered, for more than three centuries.

Since the historical trajectory of the Mediterranean is investigated here in a double register, the first part of the book is devoted to highlighting processes unleashed by the rise, consolidation, and demise of the age of the city-states and how these aquatic empires reshaped the Inner Sea according to their oftentimes conflicting needs. It accordingly tracks the acculturation in, and departure from, the Inner Sea of tropical crops, the provenance of which was the Indian Ocean, and depicts how the dissemination and egress of these crops made and unmade the fortunes of the twin city-states. Stated briefly, it accounts for the withdrawal of commercial agriculture from the shores of the Mediterranean.

The second part of the book plots the great agrarian cycle that framed the afterglow and autumn of the Mediterranean and the ramifications this long cycle had on the region's landscape in terms of settlement and cultivation patterns. It attributes a critical role to the retreat of rural settlements from malarial lowlands, to the benefit of surrounding hills. Withdrawal of commercial agriculture from the maritime plains on the one hand, and the retreat of rural settlements from afflicted low-lying lands on the other, set the parameters of change in the Mediterranean in its twilight. The second part also examines the ramifications of the dissemination of American food crops, as alternative cereals, along the shores of the Inner Sea from the late sixteenth century and demonstrates how these new crops filled the void left behind by the contraction in arable cultivation.

The city-states, rich trades, oriental crops, and commercial agriculture in the littoral plains of the Inner Sea that populated the Mediterranean landscape in part I pose a striking contrast to that portrayed in part II, a landscape dotted with tree crops and American food crops, both of which thrived on the hilly slopes of the basin. If the arrival and diffusion of sugar and cotton cultivation blessed the Venetian and Genoese men of trade and underpinned the golden age of the city-states, then the arrival of American food crops, as poor men's meat and grain, reflected the reversal in the fortunes of the Inner Sea.

Against this background, the first chapter of part I charts the symbiotic relationship among the basin's empire-building city-states and empires, and outlines the responsibilities shouldered by the celebrated city-states not only in the weaving of economic networks interlinking the Greater Mediterranean, but also in the formation and consolidation of the empires that vaulted over the basin: the Genoese *nobili* and the Castilian throne; the Serenissima and the Sublime Porte; Florence and the French throne. This it does by tracing the pivotal role the city-states played in the revival of the Inner Sea from the thirteenth to the fifteenth centuries; the perpetuation and consolidation of the interdependence between city-states and imperial polities during the long sixteenth century; the distinct yet complementary nature of the functions performed by the cities of St. Mark and St. George; and the change in the theaters and modes of operation of Venetian and Genoese men of money in the seventeenth and eighteenth centuries in tune with the economic restructuring of the basin.

The second chapter surveys the dissolution of the division of labor that the city-states successfully put in place from the twelfth century and how this restructuring created multilateral dependencies among different quarters of, and polities quartering, the Mediterranean and beyond. It dwells on the exodus of the cultivation of lucrative crops from the Mediterranean, sugar in particular, and the transformation of cotton from a plantation crop into a cash crop. The migration of these crops was complemented by the dwindling significance of commercial agriculture in the basin after the lands to the east of the Elbe were turned into the breadbasket of the Amsterdam-centered world-economy in the seventeenth century. This chapter chronicles the resultant dethroning of King Wheat, which ruled the region from the eleventh to the sixteenth centuries. It also highlights how different the Mediterranean was at the height of its power: only after it was deprived of the rich trades that the vineyards and olive orchards—that is, its renowned tree crops—came to blanket the basin's hillsides and vacated plains.

The third chapter complements the picture by dwelling on how the contraction in commercial agriculture and the departure of rich trades resulted in the return of the region's civilizational crops: vineyards proliferated and vine and raisin trade became the mother trade of southern France, Greece, Crete, and Smyrna; olive oil and soap production became widespread from the eastern Mediterranean to North Africa, not to mention the Italian and Iberian peninsulas; and livestock husbandry allowed the Mesta, the Dogana, and the tribal confederations in the Ottoman lands, along with smallholders, to extend the dominion of the "golden fleece." It also accounts for how the spread of rural manufacturing and cottage industries complemented this picture. A movement that started with the relocation, or transfer, of industries that were formerly domiciled in the Italian city-states north of the Alps, to southern Germany in particular, during the second sixteenth century, later become widespread throughout the length and breadth of the Mediterranean. The silk industry of renaissance Venice, for instance, was no longer exclusively Venetian in the seventeenth century, with first the Veneto and later Geneva emerging as new centers of silk textile production. The dissemination of manufacturing north of the Alps in the seventeenth and eighteenth centuries demanded the services, mercantile and financial, of the imperial duo of the basin's golden age: Venice and Genoa.

Venetian merchants were to be found in the region's fairs and cities, delivering fabric to Leipzig and Cracow; the banker-financiers of St. George also placed their accumulated wealth in the service of Genoa's manufacturing industries, located mostly in the city's immediate vicinity. Moreover, the growing financial requirements of states to the north and east of the Brenner Pass as a result of the change in the direction and nature of economic flows were met in part by Genoese men of money. The house of Austria profited from the return of the Genoese capital to the Mediterranean, not to mention of course France, Lombardy, and Venice. The shift in theaters and fields of operation of the Venetian merchants and the Genoese *nobili* in accordance with the diminished activity in the Mediterranean was then the third process that accounted for the direction of change in the basin's historical trajectory during the seventeenth and eighteenth centuries.

In fine, the first part of the book deals with the waning of the Mediterranean as a world-economic phenomenon. Comprised of two chapters, the second part deals with the great agrarian cycle. The first chapter of part II, the fourth chapter, outlines in broad strokes the withdrawal from the low-lying lands, particularly the plains, of rural populations, and the contraction in cereal agriculture due to an acceleration in erosion, alluviation, sedimentation, and siltation. This chapter, like its counterpart in part I that depicts the three faces of the autumn of the Mediterranean, surveys and details the dynamics of the great agrarian cycle and the traces they left on the Mediterranean landscape. This it does by dwelling on the three temporalities summarily mentioned above that contributed to the withdrawal of cultivation from the low-lying lands: the secular rhythms of the world-economy, cycles of hegemony of the world-system, and the ecological pulse of the natural world. It is in light of these triple processes that the fourth chapter traces the emptying of the Inner Sea's low-lying landscapes, maritime and inland, by the resumption of the Little Ice Age. Processes associated with this ecological volte-face, such as climatic variability, deforestation, soil erosion, sedimentation, and the march of swamps, altered the contours of the Mediterranean landscape in the seventeenth and eighteenth centuries. Withdrawal from the plains, triggered by ecological reasons, eloquently complemented the exodus of commercial agriculture from the basin, precipitated by the recentering of world-economic flows along the banks of the North Sea. By tracing the attempts to bring low-lying lands under the plow, mostly through drainage, it chronicles how the trend was reversed from the mid-eighteenth century even though the full recovery of the low landscapes from the unhealthy vapors and malarial afflictions of the Little Ice Age took much longer, well into the twentieth century. Once again, the ecological recovery was concomitant with the recentering of world-economic flows along the shores of the English Channel.

However, the evacuation of, and later return to, the plains during the great agrarian cycle did not take place in a vacuum. It demanded a spatial redistribution of productive activities, and correspondingly, of rural settlements to the benefit of uplands, given the pestiferous condition of the lowlands. The return of tree crops, for instance, favored higher altitudes and not lower. Although animal-farming utilized both the plains and mountain ranges for pasture, the expansion of their reign can largely be explained by the earlier evacuation of the plains that made these deserted lands available as winter pasture, without raising the ire of cultivators. The transition from the sixteenth-century landscape, where the plains were the main stage of agrarian production, to one where the highlands took on the task of accommodating growing numbers of inhabitants on their slopes and skirts did not happen on its own. How this process of ascent that carried rural settlements from their low-lying lands to higher altitudes unfolded is the topic of the fifth chapter.

Certainly, this was not the first time that the economic heart of the basin was removed to high ground, and life on upper elevations was not as welcoming as the warm, level, and relatively spacious plains of the low landscapes. The *differentia specifica* of the upstream migration from the late sixteenth century was the new ecological and vegetal environment that came into being as a result of the introduction of, and experimentation with, the crops of the Columbian exchange. Since these crops were originally of highland provenance, their arrival allowed the process of acclimatization of rural settlements to the higher altitudes of the Mediterranean to proceed with greater ease. In addition, lesser crops, which were ousted by the imperialist extension of wheat during the "cerealization" of the plains in the first sixteenth century, returned to the Mediterranean. The resurfacing of these lesser crops, more likely to thrive on altitudes much higher than wheat as well as less demanding than the King Crop, was equally germane to the process of ascent, for this abetted the climb of rural habitations to higher elevations.

If the rural settlements owed their mobility and flexibility to the popularity of American food crops and lesser cereals, and to the presence and versatility of temporary settlements, what still needs to be accounted for are the factors that facilitated the ascent of rural settlements to higher altitudes. Within this context, emphasis is placed on the presence of temporary and adjunct settlements as integral components of the Mediterranean landscape, and on the capability of these settlements to expand and withdraw from the arable when conditions called for doing so. After all, villages that had "slid downhill" until the turn of the seventeenth century to populate the lowlands and take advantage of the opportunities offered by the relatively sizeable plains found themselves, upon the return of the Little Ice Age, forced to retreat to their hilly abode: the number of temporary settlements (in the form of hamlets, *mazra'as*, *bastides*), tapping into the resources of the higher and lower altitudes in the hilly range where most permanent settlements were located, mushroomed throughout the basin. Systematically tapping into the resources of the uplands did not imply that the lowlands were completely deserted. The onset of the Little Ice Age and the fluvial environment it generated brought about possibilities for growing cultivation of aquatic crops—hence, the popularity of rice as a staple—and crops that were in need of irrigation, like cotton. Nonetheless, the ascent of rural settlements slowed down due to increasing demographic and ecological pressures placed on the resources of alpine regions. The revival of the plains from the latter half of the nineteenth century, the subject matter of chapter 4, was neatly complemented by the downstream migration from the hills. The confluence of these two factors opened up the plains to settlement and tillage, as in the long sixteenth century.

These are some of the avenues of inquiry that open up almost spontaneously when emphasis is placed on the unity, and forces governing the waning, of the Mediterranean—avenues left largely unexplored in individual historiographies. At the outset, it needs to be emphasized, and without any hesitation, that systematic data covering the region's twilight will never match the quality and volume of documentation that dates from the long sixteenth century. It is more than likely too that paucity of historical documentation for most of the sixteenth to the nineteenth centuries will remain with us for some time to come. Avenues offered by the afterglow of the Mediterranean, on the other hand, large-gauge and long-haul, provide us with a meaningful vantage point and a point of entry into the dim world of the seventeenth and eighteenth centuries. After all, if the heyday of the Mediterranean-centered world-economy can be best depicted from the vantage point of its epicenter, that is, from the vantage point of the Italian city-states, so should, *ex hypothesi*, its waning.

PART I

OF CITIES OF SAINTS AND
RICH TRADES

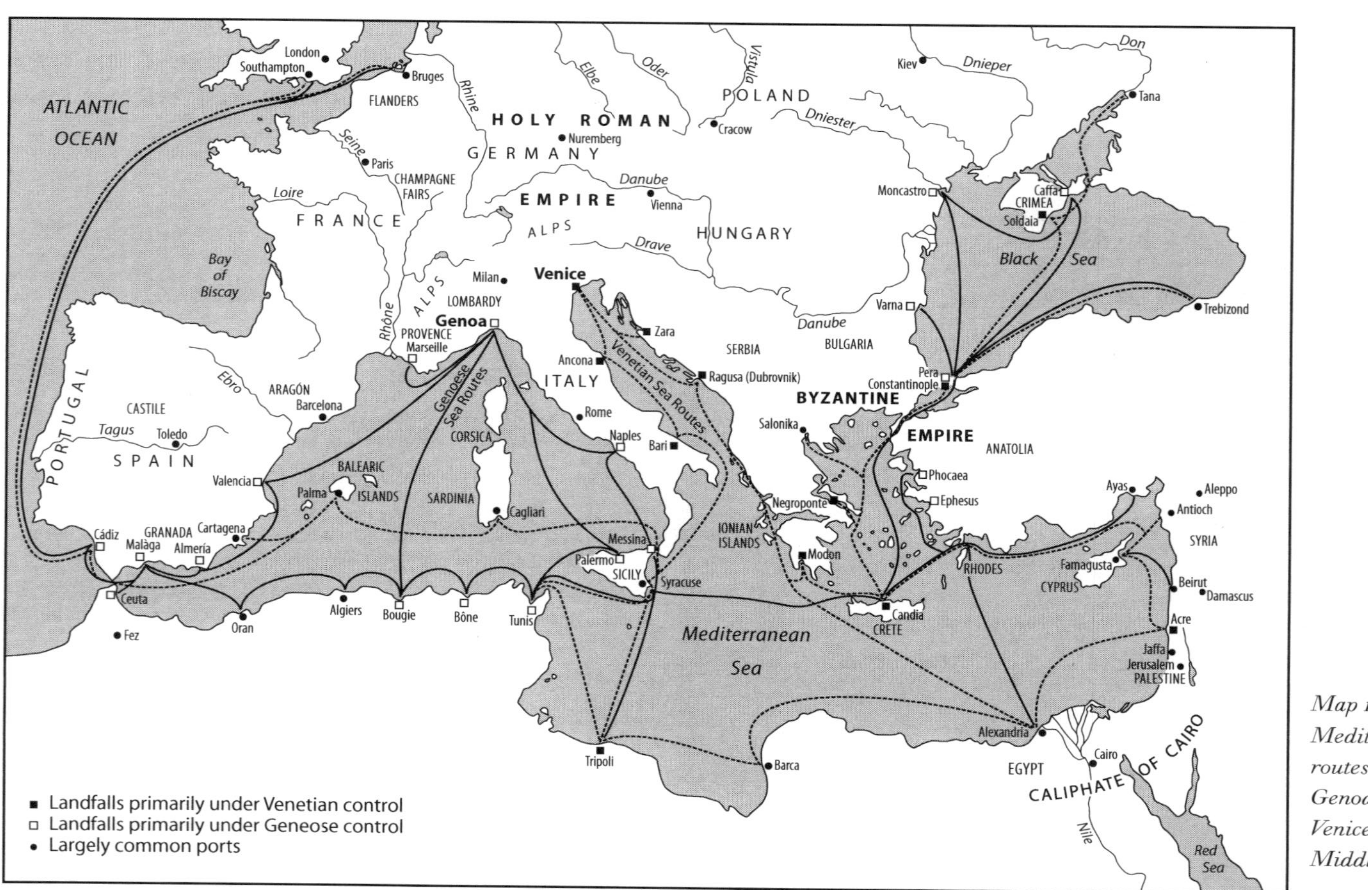

Map 1. Mediterranean routes of Genoa and Venice in the Middle Ages

CHAPTER ONE

Empires and Empire-Building City-States

The first sixteenth century commenced in the 1450s. With its onset, territorial states and imperial polities crippled by the 1250/1300–1450 downturn began to recover their strength.[1] During the tumultuous times of this drawn-out downturn, empires circumscribing the Mediterranean experienced a considerable weakening in their hold over their territorial possessions or a turnabout in their economic fortunes or, worse, both. The French throne, for one, remained engaged in the lengthy Hundred Years' War (1337–1453), and the patrimony of the house of Osman was almost shred to pieces by Timurlane's armies in 1402 and had to be rebuilt in the first half of the fifteenth century. Similarly, the throne of Aragón suffered from a steady erosion of its revenue base owing to the growing presence of Genoese merchants who eventually outmaneuvered their Catalan counterparts not only in the western Mediterranean but also in the kingdom of Castile. The Mamluk throne was beleaguered by the bullion famine of the fifteenth century as well as the decrepitude of the fleet it inherited from the Ayyubids.[2]

City-states, in contradistinction, had fared comparatively better during the downturn, partly because of the disarray in which the monarchies found themselves and principally because of the opening of the lucrative central Asian-Mediterranean trade circuit. In fact, the economic and political resilience the merchant republics exhibited during these taxing times proved to be one of the main reasons it was under the initiative and labors of a few and select city-states, Venice and Genoa at their helm, that the Inner Sea had, by the 1450s, come to house an integrated and expansive network of production and trade. Its origins deeply anchored in the twelfth and thirteenth centuries, this Mediterranean-centered system, interstitial at first, reached maturity during the fifteenth and sixteenth centuries.[3] The latticework woven by the northern Italian city-states among the port-cities that inhabited the basin as well as between lands north and south of the Alps laid the foundations for the unity and coherence of the economic life of the sea, not only along its coastlines but also deep inland.[4]

By nature, this seaborne emporium thrived on and further stimulated cross-border flows: merchants and vessels, crops and goods, currencies and ideas trespassed jurisdictional frontiers almost casually and on a continual basis despite sporadic wars and the almost ubiquitous presence of piracy. True, casting a long shadow over the Inner Sea were the twin gargantuan imperial polities of the fifteenth and sixteenth centuries, namely, the Habsburg and Ottoman empires.[5] Yet, it was primarily the Venetian and Genoese merchants who ruled the waves of the Mediterranean, singly or conjointly, the former reigning for the most part over its lucrative eastern half—east of the Adriatic Sea and the kingdom of Sicily—and the latter its western half.[6] Fittingly enough, the Castilian throne's holdings enveloped the basin from its western confines and that of the Sublime Porte from its eastern confines, in perfect accordance with the de facto partitioning of the sea between the cities of St. George and St. Mark, especially after the war of Chioggia (1378–81). And not surprisingly, it was along this line of demarcation, which cut across the strategic narrows between Sicily and the city of Tunis, that the severe frontier battles between the two dynasties took place in the latter half of the sixteenth century.[7]

Put differently, even though the basin's landscape was predominantly silhouetted by the two world-conquering empires of the sixteenth century, one jutting into the Atlantic and the other into the Indian Ocean, imperial pomp and grandeur need not overshadow the constitutive role the "empire-building cities" assumed in the shaping of the Mediterranean world.[8] For it was, in part, Genoese mercantile and financial might that allowed the Castilian throne to expand its territorial holdings across the Atlantic, paving the way for the colonization and settlement of the islands that peppered it.[9] Nor should it obscure the fact that it was the Venetian merchant fleet's solid grip on the Aegean Sea and the eastern Mediterranean that forced the Byzantine empire to relinquish its long-distance trade to the Lion of St. Mark. The existence of such an expansive commercial nexus, with the Rialto at its hub, later allowed the Ottoman dominions to partake in the riches of the Levant trade: in the closing decade of the sixteenth century, the Venetian spice trade on the Syrian shores of the empire was alone worth 2 million ducats at a time when Ottoman state revenues hovered around 8 to 9 million ducats, including revenues that accrued to the Sublime Porte from its immense *timar*, prebendal, holdings.[10] The two leading imperial denizens of the Mediterranean owed, therefore, their commanding presence in no small part to the vast maritime empires and the emporia built by Venice and Genoa, notwithstanding the differences in these merchant republics' methods of territorial acquisition, administration, and commercial practice.[11]

Towering though they may have been in stature, the Ottoman and Habsburg dynasties were not the sole imperial residents of the basin in its heyday. The third imperial denizen, the French throne, displayed, like its counterparts, a similar dependence on a city-state, Florence (and later, Milan). But the presence merchants from Tuscany (and Lombardy) commanded in the Mediterranean was in grave disproportion to their economic might. For even though the operational terrain of the Florentine men of business stretched from Constantinople in the east to maritime outposts in the Atlantic in the west, the economic presence of the city of St. Giovanni was felt more substantially to the north of the Alps than to its south, especially after the city's hold over Naples started to weaken.[12] Florence never regained the prominence it had commanded in the Inner Sea prior to the advent of the plague and the collapse of its formidable companies.[13] As such, its merchants found themselves unfavorably positioned to equip the French (or the English) throne with an economic space analogous to what the Genoese merchants provided the Castilian crown throughout the width and breadth of the western Mediterranean (and the Atlantic), and the Venetian husbandry of the rich trades offered the Byzantine, Mamluk, and Ottoman thrones in the Levant. In the process, the Signoria even earned the ill-reputation of *l'amancebada*, "the Turk's courtesan."[14] It is hardly surprising that Marseille and Aigues-Mortes, the French empire's prominent port-cities, and Lyon, where Genoese and Florentine businessmen held court, had but brief and fleeting moments of fame during the basin's golden age—unlike Cairo, Constantinople, and Seville. It was only when the French central seat of power was transferred from Paris to Bourges, home to Jacques Coeur, and the Loire valley in the 1422–1577 period, that the Gallic presence in the Mediterranean became somewhat more palpable.[15]

Three empire-building city-states and three empires, then. Of the aforementioned cities, only Florence had a feeble presence in the Mediterranean in the long sixteenth century. So, by proxy, had the French empire: the tenuousness of its presence in the Inner Sea persisted well into the second half of the seventeenth century, if not later.[16] The Ottoman and Habsburg empires, by contrast—

as the twin city-states writ large, if you will—loomed large over the Inner Sea at the height of its prosperity. Both empires had largely completed their territorial expansion by the mid-sixteenth century: the house of Habsburg under Charles V (1519–56), and the house of Osman under Süleyman the Lawgiver (1520–66).[17] In France, the reunification of the hexagon had largely been concluded under François I and Henry II (1516–59). And by the end of the sixteenth century, most of the Mediterranean coastline had come to be covered by the dominions of these colossal polities: the Ottoman littoral alone covered over 3,000 miles, more than a good two-thirds of the basin's shoreline. With large stretches of the Mediterranean littoral incorporated into its realm, the Ottoman empire was perforce fully subject to the sway of forces that spanned the Inner Sea and beyond, from Tlemcen in the west, with the Straits of Gibraltar within view, to the spice entrepôts of Alexandria and Cairo on the east, open to the winds of trade from the Red Sea and the Indian Ocean.

Certainly, the empires along the shores of the Mediterranean expanded territorially, consolidated their realms, and reached the zenith of their power during the long sixteenth century, from the 1450s to the 1650s. But the territorial aggrandizement of the empires did not necessarily create an environment that was inescapably inhospitable and perilous to the reigning city-states, Venice and Genoa. The ties that bound and interlinked the destinies of these empire-building city-states with those of the world-conquering empires were multitudinous, and the relationship between them was predominantly symbiotic. Hence, the balance of power between these two fundamentally different yet complementary polities was not instantaneously tipped against the Lilliputian merchant republics, as is usually assumed, simply because of the empires' propensity for territorial expansion.[18] The age of the Genoese, which stretched from the 1550s to the 1650s, for example, gave a new lease on life to the empires in the Mediterranean, since the Genoese bankers lifted the fortunes of the Castilian throne by providing it with much-needed gold and of the Sublime Porte by boosting the Levant spice trade. In other words, the enlargement in the territorial holdings, and the consolidation of the rule, of the houses of Habsburg and Osman did not necessarily undermine, much less erode, the conditions of existence of the city-states. Even when the key positions held by the merchants and bankers of the leading cities in imperial courts were challenged by the Portuguese money lenders or the Dutch merchants or the English trading companies, there was always need for the services of the rival siblings.[19] Historically speaking, then, the relationship between the city-states and empires was subject to change and continual revision, and these revisions were not invariably and unremittingly injurious to the former.[20]

The city-states formed, and remained much longer than usually acknowledged, as the ligaments of the Mediterranean world. More often than not, these formidable petty republics, singly and severally, set the direction and nature of change in the basin. It follows that accounting historically for how the empires and territorial states came into being, flourished, or suffered serious drawbacks becomes impossible unless the basin's history is viewed from the vantage point of the reigning city-states. Germane though the interplay between the city-states and empires was in charting the main contours of Mediterranean history, it was hardly the only factor. An equally crucial factor was the presence of acute rivalry among the city-states over the riches of the basin, not to mention among men of money of all stripes regardless of patrimony or creed. Indeed, the central role the merchant republics assumed in the Inner Sea was due neither to the concerted nature of the actions of the ruling elites of these city-states nor to commonality of purpose. Far from it. As complementary enemies, the cities of saints scarcely acted jointly despite the differential nature of the functions they came to perform in the division of labor that undergirded the Greater Mediterranean.

In the commerce of money, for example, it was Florentine, and not Venetian, merchants who operated on the Rialto.[21] Or it was in Genoese carracks that the silver the Venetian merchants utilized in the Levant left the port of Cádiz. Equally significant, this specialization was not static, but subject to continual revision: the Genoese merchants, for instance, who handled a significant portion of trade in the Mediterranean during the fourteenth and fifteenth centuries, withdrew from it and turned into prestigious banker-financiers after the 1550s, as had the Florentines before them in the mid-fourteenth century.[22] Or the Venetian merchants of the sixteenth century successfully joined the ranks of landed nobility in the seventeenth and the eighteenth centuries as did the rural notables in Tuscany.[23] The existence of a division of labor did not prevent frontal confrontations among city-states in the form of wars that pitted all against all. That their fortunes were collectively lifted by the vibrancy of the Mediterranean emporium, and that they all individually benefited from the widening and deepening of market-mediated transactions and further commodification along the shores of the basin, did not deter effrontery and unremitting warfare among the vying city-states.[24]

It is, therefore, the historical backdrop, the dynamics and the transformation over time of the relationship between the dwarf empire-building cities and expansive territorial empires on the one hand, and among the merchant republics on the other, which charted the historical trajectory of the Mediterranean. In this chapter, we discuss these two sets of relationships in four sections.

The first section frames and sketches the evolving relationship between the city-states and empires before the onset of the long sixteenth century c. 1450. It underlines how the city-states were able to extend and consolidate their dominion on the high seas as imperial and territorial warfare raged on land, on the shores of the basin. The second section then traces the transformations set in motion by the consolidation and expansion of the imperial and territorial polities during the long sixteenth century. This territorial expansion did not, however, necessarily deal a fatal blow to the city-states, for the relationship between these two polities was not antagonistic but symbiotic. How the petty city-states managed to preserve their dominion over the Mediterranean despite the territorial aggrandizement of the empires after the 1450s, and how this process was facilitated by the advent of the age of the Genoese, is the subject matter of the third section. Merchant republics, after all, initiated and furthered the westerly migration of lucrative crops, underwrote and orchestrated the spread of manufacturing across the Alps, and penetrated the emerging markets there. The fourth and final section depicts how the transformations that originated in the 1450–1650 period did not decelerate and wane, but rather went on to shape the general contours of the Mediterranean world in the 1650–1850 period and provided the city-states with precious breathing room. In sum, the chapter highlights the processes that framed the general trends that shaped the Mediterranean—and beyond—from the vantage point of the city-states.

City-States: Ligaments of the Mediterranean World

Certainly, the Mediterranean had, much before the imperial *redux* of the long sixteenth century, been a fount of, and a convivial host to, a score of empires around its shores. Indeed, the last time the basin was ringed round by the likes of empires was at the turn of the millennium, when its inhabitants ranged from the imperial offshoots of the Abbasid and Frankish empires to the politically weakened yet economically robust Byzantine empire.[25] Nonetheless, these imperial polities were, without exception, pale shadows of their former selves, and the basin was and remained effectively populated by cities that were city-states in all but name: Barcelona, Salonika, and Constantinople.[26] It was precisely during this period, when imperial polities were sparse and feeble, that a number of city-states that speckled the Mediterranean littoral extended their sway into the Levant owing to the crusades. They succeeded in establishing an economic network by underwriting and ensuring the continual and regular flow of a wide range of economic goods within and across the basin.[27] In this, the city-states benefited immensely from the presence first, of an expansive economic space that stretched from the Indian Ocean to the Canaries (the Eternal Isles to some, and the Fortunate Isles to others), and second, of a dense urban network along the shores of the Inner Sea—a string of cities that flourished in-between the not so verdant mountains that sheltered the narrow coastal strip on which these cities were built and the wine-dark sea that washed its shores.[28]

Under the watchful eyes of the city-states, imperial polities started to resurface cautiously with the economic upturn of the 1000–1250/1300 period, especially from the twelfth century onward. This was attested to in the eastern Mediterranean by the brisk revival in Byzantine rule under the Comnenian dynasty and in Seljuk rule under Alâ al-Dîn Kaykubâdh, and the emergence on the western rim of the vast Mongolian realm of the incipient Mamluk empire in 1250 with its capital in Cairo.[29] In the western Mediterranean, the rejuvenation of the Holy Roman and French empires under Frederick II and Philip the Fair, respectively, registered a similar directional change. Indispensable to the process of empire-building were the merchants and bankers of the diminutive city-states who had already established their presence even in the most distant corners of the Inner Sea.[30] The Venetian, Catalan, Pisan, Amalfitan, and Genoese merchants, among others, who had turned into the main pillars of the Mediterranean world, took over royal functions vital to the survival and expansion of the imperial polities.

Given the city-states' dominion over the Inner Sea from which the empires that bordered it also extracted a significant portion of their revenues, it is not surprising that Philip the Fair of France, for one, entrusted his finances to Musciatto Guidi, a Florentine merchant, and his navy to Benedetto Zaccaria, a Genoese merchant.[31] Alexius I, the Byzantine emperor, purchased the allegiance of Venetian merchants by granting them wide-ranging commercial privileges. The consequent breakdown in control exercised by Constantinople over its exports and the crumbling of time limitation of residence demanded of "alien merchants" benefited Venetian and Amalfitan merchants beyond measure, as confirmed by the anti-Latin riots of 1182. The ease with which silk was exported from the realm deprived Byzantium of one of its riches, its fabled textiles: artisans from Venice (and Lombardy) came to master the techniques of silk manufacturing. The hemorrhage in revenue that resulted from the loss of monopoly on silk exports further deepened the empire's dependence on the merchant republics.[32] The situation was not all that different in the case of the nascent Seljuk empire, which built its realm on lands previously held by Constantinople. It, too, was not exempt from the encroachments of Venetian merchants, as attested to by the Venetian-Seljuk treaty of 1220, which involved an exchange of chrysobulls. That the chrysobulls were issued on the one part by the sultan and on the other "not by the doge, but by the Venetian *podestà* in Constantinople," shows how stifling the Venetian presence had become in the eastern Mediterranean.[33] The notables of Barcelona, too, who had financed the city's economic expansion in the twelfth century, turned their attention, under the rule of James I of Aragón in particular, to royal finance in the wake of the acquisition of Majorca, Sardinia, and Sicily.[34]

Territorial expansion by empires therefore transformed royal finances and imperial trade into precious niches of wealth for the merchants and bankers of the city-states, and allowed both polities to expand in tandem and without considerable strife. By the turn of the fourteenth century, the process of empire-building had gained discernible momentum in the basin as evinced, among others, by the meteoric rise to power of the house of Osman, which came to administer vast territories in Anatolia and the Balkans, and by the epoch-making struggle that lasted for two hundred years between the houses of Aragón and Anjou over the control of the kingdom of Sicily.[35] At any rate, the ubiquitous rivalry among the merchant republics offered aspiring territorial rulers and dynasties an invaluable chance to extend their imperium by playing one city-state against the other. By fanning flames of rivalry, they intended to capitalize on these merchant republics' financial and commercial prowess and mastery of the sea without paying an exorbitant price—a price incommensurate with the momentary reprieve they were given from the pressing problems of the day.

Witness the attempts by the emperor Manuel I Comnenus in the latter half of the twelfth century to establish and nurture the Genoese presence in Constantinople and the ordeals he encountered in trying to protect this embryonic body of merchants from the assaults of the established Venetian (and Pisan) merchants.[36] It was due to the ubiquitousness of this sibling rivalry that, later in 1258, the supplanting of the Venetian merchants with those of the Genoese proceeded as smoothly and speedily as it did after the expulsion of the former from Constantinople once the city was restored to Michael VIII Palaeologus. Or, witness the warm reception extended to the Venetian merchants by Charles II, king of Naples, at the turn of the fourteenth century, that is, at the height of the preeminence of the renowned Florentine companies. But the policy was swiftly reversed, fittingly enough, under Robert the Wise.[37] In essence, then, the destinies of imperial dynasties and the warring merchant republics were intimately intertwined in the germinal phase of the Mediterranean world-economy. More often than not, royal finance and imperial trade had been carved up by the expansive and nimble city-states as their exclusive preserves of economic activity. That the interlinking of the fortunes of these two vastly different yet complementary polities came about during a period of economic and territorial expansion that stretched from 1000 to 1250/1300 reduced the chances of conflict between them to a considerable extent.

The expansionary economic and political conjuncture that commenced at the turn of the millennium came to a close sometime between 1250 and 1300.[38] The post-twelfth century imperial *redux* was also cut short by the 1250/1300–1450 downturn, which could not have come at a more propitious moment for the city-states, and for Venice and Genoa in particular.[39] The deceleration in the pace of economic growth fueled a fierce competition among leading city-states over the control of the Levant and the Mediterranean, culminating in the "Italian" Hundred Years' War that ended with the signing of the Peace of Lodi in 1454.[40] As truculent competition became the order of the day, the number of cities vying for supremacy thinned out during the course of the fourteenth and fifteenth centuries: Pisa, a vigorous rival, was vanquished as a sea power after the battle of Meloria with the Genoese in 1284; and Florence never recovered its former formidable strength after the collapse of the houses of Bardi and Peruzzi in the 1340s.[41] The kingdoms of Sicily and Majorca found themselves under siege by the kings of Aragón. Providentially for the cities of St. Mark and St. George, the secular downturn not only narrowed the ranks of the rival city-states. It also undermined the foundations of the imperial renaissance of the preceding period as it weakened territorial states and empires alike far more than the leading city-states. As economic turbulence and intense rivalry were settling in, the imperial cupola of the Mediterranean—Venice and Genoa—enjoyed the fruits of the stimulus given to the basin's commerce from the 1250s by the developments on the Eurasian mainland. The opening up of the terrestrial central Asian route, with Caffa as its main port of disembarkation, endowed the Genoese businessmen in the c. 1250–1350 period with great profits that were associated with the rich trades, then mainly centered in the Black Sea.[42] Later, the Venetian merchants basked in the splendor of the riches of the spice trade precisely because of the closing of the very same route, since this prompted the reconstitution, from the mid-fourteenth century, of the long-established maritime trade conduits that converged in Cairo, where the merchants of the Serene Republic remained in residence even at the height of *pax Mongolica*.[43]

The trying times of the economic slump brought about a bifurcation in the historical trajectories of the ailing empires and the thriving twin city-states, in stark contrast to what happened during the ebullient economic growth of the 1000–1250/1300 period, when both polities grew in concert. The fact that the decline in the size of the Venetian state budget during the taxing times of the

1250/1300–1450 downturn was incomparably smaller than that of its imperial contemporaries illustrates how divergent the trajectories of the two polities had become. In the fifteenth century, the English budget had fallen by 67 percent, the "Spanish" by 73 percent, but the fall in the Venetian budget was merely by 27 percent.[44] Not only did the Lion of St. Mark fare comparatively better vis-à-vis the empires, but it also forged "at home" its *stato di terra* mainly from the mid-fourteenth century.[45] The reign in Genoa in 1407 of the Casa di St. Giorgio, when "private" money-holders firmly secured the city's public finances, registered a development similar to that in Venice, only this time it was the rule of capital and not that of the state, which was firmly consolidated. The real income of the city of Florence increased elevenfold from the latter half of the thirteenth century to that of the fourteenth, from 26,000 to 945,000 Florentine lire.[46] Mention should also be made in passing that both the Ottoman empire and the kingdom of Castile suffered from the scourge of debasement during the course of the fifteenth century, between 1444 and 1481 in the case of the former and between 1429–30 and 1464–70 in the case of the latter.[47] The Mamluk empire, too, devalued its currency in 1425 under Sultan Barsbay as did the crown of Aragón.[48] Overall, then, the demanding conditions of the economic downswing not only trimmed down the number of competing city-states, but more crucially, consolidated the rule and wealth of the surviving merchant republics at a time when the empires found themselves more and more exposed to continual warfare and financial distress.[49]

The debilitating impact that the long-lasting depression had on the basin's empires thereby presented the twin cities with a precious opportunity to extend their formal and "informal" empires while holding on to their existent colonies. It was at this historical conjuncture that the hold of the city-states over the *Mare Internum* and beyond was substantially secured. Conversely, it was during these financially taxing times that the imperial polities that quartered along the Mediterranean, from the house of Osman to the house of Aragón and from the Byzantine emperor to the Mamluk sultan, were forced to rely on the seagoing city-states to an extent greater than before—on their amphibious arsenal, their command of the sea lanes, and their ability to extend significant loans without difficulty and hesitation.[50] To wit, the hold established by the Genoese over the Byzantine throne during the fourteenth century was mirrored in the fact that its customs post in Constantinople collected revenue in 1348 in the amount of 200,000 gold *solidi* whereas "the imperial customs received a mere 30,000."[51] In Aragónese Sicily and Sardinia, the downturn in agricultural fortunes that accompanied the economic slowdown and the fall in population offered the merchants of Genoa (and Pisa) an invaluable opportunity to impose their terms on the nobility who channeled the rural surplus into regional markets. The leverage the lenders established over the nobility through proverbial mechanisms of indebtedness became manifest when demand finally started to pick up: under the scrutiny of the creditor merchants, grain exports from the island, no longer reliant on the whims of the nobles or producers, skyrocketed from a level of roughly 35,000 *salme* between 1350 and 1379 to 286,000 *salme* between 1400 and 1429, when the basin started to recover from the initial devastation inflicted by the Black Death.[52] The grip of the merchants and financiers of the city-states over the rulers in the Inner Sea got tighter, thanks to the dire times of the fourteenth and fifteenth centuries.

A similar fate befell the islands of Majorca and Minorca, previously centers of transit trade which, later in the 1350–1500 period, were transformed into major wool-producers for Florence, Genoa, and elsewhere in northwestern Italian peninsula, thanks to the merchants of the city-states.[53] The success of the Genoese merchants in consolidating their hold over Sardinia, Sicily, and the kingdom of Majorca during the economic downturn was not a singular development. At the time of the fall of Constantinople to the Ottomans in 1453 and the passing of the Byzantine empire, for example, the throne's debt to Venice, "meticulously recorded to the last coin and never wavered," stood roughly at 6.5 million ducats, an enormous sum given that the Venetian state budget in the first half of the fifteenth century had reached approximately 1.6 million ducats per year.[54] Likewise, the periodic rise and decline of fairs in Champagne, Brie, and Lyon, organized and dominated by Tuscan, Lombard, and Ligurian merchants, also bound the destiny of the French empire to the pulse of the Mediterranean.

The trying economic slowdown of the 1250/1300–1450 period was not the only factor that tightened the control established by the city-states over the basin's empires in the preceding expansionary period. The ability the Lilliputian city-states acquired to successfully subdue or tame gargantuan empires during the fifteenth-century crisis can also be ascribed to the fact that the state of imperial rivalry between the merchant republics that previously had given the territorial rulers the chance to play one maritime republic against the other quickly became a thing of the past. Central to this transformation was the fall of *pax Mongolica* and the resultant eclipse of the Black Sea trade beginning in the 1350s, since this dealt a heavy blow to the Genoese preeminence in the region.[55] The dwindling of the Pontic Sea commerce, in turn, lessened the intensity of rivalry in the eastern Mediterranean as the Genoese merchants were eventually forced to relocate the heart of their maritime empire to the western half of the basin. The Inner Sea was partitioned, albeit tacitly, into two distinct spheres, the

Venetian and the Genoese. Surely, the Republic of St. George kept its presence in the Aegean and the Black Sea, at least for a while, and in a considerably diminished form. Its merchants increasingly shifted their dealings to the profitable slave trade, since demand for slaves had soared in the Mediterranean as a result of the human toll taken by the Black Death.[56] The Genoese presence lasted in the Aegean too, because the nascent Ottoman empire, in order to safeguard itself against Venetian Byzantium, extended capitulations to the warring siblings of the Signoria: the Genoese. Eventually, however, the merchants of the Serene Republic obtained their capitulations, sometime between 1384 and 1387, steadily reinforcing their grip over the Ottoman dominions thereafter.[57]

All in all, Genoa's continuing, albeit diminishing, presence in Constantinople and in the Aegean did not compensate for its sparse presence at the Mamluk court and hence in the Levantine port-cities. The closing of overland routes that terminated on the Black Sea coast reactivated the maritime trade through the Indian Ocean, which in turn boosted the fortunes of the sea-ports of the Mamluk empire and rewarded the longtime residents of the region, the Venetian traders, with the custody of the spice trade and its principal venue, the eastern Mediterranean.[58] Steadily, the Genoese merchants' share of the spice trade contracted with time, especially after the war of Chioggia in 1378–81, since the rich trades were centered once again in Cairo and Alexandria as had been the case before the rise of the Mongolian empire in earnest in the 1250s. The tapering off of terrestrial trade that terminated on the shores of the Black Sea left no option for the Genoese merchants except to rebuild their maritime empire in the western Mediterranean while keeping a steadily diminishing presence in "little Caffa" in Constantinople.[59] Merchants of the Republic of St. George were already familiar with the Iberian Mediterranean since they were the first ones to establish in 1297 a maritime connection between the Mediterranean and the North Sea and to engage in convoy navigation in order to reach, without intermediaries, consumers in the north when they were in charge of the Black Sea and, hence, the rich trades.

The ever-expanding frontiers of the kingdom of Aragón, with Sardinia, Sicily, and the Balearic islands brought within its imperium, facilitated the enlargement of the Genoese seaborne empire.[60] The city of St. George's commercial and financial presence in these islands and the Iberian peninsula and, by extension, the western Mediterranean reached imposing proportions. In the process, the waters previously traversed by the Catalan merchants came to supplant the Black Sea as the central stage of Genoa's new maritime empire sometime between the 1350s and the 1380s.[61] The apportionment of the eastern and western halves of the Inner Sea between the two rival city-states and the thickening of the demarcation line between these two distinct spheres reduced the prospect of territorial rulers pitting one republic against the other. The new power structure gave the city-states greater leverage in their dealings with the imperial polities. This is why it was during the 1250/1300–1450 period that the hold of the Serenissima over the Byzantine empire and the emirates sprouting on the shores of the Aegean Sea tightened significantly.[62]

As a matter of fact, it was the triumph in the 1340s of the land-locked Ottoman principality over that of Karası, a principality abutting the Aegean Sea, the locus operandi of the Genoese and Venetian merchants, which gave the embryonic empire a newfound economic gravitas.[63] More significant, merchants from the city of St. Mark assumed a greater presence in Mamluk ports after the empire was pounded by the advancing Mongol armies on the one hand and the onslaught of the Black Death on the other.[64] Similarly, in the western Mediterranean, as we argued above, the Genoese merchants went on to take over the commercial networks interlinking the western Mediterranean, from Majorca to Sicily. That, with the rise of the wool trade, the center of economic activity in the Iberian peninsula was transferred from its Mediterranean shores, Catalonia in particular, to earth-bound Burgos expedited the takeover of the region by the merchants of the Republic of St. George, and the Catalan merchants saw their share in the wool trade abate.[65] At any rate, the growing might of the city-states was not limited to the twin city-states of the Inner Sea. After the decline of Champagne fairs, the Florentine businessmen moved into Lyon, which served as their main entry into the financial networks centered to the north of the St. Bernard Pass.[66] Overall, the economic downturn of the 1250/1300–1450 period that mercilessly vexed the imperial polities greatly benefited the city-states by providing them with a invaluable opportunity to extend their tentacles into the far-flung corners of the Mediterranean and beyond.

The empires were, and remained, part and parcel of the Mediterranean-centered world-economy from its very inception simply because they inhabited an economic space laboriously built, jealously guarded, and patrolled by the merchants and companies of the northern Italian city-states.[67] The reliance of empires on merchant republics was magnified during the 1250/1300–1450 downturn, precisely because power struggles among the emerging territorial states, principalities, and empires were escalating with alarming speed. The empires were expanding territorially and were in urgent need of financial or commercial succor at a time when Venice and Genoa reigned supreme and remained largely unchallenged in the *Mare Internum*. The house of Osman, for one, was extending its territorial holdings from the westernmost edges of the

Anatolian peninsula to the coastal plains and shores of the Black Sea, the Aegean, and the Mediterranean. The house of Aragón was expanding south and east, incorporating most of the islands west and south of the Italian peninsula, including those of Sicily and Sardinia, into its fold. The kingdom of Castile was advancing west and south of its centuries-long homeland on the Iberian plateau, the Mesetas, to eventually reach the Mediterranean coast. The Mamluks secured Syria and Egypt, and consolidated their empire during the politically and economically challenging times of *pax Mongolica.* And despite frequent border changes in its northern latitudes, the French empire's hold over its Mediterranean provinces remained firm, even before its incorporation of Provence.[68]

As the empires were expanding their dominion on land and the city-states conveniently enough at sea, the economic networks that interlinked the shores of the Mediterranean underwent a major transformation during the 1250/1300–1450 period. Previously, the pride of place in the region's trade was held, *grosso modo,* by the pepper and spice trades. During the economic downturn, however, the range of goods traversing the Inner Sea not only widened but came to include staples and raw materials as well. The network of economic flows that interwove the Mediterranean became denser as a result. It was in large part the sharpening rivalry among the competing city-states (and the downfall of some in the process) on the one hand, and the partitioning of the Mediterranean between the Venetian and the Genoese on the other, which occasioned this transformation. Certainly, the partitioning of the basin permitted the Venetian merchants to complete their takeover of the prestigious and lucrative Levantine spice trade. More significant, the city's dependence on Romania and the eastern Mediterranean as its breadbasket and on the Levant for its cotton imports increased over time due, respectively, to the growing presence of its principal rival, Genoa, in the granaries of the basin, in Sicily foremost, and to the pivotal role cotton played in the operations of the Venetian merchants in southern Germany in the manufacturing of textiles.[69] Similarly, the sway held by the Genoese merchants over Sicily and their massive grain exports eased these seasoned businessmen's way into the trading world of the western Mediterranean. The shift in most parts of the region—starting with Majorca—to wool production for the northern Italian city-states furthered the trend toward diversification in the commodity mix in favor of staple goods; a shift not incidentally occurring concomitantly with the arrival of the Genoese in the Catalan territory. On both ends of the Mediterranean, the economic downturn thus left deep tracks by effectively forcing the merchants of the twin city-states to partake in the procurement of grain and raw materials, namely, wool and cotton.

In terms of the fortunes of the grain trade, the catastrophic blow suffered by Florence as its precocious companies bore the full brunt of the economic slow-down was of paramount import, for this inevitably obliged the city of St. Giovanni to turn "inward," or "continental" as the Venetians termed it,[70] and to place its dealings with the French throne at the center of its operations.[71] Since the ebbing Florentine presence in maritime trade was taking place exactly when the Genoese were in the process of relocating the core of their empire west and south of the Gulf of Lions, the Ligurian merchants' stifling grip over the Sicilian grain endowed them with the opportunity to comfortably penetrate into the Aragónese Mediterranean, always in need of cereals.[72] In the eastern Mediterranean, the Florentine merchants were invariably present in Ottoman cities of note, and were an integral part of the silk trade in particular, conducted through Bursa.[73] Still, their maritime presence was never a source of worry for the Venetian merchants, whose precious spice trade passed through the ports of Beirut and Alexandria, away from the sphere of operation of Florentine businessmen. Later, when the purchase of Porto Pisano and Livorno secured Florentine merchants' access to the Inner Sea in the fifteenth century, the purpose was to directly procure raw materials for, and market the products of, the cloth and silk industries, not to break into the carrying trade.[74]

Yet before the onset of the economic slowdown, Florence occupied a pivotal position in the grain, wool, oil, and wine trades.[75] The provisioning of the populous port-cities of the Mediterranean prior to the start of the 1250/1300–1450 downturn was performed by the colossal Florentine companies, some of which shouldered a large part of the basin's grain trade. In it, the prestigious trio of the Bardi, the Peruzzi, and the Accicciuoli played a preponderant part. Collectively, these companies were powerful enough to establish a syndicate in the kingdom of Naples and wrestle away from the Angevins a wide range of royal functions for themselves, such as collecting taxes and managing the grain and wine trades that were crucial for their overall operations. The volume of grain trade in the early part of the fourteenth century was impressive, even by the standards of the Mediterranean grain trade in the sixteenth century. Normally, annual exports of grain from Puglia alone—controlled by the Florentine companies and distributed by them along the Mediterranean littoral, south as well as north—averaged around 12,300 metric tons; in 1311 it reached an "astonishing total of 45,000 tons."[76] Later, however, demand for agricultural goods fell steeply owing to the onslaught brought on by the Black Death, which slashed the population drastically. The fall in demand dealt a fatal blow to the Bardi and the Peruzzi, both of which vanished in the wake of a string of bankruptcies.[77]

The lasting aftereffects of the fourteenth-century crisis were thus an abrupt demise of the expansive Florentine companies, a reduction in their share in the grain trade, and the subsequent launching of Florence's continental venture.

To be sure, similar economic misfortunes had led previously to the downfall of giant companies. But they had eventually been replaced by new and mostly larger ones. This was not the case during the fourteenth century, and the disappearance of Florentine companies left its mark on the state of the grain trade. With steady supplies of grain in peril, municipal offices—like the *Magistrato dell'Abbondanza* in Florence, the *Officium victualium* in Genoa, the *Camera frumenti* in Venice, and *l'Annonnerie* in Marseille—were placed in charge of securing the provisioning of the city-states and coastal cities as well as their territorial appendages.[78] Municipal institutions, in short, took over some of the functions previously performed by the now defunct "super-companies."[79] The great reversal in the fortunes of the stellar Florentine companies, when accompanied by the territorial partitioning of the two Sicilies between Angevin Naples and Aragónese Sicily, gave rival merchants an expected opportunity to carve out a niche for themselves in the vital grain trade.[80] This left the stage to the Venetian and Genoese merchants, who eventually had to face off in an almost unavoidable final confrontation. The battle of Chioggia solved the puzzle in favor of Venice in terms of the rich trades, but Sicilian grain trade remained the province of the Genoese merchants more than that of their Venetian counterparts, who routinely tapped into Puglia's agrarian resources to provision the city's maritime empire.

In the aftermath of the Black Death, the designated granaries of the Mediterranean—the two Sicilies, Sardinia, North Africa, Egypt, Romania, and the Crimea—were all hit equally badly, not only by pestilence but also by the vagaries of war, and were unable to provision the basin's centers of wealth in a steady and reliable fashion.[81] Egypt, for instance, a perennial grain basket, under the Mamluks suffered from pestilence, depopulation, land flight, neglect of irrigation canals, and low prices, and had to import quantities of wheat from Sicily from time to time to avert shortages.[82] Given the uncertainties surrounding the grain market after the 1350s, the city-states followed a strategy of territorial aggrandizement "at home," which brought the regions in the proximity of the city-states under the jurisdiction of the mother city with the purpose of increasing the merchants' catchment area. This, of course, was a lengthy process. The transformation of Venice into the Veneto, Florence into Tuscany, Genoa into Liguria, and Milan into Lombardy commenced during the late fourteenth century.[83]

In the short run, territorial expansion may have initially served some of the needs of the sprawling city-states, yet, as we discuss in the following chapter, finding a long-lasting solution for the mercurial grain trade was not only desirable but imperative. Even if some of the regions that had been recently annexed by the city-states were to be cajoled into accepting a division of labor that would assign them specialization in relatively less lucrative crops, such as cereals, it was certain that sooner rather than later they would have shifted to more lucrative lines of business. The *contadi* may have covered a portion of Venice's grain needs in the fifteenth century, but that proportion decreased over the course of the sixteenth century. More important, the grain trade was a strategically sensitive trade: the capability to re-export cereals to port-cities and imperial capitals deficient in bread crops provided the city-states with inestimable political leverage. As such, the grain trade was more than a source of enrichment. The volume of grain thrown into commercial circulation and placed in the hands of the merchants of the city-states could not therefore have been left to the whims of the producers or rulers in charge of channeling the produce to the marketplace. For the continual and regular provisioning of the city-states and their empires, a reliable, systemic solution was required. Yet, the expiration of the Florentine companies placed the procurement of grain at the heart of the struggle among the vying city-states. In the commercial matrix of the Genoese empire, Sicily and Sardinia came to occupy a central place as its breadbaskets as did the shores of the Aegean for the Venetian empire.[84] The rise of pastoralism in the Iberian and Anatolian (and later Italian) peninsulas during the boisterous times of the *Reconquista* and the consolidation of *pax Mongolica* further trimmed the arable and the agrarian surplus available for commercial purposes. Consequently, steady provisioning in grain remained a problem even when the economic slowdown and the loss of life resulting from the Black Death considerably reduced the demand for it.[85]

In the eastern shores of the Mediterranean, lands that had come under Ottoman rule, previously of the Byzantine and emirate realm, were already frequented and their economic resources were tapped into, if not monopolized, by the Venetian and Genoese merchants, who were indeed the masters of the Mediterranean and the Black Sea, *Hakan ül-Bahreyn*—a title coveted and zealously sought after by the rulers of the Ottoman realm.[86] The Levant trade, passive though it may have been for the Ottomans, loomed large in Francesco Pegolotti's *La practica della mercatura* long before the heyday of the rich trades. Humdrum cargoes carrying salt and cereals had traversed the waters of the Mediterranean from early on.[87] Until the opening of new frontiers off the coast of the Atlantic and the rise of the Baltic trade, the Ottoman realm—which spanned not only the length and breadth of "Romania" on which the Genoese and the Venetians had both laid claim but also the lands encircling it, and can thus appropriately be referred to as "Romania *redux*"—went on to provision Venice and its sibling city-states with victuals as much by default as by choice.[88] Venice the Dominant's sway over the eastern Mediterranean was enhanced not only by the dwindling of Genoese power in the Aegean and the Black Sea after the 1380s but also by the Signoria's growing dominion over the Byzantine empire, which was facing a serious threat of dissolution, as well as over the nascent emirates that mushroomed on lands previously of the Byzantine realm. The Serene Republic was willing to resume the commercial relations that had existed prior to the rearrangement of the political map of the region following the withering away of the Byzantine empire. So under emirate rule, wheat, wine, and soap, for example, continued to be transported across the waters of the Aegean under Venetian command.[89] Moreover, the Signoria already had Negroponte and Candia, where the Venetian "settlers" were engaged in grain production for the homeland; and after 1479, the *stato da mar* would mobilize the resources of Cyprus as well. As the share of staples in the Venetian trade increased, the immersion of the merchants of St. Mark into the Levant's networks of production and trade deepened.

The Genoese merchants were subject to similar forces in their own sphere of operation. Deprived of their Black Sea trade once the central Asian trade route turned perilous, they easily consolidated their commercial sway in the Iberian shores of the Inner Sea, from North Africa and Sardinia to Sicily, because they were already plying the waters of the western Mediterranean and the Atlantic as part of their convoy navigation. *Genova La Superba* rebuilt its empire accordingly within the economic space the house of Aragón was trying to carve out for itself, battling the house of Anjou in the process. The reincorporation of the kingdom of Majorca into the realm of the crown of Aragón in 1343 tied the fortunes of the throne to the settlement of the Sicilian imbroglio,[90] which, fortuitously for the Genoese merchants, slowed down the maritime expansion of Majorcan merchants. Moreover, the paramount role the Genoese assumed in the gradual colonization of the Canary, Cape Verde, São Tomé, and Madeira islands rendered the city valuable to the empires with an opening to the Mediterranean Atlantic: Portugal, Castile, and Aragón.[91] In most instances, political incorporation of the islands and shores west of Sicily by the house of Aragón was complemented by the subordination of these locales to the needs of the Genoese merchants: grain, sugar, and gold took the place of pride among commodities that bound the Mediterranean to the Atlantic.[92]

To be sure, it was not the grain trade alone that increased the attractiveness of the eastern Mediterranean to the Venetian traders and that of the expansive

Castilian-Aragónese realm to the Genoese merchants. The crisis in the textile industry in Venice unleashed a new dynamic. With the growing scale of production of fustians in the Swabian towns and countryside, the need for cotton from the Levant, from the Syrian shores in particular, grew in accord. The most intense period of relocation of industry into the Swabian countryside stretched from the 1350s to the 1450s—after the Black Death, that is. Fustian manufacturing went on to spread therefrom into the Rhineland, Bavaria, Bohemia, Silesia, Hungary, and Poland.[93] For the Venetian merchants, the lure of the Mediterranean as a source of cotton increased precipitously in the wake of the spatial enlargement of the cotton-/textile-manufacturing regions.[94] Concurrently in the western Mediterranean, along with the growing dependence by the Italian city-states on Iberian wool and the coeval shift to livestock husbandry, the demand for wool soared, requiring the commercial services of the Genoese merchants. The merchants of the cities of St. Mark and St. George thereby found themselves happily in charge of the provisioning of raw materials for the textile industries of the basin, cotton in the eastern half of it and wool in the western. In sum, with the inclusion of cereals and raw materials into the repertoire of Mediterranean commerce, the basin's economic structure took on a variegated and infinitely more elaborate form.

Thus, founded and shaped within the immediate orbit of the Italian city-states' ever-expanding spheres of operation, the empires had perforce a part to play in the consolidation of the Mediterranean world-economy, a process that furnished them with ample economic opportunities. For instance, access to the expanding Ottoman realm enjoyed by the port-cities of Dubrovnik and, to a lesser extent, Avlona—despite the fact that they both operated under the long shadow of Venice—bolstered the maritime industry in the Gulf of Adriatic from the 1380s.[95] Subsequently, privileges accorded to the Ragusan merchants by the Sublime Porte in 1442, 1458, and later in 1481, and the city's takeover of Genoa's maritime activities after the latter's withdrawal from trade from the 1550s, enabled the city of St. Blaise and, by extension, the empire to fully partake in the prosperity of the long sixteenth century. In the 1570s, when the total capacity of the Ottoman shipping fleet stood in the vicinity of 80,000 tons, Dubrovnik's total tonnage of shipping alone was in the vicinity of 25,000 to 40,000 tons.[96] As well, the Spanish empire remained solidly anchored in the Mediterranean in part due to the sheer momentum given by the process of *Reconquista* and in part due to its dependence on Sicilian and Apulian cereals and on the Genoese for handling of its finances and trade.

This did not mean, however, that the partitioning of the Mediterranean occurred at the expense of all but the Venetian and Genoese merchants. The

Catalan and Aragónese merchant vessels plied the waters of the Mediterranean, competing at times and cooperating at others with merchants from Venice, Genoa, and Narbonne, and had their presence established in all principal ports of call long before Andalusian merchants lost their grip on maritime trade, roughly from the thirteenth century.[97] Even after 1297, when the Genoese established direct and regular maritime links between the Inner Sea and the North Sea and when the Bay of Biscay turned into a new and complementary venue of the rich trades, the Mediterranean remained the central locus of the operations of Catalan merchants. Part of the reason this was so stemmed from the fact that the maritime activity fueled by the arrival of the Genoese, and later Venetian, merchants and financiers to the North Sea brought the Hanseatic League within the orbit of the Greater Mediterranean trade, much to the chagrin of local merchants.[98] Merchants from Gascony, Brittany, and the Basque region who had dominated this trade until the fourteenth century were consequently displaced by merchants belonging to the League. Conversely, in the *Mare Nostrum* of the city-states, thanks to the vibrancy of economic activity stimulated by the spice and pepper trades, merchants from lesser cities such as Barcelona, Montpellier, Valencia, and Marseille continued to take part in its commercial life, albeit briefly and tangentially, as attested by Jacques Coeur's undertakings in the Levant.[99] In sum, the relationship that intimately bound the fate of the empire-building city-states to that of empires and gave these cities the ability to play an integral part in the survival and expansion of the imperial polities was initially struck in the 1250/1300–1450 economic slowdown. Not merely this, the taming of the empires by the petty city-states took place as the range of goods thrown into circulation in the Inner Sea expanded to routinely include industrial and bread crops.[100]

City-States and Empires during the Long Sixteenth Century

The economic downturn that tightened the hold of the twin city-states over the basin's empires came to an end with the onset of the long sixteenth century. As economic growth picked up from the 1450s, the house of Osman regained strength after the conquest of Constantinople in 1453, and Alfonso the Magnanimous of Aragón emerged triumphant in 1458 against the house of Anjou. With imperial expansion on the horizon, it was almost certain that the Venetian and Genoese maritime possessions would come under threat, as demonstrated by the direct threat posed by Charles V to the League of Cognac, of which both Venice and Florence were members. Eventually, the territorial aggrandizement of the empires led to a changing of hands, among others, of Caffa (1475), Chios

(1566), Cyprus (1571), and Oran (1574).[101] The loss of some of the colonial holdings of the maritime empires notwithstanding, changes in the region's political landscape were not to the immediate detriment of Venice or Genoa, since both had solidified their informal empires during the 1250/1300–1450 period. When the kingdoms of Castile and Aragón brought Naples and Sicily (and Granada) under their suzerainty, this was all but at the expense of the Genoese merchants and bankers, who by that time had become so vital to the commercial operations of the western Mediterranean that they were "operating by remote control."[102] Indeed, the Ligurian merchants remained in effective charge of organizing Sicilian sugar production and conducting silk trade from the island—not despite, but because, of Aragónese overrule. Venice experienced a similar aggrandizement in its commercial realm in the eastern Mediterranean, again, not despite, but because, of Ottoman territorial expansion. The monopoly of the alum mines in Phocea, for instance, after remaining in the hands of the Genoese for close to a century, was revoked by the Ottomans in 1455, only to be turned over to the Venetians in 1463 and again in 1481.[103] More important, however ineffective the Ottoman efforts proved to be in ousting the Portuguese from the Indian Ocean, they played a part in the revival of the time-honored Levantine route and, correlatively, of the Venetian thalassocracy in the latter half of the sixteenth century.[104]

Even the loss of Constantinople, the Queen of Cities, proved to be less detrimental to the Italian city-states than it would otherwise have been. The inexorable dwindling of the Black Sea trade following the disintegration of the Mongol empire had already deflated the city's economic worth to a considerable extent. For the Genoese, the remission of taxes from Pera declined from £1.9 million in 1334 to £1.2 million in 1391, and to a mere £234,000 in 1423.[105] The contraction in trade and the fall in revenues hastened the departure of the Genoese merchants from the Black Sea. Neither did the Spanish advance in North Africa by Pedro Navarro between 1509 and 1511 dislodge the merchants of St. George from "Genoese Africa." Nor did the Ottoman conquest of Cyprus in 1571 put the Venetian orchestration of the spice trade at peril. In fact, the conquest of Salonika (1430) and Constantinople (1453) by the Ottomans, of Barcelona (1472) by Juan II of Aragón, and the incorporation of Marseille (1481)—and Provence—into the French realm ended the rule of illustrious cities that tried to mimic, albeit without much success, the commanding city-states of the era. And during the long sixteenth century, the Mediterranean remained firmly under the dominion of the republics of St. Mark and St. George despite the territorial losses suffered by the twin city-states. This was attested to by the fact that in the 1470s no one in Venice seemed to "care any longer about the

Turk," only two decades after Constantinople had escaped the clutches of the Serenissima. As long as Crete (and Cyprus) were under Venetian rule, the trans-Mediterranean trade remained as Venice's big prize.[106] It is telling that between 1479 and 1645—roughly during the long sixteenth century—the Signoria and the Sublime Porte went to war on three occasions, each military confrontation lasting just for three years, if not less. Territorially, the Ottoman empire reached its apogee in 1595, and the Habsburg empire, even at its weakest in the seventeenth century, did not witness significant loss of territory.[107] In fact, the Genoese bankers designated Seville (via Madrid) as their "eye" on the Atlantic, and the bond that tied the Signoria to the Sublime Porte got only stronger in light of the developments in the Indian Ocean.

So the territorial enlargement of the empires was not necessarily at the expense of the economic space dear to the city-states. Not paradoxically, given the degree of interconnectedness between the interests of the imperial and republican polities, even when the empires extended their sway or hold over the emporia from which the city-states derived their livelihood, the latter went on to benefit from the revamping of those emporia. The establishment under the house of Osman of a vast unified economic space that extended from the crucial maritime posts on the Aegean to those on the coast of Little Armenia, later to be expanded farther by its incorporation of the *Oltramare*, served the Venetian merchants exceptionally well. The Serenissima profited from the Ottoman empire's territorial expansion into the Indian Ocean via the Persian Gulf as well as the consolidation of its rule in the Red Sea, both of which eventually helped buttress the revival of the spice trade.[108] What is more, Venetian merchants, confined until 1552 to Alexandria, obtained the right "to set up shop in Cairo," much to their own surprise and delight, and thus were given the chance to "penetrate to the very heart of trade" with the Indian Ocean.[109] The annual customs revenue in Mecca, an important entrepôt in spice trade, was considerably higher at the end of the sixteenth century than at its beginning: in 1587, it was estimated to total 150,000 gold ducats as against 90,000 in the early sixteenth century. Spice merchants in Cairo were still transporting huge sums of silver to Mecca in the opening decades of the seventeenth century to pay for their imports, despite the dangers associated with such an undertaking.[110] Stated differently, the expansion of the Ottoman realm did not augur a slowdown in the Levant trade, nor did it harm Venetian interests. In fact, the empire's southerly expansion itself was a reflection of the slowdown in the spice trade in the first half of the sixteenth century due to the growing Lusitanian presence in Goa and Hormuz.[111] The consequent lull in trade and the decline in revenues derived from the spice trade did cost the Mamluks their empire in

1517—an empire that had withstood even the unwavering Mongol armies at their fiercest.

In a similar fashion, the Genoese merchants, as one of the vanguards of transoceanic voyages, benefited from the colonization and settlement of the islands in the Atlantic and the territorial expansion by the Iberian empires. In Seville, the city that commanded the Atlantic trade and hosted the treasure fleets, the Genoese population doubled in the later half of the fifteenth century and reached new heights in the 1500–1550 period.[112] The territorial expansion of the Habsburg empire invested the Genoese with unhindered access to Sardinia and Sicily: by the turn of the sixteenth century, grain exports from Sicily had reached 100,000 *salme* a year, and oscillated afterward between 120,000 and 150,000 *salme* per annum.[113] The oceanic voyages expanded the Genoese merchants' dominion over new plantation sites and crops, thereby lessening their reliance on the economic well-being of the Mediterranean. After the 1550s the Genoese merchants forfeited some of their maritime activities in the Tyrrhenian Sea and elsewhere, first to the Ragusans, who had extended their commercial stretch during the first half of the sixteenth century when the spice trade in the Mediterranean had dried up and when Venice was at its most vulnerable.[114] These merchants took over the transport of Sicilian wheat and salt; assumed control of long-distance voyages to Spain, the Atlantic, and the Levant; and roamed the shores of the eastern Mediterranean for grain.[115] Later, the void left behind by the egress of the Genoese was promptly filled in by the northerners. In brief, the Habsburg empire may have come to span from the Mediterranean via the Atlantic to the Philippines, and the Ottoman empire may have expanded south, incorporating the Mamluk lands as well as the shorelines of North Africa, yet the city-states remained as forces to be reckoned with during the 1450–1650 period.

The partitioning of the Inner Sea from the 1380s and the systemic inclusion of trade in staples in the region's economic flows forced the Venetian and the Genoese merchants to anchor the core of their operations in their respective halves of the basin in a much firmer fashion than before. Consequently, the Venetian merchants found themselves immersed in the established trading world of the Levant, partaking in the rich spice and pepper trades, whereas the Genoese merchants spearheaded attempts to establish a regular connection to the Baltic, ventured deep into Africa in search of gold and slaves, and, in the process, considerably widened their theater of operations. With the settlement of rivalry in the Levant after the 1380s and the clear demarcation between these rival cities' spheres of operation, the distinction between their modus operandi became particularly pronounced as well. Surely, the organizational and administrative differences in the operations of the dual maritime empires of the Inner Sea had been in place from relatively early on. Historically speaking, "in Venice, the state was all; in Genoa, capital was all."[116] The Serene Republic exercised strict control over its territorial holdings, keeping a close watch on the colonial governors it appointed to administer its overseas empire, whereas the free-for-all approach typical of Genoa placed its possessions in the hands of independent families who often acted on their own rather than in compliance with the policies formulated in the mother city.[117] In the Levant, the necessity to deal with established empires, in charge of time-honored and far-reaching commercial networks over which the merchants of St. Mark had no or little dominion, necessitated the services of, and further strengthened, the Venetian state.[118] And in the case of Genoa, the open-to-all environment of the western Mediterranean and the Atlantic in particular, shaped by colonization, settlers, and plantations, gave free rein to, and strengthened the hand of, men of money who built in this offshore environment a world in their own image.

If changes in political overlordship failed, as argued above, to fundamentally alter the hold of the city-states over the Mediterranean in the 1450–1650 period, the reasons were deeply rooted in the ongoing revisions in the world-economy's division of labor.[119] These revisions were hastened by the disruption in the spice trade during the first half of the sixteenth century, when developments along the northerly Antwerp-Augsburg axis—which spanned between the two headquarters of the Fuggers family—overshadowed those that transpired across the Mediterranean.[120] Some of these revisions had already been firmly put in place in the fifteenth century and gathered momentum during the interruption in the spice trade in the first half of the sixteenth century, when merchants of the Mediterranean were forced to look beyond the rich trades of the Levant. Over time, the spatial redistribution of production underwent revision due to the westwardly migration of commercial and relatively large-scale cultivation, inter alia, of sugar and cotton to locations outside first the eastern, and later the western, flanks of the Mediterranean; the oriental crops eventually found a permanent home in the Atlantic.[121]

The degree to which the eastern Mediterranean served as the main theater in the dissemination and acculturation of oriental crops can be shown by the fact that when the western shores of North Africa, part of the *patrie des Islams*, acquired access to these products during the thirteenth and fourteenth centuries, it was because of the dealings of the Genoese merchants (of Visconti denomination) between the Levant and North Africa and not the terrestrial conduits that reached Andalusia via the Maghreb.[122] This meant that the situation had changed drastically by the fifteenth century. By then, these shores had taken on the responsibility of serving as sugar and cotton fields as they fell under the sway of the Genoese and Portuguese merchants.[123] As a result of this westerly vegetal migration that stripped the *Ultramar* of its precious crops and altered the areal distribution of the oriental crops' principal sites of cultivation, which we outline in the following section, the loss by the twin city-states of their colonial holdings in the eastern Mediterranean was no longer as germane to their economic well-being as it would have been a century or two before, more so of course in the case of Atlantic-bound Genoa than Venice.

Yet, owing to the influx of silver into the eastern Mediterranean, demand for lucrative goods remained strong until the 1620s, if not the 1650s: the islands and the coastal plains dotting the Mediterranean went on to produce cotton and sugar. With the onset of the seventeenth century crisis in 1620/1650, the revisions in the global division of labor started to weigh in on the basin as the "sugar revolution" started to brew in earnest along the shores of the Caribbean.[124] In the sixteenth century, ships were carrying sugar from the Madeira islands to Constantinople. Challenges to the main custodian of the sea, the Serenissima, therefore came primarily from within and without the Mediterranean, essentially from Genoa, notwithstanding the overshadowing presence of the two expansive empires, the Habsburgs and the Ottomans, who confronted each other repeatedly during the course of the sixteenth century. As we discuss later, the westward relocation of these precious crops in new sites of cultivation, by revising the regional division of labor, exempted the Genoese and, to a lesser extent, the Venetian merchants from the full impact of the territorial aggrandizement of the Ottoman and Habsburg dynasties.

An additional revision to the Mediterranean-wide division of labor that proved to be of signal import to the leading city-states in their struggle to preserve power was relatively new: the process of import-substituting industrialization. The spread of manufacturing, a strategy pursued by the city-states from the sixteenth century (in the fields of textiles, glass-making, mirror-making, and chemical and metallurgical industries in Venice) altered the nature and composition of economic flows traversing the Inner Sea to the benefit of the Serenissima.[125] Import-substitution not only reduced the volume of Venice's demand for wares from "overseas," thereby improving its multilateral merchandise balance, but also gave it an edge in the realm of manufacturing, which eventually came to complement, in the latter half of the sixteenth century, its longtime superiority at sea. The onset of the seventeenth-century crisis, by sharpening competition among textile producers, eventually brought the wave of urban manufacturing to a close, undermining the city-states' industrial edge that had given them the opportunity to extend their heyday in the long sixteenth century.[126] By the end of the sixteenth century, silk, cotton, and, to a lesser extent, wool from the Ottoman seaports on the Aegean Sea and wool from the Iberian seaports of Bilbao, Cádiz, Barcelona, and Alicante had come to supply the textile industries of the Italian city-states and the Anglo-Dutch manufacturing complex.

The imperial *redux* of the long sixteenth century was not hence in any way an insurmountable obstacle for the twin cities in their search for and exercise of exclusive control over Mediterranean-wide economic flows. Granted, despite the return of the spice trade to its former venue in the second half of the sixteenth century and the assumption by the Genoese *nobili* of the task of transmitting the American silver to the Levant, an overwhelming volume of North Sea–bound pepper and spices rounded the Cape of Violent Storms after the 1630s, bypassing the Inner Sea. But the void left behind by the spice trade was filled almost immediately by silk, before long by coffee and drugs, and eventually by cotton textiles and cotton[127]—similar to the transformations the Black Sea trade underwent after the collapse of Timurlane's empire whereby the rich trades via central Asia were replaced by the commerce in local timber, grain, Caucasian slaves, fish, and caviar.[128] Though decreased in volume relative to the sixteenth century, economic flows that crisscrossed the region after the 1650s were still considerable.[129] And, as to be expected, the share of the rich trades in these flows had shrunk, both absolutely and relatively.

Inevitably, the Venetian merchants lost ground on the morrow of the northerly shift in the epicenter of the world-economy. These men of trade suffered as their rich trades that had for centuries been of exceptional import withered away steadily, yet the presence of the Venetian commercial navy in the trading world of the Mediterranean remained substantial. Of the 4,421 ships that frequented Livorno and Trieste in 1775, over 4,000 were Venetian.[130] Venice's share of the total tonnage was less than these figures would suggest due to the smaller size of its ships, but its port and merchants remained active. Both city-states recrafted their skills, changed their theaters of action, and, whenever the opportunity presented itself, returned to the Levant, as did the Venetians during the Napoleonic Wars, taking advantage of the internment of French merchants.

In a similar fashion, the Genoese bankers had to live repeatedly with the misfortunes and bankruptcies of the Spanish crown in 1595, 1607, 1627, and 1647. The bankers may have lost their ability to husband and steer the world-economy's wealth, but the city remained an important silver market. During the seventeenth and eighteenth centuries, close to 50 tons of silver was shipped each year to the trading world of Asia via the Mediterranean.[131] Hence even at the end of the seventeenth century, "huge shipments of silver were still arriving in Genoa in the holds of English, Dutch vessels."[132] And not surprisingly, the Genoese still controlled a substantial portion of the trade that passed through Seville during the later seventeenth century: their share stood at 19 percent in 1670.[133] When the occasion arose, again as in the 1670s, they carried base money, in return for silk, to Ottoman ports in want of monetary metals.[134] In the eighteenth century, the Genoese bankers financed much of the prosperous trade that Marseille conducted with the Levant.

So the Venetian merchants and the Genoese haute finance displayed adaptability and flexibility in the face of diminished economic activity and shrinking fortunes. What was of immense significance in this show of resilience was first and foremost the role played by the Genoese bankers in handling the wealth passing through the Habsburg realm from the mid-sixteenth to the mid-seventeenth centuries. The financial aptitude and ability they exhibited in channeling American silver into the basin allowed the waning of the Mediterranean to unfold in an almost orderly and leisurely fashion. However inexorable the autumn of the Italian city-states was, no single power emerged to effectively challenge them in their millpond, much less to displace them. The sway the new commercial fleets bore over the Mediterranean, even after the age of the Genoese, remained either tenuous or short-lived, as in the cases of the British and the French, or the sway of operation of the emerging powers turned out to be limited at best, as in the case of the Greek merchant marine.[135] The easterly relocation of Jewish merchants and the position they came to occupy in the wholesale trade of products especially valuable to the Venetians (such as wool, camelots, and alum) proved to be a short-term phenomenon.[136] The Safavid silk trade, on the other hand, mostly land-borne and presided over by Armenian merchants, lacked the depth and breadth the city-states' commercial operations possessed.[137] Surely, the French merchants may have attained a bigger role in the Mediterranean in the eighteenth century, and the playing field may have gotten more crowded with the rise of the renowned Greek orthodox merchant, all at the expense of Venetian thalassocracy. But the fact that *caravane*, or carrying, trade grew in volume and value also testifies to the inability of new competitors to establish a stable and enduring exchange network. After all, they were obliged to resort to offering their maritime services to raise cash in order to be able to pay for their purchases, which was hardly the case at the height of the Venetian and Genoese rule.[138] The shift in the center of the world-economy away from the Mediterranean and the economic devolution that ensued were the reasons, and excellent indicators of, why no single power was able to establish command in the Mediterranean.

In the absence of a lasting challenge, merchant republics not only kept their presence, albeit dwarfed, in the Mediterranean, they also created additional arenas of operation. As discussed in chapter 3, they crafted an extensive putting-out system, built and presided over an elaborate network of commodity and monetary flows, and assumed a new and expanded role in the terrestrial trade in the Balkans and beyond.[139] Hence, to rely solely on the volume of Venice's maritime trade or the size of its fleet in order to gauge how it fared in the twilight of its golden years fails to do full justice to its ability to diversify its operations or its significant involvement in terrestrial trade in the seventeenth century. Eighteenth-century Venice may have scarcely resembled its former self in its days of glory, yet its commercial presence in the Levant was on the wax again as all northern competitors had long left the Mediterranean to contest in the newly incorporated regions of the world-economy.[140] Even in the end of their days, the Italian city-states went on to jointly exercise considerable influence over the fate, and set the direction, of change in their *Mare Nostrum.* The unity of the Mediterranean was not fractured, as is usually taken for granted, as its autumn set in, but was recast on a new basis by the very city-states whose unchallenged reign was coming to an irreversible end.[141] The destiny of the Inner Sea cannot thus be adequately mapped unless the path set for the basin by the Italian city-states during the seventeenth and eighteenth centuries is sketched in its general outlines, if not depicted in detail.

In this frame of reference, the demise of the once powerful city-states shapes the course of Mediterranean history and provides it with a certain degree of unity and coherence, not unlike the one patterned by their rise in the fifteenth and sixteenth centuries.[142] This is another way of saying that holistic analyses of the Mediterranean, customarily reserved for its golden age, should not be restricted to its heyday only, but extended to its decline and fall as well. By the same token, however, it does not follow that developments that temporally framed the waning of the Mediterranean and fashioned the destiny of the basin should be exclusively viewed through the prisms of the twin city-states. The misfortunes of these twin city-states cannot necessarily be equated with the misfortunes of the basin at large. However vexing the arrival of northern ships and merchants may have been for the Venetians, this was not necessarily the case for all the denizens of the Mediterranean: for the Ottomans, for instance, it implied a change in the cast of characters, giving the Sublime Porte the chance to pit the northerners against the Venetians in order to break the stranglehold of the latter.[143] It was after the arrival of the northerners in the Mediterranean that the Venetians' perception and portrayal of the Ottoman court underwent definitive transformation.[144] Or when the age of the Genoese came to a close, the position of the bankers to the Castilian throne was immediately filled by the

Portuguese *marrano* bankers.[145] All the same, in plotting the region's odyssey, the significance of a holistic perspective framed with reference to the dynamics of the erosion of the city-states' dominance in the Mediterranean cannot be emphasized enough. It is therefore within this holistic frame of reference that the following section analyzes the transformations the basin witnessed during the long sixteenth century, with particular emphasis on the age of the Genoese, and shows how this age sealed the trends of the 1450–1560 period.

Age of the Genoese

The Mediterranean, at the height of its grandeur, encompassed an economic space much more expansive than the sea itself, even though it was endearingly referred to by the Venetians as *Mare Nostrum*. The Inner Sea was then part and parcel of a larger maritime economic space, the "Greater Mediterranean" à la Braudel.[146] By the mid-seventeenth century, however, the once vibrant, cosmopolitan, and expansive sea had turned into a mere millpond, a pale shadow of its former self. The spatial expansion of the capitalist world-economy from 1450 onward had reshaped the global economic space such that the tributaries of the Greater Mediterranean, originating in the Indian Ocean, the Atlantic, and the Sahara, were all rerouted away from the Inner Sea. Despite the imperial *redux* of the long sixteenth century, three processes radically transformed the economic infrastructure of the Mediterranean to the benefit of the city-states more so than the empires, despite the latter's tendency to expand territorially during the long sixteenth century.[147] Three-pronged, these processes were far-reaching and deep-seated. Even though they could be traced back to the period following the 1450s, they gained momentum during the age of the Genoese, from the 1550s.

First, the geographical mobility invested in the cultivation of oriental crops, which transported them to the islands dotting the Mediterranean where they were cultivated on a relatively more substantial scale, exempted the city-states that had remained in the forefront of this vegetal relocation from the territorial aggrandizement of the empires in the long sixteenth century. Even the loss of Cyprus, coming when it did, two decades after the spice trade had resumed, was not as deleterious as it would first seem, since an annual payment of 100,000 ducats to the Sublime Porte helped Venice to recuperate some of its possessions.[148] Second, the proliferation of cottage industries across the length and breadth of the Mediterranean boosted demand for raw materials, like cotton and silk, the commerce of which was largely controlled by the mercantile establishments of these city-states. Endowed with a plenitude of resources and skills, commercial as well as financial, merchants of the twin cities naturally took full advantage of the geographical dissemination of manufacturing into secondary towns and the deep countryside owing to the splendid reach of their operational terrain. The bypassing of urban locations and municipal regulations that governed manufacturing in urban centers widened the fields of operation of these men of trade. Cotton and silk played a key role in the formation of a wide range of textiles, and facilitated the creation of invaluable niche markets where competition in woolen textiles could be fended off to a large degree. Third, the resumption of the spice trade in the Levant, albeit only until the 1620s, and the Castilian throne's financial dependence on Genoese lending until the mid-seventeenth century engendered a new rapprochement between the city-states and imperial seats of power. The gradual and northerly shift in the economic hub of the commercial world of the region from the Levant to the Aegean and the growing presence of Istanbul in matters Ottoman on the one hand, and the rise of Atlantic trade and the establishment of Madrid's crushing presence in Castile on the other, generated two new poles of growth at opposite ends of the basin during the latter half of the sixteenth century.[149]

These three processes recast the relations between the empires and city-states, and enabled the leading cities during a time of imperial aggrandizement to evade the immediate consequences of the redrawing of the map of the Mediterranean during the 1450–1650 period. All three served Genoa better than its sibling, but this did not mean that the Mediterranean that remained in Venice's embrace failed to enjoy the economic benefits of the long sixteenth century.

The first trend mentioned above was the migration of oriental crops and commercial arable production. With the shift in the center of gravity of economic activity from the shores of the Mediterranean to the Baltic and the North Sea and, later, the Atlantic, the economic might of the Italian city-states came to be dwarfed, during the age of the Genoese, by that of Amsterdam (and the United Provinces). This meant that, on the one hand, key industries such as shipbuilding and textiles were now chiefly located in regions "beyond the olive-tree line"; the rich trades changed venue and character under the hands of northern merchants; new lines of specialization such as brewing were crafted and mastered to the north of the Alps; and it was there that the fishing and dairy industries registered their most striking advances.[150] On the other hand, labor-intensive production, such as sugar and cotton, became more and more the preserve of slave labor and of the Atlantic, where such labor was to be found in profusion and where extensive stretches of land for such soil-exhausting crops could be firmly secured owing to the territorial aggrandizement of the Habsburg and Portuguese empires.[151]

The exodus from the Mediterranean was not confined to oriental crops alone: the dynamic that allotted the commercial cultivation of sugar to the realm of coerced labor was systemic, and thus operative in cereal agriculture as well. Commercial grain production shifted away from the Mediterranean to the north, to the east of the line running from Hamburg to Venice via Vienna. Correspondingly, large-scale agricultural production was undertaken by those—mostly nobles—who had the wherewithal to successfully institute in their dominions or estates an agrarian system that did not rely on the whims of smallholders. The nobility, after all, had the capability to coax and cajole producers into serfdom.[152] The long sixteenth century, therefore, by extending the world-economy's networks of integrated production and trade into the Atlantic, effectively exposed the producers in the basin to the vagaries of the world market, now centered in the Atlantic. And the exigencies of the new axial division of labor became fully manifest during the seventeenth-century crisis. With slavery expanding its realm in the Atlantic and serfdom in east of the Elbe, the basin found itself in an extremely precarious position with its sharecroppers and peasantries.

The magnitude of the flight of oriental crops from the coastal plains was thus amplified by the northerly shift of commercial grain production to Silesia, Pomerania, and Poland. The departure of commercial grain production to places within the compass of the Dutch merchants rather than that of the merchants of the Italian city-states dealt a mortal blow to commercial cereal cultivation in the basin and its select site: the plains. Consequently, the time-honored granaries of the Mediterranean—Sicily, Apulia, Puglia, Genoese and Venetian "Romania," and Sardinia—not to mention occasional providers, like North Africa, became less reliable and irregular suppliers.[153] This was unsettling for the Genoese and Venetian merchants, not because the provisioning of the city-states was in peril of disruption, but the ability of the merchant republics to re-export lesser-quality grains imported from the Levant waned. As cereal culture in the plain of Foggia in Apulia, the plain of Alèria in Corsica, the Campidano in Sardinia, the Guadalquivir plains of southern Iberia, and the plains in Sicily contracted in acreage, the affluent and discriminating consumers of the wealthy cities of the basin came to rely heavily on high-quality and expensive grains, wheat in particular, imported from Romania and from the *contadi* of the city-states. Lesser-quality and cheaper grains that were purchased for re-exportation were ordinarily provisioned by a second rung of providers located mostly in the Levant. These lesser grains originated in the lowlands of Thessaly, Thrace, Macedonia, the eastern shores of the Aegean, the Romanian lowlands on the shores of the Black Sea, and Egypt, and were re-exported by the merchants of the city-states to different corners of the Mediterranean.[154] The shrinkage in the volume of commercial cereal flows weakened the city-states' ability to redistribute and re-export the imported grain to satisfy the needs of their far-flung empires.

Yet agrarian producers in the region, even when subjected to the vexations of the tax-farmers and landlords, lay or ecclesiastical, were not alienated from their lands for most of the seventeenth and eighteenth centuries. Doubtless, the Ottoman prebendal system, closely regimented in the sixteenth century by the imperial bureaucracy, was undermined by the rise of tax-farming, which was later transformed into lifetime tax-farming; the end of the *beau seizième* in France was marked by the growing concentration of land ownership under the hands of a few; the sale of public lands in Iberia turned lands vital to the survival of free-holding peasant households into private property; and the control the Genoese merchants exercised over producers in Sicily and Sardinia turned suffocating during the seventeenth-century crisis due to shrinking demand for cereals.[155] These measures increased the pressure on smallholders in one way or another. At any rate, agricultural producers in the basin may have turned into sharecroppers, as in Tuscany and Provence, or their tax burden may have gotten heavier than previously, as in the Ottoman and Habsburg lands. But they scarcely ended up as serfs or slaves. In the trying times of the seventeenth-century crisis, however, it was coerced labor, in the form of slavery and serfdom, which prevailed, and not unbound labor. Unarguably, the availability of coerced labor immensely facilitated the rearrangement of the global division of labor around the centrality of the Atlantic and the Baltic Sea.[156]

Ironically, the westward and northward displacement of commercial crops sown in the Mediterranean, a process that ultimately deflated the basin's economic worth and undermined the Serenissima's monopoly over it, was initiated by the Venetian merchants themselves. Their desire to establish exclusive control over the trade of lucrative crops had led them from early on to relocate the cultivation of such crops in the islands and along the shores of the Mediterranean easily accessible to, if not under the control of, the merchants of the Signoria. As long as Venice remained the organizing center of the Mediterranean economy, it naturally had unhampered access to the Byzantine *hyperperum* as well as silver from the mines of Serbia and Hungary (and gold from al-Takrūr relayed by the Genoese merchants). Once the order it built came under challenge, as in the first half of the sixteenth century, its main vulnerability—difficulty of procurement of precious metals—was then fully exposed.

The relatively easy access Venetian merchants had to precious metals until the 1515–20 period was later denied them by the loosening, with the change in spice routes, of their command on economic flows crisscrossing the Mediterranean and by the opening of new silver and gold mines across the Atlantic.[157] Since the handling, first, of German silver and American gold and, later, of silver from the Potosí mines largely escaped Venetian control, the city's mercantile establishment was obliged from early on to search for ways to reduce the volume of its dispatch of precious metals to the East. Venetian merchants' declining ability to preside over flows of precious metals became much more pronounced during the Potosí silver cycle of the 1540–1640 period.[158] As to be expected, the spread of the cultivation of crops of Indian Ocean origin into those reaches of the Mediterranean that were or came under the direct administration of the Serene Republic—Cyprus for one—initially served Venice handsomely.[159] The crop migration helped its merchants ameliorate their balance of payments due to the decrease in the volume of precious metals disbursed to pay for their imports. The colonial empire the Venetian merchants built, from Crete to Cyprus, was put to use almost exclusively for the cultivation of such crops, sugar and cotton foremost among them. As long as the dissemination of the cultivation of these lucrative crops was confined to the eastern Mediterranean, where the Venetian colonial holdings were dispersed, the Serenissima profited from it. The westerly migration gained a new character when it reached from the region that stretched from the Balearics to São Tomé island, and under the supervision of the Genoese, Portuguese, and Dutch merchants. Time and again, sugar, more often than not along with wheat, occupied the place of pride in steering the process of colonization in the recently annexed territories.[160] The benefits that accrued to Venice from the acclimatization of these crops proved to be short-lived at best, if not ephemeral.

Ultimately, the gradual shift in the production sites of these remunerative crops turned out to be more beneficial to Venice's chief rival, the Genoese, as well as the Portuguese and Flemish, merchants.[161] The city of St. George drew immense benefits from this vegetal migration, surely more than the city of St. Mark did. First, as a result of this migration, sugar and cotton (and to a lesser extent, silk) production increased considerably on the islands off the coast of the Italian peninsula as well as in the western Mediterranean and the Atlantic: these islands were opened up to commercial cultivation largely under the control of Genoese merchants and financiers. The materialization of opportunities to encourage or, in most likelihood, coax producers in these newly incorporated regions to shift to the cultivation of oriental crops in sites within their immediate grasp permitted the Genoese merchants to establish their dominion over the products of a region, the Levant, from which, commercially speaking, they had been conspicuously absent.[162] After all, sugar production took hold not only in Venetian Crete and Cyprus, but also in Sicily, Malta, and parts of North Africa, commercially under Genoese suzerainty.[163] This vegetal migration moved the commercial cultivation of oriental crops away from the eastern Mediterranean, and brought about the emptying of coastal plains where these commercial crops had been previously grown, and mostly as monoculture.[164]

Second, the Genoese merchants' ability to access and utilize slave (and Guanche) labor, easily available in the western Mediterranean owing to the ventures and raids of the Portuguese into Africa, gave them additional latitude in expanding the scale of production.[165] The Genoese Vivaldi brothers may have suffered loss of life in their attempt to reach the Indian Ocean in 1291 via the Straits of Gibraltar, the year Acre was lost to the Mamluks. Yet, thanks to the relocation of sugar and cotton cultivation, the city of St. George did not have to capture Acre by force or belabor to undo the Venetian monopoly in the Levant trade in order to establish its commercial foothold in the region to deal with, and trade in, oriental goods. All it had to do was to acclimatize these precious crops to their new habitats and "vulgarize" their cultivation.[166] In this respect, the fact that its empire was centered in the western and Atlantic Mediterranean could not have been more serendipitous.

No doubt, Venice did profit from the relocation of oriental crops in lands under its dominion. But during the sixteenth century the wealth circulating through Antwerp and Seville, managed by the Fuggers first and the Genoese banker-financiers later, largely escaped the city. The Venetian merchants, given their inability to replenish their precious metal stocks, were forced to trim down their merchandise deficit by reducing the import of oriental crops which, in turn, diminished their need for bullion.[167] In other words, the westerly shift in the production of eastern crops provided breathing room for the Venetian merchants in the short run. Over the long haul, however, the dissemination of the cultivation of oriental crops from the western Mediterranean to the Atlantic unavoidably reduced Venice's share in the north-south entrepôt trade that was of immense significance to its dealings with the cities north of the Alps. All in all, then, the migration of oriental crops further reduced Venice's pivotal commercial role in the Mediterranean over the long haul. A process thus spearheaded by the Venetian merchants eventually worked to the city's arch rival's advantage, much to the chagrin of the Signoria. Genoa may have reaped the benefits of this Atlantic-bound movement, but it needs to be emphasized that, despite the proliferation in the number of sites producing sugar (and cotton, to a limited extent, since Swabian towns obtained a portion of the cotton they needed for their fustians from Brazil), the ebullience and the riches of the 1560–1620/1650 period lifted the fortunes of most. Moreover, as we discuss in the following section, the economic vitality of the western Mediterranean due to the opening of the Atlantic world and the consolidation of Madrid's centrality in it, and of the eastern Mediterranean due to the emergence of the Aegean and the Balkans as vital economic zones and the contribution this northerly economic shift made to Istanbul's economic centrality did offset the gradual contraction of markets from the turn of the seventeenth century. When seen from the vantage point of the Mediterranean, it was only after the mid-seventeenth century, after the anchoring of sugar plantations on the Brazilian coast, that the new division of labor manifested itself in full force and displayed the structural weaknesses of the basin's economic edifice.[168] If the Genoese traders and financiers thrived on account of the part they played in the opening of the Atlantic from the fifteenth century, after the 1550s they resorted to royal finance. The move proved to be most timely because of the crowding out of the Atlantic trade after the arrival of the Portuguese, Flemish, and northern merchants.[169]

The second trend was closely linked to the first because the aforementioned westerly vegetal migration that took a toll on Venice's ability to serve as the main provider of oriental crops and spices had an adverse effect on the profits it derived from this trade. The progressively mercurial nature of the profits accruing from the Levant trade, especially during the hiatus in the spice trade and later due to the intensification of competition after the arrival of northern ships in the Mediterranean, served as an incentive for the city's patricians and merchants to invest their fortunes in urban manufactures and, toward the end of the sixteenth century, in land.[170] The strategy was at first an unqualified success and gave Venetian woolen cloth a reputation as sterling as its ducat in the late sixteenth and early seventeenth centuries.[171] In the process, the Veneto became part and parcel of the Venetian industrial landscape, at times by design due to the putting-out strategies devised by the merchants and at other times by force when cloth producers in the secondary and lesser cities and the countryside managed to assert themselves as the exclusive producers of certain fabrics and textiles.[172]

For one, the factors underlying the growing preference for investment in urban industries in the sixteenth century was not unique to Venice, but stemmed from the large-scale transformations that followed the Mediterranean's loss of centrality in world-economic flows. And similar developments took place elsewhere in and outside the basin. As the Genoese and Portuguese merchants joined their Venetian counterparts in securing the transfer of oriental crops from the basin to the tropical climes of the Atlantic, Venice was accompanied, in the industrialization drive of the sixteenth century, by the central players of the Mediterranean world. Following suit were Geneva, Lyon, and Genoa, as well as the cities of the north, Antwerp in particular, especially after the city fell from favor with the rise of Amsterdam. As a result, the producers located in the established textile centers of the time, Flanders and the northern Italian peninsula, which had historically served as hosts to vigorous manufacturing activity, were joined in the latter half of the sixteenth century by a new group of contenders. Nonetheless, despite growing competition, the addition to the Venetian businessmen's repertoire of re-exports of manufactured good that could be exchanged against spice imports, at a time when their control over precious metal flows was at an ebb, contributed to an improvement in the city's balance of payments.[173]

By the turn of the seventeenth century, Venice's textile industry had developed into the city's principal economic activity. Successful though the industrialization drive was, it proved to be relatively short-lived, very much like the success attained by the westerly transfer of oriental crops. The movement lost its initial stimulus roughly in the 1620s.[174] Confronted by fierce competition from the north on the one hand and taxed by high-wage levels exacted by the city's urban labor force on the other, Venetian merchants, unwilling to relinquish their control over manufacturing, found a solution to their dilemma by moving textile production out of the established industrial centers, and mostly into the Veneto, as confirmed by the impressive growth of sericulture in Vicenza and Verona at the end of the sixteenth century.[175] In a sense, during the seventeenth-century crisis, the Venetian textile industry took refuge in the same strategy that it had subscribed to when confronted by rising wages in the 1250/1300–1450 downturn.[176] It was not cotton or fustian industries that had to be relocated this time but the wool industry. All the same, that the city's woolen textiles encountered fierce competition from the north during the seventeenth-century crisis led to a similar outcome: some or all parts of the textile manufacturing process were relocated in the Veneto. Moreover, the Venetian merchants' ability to muster, transport, and deliver raw materials from as far away as Acre to the new production sites in Bohemia and Styria, and haul finished products therefrom back to the Levant that provided the most fitting solution to their dilemma in the earlier phase of the ruralization of production served them well in the second phase as well. Befitting the splendid compass of the Serenissima's commercial operations, its merchants were still blessed by their dominion over imports of cotton (and to a lesser degree, silk) into those regions where cotton and fustian industries were still thriving.[177] Yet, unlike the earlier phase, these regions also manufactured or finished woolen cloth in direct competition with the Venetian woolen output.[178] Naturally, the process was destabilized during the Thirty Years' War, but once the war was over, the relocation of woolen industries into southern Germany resumed without much delay. The Genoese merchants were also engaged in the cotton trade, but the cotton they controlled was of inferior quality, imported largely from Cyprus, Asia Minor, and the Balkans, and not from the Levant.[179]

If the woolen textile industry moved out of its urban niche and into the countryside, this did not benefit the Venetian and Genoese merchants, because the wool trade, more voluminous and commonplace than cotton, was not the exclusive calling of the merchants of any city-state or city. The Genoese merchants, due to their hold over Iberia, indeed handled most of it owing to their control of the peninsula's southeastern ports.[180] In both cases, continuing involvement in the cotton, silk, and wool trades may have generated revenue for the traders of the city-states, yet it was evident that neither the Genoese nor the Venetian merchants were in a position to monopolize these markets, unlike the previous phase of ruralization of manufacturing. To the contrary, the northern merchants were in the process of gaining direct and greater access to the raw materials of the Levant. In Iberia the reduction in demand for wool by the Italian city-states correspondingly reduced the share of the Genoese in the wool market. That there was no centralized marketing structure in Iberian wool was a testament to the multiplicity of forces at play, given the widespread nature of woolen industries.

All in all, the industrial renaissance the city-states underwent in the latter half of the sixteenth century initiated, or hastened, a process that facilitated the spread of manufacturing outside the established, urban manufacturing sites of the time. With the steep contraction in demand after the 1620s/1650s and the proliferation in the number of competitors, the processes of outsourcing gained further momentum. The grafting of cottage industries onto the industrial or manufacturing landscape of the Inner Sea proved to be enduring. With the onset of the seventeenth-century crisis, it became part and parcel of the basin's economic landscape. The migration of the textile industry to locations where labor was relatively cheaper and, more important, unorganized, was an option available to the Venetian mercantile establishment.[181] What is more, the rural infrastructure vital to the dissemination of production on such a scale was already prepared during the age of the Fuggers, when the first wave of invasion of American treasury percolated into southern Germany, an invasion that set the stage for the interface between rural producers and merchant companies.[182] The groundwork for the putting-out system was already put in place earlier and the employment of this existing network by the Venetian merchants was a welcome development for the region whose fortunes had started to subside with the fall from favor, in Charles V's court, of the house of Fugger; this was one of the reasons the Venetian merchant or businessman found a ready home to conduct his activities, either as a provider of raw materials, or as a vendor of Florentine, Milanese, and Venetian textiles.

Thus, the migration of the textile industry out of north Italian city-states contributed to the development of a rural industry in Ulm, Augsburg, and Nuremberg and in their vicinity or to the spread of these industries to the city-states' hinterland and via labor migration to nearby locales, as in the case of Geneva's silk industry.[183] The spread and extent of the cottage industry, based on the *verlagssystem*, reached such proportions in the eighteenth century that the advance of this largely rural-based manufacturing—a process referred to as "proto-industrialization"—has engendered a literature expressly devoted to investigating the success of this phenomenon and searching for answers as to why such a sophisticated infrastructure failed to engender a genuine "industrial revolution" in these sites of proto-industry par excellence.[184] The bypassing of the *arte della lana* and *arte della seta* and the moving of the textile industry into rural locations as well as secondary towns was a successful follow-up to the initial Venetian enterprise of industrialization: as the momentum provided by urban industries came to a halt, ruralization of production gained pace, putting the city's commercial might to full use, notwithstanding the inevitable social strains embedded in the migration of industries from their urban settings and the resistance put up by combatant urban guilds.[185] As discussed later, however, the popularity of the strategy of ruralization of manufacturing, now open to all, undermined the Venetian merchants' initial advantage. It was no longer exclusively confined to regions under the Venetian merchants' reach. More important, unlike in the case of cotton, where the supply of the fiber was under the monopoly of the Serenissima, wool was easily available, from Iberia to Anatolia.[186] Both factors were instrumental in wiping out the advantages that Venetian merchants enjoyed initially. Once again and ironically, a process initiated by Venetian merchants to preserve their hegemony worked, in the long run, in their disfavor.

To put it briefly, the industrial renaissance in woolen and silk industries in the core regions of the Mediterranean world prompted the devastation of these industries in other major textile centers of the basin.[187] The wave of deindustrialization this prompted had two end-results, both of which benefited the countryside. First, it encouraged production of industrial crops for export to Venice and the Netherlands and later to the cottage industries of the time, thus boosting the production of raw materials and industrial crops. The draining of industrial crops from the Levant and the Iberian peninsula, however, inflicted grave injury on domestic production. Second, it abetted the dissemination of industries, formerly urban, deep into the countryside across and beyond the Mediterranean into regions that were all too often poorer than those that rimmed the dorsal spine of the fifteenth and sixteenth centuries.[188] In this, the proliferation in the number of mulberry plantations from the southern shores of the Iberian peninsula to the eastern shores of the Mediterranean catered to the local as well as regional industries. Similarly, the remarkable expansion of livestock husbandry throughout the Mediterranean littoral, administered by the mighty Mesta or undertaken by the tribal and pastoral communities that tirelessly crisscrossed the Ottoman dominions, served primarily the woolen industries at home and abroad.[189] Again, however, as in the case of the exodus of oriental crops, the Genoese benefited from this second process, because their shift to royal finance absolved them of the competitive spiral into which manufacturing had been thrown, from Nuremberg to Geneva, from London to Leiden.

In a related fashion, the third and final process was the reorganization of the locus operandi of the Venetian and Genoese merchants from the second half of the sixteenth century in accordance with the diminishing significance of maritime trade in the Mediterranean for the former and with the fall of the house of Fugger, bankers to the Spanish throne, for the latter. This conjuncture strengthened the systemic integration between the city-states and empires. In Iberia, the fate of Philip II's imperial ambitions was closely tethered to the financial abilities of the Genoese *nobili vecchi* from the 1550s to the 1650s. In the Levant, the Serenissima and the Signoria established a new rapprochement once the spice trade returned to its Levantine route, a rapprochement disturbed only twice, in the 1570s over Cyprus and in the 1660s over Crete. The arrival of the northerners was facilitated by the withdrawal of the Genoese from commerce, and their venture in the eastern Mediterranean, by increasing commercial rivalry, deflated Venice's previously unchallenged supremacy.

What expedited the western-bound movement of oriental crops and the spread of textile manufacturing outside the confines of urban centers was thus the commercial sweep and financial might of the northern Italian city-states. The resultant change in the basin's geography of production, in turn, demanded a reorganization in flows of capital, raw materials, and finished goods. In concert with the extension of manufacturing networks into southern Germany, the city-states' commercial activities shifted away from their principal theaters of operation, the Mediterranean no less, and focused more on the easterly latitudes of the olden dorsal spine of manufacturing that linked northern Italy to Flanders.[190] Furthermore, the arrival of northern merchants in the Indian Ocean and the role they played in delivering the fruits of their oceanic voyages to the North Sea and the Baltic without the mediation of the Mediterranean unavoidably pruned the volume of business carried on Venetian ships.

In a twist of fate, the Mediterranean (maritime) trade the Venetian merchants were reared by and steeped in and which they had governed single-handedly for so long, was later supplemented and, at times, even supplanted by a burgeoning overland trade catering to the rich north—via the Ottoman Balkans.[191] The Venetian (and Florentine) merchants expanded their operations in the region between the Adriatic and the Baltic due partly to the wide dissemination of rural manufacturing beyond the Alps. This was the time when fairs in the region, from Cracow to Frankfurt, proliferated and thrived since they constituted the nodal points in the newly established network of collection and distribution. When viewed from the Ottoman Mediterranean, the Italian city-states' predominance, now largely in overland trade, lasted well into the closing decades of the eighteenth century.[192] The rise in the share of terrestrial trade (buttressed by riverine trade) at the expense of maritime trade remained as one of the signal features of the region boxed in by the Adriatic and the Baltic in the latter half of the seventeenth and the eighteenth centuries. This sea change registered a radical shift in established patterns of commodity flows.[193]

On the eastern flanks of the basin, overland trade between the Mediterranean and the North Sea via the Ottoman Balkans and the Adriatic, especially in leather, wax, hides, and wool, picked up in volume. Parts of this terrestrial network were closely regimented by Venetian (and at times Florentine) merchants who had by then established their presence in the region that spanned from the Adriatic to the Baltic, that is, in Nuremberg, Frankfurt, Leipzig, Hamburg, Cracow, and Lwów.[194] Wagon-trains that crisscrossed this territory may have operated under the guidance of Armenian merchants, but Venetian merchants and businessmen were actively engaged in commerce that was centered in the north of the Alps.[195] That the region already had a sophisticated commercial livestock industry provided an additional boost to the terrestrial circulation of merchandise.[196] Overall, cattle-droving, which had reached a peak at the turn of the seventeenth century, was cut back by wars, but resumed its previous pace in the eighteenth. Overland trade and the livestock industry, therefore, created a virtuous cycle: development of one automatically gave a boost to the other. The proximity of rich markets to its west and the difficulty of administering overland trade over vast stretches of frontier land sanctioned livestock husbandry's reign over the farming world: "la victoire du bétail sur la blé."[197] The rise of the celebrated Orthodox merchant to the east and south of this zone, in the Balkans, not only gave a new life to the famous Via Egnatia by rendering the connection to the Adriatic much easier, but also tightened the terrestrial transportation network, the focal point of which was to be found, not surprisingly, in Amsterdam. Only the Greek contingent of this new merchant class was sea-bound.[198]

On the western flanks of the Mediterranean, the Genoese certainly played a role of paramount significance before the 1560s in transporting Spanish wool to various Italian textile centers.[199] Yet the fact that the Genoese *nobili* had largely withdrawn from commerce after they had solidified their position at the court of Philip II reduced their further dealings in commercial ventures. Befitting their specialization in matters financial, in state loans in particular, they did proceed, however, during the seventeenth and eighteenth centuries to extend loans to a series of states, some of which were to be found in the regions where a lively overland trade flourished beyond the Alps. These states presided over lands where vibrant rural cottage industries had come into being and went on to proliferate from the mid-seventeenth century onward. Likewise, where sea-borne trade exhibited signs of revival and growth, as in southern France, there Genoese men of money were to be found.[200] In sum, the Genoese financiers and Venetian merchants, rather than function in two separate realms as they had in the sixteenth century, were now jointly overseeing the economic well-being of their *Mare Nostrum.*

To recapitulate, as the seventeenth-century crisis started to unfold around 1650, the three sets of transformations inaugurated by the Venetian (and Genoese) merchants had already affected all aspects of the economic life in the Mediterranean and redefined the Serenissima's position in it. The nature of agricultural production, for one, was transformed due to the westerly migration of oriental crops, tipping the balance in favor of producers on the shores of the Atlantic. The assumption by lands lying between the Elbe and the Oder, in Poland, Prussia, and Pomerania, of the role of the new granary of the world-economy helped transform agriculture in the Mediterranean. In a similar fashion, the region's manufacturing sector had to endure the invasion of woolen textiles, first, from the Italian city-states and, later, of its cheaper varieties from the north.[201] The eastern and western extremities of the peninsula, Salonika as well as Segovia, painfully witnessed the vagaries of this invasion. From the 1550s to the 1650s, the major urban industrial centers of the basin lived through a lengthy period of deindustrialization, be it in woolen or silk textiles.[202] Rural industry, taking advantage of the vacuum created by the decline in urban industries, built a strong foothold in and around the basin and beyond. That is, the process of ruralization of manufacturing did not disperse solely into those regions north of the Alps. Its compass was much wider than that. Manufacturing activities, then, followed the path opened up by Venetian merchants: they disseminated into the countryside. And lastly, the locus operandi of the men of money from the twin cities shifted as the need to cater to the opulent north overrode their previous specialization. In the eastern Mediterranean, the new conjuncture reordered economic flows traversing the region and favored terrestrial trade. In the western Mediterranean, increasing pressures on profits due to the multiplication of those engaged in Atlantic commerce compelled the Genoese to shift from imperial trade to royal finance. The commercial geography of the region was drastically altered during the century of the Genoese, to the detriment of the illustrious port cities of Cairo and Seville, which flourished thanks solely to the Mediterranean trade, but to the benefit, among others, of Cádiz, Livorno, and İzmir, points of embarkation and disembarkation for the merchandise destined for, and originating in, the north.[203]

The Waning of the Mediterranean

If, in the final analysis, developments associated with the age of the Genoese did not prove to be fatally injurious to the economic health of the cities of saints, this was due to the fact that the three trends summarily outlined in the previous section neither lost force nor were reversed with the closing of the second sixteenth century. Set into motion in the economically prosperous 1450–1560 period, these trends—agricultural, industrial, and commercial in nature—gathered force during the age of the Genoese, and came to set the tenor of change in the Inner Sea from the 1650s to the 1850s.[204] For one, the migration of the peripatetic, labor-intensive commercial crops came to a temporary halt with their long-term residence in the isles of the Atlantic—the center stage of the world-economy after the seventeenth century. Here, its principal sites of cultivation continued to change, but along the Brazilian coast and within the Caribbean basin. The tremendous increase in the scale of sugar cultivation following its relocation to the Atlantic attests to this. The amount of sugar produced in its Mediterranean center stage, Cyprus, and even under the rule of the Sugar King Cornaro family in the fifteenth century, reached at the most a few hundred tons. In the sixteenth century, plantations in Brazil were shipping 1,600 tons a year; in the seventeenth century, sugar shipped annually from Jamaica was around 72,000 tons; and in the eighteenth century, Santo Domingo alone "was to produce as much if not more." The mooring of sugar production in the Atlantic, with slave labor occupying the center stage, took place from the 1650s precisely when the seventeenth-century crisis was just settling in the Inner Sea and beyond.[205] In brief, the westward movement of oriental crops sustained its momentum over time, and the volume of production of these crops was amplified beyond imagination.

Commercial cereal production, too, was firmly anchored in the east of the Elbe, and the grain trade resulting from it soared in volume. The grain trade, which fluctuated somewhere between 100,000 and 200,000 tons in the sixteenth-century Mediterranean, reached a maximum of 600,000 tons when relocated closer to the shores of the Baltic.[206] This was because, first, grain production was firmly affixed in its new northerly location under the strict supervision of a landlord class, and remained so throughout the seventeenth century. The free peasantry witnessed a worsening of its lot from the sixteenth-century. And when commercial agriculture showed any signs of further relocation, it was to the north of the Elbian plains, in Sweden (Livonia, Estonia, Scania), and later in England and America, rather than to its south.[207] The gradual contraction in cereal lands along with the northerly migration of commercial agriculture precipitated the dissolution of extant structures of provisioning. During the seventeenth and eighteenth centuries, the plains encircling the Mediterranean were visibly thin of people.

The low population density that resulted from the contraction of commercial agriculture was attested to by Giuseppe Galanti's dismal description of Puglia in the eighteenth century as "without inhabitants" and "infested by a great number of serpents," Henry Swinburne's frequent references to Catalonia's "scarcity of grain," and C. F. Volney's portrayal of most parts of the Levant in the eighteenth century as "ruined" and "desolate."[208] Yet, as we will see, the void left behind by the retreat of commercial agriculture from the plains of the Mediterranean was later filled by horticulture, arboriculture, and animal husbandry, all of which stood the test of diminished economic activity during the waning of the Mediterranean incomparably better than cereal culture.[209] The import of commercial cereal agriculture along the Mediterranean coast, much reduced after the rise of northern grain trade, was given a boost only after the turn of the nineteenth century with the consolidation of *pax Britannica:* signs of an increase in the area sown to bread grains reappeared from the mid-eighteenth century, and reached considerable proportions only during the mid-nineteenth.[210] As we will see in the third chapter, the exodus of the oriental crops was followed by the recrudescence of the tree crops of the Mediterranean, the region's "indigenous" crops. Maybe not so unexpectedly, the drawn-out process of devolution provided latitude for the popularity of oleiculture and viticulture along the shores of the *Mare Internum.*

The widespread dispersal of manufacturing, too, had a long-lasting presence across and beyond the Mediterranean until the closing decades of the eighteenth century. An extensive rural industry and the *verlagssystem* remained firmly embedded in the rural reaches of the Inner Sea. The dissemination was such that, although in the sixteenth century industrial activity had essentially been concentrated along the axis running from the Netherlands to the northern Italian city-states, in the following centuries it burgeoned on both sides of the axis as well and on a considerable scale. The eighteenth century is consequently seen in the economic literature as the seedbed of "proto-industrialization," launching the careers of the great textile centers of later centuries. The broadcloth industry, for instance, spread westward into western Flanders and northern France (Arras, Cambrai, Amiens, Reims), and eastward into Brabant, Liège, and the lower Rhineland. In the Mediterranean littoral, Barcelona and Languedoc, too, followed a similar trajectory in the broadcloth industry.[211]

The longevity of the dispersal of manufacturing into the countryside was due to the long-lasting and all-out competition in woolen industries and the inevitable pressure to cut production costs, which placed a premium on the bypassing of urban guilds and weakening of labor regulations. The tendency to disperse (textile) production to those located in secondary towns or the rural sector persisted until the late eighteenth century, when the manufacturing of cotton textiles—not woolen, where the competition was toughest—came to be cloistered, once again, in specialized settings.[212] It is ironic that the processing of cotton, a crop exempt from the competitive pressures to which both wool and silk were subjected from the sixteenth century and one transformed in the eastern Mediterranean into a vital industrial crop for rural manufacturing, was eventually enclosed within the walls of new manufactories and factories in the future workshops of the world, mostly located across the Channel. Here, cotton was an industrial crop difficult to procure, especially locally. The supplies were located in the plantations across the Atlantic: it was thus beyond the reach and scrutiny of urban guilds and labor. This rendered it an ideal crop for men of money, who due to their commercial might were mostly in possession of it, since they could impose their terms on the organization of the production of cotton manufactures. The challenge posed to cotton textiles produced by dispersed, rural industry by its factory-manufactured rival was alarming yet not lethal, for the penetration of British cloth into the far-flung corners of the region during the nineteenth century was much less speedy than in urban centers.

If the agricultural restructuring of Inner Sea during age of the Genoese hastened the relocation of commercial crops previously sown along its shores and, due to reduced dependence on monocrop cultivation, diversified the crop-mix, the dispersal of manufacturing, which brought about the development of a rural industry, enhanced the trend toward diversification by increasing demand for industrial crops, animal as well as fiber. In response to growing demand for Italian woolen textiles, the commercial production and export of wool to the city-states increased, from Córdoba to Salonika, as did that of silk, from Sicily to Bursa.[213] The differential between prices fetched by these raw materials at local and export markets invariably favored the latter. Before long, demand from Italian city-states declined, but overall demand for industrial crops did not fall as steeply as it did for other agricultural goods. Manufacturing may have moved and thrived first north of the Alps, yet the burgeoning cottage industries sustained demand for most industrial crops.[214] Demand for raw materials hence remained high due to the dissemination of textile production into the countryside for most of the seventeenth and eighteenth centuries.[215] The rural dispersion of previously city-based industries rendered the administrative organization of production a trying task. In those parts of the region where mercantile groups were not as dexterous as the Venetians in furnishing rural producers with raw materials and where procuring industrial crops remained the responsibility of the producers themselves, the dissemination of industrial crops and their easy availability to rural artisans became of cardinal import.

The draining of raw wool and silk from the local markets and the fierce competition put up by well-supplied producers in the *arte della lana* and *arte della seta* encouraged the cultivation, especially in the eastern Mediterranean, of cotton for local industries. Cotton cultivation expanded significantly and almost steadily in the seventeenth and eighteenth centuries. Wherever possible, it was produced in plantation-like units but mostly by petty producers despite the ecological demands of the crop. Overall, then, sustained local demand ensured the survival and extension of industrial crops. The changing morphology and geography of manufacturing hence served to diversify agricultural production, and in the process facilitated the spread of cotton cultivation as a crop which, unlike silk and woolen textiles—the former an item of luxury and the latter of necessity—was not subject to fierce competitive pressure.

The return of cotton cultivation to the Mediterranean signaled a dramatic change in its character: it was no longer a plantation crop, nor an exclusively commercial crop. It was produced for consumption by rural industry.[216] In other words, after the migration of cotton plantations to the shores of the Atlantic, in the eastern Mediterranean it became another addition to the petty producers' reserve, cultivated in smaller fields and mostly for local and regional markets. It lost its patrician character, and thereby assumed a more modest standing as a plebeian crop and underwrote dispersal of manufacturing. As a plebeian crop, however, it was not systematically irrigated or carefully tended, since it resided in the basin's low-lying plains, which were mostly devoid of permanent human habitation. If cotton cultivation expanded, this was partly because of the amphibious environment created by the Little Ice Age, an environment conducive to irrigation, as we discuss in chapter 5, and partly because of the demand created by the dispersal of manufacturing.[217] Cotton's reappearance in the Mediterranean stood in stark contrast to the inexorable disappearance of sugarcane from the region—with the exception of Egypt, where it was still produced but on a smaller scale than before.[218] As sugar production soared in the Atlantic, the sugar industry in the Mediterranean, from Sicily to Cyprus, declined precipitously. Just as in the case of the egress of oriental crops, which was followed by the popularity of tree crops, the percolation of manufacturing into lesser centers and the countryside swelled the demand for industrial raw materials, linen in southern Germany, hemp in the Italian peninsula, cotton in Ottoman lands, woad in Languedoc, and madder in parts of the Mediterranean. The basin, as a result, became more self-reliant than previously.

Transformations in the realm of commercial infrastructure mirrored transformations in the agricultural and manufacturing landscapes of the Inner Sea, both of which possessed long historical breadth and whose underlying processes only got stronger. The pattern of economic flows within and across the Mediterranean that took shape during the second sixteenth century and tilted the balance toward terrestrial trade was not reversed in the following centuries either. This was in part because Amsterdam and, later, London kept the economic center of gravity of the world-economy in the north, and these regions' centrality in economic flows naturally attracted goods from as well as via the Mediterranean basin.[219] With centers of accumulation fixed in the north, overland trade was conducted via Astrakhan, the Ottoman empire, and the north of the Brenner Pass and extended into the deep north and, via the Iberian peninsula, to the textile towns in Flanders: the terrestrial connection remained a constant during the seventeenth and eighteenth centuries. Only by the end of the eighteenth century did the westerly conduits of the Via Egnatia start to founder, and they fell into in a state of disrepair.[220] Owing to the vitality of northerly conduits, Italian merchants recovered some of the ground they had lost in maritime trade, even though scores of merchants, Armenian, Jewish, German, and Polish, also took part in this terrestrial trade.

The growing attraction of northerly conduits aside, the erosion of rich trades forced those engaged in them to diversify their operations. This had the consequence of merchants extending their tentacles into the countryside and financing the production of raw materials, as in the case of Cairene merchants who, with the collapse of the spice trade, started to underwrite the production and commerce of linen and sugar, both of which found their way, via Venice, to the northern shores of the Mediterranean.[221] The bankers of Lyon, after the fairs held by Genoese bankers moved to Besançon, extended their credit operations into the city's hinterland and underwrote the manufacturing of silk textiles.[222] The Genoese merchants, on the other hand, established a monopoly on the transmission of raw silk from the peninsula to the hexagon.[223] By increasing the amount of goods thrown into circulation and by orchestrating and underwriting their production (employing mostly a putting-out system), the bankers and merchants of the region broadened the expanse of dispersal of manufacturing. The basin's economic structure, that is, accommodated changes that stemmed from the diminishment in Mediterranean trade and the commercial environment that came into being, opened up new avenues of economic activity to merchants of all sizes and shapes, including those of the sibling city-states who still possessed certain advantages over their new competitors.

To wit, until the arrival of the steamship and the advent of *pax Britannica,* the Mediterranean trade, deprived of the commercial and financial infrastructure put in place by its leading city-states during the golden age of the Mediterranean, largely remained bereft of a comparable transactional network. At the height of their power, Dutch merchants operated mostly in the Indies and the Atlantic rather than the Mediterranean: this left the Inner Sea in the hands of lesser powers, the British and French empires, which never had a firm command over silver or gold flows as had the Venetian or the Genoese men of money in their heyday. The difficulty the lesser powers encountered in marketing their goods in the Inner Sea and their need to generate income through coastal shipping or by dumping lesser coins in Ottoman ports of call did not constitute strategies capable of generating a long-lasting exchange network that would nourish an ebullient economic environment. In the absence of such an expansive and tightly woven web of money and credit, the Venetian ducat continued to command high respect in the Levant, even in the eighteenth century. Loans extended by the Genoese bankers to the states in the basin became one of the most important sources of finance.[224] It was only in the nineteenth century, with the adoption of the gold standard and the establishment of a solid multilateral merchandise flow and payments system, that such an environment came into being. In the absence of such networks, local and regional networks woven by the merchant communities of the Inner Sea functioned without much difficulty but on a limited scale. At any rate, as long as opulence resided on the shores of the North Sea, all maritime and terrestrial trade routes ran, by design, into and out of it, with the Ottoman Balkans, the Iberian peninsula, and transalpine Italy serving as its overland arteries.[225] Thus processes affiliated with the afterglow of the Mediterranean during the 1560–1650 period did not lose their vigor with the onset of the seventeenth-century crisis. It was the establishment of *pax Britannica*, which reorganized the global economic space, that ended the reign of all three processes.

In all three instances, the trends inaugurated during the second sixteenth century did not come to an end with its closing in about 1650. Rather, these trends were consolidated and, better, deepened during the following two centuries due to world-economic developments. Lasting was the impress these three processes left on the tracks of Mediterranean history and, in particular, on the basin's agriculture and rural landscape, as we cover in part II at length. Suffice it to state here that the relocation of oriental cash crops and commercial bread crops, the widening stretch of dispersal of manufacturing, and the growing weight of terrestrial trade led, in unison, to the long-term retreat of commercial agriculture in the region. This retreat did not take place *in vacui.* The spatial corollary of the retreat was the desertion of lands reclaimed during the expansionary period that stretched from the eleventh to the mid-sixteenth centuries.[226] In the Mediterranean, lands that had been reclaimed and tilled during this period were those located in the lowlands and plains: these were vacated during the following two centuries. The early signs of a reversal in this trend that brought more land under the plow became manifest, albeit diffidently, from the latter half of the eighteenth century.[227] The arrival at the turn of the nineteenth century of Russian grain at the port of Livorno, northern merchants' stepping stone into the Mediterranean in the sixteenth century, signaled a decisive turn in the fortunes of commercial agriculture in the basin.[228] The repeal of Corn Laws in Britain and the ensuing division of labor this unilateral move induced opened up the possibility of organizing agricultural production along the shores of the Mediterranean on a commercial basis yet again, with the prospect of recovering ground, mostly on the plains, lost during the seventeenth and eighteenth centuries. The transfer, therefore, of the Mediterranean's economic epicenter from its plains to its hillsides and mountains, a corollary of the retreat of commercial agriculture, started with the second sixteenth century and lasted until the latter half of the nineteenth.

To recapitulate, all three processes that framed the eclipse of the Mediterranean were endowed with enviable longevity and had a lifespan of over three centuries. Even though the dynamics underlying these trends changed in the long run, the directionality of change remained stubbornly stable despite the twists and turns caused by world-economic and political developments and despite the divergence of interests between the rivalrous city-states. From early on, the Venetian and Genoese merchants may have cooperated in the eastern Mediterranean and the Black Sea and lived side by side as complementary enemies, but Venice's stronghold over the Levant turned asphyxiating for the Genoese from the 1380s on. Not only were the spheres of operation of these twin city-states clearly demarcated, but their strategies and theaters of expansion varied substantially. The Serenissima was given a new lease on life, from the mid-sixteenth century to the opening decades of the seventeenth, owing to the resuscitation of the Levant route and to its success in adhering to an import-substitution strategy. Genoa's heyday lasted longer, stretching into the latter part of the seventeenth century, owing to the accomplishments of its banker-financiers. Until then, the Mediterranean may have marched to two different tunes, but from the latter half of the seventeenth century, with Venetian merchants taking up their new posts in Cracow and fairs of Poland and Genoese merchants taking over Liguria and Marseille,[229] the differences between the strategies of these fierce competitors narrowed. Although on the shores of the Atlantic, the rise of Amsterdam may have flagged a transition from the era of empire-building city-states to that of territorial states, in the Mediterranean, it was Genoa that led the unification of the kingdom of Italy in the mid-nineteenth century.[230]

In light of the pivotal and constitutive role assumed by the cities of saints, the picture we have of the Mediterranean as a historic divide between rival imperial polities and, worse, religions does not hold water. When seen through the kaleidoscope of the petty city-states, the Inner Sea ceases to be the central stage of an epoch-making, civilizational confrontation between the world-conquering Ottoman and Habsburg empires. Nor can it be envisioned as the center stage of the confrontation between the worlds of Islam and Christianity, a confrontation with a timeless and ecumenical hue. What determined the tempo of change in the basin was a relatively innocent, this-worldly, and "secular" rivalry between two empire-building city-states: Venice and Genoa. The second sixteenth century, in fact, deepened the linkages between the Sublime Porte and the Signoria on the one hand, and the *nobili vecchi* and the Castilian throne on the other. As the historical account given above has underscored at length, the political, economic, and social properties that set Venice apart from Genoa did not originate in the era of Philip II or Süleyman the Magnificent. Their pedigree can be traced farther back. The diverging historical trajectories of these two city-states may have added firmer, sharper lines to their features after the Genoese

merchants-cum-bankers laid claim on the fabled treasures of American provenance and after the spice trade returned to its olden Levantine route.

If in this mirror Venetian capitalism seems less adventuresome and less
creative than its Genoese (or by the same token, Florentine) counterpart(s), it
was not only because the Serene Republic was offered the comfort of secure
access to the resources of the Byzantine, Mamluk, and, later, Ottoman empires.[231] It was more than that. Nor can the differences between the Venetian
and Genoese empires' modus operandi be simply reduced to the time-honored
distinction between the salience of statal regimentation on the one hand and
that of private enterprise on the other.[232] The difference was of a larger order
into which the Mediterranean itself dovetailed: the Inner Sea derived its main
features from the historical attributes and qualities of two disparate worlds of
which the city-states were a part. One existed on its easternmost edges "with a
well developed and rich money economy" into which the Venetian merchants
were drawn. The other, where "the development of a money economy had
hardly begun," was a world the Genoese opened up and shaped in accordance
with the requirements of the moneylenders and capitalists who led the colonization.[233] If Genoa had the luxury of structuring and restructuring this tabula
rasa in the image of the world it created in the western Mediterranean, this was
a luxury Venetian men of money never had.

Indeed, Venice the Dominant, as it were, stood at the westernmost latitudes
of a circuit that was centered in the Indian Ocean, the Arab Seas, and Ming
China. This was a sophisticated and wide-spanning circuit, built over time, with
its financing and shipping mechanisms put in place by a wide and complementary array of merchants, shippers, bankers, and rulers.[234] In this network, even
the Mongol empire, which is always seen as a mere conduit, itself was a significant market for horses, and more important, textiles, because this was an empire
draped in textiles, from massive tents to heavy clothing for the violent temperature changes of the inner steppes of Asia.[235] The Venetian businessmen had
little control over this distant world, though by the first half of the sixteenth
century, thousands of Venetian families had managed to make serious inroads
into it via their growing presence in the Safavid lands. The network these
merchants put in place stretched from Aleppo to Diu. Yet, eclipsed by the long
shadow of the Florentine bankers at home and by those of indigenous bankers
and merchants in the Indian Ocean, Venice was, very much like Genoa at the
time when the city of St. George was in charge of the Pontic Sea, smoothly
integrated into a world that was neither of its making nor even aware of its
machinations, if it had any. The Genoese, on the other hand, built an aquatic
empire in the western latitudes of the Greater Mediterranean, extending into

the *Mare Tenebrosum*, "the Sea of Darkness." In this Genoa was aided by the Spanish, Portuguese, Flemish, and at times Florentine businessmen, yet it was its commercial and financial dexterity that paved the ground for the empire's extension into the Atlantic, a world that was not commercially integrated into the world-economic networks, the opposite—in this sense—of the Indian Ocean–centered system. In the Iberian peninsula and the western Mediterranean, the Genoese found not only a "stronghold to which to retreat" but also "an outpost from which to advance."[236]

Venice and Genoa may have inhabited the same peninsula and coexisted in eerily close proximity. But they were worlds apart in the 1450s. Venice symbolized and mirrored a world that dazzled even the most prudent. *Genova La Superba*, on the other hand, mirrored the stark world of the "offshore" Mediterranean in the fifteenth century, and in that image it cast the Atlantic world. The Mediterranean may have been outcompeted by the relocation of the center of economic flows in the North Sea and the Baltic. But the offshore world, the world on the move as we depicted it, the world capital built de novo, had a long and lustrous future before it. A tale of two cities maybe, but more so, a tale of two worlds as reflected in the shallow waters of the inner pond of the Mediterranean.

CHAPTER TWO

City-States and the Inner Sea

The second sixteenth century commenced in the 1560s. By then, Antwerp's greatest merchant-bankers and merchants, Gasparo Ducci, Luis Perez, and Erasmus Schetz, to name a few, had already risen to prominence as grain-brokers for Lisbon and Castile. Inundated by the wealthy *marranos* from Iberia, Antwerp had established its command of the licit and at times illicit flow of Baltic grain south of the Channel. Fittingly, the city on the Scheldt was located at the crossroads of the grain (and bulk commodities) trade from the Baltic, the Flemish textile trade, and the rich trades from the Mediterranean and the Atlantic. It was also in Antwerp that the easterly flow of West Indies gold converged with the westerly flow of German silver.[1] The estuary formed in Antwerp by these monetary flows gave merchants of the city the ability to finance the grain trade with enviable ease.[2] Via Antwerp, the Baltic grain found its way, first, to the Cantabrian and Atlantic shores of Iberia, Lisbon in particular, where grain production had given way to the cultivation of lucrative crops and where vineyards were spreading with celerity; second, to New Spain, where settlers were growing in number and in want of the bread grains they were accustomed to consuming in their natal lands; and third, to the Italian city-states, where a ubiquitous demand for cereals was mounting because of the looming grain deficit in the Mediterranean.

The city of St. Giovanni, for one, was struck by a long string of acute food shortages at the turn of the seventeenth century.[3] By then, the Ximenez of Lisbon and Antwerp, in charge of arranging deliveries for the grand duke of Tuscany, had, along with other wealthy Portuguese merchants dealing in grain, taken up residence in Florence.[4] These merchants were, in fact, among those who had invested in—or enclosed—land from the final decades of the sixteenth century to profit from the hike in cereal prices that arose from the relentless population growth across the Mediterranean basin, from Valladolid to Aleppo. These were the times of the return of capital to land and the "defection of the bourgeoisie." The sale of public lands, *tierras baldías,* in Habsburg Spain by Philip II in the 1570s; the attempts by Ottoman authorities from the turn of the seventeenth century to arrest the abandonment of the arable by insisting on the collection of land-breaking tax, *qasr al-faddan* or *çiftbozan,* and to enforce the return of fleeing peasants to their homesteads; and the sale of Church property in France under Charles IX and Henry III proved equally inefficacious in reversing the retreat of land under cultivation.[5] Yet, even the return of significant sums of money to land and the land bonification that followed, admittedly only in certain parts of the Italian peninsula (and Sicily), naturally failed to arrest the general fall in agricultural production. The Baltic grain trade, by contrast, went on to grow in volume and significance. After the fall of Antwerp to Alexander Farnese in 1585, merchants from Amsterdam instantaneously assumed their predecessors' pivotal position in this staple trade. Already by the beginning of the seventeenth century, they had put in place an impressive network of grain collection, storage, and redistribution that enabled them to establish a separate Corn Exchange in Amsterdam.[6] In the latter half of the sixteenth century, in short, the Baltic grain trade expanded its sweep, gradually yet steadily, to encompass not only the port-cities presiding over Atlantic Iberia, but more important, those at the very heart of the Inner Sea.

The inability of agricultural production in the Mediterranean to keep pace with the demographic growth of the long sixteenth century and the growing dependence by the city-states on grain that originated, first, in the Levant, later, in their own hinterland (*contadi* and *stato di terra*), and, finally, in the Baltic for their provisions, were not necessarily manifestations of trying times. Quite the contrary. The economic misfortune that befell the Mediterranean from the turn of the sixteenth century, when the spice trade came to be recentered via Lisbon and Antwerp on the shores of the North Sea, was promptly reversed by the 1540s, and the recovery was firmly secured by the rerouting by the Genoese of American silver to the Levant. The resultant monetary influx was one of the principal factors, along with rampant urbanization, that eventually fueled the famous price inflation of the last quarter of the sixteenth century.[7] North and south of the Alps, prices kept rising, reaching new heights in the closing decades of the century. Meanwhile, the rate of growth of agricultural production, which had kept up with and at times even surpassed the pace of population growth in most parts of the Inner Sea for more than a century, from 1450 to 1560, started to lag behind, often with disastrous consequences.[8] The relative scarcity of bread grains resulted from the inability of agricultural production to keep pace with the dizzying rates of demographic growth. And the resultant paucity was further aggravated by the shift away from cereal husbandry to more lucrative crops, which exerted upward pressure on grain prices until the mid-seventeenth century.[9] The slowdown in population growth during the seventeenth-century crisis relaxed the pressure on prices, if not completely lifted it.

The repercussions of the inflationary movement were felt throughout the Mediterranean, since the steep rise in prices laid the groundwork for the fiscal crises that taxed the basin's states in its wake, mostly in the closing decades of the sixteenth century. True, the empires suffered more than the smaller states. The Castilian crown's descent into bankruptcy in 1575 and again in 1596, the Sublime Porte's ballooning budget deficits from the 1590s, and the intensification of peasant revolts in France between 1580 and 1595 reveal how the changing economic climate tested the capabilities of the imperial bureaucracies.[10] The Italian city-states may have suffered less in the short run under the fiscal pressures of the late sixteenth and the early seventeenth centuries. A rapid growth in the proportion of manufactured goods in their economic transactions with the Levant exempted them from the travails of the times until the 1620s, and perhaps until the start of the Cretan war in the 1650s.[11] In the long run, however, the merchant republics were beset by a development that was much more grievous, if not fateful, than the fiscal turmoil that shook the empires to the core. This was the erosion of the Mediterranean-wide division of labor that the city-cities, led by the Serenissima, had striven so long and hard to fashion and maintain.

The new bout of silver inflows from the Americas in the mid-sixteenth century led to the dissolution of the existent division of labor, since this underwrote "the spread of luxury."[12] The metallic invasion replicated in the eastern Mediterranean the scenario that had unfolded in the basin's western extremities in the first half of the century. Earlier, the flood of German silver and West Indies gold via Antwerp had induced a shift in economic specialization on the Atlantic shores of the Iberian peninsula toward lucrative crops. This directional change was manifested, among others, by the retreat of arable farming to the benefit of viticulture in Atlantic Iberia (where ecological conditions for it were more than wanting) and oleiculture in Andalusia.[13] After the 1550s, the flow of

Potosí silver into the Levant occasioned a comparable outcome in the eastern Mediterranean, where increasing crop diversification and rapid urban growth worked to the detriment of cereal agriculture. That is, as the influx of German silver and West Indian gold during the age of the Fuggers (c. 1460–1530/1540) accelerated the tempo of commercialization and commodification in Atlantic Iberia at the expense of low-yielding grains, so did the influx of silver from New Spain into the Inner Sea during the age of the Genoese (c. 1560–1620/1650).[14] By the last quarter of the sixteenth century, the waves caused by the white metal's *drang nach osten* had finally reached the shores of the eastern Mediterranean, three-quarters of a century after the arrival of Caribbean gold in Seville.

Grain prices skyrocketed from the 1590s to the 1620s, but the overall inflationary upsurge lifted the prices of most, if not all, crops and goods. When accompanied by dizzying rates of urbanization and soaring urban demand for a new and widening range of goods,[15] arable production in, and subsequently exports from, the Ottoman lands diminished, as happened in parts of the Iberian peninsula from the turn of the sixteenth century. Equally important in triggering this outcome was the deceleration in the tempo of arable reclamation due to the exhaustion of lands that could easily be converted into cereal husbandry or due, at times, to the declining fertility of relatively marginal lands that were placed under the plow during the frenetic phase of economic expansion in the 1450–1560 period.[16] As a result of the gradual fall in agricultural output per capita, crop failures and famines that came to plague the region surfaced, first, in the western Mediterranean, in Castile and Andalusia from the 1530s. Similar shortages appeared later, in the closing decades of the century, in the eastern latitudes of the Inner Sea, in the Ottoman dominions as the intensity and frequency of crop failures and famines reached new heights in Iberia.[17] Thenceforth crop shortages, sporadically coupled with famines, became a recurrent phenomenon in the grain-producing regions of the basin and the ramifications of this shortfall were felt even in the most far-flung quarters of the Inner Sea. To remedy the situation, major cereal producers tried desperately to place restrictions on, or curb, seaborne grain exports. After all, what was seaborne grain, *grano del mare*, to the Venetians was conversely sea-bound grain to the Ottomans, *deryaya giden tereke*—grain lost to sea, to be precise.

As the basin's cereal-producing regions struggled to monitor and regiment the grain trade, the merchant republics found themselves obliged to rely on a provisioning strategy which, by nature, was an untenable one, even in the short run. Faced with dwindling supplies of seaborne grain, the city-states, in order to ensure steady and adequate provisioning of their dominions, adopted a strategy of tapping into the resources of their immediate hinterland, Terraferma or *contadi.*[18] Yet the strategy of "self-reliance" in light of the decline in "grain from the sea" offered a fleeting solution at best, especially considering that these cities could normally survive on the produce of their *contadi* for a few months a year—in the case of Florence, for instance, five months a year.[19] More damaging, the city-states' *contadi,* that is, the ambient shadow cities and their hinterland, inhabited an economic space tightly integrated into the division of labor centered around the needs of the mother cities. As such, the immediate hinterland of the city-states, as the underbelly of their mighty brethren, not only housed populations who commanded relatively higher wages than in the grain-producing zones of the basin, but also shouldered a heavier burden in the late sixteenth and early seventeenth centuries in textile manufacturing when the competition in woolen production reached its crescendo.[20] The *contadi* could not therefore be easily given over to arable agriculture—except in periods of acute emergency. Even then, the supplanting of grain from the sea, *blé de mer,* with "home" grain all but averted the grain crises of the 1590s.[21] It was only in the mid-seventeenth century, when the city-states' industrial renaissance was cut short, that their hinterland was "assigned" a greater share in the provisioning of the mother cities. Conveniently enough, it was then that more money was invested in land and more land was allotted to cereals by the city-states' prosperous patricians as fortunes they reaped from long-distance commerce started to dwindle.[22]

Overall, across the Mediterranean the planting of the arable to commercial crops and the conversion of lands previously under cereal cultivation into vineyards, orchards, and pastures, all of which had already taken land away from grain, proceeded unhindered as if unaffected by the turmoil generated by the endless string of crop shortages. It is telling that of the measures put into effect to remedy the soaring deficit in grain production in the Inner Sea, only those introduced in the Italian city-states and Sicily were designed to extend the arable and increase agricultural output under the iron fist of aristocratic and baronial families.[23] Not to anyone's surprise, men of money from the Italian city-states—the Venetian patriciate in particular—invested in farming when grain prices were skyrocketing.[24] There were even attempts to recover land from scrub or swamp. And some new land was created, thanks to the draining of water-logged ground.[25] Still, success proved to be elusive despite the enormity of resources, financial or otherwise, the merchant republics commanded. In any event, investments in land came to a halt when agrarian prices started to stagnate and declined sometime from the 1620s/1650s. Soon enough, the grasslands of the Veneto were put to work toward livestock and meat production and not arable agriculture, and this specialization extended well into, and lasted past, the age of enlightenment.[26] In this, the Venetian patricians' experience was not too dissimilar from that of the Antwerp merchants who, too, bought up land on the banks of the Scheldt in the sixteenth century. The reason behind their return to land was that they were able to borrow money on the strength of their landholdings. For others, purchasing land was a way of keeping up with inflation.[27] Given the palliative nature of measures taken by states or landowners in the grain-producing and -importing regions, the trend of declining arable production remained entrenched and could not be reversed. By then, cereal production had by and large ceased to be an economic activity worthy of the efforts of the denizens of the Mediterranean's rich quarters. The onus was thus placed on the shoulders of the rural producers in the eastern shores and, later, outside the basin, in the east of the Elbe.[28]

That these circum-Mediterranean transformations were deep-flowing and structural is corroborated by the fact that neither the frequency and pervasiveness of crop shortages and famines, nor skyrocketing prices at the end of the century, were able to decisively alter the main tenor and direction of agrarian change and boost grain production. The severity of the crises was, in itself, a telltale sign of how successful the conversion of the arable to vineyards, orchards, and pastures as well as the shift away from cereal agriculture had all been. For those who toiled on the land, switching back to grain farming at a time when part of the Potosí silver was flooding the Mediterranean basin was no longer a tempting prospect. This was precisely because this metallic invasion, by propping up regional demand for lucrative crops and goods, rewarded crop diversification and disfavored a return to the tyranny of low-yielding grains. The former granaries of the Italian city-states consequently devoted less land to cereal cultivation, or to put it in another way, the volume of agricultural produce that found its way to the Mediterranean markets shrank considerably in the face of human plethora.[29] During the long sixteenth century, therefore, the division of labor that characterized the Mediterranean at its height came to be undone by and by, first, in the western and, then, in the Levantine shores of the basin.

Essentially, the process involved, in both instances, a gradual disappearance of cheap and easily accessible grain accompanied by a contraction in the arable due in part to crop diversification as well as rapid demographic growth. Again in both instances, the shift away from cereal agriculture came about in periods of infusion of precious metals. With the inflow of German silver and gold from the West Indies during the 1460–1530/1540 period, Atlantic Iberia shifted away from arable culture.[30] The high mobility of rural labor within Iberia due to the concentration of land under few hands during and after the *Reconquista* and the overwhelming presence of agricultural laborers, *jornaleros,* when accompanied by the emergence of transatlantic migratory movements, facilitated the adoption of market-sensitive strategies on account of rising prices. In the eastern Mediterranean, by contrast, smallholding peasantry endowed, or burdened, with inalienable rights to land did not have a choice except to increase and diversify agricultural output.[31] Consequently, as far as the provisioning of the wealthy quarters of the Inner Sea was concerned, the eastern Mediterranean came to take over more responsibility in the sixteenth century than before. Grain shipped from the Ottoman dominions was dispatched as far away as Lisbon.[32] Given the centrality of the Levant in the provisioning of the city-states, the infusion of American silver into the basin in the 1560–1620/1650 period proved particularly perilous, for it unleashed a process somewhat akin to that which affected the western Mediterranean earlier, but this time with more serious consequences.[33] By then, the eastern Mediterranean had assumed the task of provisioning the rich zones of the Inner Sea, the Italian city-states first and foremost. This happened at a time when the steady provisioning of the cities of the western Mediterranean was already a source of disquiet, from Lisbon to Venice, even before the dissolution of the basin's division of labor.[34]

Crucial in this scenario was the dwindling of commercial grain trade in the Inner Sea exactly when the Atlantic was becoming not only the principal conduit of the grain and spice trades but also the main habitat of sugarcane (and, to a lesser degree, cotton) cultivation. Collectively, the rerouting of the rich spice trade to the benefit of Lisbon via the southernmost maritime route of the Indian Ocean, the relocation of sugarcane (and cotton) cultivation in the Atlantic isles, and the newly gained centrality of the Baltic grain in the world-economy's division of labor in the 1450–1650 period—the three goods that undergirded the city-states' command over the Inner Sea as we highlighted in chapter 1—inescapably punctured the economic fabric of the basin. The transformations associated with the emigration of the staples of the rich trades took deeper hold in the Iberian Mediterranean from the 1450s and accelerated under the age of the Fuggers; the Ottoman Mediterranean followed suit from the 1550s under the age of the Genoese. The void left behind by the departure of sugarcane from the coastal plains and valleys of the basin was later filled in by the march of the tree crops of Mediterranean fame: the olive and the vine started to blanket the hills and, when possible, vineyards even took the place of fields of grain.[35] Small livestock and wheat, the other two members of the Mediterranean triad, saw their fortunes reversed from the turn of the seventeenth century after having led the colonization of the highlands and the lowlands of the basin, jointly on occasion and oftentimes in competition.[36] Smallholders' dominion over sheep husbandry increased steadily, albeit slowly, in the following two centuries to the detriment of large herding associations. In numbers of sheep, the ceilings reached during the reigns of Louis XII and Francis I, the golden age of stock-raising, had become a dim memory in 1700.[37] Arable agriculture encountered a similar fate: the share of wheat diminished in gross product, since it no longer was the undisputed king of the basin's low landscapes. The reclamation of new lands of which it was put in charge came to a halt. Wheat was now grown "in the well-drained soils of the valley bottoms and in the floors of the intermontane depressions" rather than in the low-lying plains,[38] as slopes hosted vineyards and orchards and the high landscapes were put to use by livestock as summer pastures. And the coastal valleys and plains, previously sown with sugarcane, cotton, and wheat, were now given over to transhumant herds and tree crops. Arable farming was also limited by the impressive dominion of livestock husbandry, which went on to expand by fits and starts until the closing decades of the eighteenth century.[39] A new ecological and economic order arose, an order that removed the vestiges of the city-states' golden age with striking and incredible speed, as the staples of the Mediterranean ecology came to supplant the staples of the rich trades. The transition proved to be a smooth one and was tightly sealed as of the mid-seventeenth century at the latest. How the Inner Sea, at its height, was shaped by the rich trades—the exclusive realm of the triumphant city-states—which eclipsed the basin's own staples, and how the latter started to recover from the 1450s the territory it lost, is the subject matter of this chapter.

To account for this long-term transformation in the agricultural landscape of the basin, we first trace how the staples of the Mediterranean trade came to lay the foundations of the maritime empires put in place by Venice and Genoa. Here, we emphasize that the spice trade was but a component of the city-states' fabled wealth. Not only that, but the rivalry among the sibling city-states over the control of the rich trades of the Levant was the driving force of economic and social change and it led to the hollowing of the eastern Mediterranean. The second section outlines why in the heyday of the Inner Sea, the triad of Mediterranean crops—the vine, the olive tree, and cereals—failed to partake in the commercial buoyancy of the era. Of the three, the cereals presented a lingering problem for the provisioning of the city-states and their maritime empires. The third section highlights the agrarian transformations associated with the long sixteenth century, especially with the age of the Genoese. It was then that the crops of the Mediterranean started to become popular once again and the land devoted to them expanded. After sketching the restructuring in the basin's division of labor and the place occupied in it by its Levantine quarters, the section chronicles the dwindling of the agrarian surplus in the latter half of the sixteenth century. As the reclamation and colonization of the new lands came to a grinding halt in the Inner Sea, wheat inescapably lost the pioneering role it had previously played.

The Moving Feast: Oriental Crops

As discussed in the previous chapter, the Mediterranean had, from early on, an elaborate division of labor that interwove and internested the basin's port-cities—and their areas of catchment, stretching at times deep into inner Asia and sub-Saharan Africa. Put in place and diligently policed by the ruling city-states, this areal division of labor was, by nature, dynamic. It was subject to change owing in large part to the omnipresent rivalry among powers vying for hegemony, Venice and Genoa for the most part. Indisputably, the rich spice and pepper trades were the life-blood of, and the most sought after trades in, the Inner Sea—bringing in, along with the Constantinople trade—untold fortunes to the merchant republics.[40] Following the anchoring of the crusader states in the Levant after 1099, a prolonged and all-out war was fought over these lucrative trades from early on among the contenders to the Mediterranean throne: Pisa, Barcelona, Amalfi, and Florence, too, initially featured among the select few. At the onset of the crusades, the maritime route that transmitted the spices from the Indian Ocean to the Mediterranean reached the Inner Sea via the Persian Gulf and the Red Sea, giving the basin's eastern shores from Beirut in the north to Alexandria in the south a considerable share in this prosperous commerce. To wit, in the thirteenth century, "the revenues of the single city of Acre from the transit spice trade were about the same as those of the Kingdom of England."[41] Yet, from the mid-thirteenth century or so, the unification of the Eurasian landmass under the Mongolian administration and the formation of a unified and well-protected commercial domain stretching from the South China Sea to the Sea of Azov diverted the westerly flow of oriental merchandise from maritime to terrestrial routes. What is more, goods originating in the Indian Ocean changed their principal itinerary exactly when a political sea change was taking place in Constantinople, the Byzantine capital located on the only debouchment of the Black Sea. The modification in trade routes turned out to be a great blessing for the city of St. George, much to the chagrin of the Signoria.

What provided the Genoese merchants this prodigious opportunity was the commercial policy pursued by the Byzantine throne after the recapture of Constantinople by Michael VIII Palaeologus in 1261. Punitive by design, the policy followed by him was intended to inflict ultimate injury on Venetian interests, much to the delight of its perennial enemy, the Genoese.[42] The former, after all, had led the sacking of Constantinople during the fourth crusade in 1204, paving the ground in the process for the subjugation of Romania and the Dalmatia to the needs of its emerging maritime order. The banishment, however, of the merchants of the *città galante* from the Black Sea, even when these indefatigable merchants eventually managed to carry on some of their operations from their post in Tana, turned out to be a punishment infinitely more severe than originally envisioned by the Byzantine throne due to the tectonic change in the nature of trade routes interlinking the Indian Ocean to the Mediterranean. The "orgy of capitalism" that followed the fourth crusade would cost the Venetians dearly.[43] Providentially for the merchants of the republic of St. George, the concordance of these two events—the opening of the central Asian route on the one hand and the enthronement of the Paleologus dynasty in the Byzantine empire on the other—granted them the keys to the Black Sea and *il deserto dei Tartari*, the destination point of the valuable merchandise that originated in or transited through the Mongolian dominions.[44] Once the Genoese traders found themselves in charge of the Pontic Sea (and the Danube), the trade conducted through the city of St. George generated three times more revenue in 1293 than the kingdom of France under Philippe le Bel.[45]

As can be surmised, the rise of the central Asian terrestrial route and the intensification of trade along the Black Sea littoral worked to the detriment not only of the Venetian merchants, but more crucially, to the Levant trade in general. Cairo was one of the two major nodes that connected the Indian Ocean to the Mediterranean. And under the Mamluks, it remained beyond the ambit of Mongolian armies unlike its rival, Baghdad, which was laid to waste in 1258. The inability of the empire of the steppes to subjugate Cairo failed to prompt the Levant trade, however. Undoubtedly, the city of the Caliphs stood unrivaled after the ouster of Baghdad from the post of the main port of disembarkation for western-bound oriental goods. Yet in the absence of a vibrant thalassic trade, what it controlled was merely a bridge that spanned over a largely desiccated river. By the same token, the fall of the crusader states in the 1290s and the inability of these states to reestablish their rule in the Levant during the fourteenth century despite the steady inflow of fresh supplies of pilgrims was due, in no small measure, to the diversion of the rich trades away from the Levantine shores of the Mamluk empire in favor of the Mongolian emporium.[46] The consequent economic slowdown that deprived the crusaders of the lucrative profits associated with the rich trades rendered the preservation of these states inordinately expensive for the settlers.

The extension of Mongol rule into central, highland Anatolia certainly opened up a conduit to the Inner Sea through the Armenian kingdom of Cilicia and its port-city, Lajazzo (Ayas). This leg of the trade was handled by merchants based in Famagusta.[47] But, on the whole, the share of the spice trade that followed this itinerary remained relatively modest, declining further after Lajazzo, the Mongol window on the Mediterranean, was captured by the Mamluks in 1322.[48] The final outcome of the crusades, however, was one of the ultimate ironies of history, since the crusaders had intended to squeeze out the commercial intermediaries in the Levant by allying themselves with the Mongol throne, the patron-saint of the rich trades, even if briefly. Eventually though, they themselves were driven out of the Holy Land by the Mongolian empire's planned marginalization of the Levant route. Some of the staunchest supporters of the crusades, the Genoese merchants ended up reaping the fruits of *pax Mongolica* in a place far and away from the Holy Land, with their gaze fixed on the wide open steppes of "pagan," "shaman" inner Asia rather than on the Holy Sepulcher.

The conjuncture that gave its blessings to the Genoese traders lasted until the downfall and dismemberment of the Mongol empire beginning in the mid-fourteenth century. The resultant deterioration in security along terrestrial routes following the fragmentation of the empire served to reactivate the time-honored southerly maritime route that connected the Indian Ocean to the Mediterranean via the Cairo-Alexandria isthmus and the Damascus-Beirut route.[49] That is, the closing of the Black Sea route was followed by the reopening of the Red Sea route. And the change of venue rekindled economic activity on the shores of the Venetian Levant. This turn of events rewarded the Venetian merchants handsomely, since these men of business had not pulled back from the region when the basin had ceased to be the main conduit of the spice trade and when the share of the city of St. Mark in the rich trades had diminished in volume and value. Instead, the Venetian merchants kept their mercantile operations in the eastern extremities of the basin, and the array of goods they handled broadened to include sugar and cotton (also salt and copper).[50] Arguably, the waning of the spice trade in the eastern Mediterranean from the 1250s to the 1350s when the Genoese plied the waters of the Black Sea, in effect, cajoled the Venetians to better canvas the shores of the Levant and cast their commercial net considerably wider than previously. In this they were helped by the extension of sugar and cotton plantations in the Mamluk lands on account of the dwindling spice trade.[51] When the rich trades returned to the Levant from the 1350s, the array of goods dealt by the Venetian merchants did not subsequently decrease in number. Sugar and cotton had by then become indispensable to the operations, and amply enriched the commercial arsenal, of the Venetian merchants.

In the Aegean as well, a similar development unfolded contemporaneously: the merchants of the Serenissima, whose presence in the Black Sea turned tenuous after the 1260s, chose instead to tap into the extant silk and cotton supplies of the region in the thirteenth and fourteenth centuries at the expense of the Genoese merchants who had initially led the way in the acquisition of raw materials therefrom.[52] The Genoese merchants, in charge of the Pontic Sea—at the fount of the rich trades, including silk from the Caspian—went on to deal in *seta de Romania* and cotton from their most valued outpost, Chios. But Venetian rule over Crete, Coron, Modon, and Negroponte furnished the city's merchants with ample opportunity in the Aegean for the procurement of raw materials.[53] The Venetian empire, after establishing a base at the heart of the Constantinople-Alexandria commercial axis, that is, in Crete—"the nucleus and strength" of its empire—perfunctorily consolidated its command over the Byzantine-Levant commerce.[54] Genoese merchants, especially after relocating their empire in the western Mediterranean, gained greater access to the Sicilian, Calabrian, and Granada silks, which they distributed in the textile centers in the Italian peninsula, Catalonia, and Montpelier.[55] The Aegean and Romanian silks were hence left largely in the hands of Venetian merchants. At any rate, silk production, partly because the Latin Aegean was within the easy grasp of Venetian merchants, from Zara to the Morea, and partly because of the mulberry tree's ecological demands, did not disseminate as widely as sugar or cotton, with the notable exception of the Po valley. Moreover, whatever was available in Granada and Sicily could not compete, in terms of quality, with silk dispatched from the shores of the Caspian Sea and farther east.

The team of sugar and cotton (and silk) thus secured a shelter for themselves in the commercial haven offered by the Venetian men of trade. Always in demand, cotton found expanding markets in the twelfth and thirteenth centuries in the budding textile industries of Catalonia, Provence, and the city-states of the northern Italian peninsula. So did sugar, since it could be procured "even in remote towns" in England in the thirteenth century.[56] Crucial, however, was the relocation of the cultivation of these profitable cash crops to new sites, in Cyprus, Crete, and other Venetian and Frankish possessions across the Mediterranean. The presence established by the early crusaders on the shores of the Levant proved to be an invaluable asset in launching this vegetal dissemination. Fortuitously for the Frankish rulers and businessmen, the occidental movement of the profitable oriental crops was accelerated by developments that hastened the fall of the crusader states in the 1260s. The Mamluks, fearful that crusader attacks might resume momentarily, evacuated parts of the Levant, the very sites of commercial agriculture in the region in which sugarcane and cotton cultivation had long taken up residence (sugarcane in Sidon, Tyre, Acre, and Tiberias, and later in the Jordan valley, and cotton in Palestine, northern Syria, and the Jordan valley).[57] The ports of Tripoli and Beirut, located away from Jerusalem and Acre, firmly fortified, served as the commercial outlets of the region.[58]

The catastrophic blend of the atmosphere of insecurity that overhung the coastal regions of the Levant and the shift in trade from the maritime Indian Ocean route to the terrestrial central Asian route in unison put the newly constituted Mamluk polity in a difficult bind. Its revenues from the transit spice trade shrinking, the throne found itself in charge of the disagreeable task of increasing customs dues exactly when the coastal regions where sugarcane and cotton were cultivated found themselves thinning in population. Later, one-third of the population of Mamluk Egypt and Syria was wiped away by the Black Death.[59] True, the crusader assaults debilitated the economic health of the Syrian coast, but not that of the Nile valley, where new villages were founded and irrigation projects were constructed during the lull in the Levant spice trade in the 1250–1350 period.[60] Nonetheless, attempts to revitalize Mamluk Egypt could not remedy the malaise that plagued cotton cultivation in the Levant, given that most of the highly valued varieties of it were grown along the Syrian littoral, in Hama, Aleppo, and Cilicia.[61]

The policy adhered to by the Mamluks, then, contributed to the acceleration and consolidation of a movement that was to constitute the underlying dynamics of change inside and outside the basin in the centuries to come: namely, the relocation of the cultivation of oriental crops in new, westerly locations. As the coastal regions of the Levant, home to commercial agriculture, were vacated, oriental crops acclimatized in the eastern Mediterranean embarked on their long-lasting and far-reaching westerly journey, under the aegis of Venetian and Genoese traders. The migration of oriental crops also signaled the beginning of a lengthy process of economic devolution that would eventually culminate in the decline of the sugar industry in, and its disappearance from, the Levant from the fifteenth century on.[62] Cotton as well encountered a similar fate, yet with a considerable and precious time-lag; its commercial significance did not diminish as hastily as that of sugar, as we discuss shortly.[63] In the Atlantic, cotton was transformed into a plantation crop; in the Balkans, Anatolia, and the Levant, the crop turned into a peasant-produced "cash crop."[64]

The crusaders had firsthand experience in the cultivation of cotton and sugarcane and the production of sugar in the eastern Mediterranean—the Venetians in Tyre, the Teutonic Knights in Tripoli, and Baldwin II, king of Jerusalem, in Acre. Concomitant with the fall of the crusader states, sugarcane and cotton cultivation was rekindled in locations where it had been introduced earlier, and acclimatized in others, such as Cyprus, Crete, Rhodes, and Malta, which were all insular, and were placed under the jurisdiction of Frankish lords or dynasties.[65] Cotton cultivation was officially encouraged in Venetian possessions in the Aegean, Crete, the Morea, and Negroponte. Under the rule of the Lusignan dynasty, sugar production expanded in Cyprus owing to investments by the Knights of St. John, the Ferrer family from Catalonia, the Cornaro family from Venice, and the Banco di San Giorgio, and sugar became one of the major crops of the island.[66] If monies invested in sugar production in these isles originated in the wealthy cities of the western Mediterranean, some of the labor put to work in these new sites of cultivation also came from afar. The opening of the Black Sea in 1204, when Constantinople fell into the hands of the Venetians during the fourth crusade, increased the number of slaves available to sugar plantations, breaking the reliance by the enterprising, insular dynasties on enslaved war-captives alone. Under the aegis of the Latin empire of Constantinople, the slave trade flourished in the Levant between 1204 and the 1260s, and maintained its vitality thereafter.[67]

Evidently, the employment of slaves as agricultural hands did not materialize all at once and even when it did, given the relatively low numbers of slaves imported initially, the overall impact of the newly arriving hands on production costs must have been less than modest. Nonetheless, the imposition of heavier customs duties on sugarcane and cotton cultivated in the Mamluk lands just as slave labor was making its debut in lands under Frankish rule must have given the rivals of Levantine sugar a competitive edge, slight though it may originally have been. This was hardly the only benefit derived by Venetian merchants from the command they had established over the Black Sea slave trade. From their post in Tana, they went on to supply the Mamluk empire with slaves. In the fourteenth century, the Venetian merchants' ability to help Cairo replenish its slave army was so highly deemed by the Mamluks—"the empire of the slaves" no less—that the merchants were exempted from fiscal dues from Aleppo to Ghaza, and gained greater access to the region's local markets and goods.[68]

Certainly, the employment of plantation slave labor remained limited in scale, but the cultivation of sugarcane in the islands of the eastern Mediterranean was not necessarily organized around the centrality of slave labor per se, but of *corvée*, coerced labor that was part of the idiom of the feudal order and thereby available, with some restrictions, to lay and ecclesiastical lords. Coerced labor was put to full use in sugarcane cultivation on demesne lands in the isles for good reason. Offshore and not subject to the strict codes of the feudal order that governed the usage of demesne land and *corvée* labor, these islands offered the Frankish nobles who had invested in the cultivation of oriental crops the chance to rely on coerced labor more freely than those at the heartlands of feudalism where the system's growing crisis from the 1250s was placing growing restrictions on the use of bound labor. After all, the concurrent collapse of the Church and the state and the "crisis in seigniorial revenues" à la Bloch, compounded later by the scarcity and growing bargaining power of laborers after the Black Death,[69] must have played their part in stimulating offshore ventures. Offshore, the marriage of sugarcane cultivation with coerced labor may have started with *corvée* labor first, but slave labor came to supplant it as the crop's westerly odyssey gained velocity.[70] So, owing to the presence in the Mediterranean islands of coerced labor, initially the employment of slaves proceeded tardily, and the reorganization of production did not undergo ostensible restructuring until the fifteenth century—until new and larger supplies of slave labor, forced to populate previously uninhabited islands, gave the western Mediterranean a competitive edge over the Levant. Not coincidentally, it was before the fifteenth century, before the heyday of the sugar industry in the western Mediterranean, that the Cornaro dynasty, the sugar kings of Cyprus, reached the apogee of its power.[71]

Accompanying the westward movement of sugar and cotton was the transformation in the nature of the primary function performed by the Venetian colonies: from providing victuals to the city's fleets into hosting the culture of oriental crops for the merchants of the Serenissima. As mentioned above, these lands, unlike the heart of the feudal world, were not bedeviled by the growth in the bargaining power of rural producers. This is why the Frankish Peloponnese along with Cyprus, Crete, Negroponte, Malta, and Rhodes featured prominently in matters Mediterranean from the 1250s. Witness the prestige commended by, and reverence shown to, Peter de Lusignan of Cyprus at the 1364 convention of nobles in Cracow.[72] Moreover, the survival of arrangements that governed bound labor and the availability of slave labor meant that population loss in these offshore locations could more easily be compensated. This enhanced the economic centrality of the isles dotting the Mediterranean. In Cyprus in the sixteenth century, for example, more than one-third of a total population of two hundred thousand were serfs laboring on crown lands.[73] The import of coerced labor in this context was that, by eliminating the whims of producers, it gave market-mediated transactions a big boost and much desired continuity.

The conjuncture changed dramatically after the 1350s, for two reasons. First, the occidental movement of oriental crops accelerated significantly as a result of the growing Genoese presence in the western Mediterranean after the demise of the Pontic route.[74] In the decades that followed, sugar became the *primum mobile* in the colonization and settlement of the "fortunate isles" that stretched from Madeira in the north to São Tomé in the south, not to mention the cane's spread in the south Atlantic—east of the Tordesillas line, to be precise. Second, with the growing dearth of labor following the Black Death, slaves came to be used in growing numbers in the production of sugar throughout the basin. Yet, precisely when slave labor was becoming more easily available and commonplace in the cultivation of high-value crops in the western latitudes of the Greater Mediterranean, it was becoming less common in the Levant, especially after the closure of the Black Sea to the Frankish trade, an important source of supply for slaves in the decades that followed the fall of Constantinople to the Latins. Moreover, the fissiparous development of offshoot empires on the outer fringes of inner Asia following the demise of the Mongol empire and the attendant demand for slaves for manning the military and bureaucratic corps of its imperial descendants must have diverted the region's slave trade away from its westerly outlets. Not surprisingly then, in Cyprus, where the last years of the Lusignan rule and the subsequent Venetian period were marked by a decline in the number of slaves, new sources of supply were found in the port-cities of the Maghreb, rather than the Black Sea.[75]

The burgeoning slave trade in the western Mediterranean and the relatively easy availability of slave labor transformed the sugar industry in this part of the basin, since it gave this lethargic industry a newfound life.[76] Earlier, sugar may have followed the Koran, and its spread in the western Mediterranean may have been completed by the ninth or the tenth century—Sicily, Tunis, Morocco, and Andalusia all had sugar industries dating from the ninth century.[77] But given the new symbiosis between slave labor and sugarcane "born-again" in the baptismal waters of the Genoese Mediterranean, it is not surprising that the sugar industry blossomed in the western shores of the Inner Sea. The expansion of the slave trade accelerated under the stimulus of the Genoese (as well as Portuguese and Florentine) merchants, who built their maritime empires owing to the profits fetched from the sugar trade.[78] Sugar production in Sicily, for instance, picked up in the fifteenth century.[79] The launching of trans-Saharan and, later, Atlantic excursions into Africa for slaves gave sugar production a boost in the Iberian Mediterranean from the turn of the fifteenth century. In Portugal, Algarve, and even a location as northerly as Coimbra had sugar industries that dated back to the tenth century. Yet these sleepy industries were given a new lease on life when slave labor became the golden currency in the western Mediterranean, circulating within, and manning the commercial nodes and outposts of, the incipient Portuguese maritime empire.[80]

During the sixteenth century, the exclusive devotion of the islands on the western edges of the Greater Mediterranean to sugar production started to pose a veritable threat to sugarcane cultivation on the basin's eastern fringes. Madeiran sugar made its debut along the shores of the Atlantic, in Antwerp after 1450. By 1550, it had reached markets as far east as Constantinople and, not surprisingly, Chios—Genoa's remaining outpost and entrepôt in the Aegean off the coast of Asia Minor. Madeiran sugar's majestic sweep, from the Atlantic to the Aegean, put on display the Genoese merchants' awesome geographical reach.[81] Sugar from São Tomé followed suit from the 1490s, and finally, Brazilian sugar began to arrive in Antwerp and elsewhere in the 1530s and 1540s.[82] While sugar production was spearheading the colonization of the Atlantic islands and the coastal regions of Brazil, and competitive pressures were building up in the Levant, sugarcane cultivation and milling in the insular nerve centers of the Venetian empire were accordingly placed under close administrative scrutiny. It was then that the Venetian colonial possessions received munificent attention from the Signoria. Thus, efforts to repopulate Crete to bolster its economy after the pacification of the Greek and Venetian magnates' revolt in 1363 started in earnest in the late fourteenth century.[83] Not merely this, the competition over Cyprus between Genoa and Venice flared up in the fifteenth century. The city of the Doges took effective control of Cyprus in 1479, as was acknowledged by the handover of power in the following year by the Lusignan dynasty. The cultivation of cash crops was also closely regimented and codified with renewed vigilance.

In spite of the measures introduced by the representatives of the Signoria, in the final analysis it was almost impossible for the Venetian colonies to successfully compete with the newly reclaimed and slave labor–populated Atlantic islands. In Crete, sugarcane cultivation was supplemented by viticulture, and in Cyprus by cotton cultivation. Yet, however unsuccessful or inefficacious these measures turned out to be, sugar from the Atlantic failed to shake the very foundations of the Mediterranean sugar industry for quite some time. Even in the late sixteenth century, al-Mansur, emir of Morocco, was financing the construction of a number of sugar refineries in the southern Sous region to lift the economic fortunes of his realm.[84] During the long sixteenth century, "there were few Mediterranean coastal valleys or plains with water for irrigation where sugar was not cultivated."[85] The full scale of the transformation and the strength of the Atlantic enterprise became manifest only at the end of the sixteenth and at the turn of seventeenth centuries. Until the onset of the seventeenth century crisis in 1620/1650, due to the splendors of the age of the Genoese, sugar (and other lucrative goods) produced in the Mediterranean as well as the Atlantic went on to find markets in (and outside) the basin. Yet during the lean times of the seventeenth-century crisis, the contraction in trade placed severe restrictions on the ability of all sugar producers to market their goods, to the injury in particular of those in the Mediterranean.[86] The Levant, which was once "fragrant with sugar," was evidently less so. When Evliyâ Çelebi visited different quarters of the Ottoman empire in the mid-seventeenth century, the number of places in which he spotted sugarcane cultivation was few and far between.[87] Although sugarcane had been cultivated and milled since the mid-fourteenth century, albeit on a small scale, on the Mediterranean shores of Anatolia (as witnessed by al-'Umarī), by the sixteenth century, the volume of production was insufficient to satisfy the needs of the Sublime Porte in the sixteenth century, and the shortfall was covered by shipments from Egypt and Cyprus and, later, the Atlantic.[88]

No doubt, Venice suffered from the westerly movement of the oriental crops, but more so in the case of peripatetic sugar than cotton despite the fact that cotton, too, followed an itinerary similar to that of sugar: it migrated west and reached the Atlantic shores of the Marinid empire.[89] It, too, was grown on the Atlantic islands, in São Tomé, for instance. Yet, of the goods the Portuguese and Genoese merchants had on offer in return for their purchases of malaguetta pepper and slaves in Atlantic Africa, it was cotton textiles that proved to be in high demand among the peoples of the region. Cotton grown offshore on the Atlantic isles was therefore woven into textiles in nearby sites to suit the tastes of the region's inhabitants, with a view to facilitate the Portuguese empire's commercial penetration into what was to become Lusitanian Africa. This in situ employment of cotton placed strictures on the crop's availability in the Mediterranean.[90] It was only at the end of the seventeenth century that cotton shipments from the Atlantic started to reach manufacturing centers on both sides of the English Channel, later than sugar, and even then in much more modest quantities.[91] Cotton trade in the eastern Mediterranean maintained its vivacity, because the quality of cotton produced in new locations outside the basin was not only considerably lower than that of cotton produced in the Levant, but was also shorter in fiber. Given that cotton went on to be used on both sides of the Alps for the manufacture of fustians made it an integral part of the Venetian, hence Levantine, trade. Cotton's longevity thus depended on the success and spread of textile industries in northern Italy and Swabia, a process that lasted well into the turn of the seventeenth century.[92]

Earlier, too, in the twelfth and thirteenth centuries, demand for cotton had grown in tandem with the expansion of the incipient fustian industry. Even during the early part of the economic slowdown from the 1250/1300–1450 period, the industry managed to survive owing to its ability to relocate parts of the production process into the immediate vicinity of the city-states where guild regulations were less binding than in Venice or Florence. Later, when labor rapidly turned dear in the aftermath of the Black Death, textile manufacturing was relocated to places where labor was relatively abundant, unorganized, and cheap—across the Alpine passes to southern Germany. That is, demand for cotton continued unabated notwithstanding the economic slowdown. With such persistent demand for it, cotton was appropriately called, even in trying times, the "plant of gold," as was the case in Cyprus.[93] The Venetian merchants' orchestration of raw cotton trade between the main sites of cultivation in the Levant and the cotton textile manufacturing regions in and across the northern Italian peninsula proved to be germane to the survival of the industry and, by extension, the cotton trade.[94] With sugar on the move in the Atlantic, cotton emerged as the ideal candidate to take the commanding role from, if not supplant, sugarcane. In the fifteenth and sixteenth centuries, Cyprus, host previously to lush sugar plantations and a vibrant sugar industry, came to be known for its cotton, and not for the marzipan confection for which it was once famous and which, as an exotic culinary offering, had created such a stir in the convention of nobles at Cracow in 1346.[95] After the island changed hands from the Lusignan dynasty to the Venetian authorities, cotton production increased threefold.[96] When the Ottomans captured Cyprus in 1571, the island was still famed for its high-quality cotton.[97]

Stated briefly, the Frankish and Venetian merchants' involvement in cotton and sugar trade intensified after the diversion of the spice trade away from the Levant in the mid-thirteenth century. Concurrent with the relocation of the Genoese maritime empire in the western Mediterranean after the 1350s, sugarcane trespassed rapidly into the same territory, and its Atlantic-bound march eventually took a heavy toll on sugar production in the Levant. The economic consequences of sugarcane's egress were mitigated, however, by cotton's new-found popularity in the eastern Mediterranean. As profits derived from the sugar industry helped underwrite the colonization of the Atlantic islands, in the Levant, cotton took the lead as the staple of the ascendant manufacturing industries. This spatial order was not without its ramifications on the commercial structures of the eastern and western halves of the Inner Sea. The dependence of Genoese merchants, patron-saints of sugar cultivation, on slave (hence specie) trade gained in intensity.[98] Likewise, the Venetian merchants' addition of cotton and silk to their hefty spice imports from the Levant amplified their susceptibility to the vagaries of the specie trade. After the 1350s, then, the slave and sugar trades drove Genoese traders west, and the spice, cotton, and silk trades tightly anchored their Venetian counterparts in the Levant.

Finally, the Genoese merchants' involvement in the Mediterranean-bound wool trade, which catered to the north Italian city-states' woolen industries, added a new flavor to the areal distribution of economic activities across the basin. The ventures of an animal fiber in the westerly half of the basin in the form of wool trade reflected a development eerily similar in nature to what was transpiring in its easterly half with regard to a vegetal fiber, cotton.[99] The occidental movement of the merchants of St. George was accompanied by the inexorable growth, especially in Iberia, of livestock husbandry and the wool trade.[100] As the Levant was turning into the chief supplier of the raw material, cotton, to the fustian industries dotting southern Germany, wool from Iberia started to find its way, via Burgos, to the textile centers in the north, Flanders and Brabant. Thanks to the Genoese merchants' established presence in Valencia, Murcia, Alicante, and Cartagena, it also catered to the northern Italian peninsula.[101] The centrality of the merchants of the city of St. George in the procurement of raw wool for the northern Italian city-states mirrored the Venetian merchants' procurement of raw cotton for the very same cities' textile industries. In effect, the relocation of the Genoese empire in the western Mediterranean and the success registered by the city's merchants in outmaneuvering local traders in the kingdom of Majorca facilitated the devotion of most of the island's arable to sheep husbandry at the expense of cereal cultivation. The conversion was facilitated by the Genoese merchants' comfortable access to the grain supplies of the kingdom of Sicily.[102] It needs to be emphasized that the expansion of livestock husbandry was already underway in the Iberian peninsula due in part to the tumultuous times of the *Reconquista*. In times of almost perpetual warfare, investing in livestock husbandry, which was mobile in nature and hence not subject to easy seizure, was more preferable than tying one's fortunes in immovable repositories of wealth, such as land, given the uncertainties about whether the land would change hands or jurisdictions, or be laid to waste.[103] Thus, in both halves of the Mediterranean, the immersion of the triumphant city-states in their respective spheres deepened after the 1350s as their dominion over oriental crops and raw materials became germane to the survival of their overall imperial enterprise.

The epic struggle over the control of the rich trades that had pitted Venice against Genoa was ultimately settled in the period that stretched from the 1350s to the 1380s in favor of the former. Over the course of the fifteenth century, the Levant trade reached new heights under the aegis of Venice and remained the envy of those who struggled to increase their share in it. No surprise then, a Catalan text, penned in 1453, the year Constantinople changed hands, referred to it as "the head and principal of all commerce."[104] The Serenissima's Levant trade ranged from about 580,000 to 730,000 ducats at the end of the fifteenth century whereas that of Genoa remained in the range of 130,000 to 150,000 ducats.[105] The share of Venetian merchants in the total spice trade, including sugar, increased to 60 percent in 1500, up from less than 45 percent a century earlier, as the share of pepper in the overall spice trade decreased relatively in favor of other spices.[106]

Nonetheless, the return of the spice trade to its erstwhile route from the 1350s did not erode the standing attained by oriental crops in Venetian merchants' commercial operations in the preceding century or so. The post-plague depression, when compounded by the scarcity of precious metals from the 1380s, pushed the geographical stretch of oriental crops in both halves of the Mediterranean. The bullion famine of the fifteenth century elevated cotton's stature in the Levant to that of a golden crop. Cultivated in Venetian-held territories, always in demand in southern Germany and elsewhere, and marketed at premium prices, cotton came to occupy a prestigious place in matters Levantine. In the western Mediterranean the situation was similar: sugar, as the chief navigator of westward explorations, helped the Genoese to rebuild their maritime empire along new shores and under new skies, and assumed a vital role in the shaping of the Atlantic.[107] The adverse effects of the bullion famine became more pronounced on account of the newfound vigor of the spice and slave trades, since a lack of specie simply placed the onus of procuring return merchandise on the shoulders of the Venetian and Genoese merchants. To the chagrin of Venetian and Frankish merchants, the merchandise shipped east covered only a modicum of the imports from the Levant, even at the end of the fifteenth century. The difficulty of procuring precious metals intensified the warring rivals' involvement in the production and distribution of precious oriental goods (and raw materials such silk and wool), which for all practical purposes, facilitated the maintenance of these cities' precarious multilateral merchandise balance.

The elevation of oriental crops to the commanding heights of the Mediterranean economy was not the only outcome of the bullion famine. Under the taxing circumstances of the post-1350 period, the city-states found themselves compelled to reduce the level of their imports not only by relocating the cultivation of oriental crops within their imperium but also by assuming the manufacturing in loco of merchandise erstwhile imported. The geographical distribution of economic activities within the basin was revised to the benefit of the city-states where most of these activities later came to be cloistered.[108] Correspondingly, the economic structure of the city-states became ever more sophisticated, endowed in their heyday with a solid manufacturing base not only in textiles but also in paper-, soap-, mirror-, and glass-making.[109] Along the banks of the Po valley, for instance, mulberry trees witnessed their dominion extend, boosting the silk industry. As a result of these transformations, there had been significant change by the 1450s in the basin's agricultural landscape, "particularly in Lombardy with the development of forage crops, irrigated fields and new plants (rice and mulberry trees), while important progress had been made in livestock farming."[110] The city-states hence managed to ameliorate their balance-of-payments deficit by resorting to import-substitution and by decreasing the inflow of manufactured and agricultural goods, now produced and cultivated within their imperium. The rival city-states concomitantly extended their terrestrial holdings to the detriment of the lesser cities in the first half of the fifteenth century, as discussed in the previous chapter.[111] In sum, the period commencing in the 1350s proved to be pivotal in the formation of the new contours of the Mediterranean economy. The arrival of the Black Death and the ensuing labor shortage facilitated the wedding of sugar to slave labor; the ousting of the Genoese merchants from the Levant hastened the occidental migration of oriental crops; the mounting demand for cotton and silk compelled the Venetian merchants to tighten their grip on the Levant trade and the Genoese merchants to strengthen their sway over the sugar and wool trade. The triumphal occidental march of the oriental crops that blessed the cities of saints was not without its ramifications for the economic fabric of the Inner Sea, however.

The Trio of Mediterranean Crops at a Commercial Ebb

The reordering of the Mediterranean by, and its partitioning between, Venice and Genoa may have been partly based on the dissemination of the cultivation of lucrative oriental crops throughout the basin on the one hand and the concentration and centralization of the spice trade under the hands of the Serenissima for longer than a century (c. 1380–1490) on the other. Yet, this was only part of the story. The last leg of the merchant republics' awesome might was owed to their ability to procure and peddle a commodity much less glamorous and much less lucrative than either the spices or the oriental crops: bread grains. One of the oldest denizens of the Inner Sea, bread grains, specifically wheat, had never been grown in sufficient quantities, if at all, in the dominions of the leading city-states. Venice, secluded by, yet also trapped in, the salty lagoons of the Adriatic had no arable worthy of the name,[112] and Genoa, with its narrow coastal plains squeezed between the Apennines and the Gulf of Lions, was almost always in need, and in search, of bread crops. Florence, simply by virtue of its geographic location, fared better but not by much.[113] Not coincidentally, these city-states started their careers as maritime powers by successfully acquiring the capability to tap into and transport grain from Romania, *granum de Romania*, and the Black Sea not to mention Puglia and Sicily from very early on. Since these cities were principally and habitually provisioned by *blé de mer*, grain prices in Genoa and Venice, unlike in other Mediterranean port-cities, exhibited unmatched long-term stability on account of the steady and continual arrival of cereal imports.[114] Initially, the two Sicilies, Romania, and the Black Sea constituted the bread lands of the Italian city-states and the Mediterranean at large: at times, Sicily alone, "a sort of sixteenth-century Canada or Argentina," dispatched a quarter of its gross agricultural product.[115] Since a significant portion of the *grano ciciliano* was drained by Florentine companies before the 1340s, the Genoese and Venetian empires, deeply embedded in the Aegean and the Black Sea, had access to grain originating therein: Thessaly, Macedonia, or the Venetian colonies in Negroponte and the Morea.[116]

Unlike the Portuguese settlers who devoted, almost by instinct, large tracts of the Azores and Madeira, which were uninhabited prior to their arrival, to wheat cultivation, the Venetian colonialists established their jurisdiction in lands that were long settled and where the arable had long been sown under bread crops.[117] Channeling of the agrarian surplus into Venetian hands was secured through taxation and the revamping of land tenure to the benefit of the newly arriving lay nobles of Latin stock and ecclesiastic bodies of Catholic denomination.[118] Part of the appropriated grain was dispatched home, part of it was earmarked for victualing the mariners as well as the pilgrims on their way to, or back from, the Holy Land. However, given how unsteady available supplies were in each of these locales, casting a net in the basin as wide as possible was the most effective way of exempting the city-states (and their empires) from the vicissitudes of crop shortages. The initial benign neglect shown by the Venetian authorities in their colonial holdings, which were given over to cash crops rather than to staples, was a reflection of the ease with which agricultural produce was appropriated along the Mediterranean coast in the twelfth and thirteenth centuries.[119] But the favorable conjuncture proved to be fleeting.

One of the reasons the state of grain trade looked abysmal between 1250 and 1350 was due to the closing of the medieval frontier in the Iberian peninsula and the Levant.[120] At both ends of the Mediterranean, the state of turbulence that forced the closure of the medieval frontier was more supportive of livestock husbandry and hence injurious to arable agriculture. In the east, the political uncertainty that resulted from the westerly march of the Mongol empire engendered an environment hospitable to nomadic existence and livestock agriculture. In Byzantium as well, economic recovery of the twelfth century was cut short, stimulating livestock husbandry and transhumance in the Anatolian plateau and the Balkans.[121] Likewise, the disintegration of the Seljuk realm, which at its height enveloped four hundred thousand villages (only thirty-six thousand of which lay in ruins despite all the incursions), gave way to an expanding "marchland."[122] In all, the growth of livestock husbandry in most parts of the Levant led to a gradual withdrawal of grain cultivation.

In the western ends of the basin, the relentless settlement of the Mesetas was eventually checked by the growing force of the Mesta. The disruption of links that tied whatever was left of highly commercial Granada to the cereal-producing Mesetas as a result of the devotion of the latter to sheep farming compelled the Genoese, in search of new grain supplies, to extend their commercial operations into the Morocco plains.[123] In like manner, as the Black and White Sheep Turks expanded their realm in the east and Iberia's golden fleece claimed huge expanses of land in the west, the Danubian and Thracian plains as well as the south Russian plains took over the task of provisioning the Levant, and the Moroccan plains were gradually turned into grain providers for parts of the Iberian peninsula.[124] Concomitantly, then, the city of St. George incorporated the Maghreb into its maritime emporium as its grain lands, as did the Venetian merchants' hold over the Aegean tighten as their bread lands.[125] The galleys of Romania were as elemental to the functioning of the Venetian empire as were the Venetian *muda* that transported spices and cotton from Alexandria.

The search for new sources of grain was further intensified as the maritime republics of the sibling cities grew in size and the arable in their colonial holdings was increasingly, but never exclusively, allocated to cash crops. This had the unwarranted but not unexpected outcome of increasing the vulnerability of these aquatic empires to bottlenecks in supplies of bread crops as well as to the fluctuations in climatic conditions. Equally consequential for the survival of the thalassic empires was the city-states' ability to tend to the needs of their colonial possessions with a view to maintain an imperial order and, if need be, to provide their holdings with basic cereals in times of emergency.[126] Given that a shift to cash-crop production had become the norm in the twin merchant republics' territorial holdings throughout the Mediterranean, often to the detriment of subsistence production, the provisioning of colonies required circulation of bread grains within the maritime empires—in addition to the radial flow of cereals to the mother cities at the center of this expansive commercial web.[127] A significant portion of the imported grain was then re-exported. A caveat is in order here. Transporting grain to coastal cities by sea was oftentimes cheaper than provisioning them from the interior.[128] This meant that the volume of Mediterranean grain trade had little to do with that of arable production inland.[129] But this did not take away from its centrality in the survival and upkeep of these maritime empires, for these empires were built on webs interlinking first and foremost the coastal regions of, and the islands speckling, the Inner Sea.

Given the precarious nature of supplies due to intemperate weather and harvest fluctuations, the persistence of low yields, and the unending political turbulence, not to mention the frequent visitations of pestilence, the city-states extended their catchment areas into the distant parts of the basin and mobilized grain surpluses from Negroponte to Provence.[130] The fluctuation in grain shipments was best attested by changes in the provenance of imports into the cities of Genoa and Valencia. In 1402 the city of St. George received 70 percent of its imports from Provence and only 2 percent from Sicily. In 1405, however, Provence was the provenance of a mere 0.5 percent of the imports of the city whereas Sicily's share had increased to 46 percent. In the case of Valencia, in 1475 the city imported 8 percent of its grain from Sicily; in 1486 Sicily's share stood at 56 percent. Or, Catalonia provided less than 3 percent of its grain imports in 1479, but over 45 percent a decade later.[131] Given the mercurial nature of the grain trade, the web formed by grain flows linking the basin had to remain, and remained, densely woven.[132]

Wheat itself may have been a humble and quotidian cereal, easily available in local markets but not necessarily in regional and supra-regional markets. As such, the grain trade belonged to the distinguished family of *grand commerce*, along with that in wine and woolen textiles from early on.[133] Unlike the wool and wine trades, however, where bills of exchange and other forms of payment dominated, in the grain trade the "attraction" (or the "bait") was payment in cash.[134] Moreover, the grain trade did not merely serve to provision the city-states or their colonial empires. Additionally, grain served as an invaluable merchandise that gave the merchants the opportunity to establish a permanent foothold in lands that offered them the keys to the rich trades, like the cities of North Africa, gateways to the gold trade, and the Mamluk empire, high temple of the spice trade.[135] Egypt, once the granary of the Roman empire, true to the wealth accruing to it from the spice trade and from its control over the gold of the Sudan, imported grain under the Mamluks until the lucrative trade in spices and pepper suffered disruptions.[136] Reflecting this, cereal prices in the eastern Mediterranean remained higher than elsewhere in the basin.[137] Similarly, the

North African littoral, which remained, until the early fourteenth century, an overland conduit for trans-Saharan gold that flowed into the Italian and Iberian peninsulas, imported grain, mostly from Sicily, befitting the wealth it commanded. Its situation changed, however, in the fourteenth century, as mentioned above.[138]

Put differently, the provisionment of these aquatic empires, where the cultivation of cash crops was more highly valued than that of subsistence crops, assigned a prominent place to the grain trade. Initially, the availability of the *granum de Romania* and the ability of Venetian colonial authorities to coax, if need be, their subjects to plant wheat, rendered the provisioning of the empire a manageable task. In the case of Genoa, its merchants' formidable grip over Sicilian and Sardinian wheat, and the allocation of lands in the Atlantic islands—at least at the outset of their colonization—to cereal agriculture by settlers who took up residence therein, prompted the grain trade. Yet sooner rather than later, the islands throughout the length and breadth of the basin were eventually consigned to cash-crop cultivation. Venetian colonial authorities, by encouraging the planting of vines and cotton, and the Genoese merchants in the Atlantic, by overseeing the expansion of sugar production, gave cereal agriculture elsewhere a considerable boost. Between 1350 and 1450, in matters pertaining to the provisioning of the north Italian city-states themselves and of their empires, both Venetian and Genoese traders had to significantly widen their catchment area.

What is more, in the western Mediterranean, the building of the Genoese commercial empire was concomitant with the expansion of livestock husbandry in Iberia and Majorca. Given that sugarcane's lead in the colonization of "new" lands in the western Mediterranean hastened from the 1350s, the need to provision these sugar islands as well as lands devoted to sheep husbandry regularly gave the search for new grain supplies a new urgency.[139] The inclusion of Sardinia and Sicily into the kingdom of Aragón after the 1380s compelled merchants of the city of St. George, in conjunction with the Portuguese, to extend their locus operandi to the shores of the Sea of the Atlas Mountains: the plains of Morocco. That is, as Barcelona turned to the kingdom of Sicily for its provisioning, Genoese merchants correspondingly turned south for fresh bread grain supplies on account of soaring demand.[140] In the process, the North African littoral, which remained a recipient of Sicilian grain until the late fourteenth or early fifteenth century, found itself in a position where it had to start exporting grain to the Iberian peninsula.[141] Overall then, when the rich trades traversed the waters of the Inner Sea, they were accompanied by the not so profitable yet ever so strategic grain trade.

By contrast, the time-honored tree crops of the basin, the vine and the olive, did not have a palpable presence in the economic life of the region in the fourteenth and fifteenth centuries. In the case of the former, the devastation left behind by the Black Death, which reduced hands available to toil on the land, single-handedly induced the retreat of vineyards from the 1350s to the 1500s.[142] From the tenth century, vine-growing and commercial viticulture had spread its tentacles in a northerly direction toward the English isles, the Rhineland, the Brabant, and the Tokay, Hartz, and Paris regions, to name a few. Viticulture expanded in the Mediterranean as well, as attested by the pervasiveness of its consumption in Andalusia the wines of which, from Malàga to Jerez, were also "renowned in the east."[143] Vineyards expanded at the expense of the wasteland, mostly the *palus*, marshes, in the Bordeaux region and the English woodlands. It was encouraged not only by Edward I but possibly more so by the trend of global warming that set in at the turn of the millennium.[144] The septentrional advance of the vineyards prepared the heyday of Atlantic wines from the fourteenth to the sixteenth centuries. Gascon wine became a staple of the English market and was buoyed by the fall in cereal prices during the fifteenth-century crisis. The trend was reversed in the sixteenth century, partly to the benefit of the Mediterranean, as we discuss later.[145]

Because viticulture required intensive labor, the scarcity of labor necessitated the employment of this precious commodity in a judicious manner in the demanding times of the post-plague period. Vineyards' phenomenal expansion along the shores of the Inner Sea from the turn of the millennium was thus reversed, for land devoted to the vine went on to shrink in acreage until the turn of the sixteenth century. The contraction in land devoted to viticulture was counterbalanced in some measure by another development: the surviving vineyards took over valuable, fertile land from the former plow lands located in the plains, leaving the stony soils of the surrounding hills behind.[146] An opportunity was presented by the thinning of populations in those parts of the basin hit by the pestilence. The combination of falling production and the establishment of new vineyards in prime lands led to the emergence of regional, specialized *terroir* wines, rendering obsolete the generic wines produced in the preceding period.[147] Those varieties of wine that acquired a long-lasting reputation as *grands vins* did so in the taxing times of the post-1350 period, but the emergence of these wines should not obscure the fact that vineyards on the whole lost their pivotal place in the Mediterranean landscape. In the early fourteenth century, Bordeaux exported 850,000 hectoliters of wine; in the mid-sixteenth century, the Loire region's exports did not exceed 300,000 hectoliters.[148] Overall, viticulture remained under the dominion of four major regions: Bordeaux

(Gascony), Seville (Andalusia), Naples (Liguria), and Candia (Crete).[149] The *monemvasia* vines, transplanted from the Aegean islands, came to produce the highly esteemed malmsey of Madeira, which in the 1460s was considered to be the "best wine in the world," surely to the detriment of the Mediterranean.[150] By the sixteenth century, viticulture had largely lost the standing it commanded in the thirteenth, when wine was considered as valuable as "gold dust."[151]

It was not the vineyards alone that became relatively scarcer along the coastal plains and hills of the Mediterranean from the fourteenth century. The same fate was also shared by olive groves. Olive oil production and trade, which had contracted during the Mediterranean's "dark ages," started to pick up, especially in Iberia and Apulia, from the turn of the millennium. The olive tree was also widely cultivated on the islands of the Aegean Sea, the shores of Asia Minor, and the Peloponnese. The resumption of olive cultivation remained sectoral, though.[152] In the Byzantine and North African realms, the arrival and proliferation of pastoral nomads reduced the popularity of olive oil. Even Mamluk Syria, normally one of the major producers of olive oil (and soap), was not only incapable of supplying Cairo, but also had to import it, primarily due to the spread of oriental crops.[153] When oil was needed, it was dispatched from Apulia or Andalusia to Tunis, Alexandria, Acre, and Jerusalem, and at times to Asia Minor.[154] Again, the taxing times of the 1250/1300–1450 period and the political tumult and labor scarcity that accompanied it provided the backdrop against which the change in the fortunes of olive groves needs to be assessed. For the encroaching compass of livestock husbandry at both ends of the basin had a perverse effect on the consumption of olive oil. The increasing popularity and easy availability of animal fats at the expense of vegetal fats prompted a contraction in the everyday usage of oil for cooking purposes, from Provence to Asia Minor.[155] If the salience gained by the Mesta attests to the spread of stock-raising in the western extremities of the Inner Sea, the loss by Byzantium of Asia Minor and the resultant growth of livestock husbandry in the Balkans generated a similar effect in its eastern extremities.[156]

To be sure, olive oil had other uses besides cooking. It was used as a preservative, for oiling wool for the manufacture of cloth, and for tawing leather.[157] As the pace of manufacturing accelerated due to the initiatives of the city-states, Italian soap-makers managed to take over the art of soap-making from the Castilians, who had succeeded in developing a highly prized white, olive oil–based, soap alongside the traditional black soap used in manufacturing.[158] The partitioning of the Mediterranean in two placed the main oil-producing regions under the command of the reigning city-states. Olive oil was largely produced in southern Iberia, and to a lesser degree in Apulia. In the mid-fifteenth century, oil from Iberia, mostly from Andalusia, and Djerba, was being shipped, mostly by the Genoese (and at times by the Catalan) merchants, to the Levant, Anatolia, Chios, Phocea, and Flanders.[159] The regions in which Venetian presence was pronounced were all or became producers of olive oil, from the Morea to the Peloponnese and western Asia Minor, not to mention Apulia.[160] As olive oil production in Apulia advanced, the region's oil was shipped primarily to North Africa.

It was during the long sixteenth century that the Mediterranean trio gained wider commercial appeal.[161] Before the 1450s, the commercial presence of wine and olive oil was therefore relatively less tenuous than that of the oriental goods that constituted the staples of the Mediterranean commerce: sugar and cotton. Wine and olive oil, of course, crisscrossed the basin in the fourteenth and fifteenth centuries, yet the volume of this trade was comparatively limited.[162] At its zenith, the city-states' rule was based on the procurement and distribution of "scarce" and "exotic" goods rather than the main staples of the Inner Sea— easily available to most where ecological conditions permitted.

The unity of the Mediterranean was thus solidly embedded in the division of labor put in place by the reigning city-states between 1250 and 1450, during a period of intense rivalry and competition. It did not merely emanate, as argued or intimated at times, from the ecological attributes of the basin, as captured by the low profile kept by the tree crops of the region. Nor was this division of labor ahistorical. Rather, the cultivation of highly gainful crops that undergirded and embodied the Venetian and Genoese merchants' unmatched sway over the economic flows of the Inner Sea was either revived in, or disseminated into, new and occidental locations within and without the basin by the very same city-states. By the 1450s, the basin's economic structure had become much more sophisticated and was more tightly integrated than in the twelfth century. Not only was the Iberian peninsula's central role, mirrored in the dominance of the entrepôts of Malàga and Alicante, undermined by the rise of the northern Italian city-states. But also, and more important, was the range and scale of goods that traversed the Inner Sea and entered into commercial circulation.[163] At the start of the long sixteenth century, the crop trio, comprising the spices, oriental goods, and grain, constituted the foundations upon which the commercial empires of the Mediterranean were based. The attempts to relocate their cultivation in new sites or, as in the case of the spices, to find an alternate route for their transmission played a crucial part in the basin's history.

If the partitioning of the Mediterranean hastened the westerly migratory movement of oriental crops, the fifteenth-century crisis sealed the subjugation of the trio of Mediterranean crops to the cupola of "golden crops." When gold fetched from Africa and silver extracted from the mines in Kutná Hora largely dried up after the 1380s, cotton and sugar played a pivotal role in obtaining silver or gold. Venice, given its sway over the Bosnian and Serbian mines, suffered less, but overall supplies of silver remained "erratic and precarious" until the 1460s.[164] In the fourteenth century, the merchandise deficit was still made up by bullion shipments, despite the so-called commercial revolution of the good thirteenth century.[165] Similar problems were simmering in the western Mediterranean as well (and went on to do so until the 1560s). Largely divested of the spice trade, the Genoese merchants who had invested their fortunes in the slave and sugar trades found themselves, like their Venetian counterparts, encumbered by a lack of specie. Genoese expeditions into North Africa and farther south generated occasional success in spotting oases of gold, yet failed to solve the problem in a definitive fashion.[166] For the merchants of St. George, lack of specie posed a more formidable problem than that in the Levant, because here the crop that underwrote the process of colonization, sugarcane, was closely associated with and blossomed in the company of, slave labor. The elective affinity between sugarcane cultivation and slave labor, which had been established recently and served the Genoese merchants handsomely, was bound to falter in times of scarcity in precious metals, since resulting difficulties in the acquisition of slaves found their manifestation in the profitability of sugar production. The predicament was momentarily overcome by the enslavement of the *guanches* and the inhabitants of the newly colonized islands.[167] But in the medium run, the drying-up of precious metal flows exactly at a time when the spice trade in the east and the slave trade in the west were picking up speed compelled that goods be dispatched in return. Stated differently, the end of the bullion famine may have lessened the pressures on the twin city-states, but it did not alter the tendencies of the previous period.

Since Venetian and Genoese businessmen both suffered from the same ailment, the bullion famine, and the reigning city-states resorted to comparable remedies, that of boosting their manufacturing and re-exporting activities to the benefit of finished cloth, both halves of the basin underwent analogous and parallel transformations.[168] The Milan-Florence-Venice-Genoa quadrilateral thereby served as a magnetic pole for raw material flows, namely, silk and dyes from Romania, Sicily, and Granada, cotton and alkali from the Levant and Romania, and wool from Iberia (and, to a lesser degree, the Levant). In other words, the launching and maturation of the woolen and silk industries in, and outsourcing of cotton textiles by, the cities of the northern Italian peninsula led to a more comprehensive canvassing and mobilization of the raw materials within and without the region, as confirmed by the centrality of Florentine merchants in silk markets who positioned themselves between main silk cloth producers and principal silk-growing regions.[169] The rejuvenation of silk and cotton cultivation in the occidental latitudes of the basin was therefore complemented by the extension in the realm of wool throughout the length and breadth of the Mediterranean.[170] As sugarcane sailed erratically from one isle to the next and reached westernmost Madeira, cotton and silk cultivation was given a new lease on life as a result in the Genoese-controlled half of the basin: silk in Granada and the kingdom of Sicily and cotton in North Africa.[171] Even so, these new sites could not effectively challenge the preeminence of either Caspian silk or Levantine cotton. The economic vitality of the occidental Mediterranean, however, was not negatively affected by this, since the rise and centrality of the wool trade in Iberia and in Majorca effectively added a new dimension from the fifteenth century to the economic edifice of the basin. Wool and sugar in the west and cotton and silk in the east saw their lot consolidated with the acceleration in the pace of manufacturing.

As the economic reticule woven by the flow of raw materials, textiles, and finished cloth became more complex and expansive, adding new layers to the commercial structures put in place from the thirteenth century at both ends of the Inner Sea, the grain trade experienced a different fortune: its destiny was closer to that of wine and oil than to that of the oriental crops. For one thing, the grain trade suffered a veritable drawback in the latter half of the fourteenth century in particular, because the basin's breadbaskets were hit by the Black Death and the political uncertainties of the period. In the east, Ottoman excursions into Thessaly and Salonika and the resulting atmosphere of uncertainty, especially around Salonika from the 1380s to the 1430s, when the city changed hands between the Ottomans and the Venetians, had already induced a fall in grain exports from Romania, with Salonika losing its role as a main port of trade.[172] From the 1480s onward, the situation turned alarming. The imperative to provision Constantinople, the population of the capital city soaring to new heights under Ottoman rule, and the attendant closure of the Black Sea to Frankish merchants, siphoned off large volumes of commercially available grain.[173] This problem had not arisen previously, simply because the city's population had fallen to eighty thousand in the first part of the fifteenth century. Its population, however, ranged between six hundred thousand and seven hundred thousand in the course of the sixteenth century.[174] The provisioning of this demographically colossal capital city principally relied on shipments from the Black Sea and the Balkans.[175] Venetian traders, always resourceful, had already expanded their areas of catchment to include the principalities on the Aegean shores of Anatolia beginning in the fourteenth century.[176] They went on to ply the waters of the Aegean, unperturbed by the loss of Salonika to the Ottomans in 1430 and Negroponte in 1470. All the same, the deficit created by the diversion of Romanian grain to Constantinople was a serious blow to the provisioning of the Venetian empire.[177] The access that the Frankish merchants had to the Pontic Sea from 1204 ended with its closure in the 1480s and undermined the bases of the previous strategy of provisioning.[178]

The Genoese Mediterranean was not exempt from these misfortunes, even though the reasons were slightly different. Its granaries recovered very slowly from the ghastly losses inflicted by the Black Death. In Sicily, the demographic decline reached its nadir around 1400. In 1282, the island's population was around 850,000; in 1501, after the recovery of the fifteenth century, it stood in the neighborhood of 550,000 to 600,000.[179] Although the island's economy had become less diversified in 1450 than in the thirteenth century and its dependence on the Genoese financiers and traders had grown stronger, the volume of its grain exports stagnated in the fourteenth and fifteenth centuries at best, and decreased in great likelihood.[180] Sardinia, for one, had a population of 160,000 in 1485, down from 300,000 to 400,000 early in the fourteenth century. The early figure was not reached again until after 1750, having gone through a second slump in the latter half of the seventeenth century.[181] As the number of producers went on to shrink, only to reach their former levels in the eighteenth century, two new developments compounded the demographic loss. On the one hand, with Genoa's relocation of its empire west of the kingdom of Sicily, Barcelona was drawn into the orbit of Sicily after the 1380s. Unlike Andalusia and Portugal, Catalonia and Majorca (and at times Castile) were secured access to the Sardinian and Sicilian grain owing to the reach of the Genoese merchants. Starting from the first half of the fifteenth century, merchants from Lisbon and Genoa and, later, Seville started to show up more often, if not regularly, on the shores of Morocco in order to secure grain imports.[182] On the other hand, that Naples, the second most populous city in the Mediterranean after Constantinople until the eighteenth century, went on to rely on grain from Sicily and Puglia, contributing to the further dwindling of grain supplies. Supply was getting ever tighter in the basin's former bread lands, undermining the ability of the Genoese and Venetian merchants to provision the basin with grain and victuals.

If the fifteenth-century conjuncture proved to be vexing for the sibling city-states, it has to be underlined that the territorial expansion they pursued via the enlargement of their Terraferma and *contadi* in the first half of the fifteenth century must have slightly brightened the bleak picture that emerged in the Levant with regard to grain. Starting from the turn of the fifteenth century,

Venice incorporated Brescia, Verona, and Bergamo, and Pisa fell to Florence. By the time of the Peace of Lodi in 1454, Venice was still in the process of building up its Terraferma, Milan came to tower over Lombardy, and Florence over what became Tuscany.[183] These new political entities were in effect city-states writ large, and met the needs of their mother cities in manufacturing and at times in agriculture, serving as a buffer zone. Despite the aggrandizement in the city-states' "home" territories, it should not be automatically assumed that the newly incorporated lands came to take over the duties or the functions of the offshore colonies and territories. They did so momentarily, in the latter half of the sixteenth century, and not very successfully. The functions undertaken by these cities and their associated hinterlands proved to be more significant in the realm of manufacturing, which gained pace during the sixteenth century, than in agriculture.[184] Besides, as stressed above, provisioning the needs of the city-states constituted but one component of the demand for grain. That it was mostly "home" grain that served the city-states still necessitated that other supply sources in the basin be tapped to provision the empire at large.[185] Providentially for the city-states, the travails of the grain market came to an end due to completely unforeseen circumstances, owing to the economically prosperous *beau seizième,* as we discuss shortly.

Here, note should be taken of the fact that search for new sources of grain, when supplemented by the growing demand for raw materials at the economic heart of the Italian peninsula and elsewhere for flourishing textile industries, perforce extended the territorial sway and command of the twin city-states in their respective spheres of operation. What solidified the relations between Habsburg Castile and the Genoese *nobili vecchi* on the one hand, and the Sublime Porte and the Signoria on the other, was the mutation in the direction of commercial and financial flows, as we will see below. The coming of German silver (and West African gold), drained first via Venice (and Tunis and Tripoli) to the Mediterranean and later via Antwerp (and Sijilmasa) to the Atlantic, eased the bullion shortage, but fell short of satisfying the needs of the Levant and Atlantic trade.[186] Certainly, given the centrality and the steep rise in the volume of the spice trade in the Levant and the slave trade in the Atlantic Mediterranean, German silver failed to radically alter the status quo. Yet when the silver from German mines was complemented by the apparition of the American treasure in Seville, the western tier of the basin became increasingly subject to the rhythms, *pace* Chaunu, of a movement that was centered in the Atlantic.[187] The confluence of the pepper, sugar, silver, and Baltic grain trades first at Antwerp and later at Amsterdam deprived the Mediterranean of its rich trades. Still, if the cultivation and distribution of the golden crops of the region, from cotton and silk to wool and sugar, maintained their salience until the end of the age of the Genoese in the mid-seventeenth century, the same cannot be said of the grain trade, or at least in the Atlantic Mediterranean. The unmaking of the Mediterranean division of labor that was structured in the 1250/1300–1450 period started to come undone because of it after the 1450s, not all at once but in a steady yet inexorable manner. The gradual diminution in commercially available agrarian surplus, first in the western Mediterranean and later in the Levant in the long sixteenth century, shook the foundations of the division of labor that undergirded the Serenissima's sway over the Inner Sea.

End of the Imperialism of Wheat

The thickening of the demarcation line between the Venetian and Genoese spheres of operation and the diverging trajectories of the rival city-states had their political repercussions. In the western tier of the Inner Sea, the relationship struck between the Castilian throne and the merchants of the city of St. George, the latter in charge of the Sicilian silk and grain trade and transatlantic explorations, came to be built on a new foundation from the 1520s, when Genoese merchants started to extend loans to Charles V.[188] The availability of royal finance that had previously been earmarked for the house of Fugger allowed the Genoese men of money to diversify their activities between mercantile and financial operations.[189] The new modus vivendi between the Genoese *nobili vecchi* and Philip II was neatly complemented by developments transpiring in the eastern Mediterranean, where the diversion of the spice trade away from the Levant not only summarily hastened the fall of the Mamluk empire to the house of Osman, but also deepened the relationship between the Signoria and the Sublime Porte. The territorial expansion of the Ottoman empire did not necessarily sap the vitality of the Venetian empire. The formation of a vast economic space under Ottoman rule, from Salonika and Istanbul to Cairo and Baghdad on the one hand, and the emergence of the Arabian Seas as a node of lively economic activity on the other, benefited the merchants of the Serenissima substantially: witness the growing presence of Venetian merchants in Hormuz and Isfahan.[190] In some measure, the salience of the Arabian Seas stemmed from the Portuguese merchants' inability to fully redirect the spice trade toward the Atlantic: they were forced to sojourn in, and conduct business via, the Persian Gulf. Pivotal was the restoration of the spice trade in the Indian Ocean and the Persian Gulf under the aegis of local merchants, even though Lusitanian sea power partly circumscribed their activities.[191] The opening of the Atlantic and a new Genoese-Habsburg rapprochement was thus neatly paralleled by the unification of the eastern Mediterranean under Ottoman rule, giving a boost to the Venetian enterprise in the Levant. In fact, it was under Ottoman rule, in 1552, that Venetian merchants were given the right to conduct their operations in Cairo; until then and especially when the city was under Mamluk rule, these merchants were confined to Alexandria.[192] Relations between the Genoese men of money and the Habsburg throne and the Venetian merchants and the Sublime Porte took on a new hue during the century of the Genoese.

As to be expected, the Atlantic-bound movement boosted the vitality of the western Mediterranean. It tethered the fortunes of the Genoese traders' expanding emporium and the continuation of the African slave and pepper trades to their operations in the basin's main ports of call, whence they acquired textiles and goods. After all, the Atlantic-going Genoese gained access to the precious American treasure by dealing in textiles, marketing them in Lisbon and Seville, and thereby obtaining gold from former and silver from the latter. Besides, the devotion of the Atlantic isles to sugar cultivation and the centrality of wool production in the Iberian peninsula and in offshore sites were both dependent on the availability of grain, thanks to the Genoese merchants and their impressive fleet, which reached its zenith in 1556–58 with 29,000 tons. At the end of the fifteenth century, the size of the fleet varied between 12,000 and 15,000 tons.[193]

In the Levant as well, Venice's hold over the Mediterranean circuit may have decreased in the first half of the sixteenth century as a result of the diversion of the spice trade, but this did not necessarily translate into declining commercial activity. The Signoria was still able to capture a large portion of the remaining spice trade: it made allowance for spices to be transported on any vessel, not exclusively on ships belonging to the *galere da mercato*, as was the case previously. It waived customs duties on the entry of spice-bearing ships to Venice, and forced the merchants of the towns in the Terraferma to travel to Venice for all their purchases.[194] Thanks to the vitality of the Hormuz connection, spices still made their westward journey, not via maritime routes, but via overland routes that traversed Caucasia, Anatolia, and the Balkans, and reached the debouchment points of Akkerman, Lwów, and Brašov.[195] That these overland routes proliferated at a time when maritime spice trade ebbed was not coincidental, for it was via terrestrial routes that the oriental goods eventually found their way to the well-established markets north of the Alps.

Raw materials that fed the industries of the city-states—cotton, wool, and silk—also followed a similar terrestrial itinerary and reached the Adriatic via Ragusa and Spoleto. The maritime trade was still in place, albeit diminished:

the rise of the port of Ancona, which served as an outlet for the woolen textile manufacturers of Florence and the peninsula at large, was part of this development: silk, cotton, dyes, and spices constituted over 80 percent of the city's mercantile activity during the first half of the sixteenth century.[196] That is, the "local" content of the Levant trade grew during a temporary abatement of the spice trade. The number of small ships and barges that were engaged in coastal trade and transported wool (and grain) multiplied. A new fleet grew up, specializing in the transportation of bulky goods: these fleets, mostly Venetian and Ragusan, were comprised of ships that could carry over 14,000 hectoliters of burden.[197] In sum, as the basin's western shores were being reshaped under the kinetic "synergy" of the sugar and slave trades, in its eastern shores, the local content of the goods that crisscrossed the basin was increasing, in the form of grain, cotton, and silk.

Both city-states were thus bound by the same set of constraints, for the manufacturing drive of the fifteenth and sixteenth centuries impelled them to intensify their dealings in the procurement of cotton (and silk) in the Levant and wool (and silk) in the Genoese Mediterranean due to the precipitous rise in the share of textiles in their merchandise exports. While the market for industrial crops thrived, the precarious state in which the grain trade found itself in the 1350–1450 period, as examined in the previous section, deteriorated further, especially in the westerly quarters of the basin. Two long-term developments played a part in this. First, the assumption by Habsburg Spain from the late fifteenth century of the role of the principal wool supplier to Flanders and to the Italian city-states, as the manufacturing of woolen textiles took the center stage, opened up the arable to competition between the plow and Merino sheep, and turned the vastitudes of Iberia and Puglia into grazing fields. The plow may have "moved inexorably forward," but Charles V issued decrees at various times urging his subjects to return to pasture all new lands reclaimed during his reign.[198] Loss of arable to grazing was particularly deleterious in the Mediterranean, because wheat took up large areas of the basin's corn lands given the low level of yields and the need to leave the sown land fallow the next year. In Ottoman Anatolia, for instance, the yield rate was roughly 3–4 to 1; in Iberia 5 to 1 (4.5–6 to 1 or more in Andalusia); and 3–4 to 1 in Provence.[199]

Second, the earmarking of the newly reclaimed islands for sugarcane cultivation reduced the chances that cereal agriculture would be established successfully in these places. Provisioning the settlers, sailors, and the Iberian urban centers increased the need for grain at the same time that vast tracts of land in the western Mediterranean were now given over to livestock husbandry. Even Andalusia, which followed a path different than that of the Mesetas, for it served as a grain-basket for Castile until the mid-sixteenth century, ceased to export grain, and even when it did, it was in ever tapering quantities. "Flour from Andalusia was no longer sufficient to make the biscuit for the fleets."[200] The influx of American gold and the availability of Baltic grain, which had made its appearance in the region as early as the fourteenth century,[201] sealed the trend in the western Mediterranean toward livestock husbandry, restructuring the division of labor in the basin. At any rate, the Atlantic and Mediterranean Iberia, thanks to their surpluses in wine, oil, and American silver, could afford to import grain.

If the above-mentioned twin developments took land away from cereal husbandry, there was also a countertrend that eased the bottleneck in grain trade, however briefly. From the mid-fifteenth century, economic growth picked up and the basin started to recover as new lands were opened up to cultivation, new rural settlements proliferated, and urban centers grew at a dizzying pace.[202] This was by no means Chaunu's "full world," which was reached in the mid-fourteenth century. Nonetheless, recover the Mediterranean did during the course of the long sixteenth century, and the consolidation of the empires encouraged the settlement and exploitation of new lands. In almost all quarters of the basin, from Andalusia and Provence to the Ottoman dominions and North Africa, the new era of expansion found expression in demographic and agricultural growth. Large sections of the arable in the Habsburg and Ottoman dominions were brought under the plow, at times at the expense of, or in competition with, livestock husbandry. That animal husbandry had gained salience during the politically tempestuous times of the period prior to the 1450s and was later stimulated by the efflorescent wool trade turned the competition between bread lands and pastures into an epochal confrontation between cereal and livestock husbandry.

The proverbial struggle between Abel and Cain that put the Catholic kings in an uncomfortable position from the 1480s—after the termination of the civil war—to the end of the sixteenth century neatly paralleled attempts to settle the nomads in the Ottoman empire at the end of the fifteenth century.[203] Notwithstanding the fierce competition between livestock husbandry and arable agriculture over land, large tracts of previously uninhabited land were brought under the plow. The period of growth involved a fast demographic recovery and a large-scale reclamation and settlement of lands previously left untilled.[204] In certain parts of the Inner Sea, the recovery may have started later, but the overall direction and tenor of change was unmistakably similar: demographic and urban growth colored, almost in a pointillist fashion, the shores encircling the basin. The trend was manifest in Castile, Andalusia, Provence, Anatolia, the Balkans, and the Levant—throughout the length and breadth of the Inner Sea.[205]

Cereal production, now expanding on lands that had been given over to nature after the Black Death and hence were well-rested, allowing for good yields, rose swiftly.[206] Where marginal land abounded, as in lands attended to by the nomads on a periodic basis in inner Anatolia, or where the forfeiture of the arable was extensive, as in the case of the uprooted Moriscos in Andalusia, the tempo of reconstruction was especially rapid. Broadly speaking, the recovery was more pronounced in the latter half of the fifteenth century, after the conclusion of the civil war in Iberia and the takeover of the Pontic shores by the house of Osman. The reconstruction in the Levant started later, after the fall of the Mamluk empire to the Ottomans in the 1510s. Judging by the rise in deliveries in tithes, the proliferation in the number of new villages and *latifundios*, and the intensity in the activities of *waqfs*, pious foundations, in the colonization of new lands, this was a period of plenty.[207] In fact, owing to the large nomadic populations, the settlement of new lands and the relentless engagement of these peoples in agriculture as an ancillary activity to their animal husbandry deepened their agro-pastoral specialization.[208]

Developments in the western extremities of the basin differed from those in the Levant. The Atlantic coast of Iberia, now territorially larger owing to the settlement of lands from the Azores to Cape Verde and by the tentative incorporation of post-Marinid Morocco, followed a different trajectory.[209] Given that initially the newly conquered lands were devoted to grain cultivation, it became easier for the inhabitants in the coastal regions of Iberia to devote more of their land to lucrative cash crops, such as wine and olive oil. Madeira, the Canary islands, the Azores, and the wide Moroccan plains all supplied, at one point or another, Portugal and coastal Iberia with grain. Queen Isabella's expectation that the Spanish settlers in newly acquired lands would exist as farming communities and live alongside the natives reflected a need.[210] Portugal, more heavily invested in these explorations, displayed the scale of transformation in a crystal-clear fashion. Previously exporting grain to England, it had built its imperial policies, from the late fifteenth century, around "grain imperialism."[211] It was not the expanding reach and stretch of merchants of the western Mediterranean alone that was at the root of this transformation.

If the presence of Baltic grain and the expansion of livestock husbandry hastened the conversion of large tracts of the arable in the occidental Mediterranean to the cultivation of lucrative crops, two additional factors reinforced the trend toward diversification. First, the arrival of Caribbean gold pushed up prices. The years from 1506 to 1510 and from 1521 to 1525 registered the highest price increases in Spain. What is more, increasing liquidity rewarded some crops more than others. Provisioning Spanish settlers during the sixteenth century, as late as about 1580–90, placed a premium on cargoes of grain, biscuits, wine, and oil.[212] In line with this development, wine and oil prices increased faster than that of wheat. Taking 1511 prices as the base 100, the price of wine climbed more than others: in 1539, it stood at 350 as opposed to oil at 297.5, and grain at 264.4. The tendency remained intact in the decades to come. Between 1511 and 1559, the price of wine in Andalusia rose nearly eightfold, a much higher rate of increase than for grain and olives.[213] Cultivation of vines and olives accelerated rapidly, as in Jerez and Jaén. And the commercial circuit Indies–Seville/Lisbon–Medino del Campo was extended to include Barcelona and Perpignan.[214] In Mediterranean France as well, "fairly important olive plantations were established" from 1500 and 1570. Commercial production in the Garonne, the Loire, and the Seine was encouraged by the Atlantic trade.[215]

Second, the ascent of Antwerp, with its comfortable access to the Baltic grain, only enhanced the direction of specialization. In the opening decades of the sixteenth century, the rise of Antwerp and with it the opening of the Scheldt markets, when accompanied by the growing presence of Iberian merchants in the Low Countries, had animated the northerly flow of oil and wine, not to mention sugar.[216] With this came, of course, Atlantic Iberia's dedication of its lands to orchards and vineyards. By the mid-sixteenth century, the process had progressed to such a degree that even a region as rich in grain as Andalusia had started to import grain as the city's olive orchards were spreading at a rapid rate. The economic buoyancy generated by the centering of the spice trade on the Atlantic coast brought about the growing inclusion of Andalusia into the commercial networks interweaving Castile, hence the consequent shift from the Burgos–Medina del Campo–Bilbao axis to the Burgos–Medina del Campo–Sevilla axis.[217] This did not rule out the assignment of larger areas of inland Iberia—the Mesetas—to grain production, but it did not alter the situation fundamentally.

Put simply, however anemic the influx of "Aztec gold" may have been at the outset, the resultant acceleration in economic activity along the banks of the Atlantic attracted scores of settlers to the outposts of the maritime Iberian empires. Demand in these newly incorporated regions for the staples of the Mediterranean—olive oil and wine—buoyed economic life in Andalusia (and Iberia), the gateway to the Atlantic.[218] Provisioning settlers in New Spain and the islands was a task cherished by the merchants of Amsterdam, for it provided a timely and convenient outlet for Baltic grain. Because the volume of incoming bullion, albeit mercurial, was on the rise and commercial grain within easy reach, the growth of livestock husbandry proceeded unfettered. The developments along the shores of the Atlantic Mediterranean were accompanied by and closely intertwined with the migration of the spice trade away from the Inner Sea after 1500.

Diversification of agrarian production in Iberia was not the only result generated by the inflow of American gold and German silver. Along with Iberia, hub of the silver trade, regions north of Augsburg experienced an inflationary upsurge owing to an increase in the volume of bullion passing through the hands of the Fuggers. Precious metals flowed between Augsburg and Antwerp, leaving their tracks north of the Alps.[219] The concomitant confinement of inflation to the shores of the North and Baltic seas caused a disparity in prices to the north and south of the Augsburg-Antwerp axis, to the advantage of the latter. Stimulated by the increased output of German silver mines under the aegis of the Fuggers, prices in general and grain prices in particular to the east of the Elbe registered their sharpest increase between the 1530s and 1550s, rendering Ottoman grain relatively cheaper by default.[220]

Of the 100,000 to 200,000 tons of grain thrown into circulation per annum during the mid-sixteenth century in the Mediterranean, more than 50,000 tons originated in the Ottoman lands, mostly in the former Venetian and Genoese Romania.[221] The commercial availability of Ottoman wheat in the Mediterranean markets was boosted by demographic trends of the era. From the 1530s to the 1570s in particular, the population of the empire grew by significantly more than 0.07 percent per annum (reaching more than 2 percent in certain places, especially during the early half of the century), much higher than the median rate of demographic growth in the basin.[222] Nonetheless, given the pace of the empire's territorial expansion, there was plenty of land to be reclaimed, and lands that reverted to nature were naturally invested with relatively high productivity. That the volume of production was not outstripped by demographic growth was confirmed by the fact that food prices in Istanbul registered a 31 percent rise in grams of silver, between 1489 and 1573, "indicating a modest rate of silver inflation." Relatedly, the weight and silver content of the Ottoman currency, the *akçe*, remained unchanged from 1481 to 1585.[223] On account of the level of taxation, which in the first part of the century was just a fraction of what it became at its close, the favorable conjuncture of light taxes and high productivity allowed the *reaya*, peasants, to retain 50 to 65 percent of their total production.[224]

In the Balkans, for instance, the lands the Ottomans inhabited were confined mostly to the low marshy plains, river valleys, and the lower altitudes of ambient hills: Thrace, Thessaly, the lowlands of Macedonia, and Danubian Bulgaria. These lands were referred to as *magna valachia,* and served as grazing grounds for the flocks of the Wallachians.[225] Towered over by a chain of castles erected on ambient higher ground, these low-lying plains were thinly populated. They were vacated, in part, before the arrival of the Ottomans due to the escalation of lordly coercion during the bullion famine of the fifteenth century that popularized labor rent and placed a premium on labor services.[226] They were vacated, in part, after the arrival of the Ottomans by the exodus of its remaining inhabitants onto higher altitudes. The *yürük,* nomadic, populations who were in charge of the colonization of these lands were stock-raisers practicing transhumance and found themselves on a terrain perfectly suited to their winter pasture needs, with an abundance of grazing ground. Summer grazing on higher altitudes was a different matter, since the hills and mountains offered a sanctuary to those who had vacated the plains. An increase in the populations of these altitudes after the desertion of the plains diminished the area available as pastureland, not to mention the fact that these pastures were located in hostile territory. The diminishing capacity of newly arriving nomadic populations to perpetuate their peripatetic existence was one of the factors that gave a boost to sedentarization.

A second boon to the process of settlement was the awarding of large estates to military cadres. Often, these estates were eventually transformed into religious endowments that were, in comparison with their counterparts elsewhere in the empire, considerably larger. In 1530, for instance, when the number of prebendal villages in Rumelia was 2,709, in the Pasha *sancak* alone there were more than 750 *waqf* villages.[227] The pace of colonization was swift, then, thanks to the fact that the lands taken over by the invaders were sparsely populated lowlands and hilly sites overlooking them, home previously to medieval castles: most of the early settlements carried the name *hisar,* castle, as well as the suffix *âbâd,* won from the wild—heathland or thicket. That the latter term was used interchangeably with *ova,* the plains, registers plainly where the process of recolonization was first at play.[228]

Territorial expansion along the Syrian coast was likewise largely confined to the lowlands of the Fertile Crescent, in a geographical setting similar in character to its Rumelian counterpart: in 1520 when the Ottomans made their first survey of Tripoli, they found only 800 out of 3,000 villages still revenue-bearing.[229] And these lower landscapes were the first to be resettled and repopulated; as a result, the region regained its economic vitality in a relatively short span of time. From 1521 to 1569, the number of villages in Damascus province increased from 844 to 1,129.[230] Here, too, the castles overlooking the Cilician and eastern Mediterranean plains fell into disuse as soon as their function of overseeing the lowlands lost its relevance as these lands were recolonized and inhabited.[231]

Overall, then, the commercialization of cereal and agrarian production increased in the Ottoman Mediterranean whereas in the Genoese Mediterranean, it was mostly the "traditional" tree crops of the Mediterranean, hence the vineyards and olive groves, which grew to the detriment of cereal agriculture. Certainly, the pace of recovery varied from place to place, but throughout the expanse of the Mediterranean, it was the coastal regions which, *grosso modo*, recovered earlier than the inland provinces, save for Catalonia. The vitality of the Cantabrian and the Galician provinces of Iberia from the fourteenth century on gave these regions the opportunity to shift their economic specialization accordingly. Northern, northwestern, and Atlantic Iberia, despite the fact that they never were ideal lands for viticulture in terms of soil and climate, gave over significant stretches of their land to vineyards, furnishing the rural producers with substantial income from market transactions.[232] The coastal regions of Iberia, which were commercially reshaped by the economic pulse of the Burgos and Medina del Campo fairs, later were tuned to the developments in Antwerp. After all, the wool trade, centered in Burgos, boomed in the late fifteenth century and its prosperity remained in place until the 1560s. It was then that difficulties appeared in the plow lands of the Iberian peninsula, primarily because that was when the expansion of the central Meseta came to an end. Andalusia, therefore, in stark contrast with Castile, was able to reallocate its arable in favor of vineyards and olive orchards. Murcia followed a similar trajectory, because in this instance, the cash crop that led the region in a different direction was silk which was sent to the commercial centers of the interior, to Toledo in particular: "So dominant was the position of silk that the mulberry ousted wheat as the principal crop."[233] The situation in Valencia was not all that different.[234] The Baltic's supremacy was also confirmed by the movement of northern (and English) grain via Livorno.

Capturing this ongoing process fully was the appearance of grain shortages that started to plague the western Mediterranean. The flow of *blé de mer* into the region, which was almost always guaranteed, experienced frequent disruptions. In the core zones of the basin, land sown under grains shrank systematically as its peripheral zones took on the burden of cereal husbandry, but not for long.[235] During the course of the second sixteenth century, this division of labor came undone as cultivators in Ottoman Romania and Sicily, among others, who hitherto had shouldered the burden of cereal agriculture, chose to switch to the cultivation of more profitable crops, with disastrous consequences for the basin.[236] Growing demand for industrial crops on the one hand, and for meat and dairy products on the other, reallocated some of the land sown under bread crops to more profitable crops and uses.[237] As a result, the Ottoman lands, too, started to suffer from occasional shortages. Increasingly then, the amount of land devoted to the cultivation of grains along the banks of the Mediterranean contracted to the benefit of competing and more lucrative crops, vegetal or animal. It was partly this retreat from cereal husbandry in the former granaries of the basin that prepared the way for the arrival of Baltic grain in its heartlands.[238] This was a major sea change for the ports of the western Mediterranean, although the northern tier of the Iberian peninsula had long been provisioned by Baltic grain, thanks to its ports on the Atlantic. Following the diversification of the basin's crop-mix, its corn lands shrank and its cereal production declined. The process accelerated during the age of the Genoese. The diverging yet complementary trajectories of the twin city-states that had underwritten the prosperity of the *beau seizième* by, in effect, dividing the basin into two functional halves would be undermined by the very age of the Genoese.

It was within this modified and diversified landscape that bread grains yielded to the march of lucrative crops and pastureland. The modifications in the division of labor that took place after the 1550s and allowed the denizens of the western Mediterranean to divest themselves of grain production from the 1450s started to generate strains toward the end of the sixteenth century. In Crete, for example, the Venetian rulers resorted to uprooting vines and cajoling local producers to plant wheat in order to ensure the steady flow of grain to its navies. The Sublime Porte, for its part, periodically placed restrictions on, or banned, grain exports from its dominions.[239] High grain prices briefly tempted the wealthy bourgeoisie who invested in land to try to take advantage of the bottleneck,[240] but failed to stimulate petty producers' interest. Even in the case of the landlords, their greatest remuneration came, as in Iberia, not from the lease of lands but from rights such as "the use of baronial flour mill, wine and olive press, and in some cases, baronial shops."[241] Looming crop failures and shortages notwithstanding, the expansionary thrust of the second sixteenth century helped preserve the momentum of crop diversification in the eastern Mediterranean at the expense of bread cereals. The return of bullion and spices to the basin after a hiatus of a half-century thus left its imprint on its landscape.

As was to be expected, the retreat from agriculture reached unsettling proportions in the western Mediterranean. In Spain, the Cortes started petitioning the emperor from 1579 to restrict the planting of vineyards because more and more land was being devoted to viticulture. In fact, in early sixteenth-century New Castile, viticulture was much more widespread than it is today. And Andalusia had become a regular grain importer by 1560 at the latest. In Languedoc, the loss in 1535 of more than half the flocks of sheep due to epizootic diseases did not translate into a boost for arable production due to the availability of more land: there, too, a ceiling had been reached by 1560 in tithes and agricultural production, and the situation was made worse intermittently by the Wars of Religion. Subsistence crises intensified between 1560 and 1575 and again between 1584 and 1595.[242] Provence, Languedoc, and Catalonia escaped the depressing times after 1650 by devoting more land to viticulture.[243]

The situation changed decisively once the responsibility of handling the wealth of the world-economy passed into the hands of the Genoese bankers, who supplied the Inner Sea with large quantities of silver and underwrote the return of the spice trade to its time-honored route. The infusion of silver monies, *reales de a ocho* in particular, into the eastern Mediterranean in the last quarter of the sixteenth century was not without its consequences. As attendant inflation started to plague the basin, it was not only the price differential between the Baltic and the Levant grain that shrank to the detriment of the latter.[244] Crucially, the inflationary environment generated by the influx of silver lured enough producers in the eastern Mediterranean to switch from grains to more lucrative crops or goods.[245] A development that had picked up pace in the central zones of the Mediterranean world between 1450 and 1560 was replicated, from the 1550s, in its eastern extremities, the Ottoman dominions. Growth continued in this period, but was mercurial and torn by political trauma that shook most quarters of the Mediterranean (and beyond).[246]

The contraction in grain production consequently exposed the Mediterranean to the vicissitudes of potential crop failures and shortages. This was amply demonstrated by the onset of a series of crop shortages that had started earlier at the very heart of the millpond—in Venice in the 1530s and 1540s and in Spain in the 1560s—and now began to affect its outer rims, the Ottoman dominions, from the mid-1560s to the 1590s.[247] The deceleration in population growth in the Italian and Iberian peninsulas after the 1570s prompted landlords to plant vines and olives and to convert the arable to pasture. In the Ottoman countryside, too, the crop-mix became more varied and market-sensitive in concert with the growth of urban centers and handicrafts within and across its borders, boosting demand for raw materials.[248] Conversely, the growth registered by grain production frequently remained below that of other crops. More often than not, it failed to catch up with demographic growth, if it did not decline outright.[249] The second sixteenth century was thus not conducive to the imperialism of wheat, which had colored the Mediterranean more or less since the turn of the millennium, with the exception of the 1350–1450 period. It was precisely because grain production in and around the basin was losing ground vis-à-vis lucrative crops as the population was growing ceaselessly that the grain crisis of the 1590s proved to be more devastating than before, in Castile as well as in Damascus. The timing of the Ottoman wheat boom, which lasted from the 1540s to the mid-1560s, provides solid testimony to the fact that, in part, the boom was rendered possible by the inflationary upsurge in the northern latitudes of the Augsburg-Antwerp axis in the early half of the century. After the 1560s, the empire, too, suffered from frequent grain shortages.[250]

Even though the Ottoman empire's territorial expansion commenced in the mid-fourteenth century, the process took on a new form from the mid-fifteenth century as the pace of settlement accelerated. The establishment of Ottoman dominion over the Anatolian peninsula and the Mamluk lands, completed respectively in the 1450s and the 1510s, coincided with the economic expansion of the 1450–1560 period, allowing the process of settlement and reclamation to take place much easier and faster than otherwise would have been possible. The settlement of nomadic populations, slow or sudden, circumscribed the area left to use by nomadic populations, thereby compelling further sedentarization. In the low landscapes of the empire, the colonization and sedentarization processes advanced in a speedy manner in the fifteenth and sixteenth centuries. Changes in the nature and extent of pious foundations' rural holdings attest to this. In most deeds, villages and *mazra'as*, subsidiary settlements, appear with greater frequency from the turn of the fifteenth century.[251] Early deeds are filled with acts of endowment that place a wide array of rural revenue-bearing sources at the disposal of the foundations. What is striking in these deeds is the conversion, time and again, of entire villages—with their auxiliary fields, near and afar— and uninhabited fields—used for farming or pasture—into mortmain. Over the course of the sixteenth century, the composition of revenue-bearing sources shifted in favor of urban assets. The turnabout in the fortunes of pious foundations, whereby revenues flowing from urban assets surpassed that of rural assets, eloquently registers the changing balances between rural and urban fortunes.[252] Rural revenue sources endowed for pious purposes were not as generous in the latter half of the sixteenth century: only shares, indeed fractions, of villages or *mazra'as* were earmarked as mortmain. This demonstrates that most of the prime land that was easy to reclaim and cultivate was already under cultivation and its rights of usufruct granted.[253]

The juxtaposition of these two trends signaled the inevitable. As the colonization of agricultural land was slowing down, urban growth was picking up in pace. The tempo of urbanization was as determining a factor in the eastern Mediterranean as it was in its western half in setting the overall tone of agricultural growth. No doubt, the density of population in the Ottoman countryside was not as high as it was at the economic center of the Mediterranean. Nor was the land under cultivation as expansive, partly due to the absence of wide plains that so eminently marked the northern latitudes of the Alps, from Hungary to the shores of the Atlantic. Nonetheless, the empire not only contained cities big and populous by any standard, but also the percentage of its urban populations was and remained relatively high.[254] As a result, the expansionary thrust of the period encouraged producers to venture into new territory by cultivating lands previously left untilled or deemed unsuitable for permanent tillage, however fractional they were.

Coastal and inland plains that had been devoid of permanent settlements but periodically tilled came to be cultivated regularly. During the sixteenth century, the pace of conversion of temporary settlements into villages accelerated as new and additional temporary settlements were founded. Most of the settlements classified as *mazra'a*s in land surveys conducted in the first half of the century were later registered as villages "proper." One of the indications of temporary fields turning into cultivated fields was that, instead of being farmed out for a lump-sum payment, in later land surveys, they were registered in some detail, including a breakdown of their agricultural output, which was comprised over-whelmingly of grain crops.[255] As nomadic populations settled down or extended their stay in their winter quarters, they shifted their long-range migration patterns to short-range. Newly established villages mushroomed on or in the proximity of routes of transhumance or migration, inescapably interfering with the movement of flocks. The emerging pattern of settlements hence generated strains between settled and nomadic populations, especially in the coastal plains, where the nomads took refuge from the harsh winters of the plateau. The cultivation of the plains in Cilicia and Pamphylia on the Mediterranean, on the banks of the river Meander on the Aegean shores, in Thessaly, Akkar, and Nablus, deprived the nomads of valuable grazing ground.[256]

The plains of Konya and southern Syria best exemplify the vigor and achieve-ments of this wave of colonization.[257] What was impressive about the reclama-tion and recolonization process of the sixteenth century was that both plains were more densely populated at the end of the sixteenth century than at the end of the nineteenth. To put it in perspective, these plains were still not as densely populated even after the institution of a new Land Code in 1858 that facilitated the establishment of quitary claims over vast stretches of land, in particular of vacant land, even after massive waves of migrants were strategically settled in scarcely populated parts of the empire from the mid-nineteenth century; and, equally important, even after agricultural production was fueled by the mid-

Victorian boom. The extent and scale of colonization achieved during the sixteenth century was plainly impressive.

Using an index with a base level of 10 in 1475, the extension of cultivation in around 1575 lingered at 12 whereas population growth reached 17; the ratio illustrates the growing discrepancy between the two.[258] Thus, despite the gigantic task of turning vast tracts of land into arable and the impressive progress registered in the process, the second sixteenth century witnessed a slowdown in the pace of increase in agricultural output. From the lush valleys of the Meander to the calcareous lands of Aleppo, from the plain of Konya to the riverine fields dotting the banks of Yeşilırmak and Kızılırmak, agrarian production during the latter half of the sixteenth century failed to catch up with the impressive rise in population. On a per capita basis, production of bread grains fell, in most places, almost by half, if not more.

In Manisa, for instance, located on the fertile banks of the river Meander, cereal production failed to catch up with the pace of demographic growth: wheat production per capita fell from 270 kilograms in 1531 to 240 in 1575; so did barley, from 225 kilograms to 210.[259] In Antep, production increased, as elsewhere, until the 1560s; thereafter, it trailed behind population growth. The gains registered during the first half of the century were mostly wiped out during the latter half: wheat production per capita in 1543 was over 500 kilograms per person, and barley over 250; yet as of 1574, both had fallen, wheat to 330 kilograms and barley to 175, a fall by one-third.[260] The situation was not any different in other regions of the empire. In the *liva*, province, of Çorumlu in northeast central Anatolia, for example, the rate of population growth in the three districts of the province between 1520 and 1574 was in the neighborhood of 90 percent. However, the region's wheat production went up by 21 percent, and barley by 48.[261] The situation only got worse toward the closing decades of the century: in some villages around Kayseri in the 1580s, grain production per household was 55 percent below its level in 1490, and Aleppo did not fare any better.[262] As agricultural production was sliding, the average peasant holding had fallen from a *çift* to one-third and even one-quarter of a *çift* by the end of the period, not to mention the surge in numbers of landless peasants.[263] The inability of production to catch up with demographic growth was also indicated by the growing differential in administered prices. Whereas the price of wheat was 60 *akçe* per *mud* in 1487 in the province of Hüdavendigâr, it fluctuated mildly in 1521 between 70 and 80 *akçe;* yet fluctuations widened notably in the 1570s, and ranged from 100 to 300 *akçe,* even before the dawn of the price inflation of the turn of the century. The situation in the neighboring provinces was no different: the price range broadened sharply from the 1570s.[264]

What needs to be mentioned here is that grain production per capita in the empire, befitting its role and position in the Mediterranean division of labor, was significantly higher in the eastern half of the Inner Sea. In the 1570s, grain production per capita in Manisa was around 500 kilograms; in Castile, it was less than 250 kilograms.[265] This situation was not specific to the Iberian peninsula, but the norm in the western Mediterranean. Rural producers in Castile and Valladolid drove a larger proportion of their incomes from viticulture and livestock breeding.[266] In fact, gross agricultural product and agricultural surplus were higher in the Balkan provinces, a significant share of which served to provision the city-states, especially when a series of crop failures and threats of famine lingered over the Mediterranean in the mid-sixteenth century.[267]

There were similar crop failures on the eastern shores of the Mediterranean, most notably in Syria in the 1580s, and in parts of Anatolia as well, from 1585 to 1595.[268] Yet, even at the height of the *celâlî* rebellions at the turn of the seventeenth century, the indications are such that, though lessened in intensity and reduced in volume, exports continued well into the 1650s.[269] The edicts actively prohibiting the export of grains, copious before the 1620s, decrease in number from that time on. So did the quantities involved. If Ottoman grain exports continued for a while in the closing decades of the sixteenth century, it was because of the massive devaluation of 1584, which made exports temporarily cheaper. The inflationary spiral the debasement provoked taxed the central bureaucracy and timar-holders, paving the ground for waves of social strife to shake the foundations of the empire until the mid-seventeenth century.

It was mostly bigger estates, held either privately or, more generally, by large *waqf*s, which profited from the boom.[270] They were able to mobilize the rural surplus on a large scale after all. The mortmains in the Balkans and elsewhere in the empire amassed huge fortunes from marketing the grain they collected from their *waqf* villages in the form of tithes and from their sharecroppers in the form of one-half to one-third of the produce. Sultan Bayezit's foundation acquired 337,136 *akçe*s by vending cereals; another, Şehabettin Vakfı, 51,456; and Murad II's foundation, 50,000. These foundations were all located in the Balkans, that is, in close proximity to the western markets and within the easy reach of Venetian merchants. But even in Konya, where conditions in the city's hinterland deteriorated rapidly during the *celâlî* rebellions, the famous Mevlevi Dergahı enjoyed surpluses, ranging 119,000 *akçe*s in 1599–1600 to 34,000 in 1608–9. A deceleration in the price rise after the 1620s lowered the foundations' grain sales, and their surpluses dwindled precipitously. So when prices reversed course around 1650, most of the foundations had already become familiar with chronic deficits. What is more, the increase in the share of urban assets in their holdings, which started timorously in the latter half of the sixteenth century, speeded up from the seventeenth century, exhibiting the ebbing significance of agricultural pursuits within the realm. To wit, the average size of peasant holdings rose, as a result of abandonment, from 30 *dönüm*s in the sixteenth century to 60 or more *dönüm*s in the seventeenth century.[271]

As commercial grain husbandry lost its appeal, specialization in, and monoculture of, lucrative crops spread along the shores and in the islands of the Mediterranean in response to the demand fueled by the arrival of American silver, but once again, for not too long. The seventeenth-century crisis arrested the commercialization drive of the post-1560s period and eroded the basis of both specialization and monoculture. In their stead, interculture, *coltura promiscua*, which combined the renowned triad of crops sui generis to the Mediterranean, returned with force. Tree crops and the growing ownership of small livestock came to complement wheat, whose role was now much diminished. There were two regions, however, where this dynamic did not hold and, surprisingly enough, they both were outside the Mediterranean climatic zone, an issue that we discuss in detail in chapter 5. These two districts, which were previously renowned as the breadbaskets of the Roman empire, Nilotic Egypt and the plains circumscribing the Black Sea coastlands, resurfaced to fill in the gap left by the decline in commercial cereal production. It was these two districts that served as the Sublime Porte's breadbasket, almost exclusively until the latter half of the eighteenth century.[272]

The withdrawal of commercial agriculture had a salient ramification. The fading away of commercial opportunities in the countryside lifted pressures on petty producers in that efforts by the politically powerful to enclose large tracts of the arable lost its initial appeal. The first forceful appearance of large estates in Ottoman history occurred during the sixteenth century, when *çiftlik*s proliferated in number.[273] Even so, these estates proved to be short-lived simply because of the onset of the seventeenth-century crisis and the pursuant revision in the world-economy's division of labor. Of the large estates that came into being during the seventeenth and eighteenth centuries, the *çiftlik*s remained limited in compass and less than modest in size. With the exception of the 'Azm family of Damascus, the notables who reigned in the eighteenth century derived only a tiny fraction of their income from cereal-related activities. After the sixteenth century, the pivotal economic activity in most of the *çiftlik*s was cattle breeding and livestock husbandry.[274] The structural imperatives of this livestock-centered organization to radically transform the lives of millions of petty producers were different from those of their Polish and Bohemian counterparts. The facility with which large estates acquired bountiful grazing fields on deserted, *mawât*, lands left the peasants largely unaffected by their expanding sphere of operation. Even when the Sublime Porte found itself weakened vis-à-vis the provincial notables after the sixteenth century, the absence of large-scale alienation of peasants from their lands disallowed the would-be contenders to fashion and impose an agrarian order to their own liking.[275]

By the time the seventeenth-century crisis dawned on the world-economy sometime between 1620 and 1650, the long sixteenth century had changed the Mediterranean landscape beyond recognition. For one, its division of labor became infinitely more elaborate as the basin recovered from the near mortal blow inflicted on it by the Black Death, and urban growth had accelerated with a corresponding rise in demand for agricultural goods. From the mid-fifteenth century, large stretches of the Mediterranean countryside were opened to cultivation; marshlands, woodlands, and lands taken over by nature were turned into arable. The process of colonization was to a great extent synonymous with the expansion of the area sown to grains. Yet, as the recovery picked up and turned into a full-blown expansion, sowing croplands with bread grains lost its initial fervor, more so in the western Mediterranean than elsewhere. As producers in the Italian city-states and Iberia converted more of their arable into vineyards, olive groves, and orchards, the production of bulky and low-value goods such as cereals and crops demanding intensive labor such as sugar and cotton were migrating to, and their production was expanding on, the outer skirts of the Inner Sea. The areal division of labor that had colored the Mediterranean at its heyday was coming apart after centuries of existence.

The westward and Atlantic-bound spread of the oriental crops then took place in an expansionary economic environment until the start of the seventeenth-century crisis. The world-economic upswing helped accommodate an absolute growth in the volume of production, at least until the 1620s and possibly until the 1650s. Despite a proliferation in the number of production sites, the migration of crops did not necessarily dislodge established centers of production—which were located in the Mediterranean—from their entrenched positions. Since the cultivation and trade of these crops in the Mediterranean basin went uninterrupted, the Levant trade remained largely unaffected by the extension in the world-economy's spatial compass. As we document in the following chapter, the seventeenth-century crisis altered the rural landscape along the shores of the Inner Sea, to the benefit its staples, the horticultural and tree crops specifically, and at the expense of the staples of the rich trades that had colored the Mediterranean for more than four centuries.

CHAPTER THREE

Eclipse of the City-States and the Resurfacing of the Mediterranean

The second sixteenth century commenced in the 1560s. By then, Jewish populations expelled from the kingdom of Castile had taken up residence in distant corners of the Mediterranean, including Salonika and Safed on the eastern shores of the basin, in the Ottoman lands.[1] In the new lands they settled, these migrant populations went on to practice, more often than not, the trades they had excelled at in their natal lands before they were uprooted by the edict of expulsion issued by Ferdinand of Aragón. Among the trades the Jewish populations mastered in the land of the Golden Fleece was, as to be expected, the manufacture of woolen textiles.[2] And not coincidentally, some of the popular points of destination in the easterly migration of the displaced artisan populations were the preeminent woolen manufacturing centers of the Ottoman Mediterranean—Constantinople and Salonika, to name two.[3] In Salonika, for instance, where the city's woolen industry was charged with supplying the large Ottoman infantry with *çuha*, woolen cloth, the newly arrived artisans found themselves, along with the local practitioners of the trade, taking part in an enterprise deemed militarily sensitive by the Sublime Porte.[4] Less than two decades after their expulsion and in a city largely devoid of Jewish inhabitants in the 1480s, artisans from Iberia established themselves in Salonika as one of the principal manufacturers of woolen cloth for the janissary army. The relative ease with which these migrants went on to practice their habitual trades at the opposite end of the Mediterranean, in the *Oltramare*, was owed in large part to the expansionary élan of the 1450–1560 period. That the Inner Sea was laid under siege from without by the Portuguese merchants in the Indian Ocean—a maneuver that temporarily weakened the Serenissima's commercial hold over the spice trade—facilitated the anchoring of Jewish merchants in the Levant. The involvement of immigrant Jewish merchants in wool, cloth, and camelot trades at this critical juncture perforce turned them into indispensable middlemen in the Venetian Levant trade.[5]

Throughout the length and breadth of the Inner Sea, the woolen (and silk) textile industries were buoyed, as elsewhere, by the steep demographic surge and the urbanization wave of the *beau seizième*. By the same token, however, a proliferation in the number of textile manufacturing centers and the ensuing ferocious competition among them rendered these industries increasingly subject to the vagaries of the marketplace starting from the 1560s. The ceaseless upsurge in demand for raw wool over the course of the sixteenth century turned the acquisition of this animal fiber into a vexing enterprise. The woolen industry in Salonika as elsewhere was not exempt from the stiffening competition in the procurement of raw materials, a competition brought about by the growing needs of the thriving industries of the opulent city-states, most notably Venice.[6] The underlying and accumulating strains in the manufacturing sector, which flowed from soaring raw wool prices, surfaced during the visitation of the plague in 1568–69, when Salonika was briefly evacuated. Some of the artisans who had moved to nearby towns during the plague were less than willing to resume their vocation in Salonika when the city returned to full health.[7] Indicative of the exodus of textile craftsmen from Salonika was the steady decline in revenues of the city's stamp tax-farm: first in 1560 and 1566, and later in 1572, 1575, 1581, and 1583. Unarguably, the fall attested to a decline in the volume of production.[8]

The acceleration of industrialization in Venice in the sixteenth century—and in Genoa in the seventeenth—was central to the rising demand for industrial raw materials. The economic slowdown and lull in the Levant trade that followed the diversion of the spice trade away from the Mediterranean since the turn of the sixteenth century triggered immediately a process of readjustment and renovation in the Serene Republic. The city's output of woolen cloth skyrocketed from two thousand cloths a year in 1516 to twenty thousand in 1566.[9] The city of the Doges was not alone in adhering to this strategy, however. In a span of few decades, it was joined by Lyon, Antwerp, and Geneva—the primary beneficiaries of the age of the Fuggers—which all adopted a strategy of promoting manufacturing during the latter part of the sixteenth century.[10] For the centralization of American treasure in the hands of Genoese bankers prompted the urban centers of the bygone Fugger era to diversify their economic activities in conformity with the loss of the key financial role they had played pursuant to the attenuation in silver flows from Germany.

As a result, between 1560 and 1650, soaring demand for industrial raw materials—specifically, wool and silk as highlighted in the previous chapter—intensified competitive pressures on the supplies of animal and vegetal fibers across the Mediterranean. Easily accessible by sea and within the orbit of Venetian and Frankish merchants, raw wool from Salonika started to find its way to more lucrative markets, where it fetched significantly higher prices than in the Ottoman dominions. In conditions of soaring prices, measures implemented by the Sublime Porte to prevent an exodus of industrial fibers from the realm proved futile.[11] To combat the rerouting of raw materials away from the major manufacturing centers of the empire, the central bureaucracy accorded, by fiat, the Jewish artisans privileges that guaranteed that they receive raw wool at beneficial and administered prices. Additionally, they were offered exemptions from certain taxes.[12] Nonetheless, given the pace of industrialization in Venice, these measures fell short of offering lasting relief to those in the textile industry. The debasement of Ottoman currency in 1585–86, by reducing the value of the *akçe* vis-à-vis the ducat, made matters only worse by lowering the prices of raw materials to the benefit of ducat-holders.[13]

Squeezed between the imperial bureaucracy's unwavering determination to ensure that the infantry was provisioned with the necessary garments at administered prices on the one hand, and the unavoidable leakage of raw wool toward more profitable markets where it fetched prices worthy of the wealthy centers of the Italian city-states on the other, the artisans of Salonika found themselves subject to two irreconcilable sets of forces—both of which were hopelessly beyond their control.[14] What followed next was a new round of migration for Salonikan artisans, this time shorter in distance but significant nevertheless. The Jewish artisans started to move away from the empire's primary centers of production to new locations where raw wool was less susceptible to the Mediterranean-wide economic oscillations and price hikes. Some moved to smaller urban centers in the vicinity of Salonika, such as Serez (Serres), Manastır, and Üsküp. Others made their way to the Aegean shores of Anatolia, to the relatively unknown towns that dotted it, precisely because these locations, frequented by nomads, had easy access to wool clip and, more significant, were not under the immediate sway of the Frankish merchants. Initially, the primary destination of the artisans was therefore not İzmir—a city that was in the process of gradually rising above its competitors, which meant that its economic life was becoming subject to the same set of pressures that marked Salonika.[15] More attractive was nearby Manisa, former Magnesia. Not only was Manisa relatively exempt from the pressures that taxed the established urban manufacturing centers of the time, from Segovia to Salonika. Crucially, it was host to a princely court—hence, Şehzade Sancağı—with a vibrant demand for textiles.[16] The presence of a captive market, coupled with the easy availability of wool at affordable prices, must have offered a temporary haven to artisans who had left behind an environment where they were not allowed to pass the rising costs of production on to their regal customers. By the opening decades of the seventeenth century, this migratory movement had caused considerable consternation in Istanbul so as to force the Sublime Porte to issue a series of edicts to ensure that the artisans who had relocated to Manisa and, later, to İzmir and its environs, be dispatched back to Salonika.[17]

Artisans in the woolen industry, trapped at the intersection of administered and market prices, were not the only ones who suffered, however. The new environment proved to be suffocating for those employed in the cloth industry in general, because the availability of Dutch and English "new draperies" at relatively affordable prices prevented most artisans, be they employed in the woolen, silk, or cotton industries, from increasing their prices in accordance with the hike in raw material prices. Overall, then, the itinerary taken by the Jewish artisans—who moved away from the established centers of manufacturing into lesser towns—eloquently captures a movement that left a deep imprint on the Mediterranean between the sixteenth and the nineteenth centuries, not only in the Ottoman dominions but in other quarters of the Inner Sea as well. In Segovia, another major woolen cloth producer, the number of looms decreased from 600 in the 1580s to 159 in 1691, and the town was "ruralized."[18] As the villages surrounding Bursa specialized in silk twisting, the putting-out system became widespread from the seventeenth century.[19] Similarly, in Lyon, the city's main industry—silk—was hollowed out as production spread in the seventeenth century, especially under Louis XIV, to the city's hinterland: St. Étienne, St. Chamon, and Vivarais.[20] Cloth manufacturing, from low to middling quality, was also widely diffused throughout Languedoc.[21] In other words, the geographical mobility of the Jewish artisans, from Salonika to Manisa, was emblematic of a basin-wide relocation of manufacturing activities away from the established, urban, and guild-regulated centers to locations where restrictions on the organization of production could easily be skirted or where labor markets were, to use a present-day euphemism, relatively "flexible."

Despite the fact that the relocation of manufacturing was a Mediterranean-wide phenomenon, the underlying reasons for it were of different provenance, albeit in essence intimately interrelated. When seen from the vantage point of the well-established and distinguished industrial centers located to the south of the Alps, which were all pitched against the Anglo-Flemish producers in the north and the *fustignari* producers in southern Germany, subscribing to a strategy of relocating manufacturing operations into towns and villages, where industrial organization à la city-states was not the norm, was a weapon in their arsenal of competition.[22] Not only the rural hinterlands of Lombardy, Piedmont, and Liguria, but the *contadi* of the city-states too—in the case of Veneto, for instance, Bergamo, Verona, and Padua—registered notable success in developing their woolen (and silk) industries as a result.[23]

If the core, wealthy regions of the Mediterranean embraced, under intensifying competition, the strategy of shifting manufacturing into comparatively cheaper zones of production, the dynamic was different outside the northern latitudes of the Italian peninsula. The continual hemorrhaging of raw materials from the Iberian peninsula, North Africa, and the Levant to the northern Italian city-states arrested the development of urban manufacturing in these very regions. Not surprisingly, the ability of urban industries located in the basin's outlier zones to compete with the wealthy city-states over the procurement of raw materials, which had suddenly turned dear due to the lure of lucrative Mediterranean markets, was not unlimited.[24] The hardening competition compelled the migration of unemployed or displaced textile workers to lesser manufacturing centers where the threat of competition did not loom as large on the horizon as it had in the bigger urban centers.

All the same, in both instances there was a growing dissemination of industries previously located in the basin's urban centers into their hinterland, immediate or distant, or farther into the deep countryside. When the Grand Vizier Damad İbrahim authorized the construction of new woolen mills in the opening decades of the eighteenth century, the heart of manufacturing activity in Thessaly had long since moved away from urban centers like Salonika, to be relocated in the mountainous regions overlooking the plains. Concordantly, cotton textiles had largely supplanted woolen textiles,[25] and cotton cultivation itself had undergone major expansion in the eastern Mediterranean during the course of the eighteenth century.[26] Contrary to the expectations of Charles V, the Spanish woolen industry, led by the ill-fated royal woolen textile factories, failed to take off in Guadalajara in the mid-eighteenth century exactly when the calico—cotton—industry was experiencing a boom in Barcelona, whence it spread into the deep recesses of the province of Catalonia.[27] With the diffusion of manufacturing into the countryside in Languedoc, for most households rural industry was added on to the produce of the soil as the "second providence."[28]

Cottage industries had taken deep roots along the shores of the Inner Sea by the eighteenth century.[29]

Of course, woolen cloth was not the only industry to be adversely impacted by the ongoing concentration of manufacturing activities within the dominions of the city-states. The silk industry, as exemplified in the cases of Murcia and Bursa, had to cope with mounting demand for raw silk in concert with the industrial renaissance to the south of the Alps. Less easily available than wool, silk appealed to a smaller and relatively well-off group of consumers. Hence, however exacting the dwindling of supplies of silk may have been for the producers, the pace of the demise of silk industry proved to be less precipitate than that of the woolen industry, which was subject to cut-throat competition. Even so, this did not mean that the hemorrhaging of local supplies failed to place similar strictures on the ability of silk textile producers to carry on with their operations as before. Ailing silk industries eventually suffered a fate depressingly similar to that of their woolen brethren.[30]

The industrialization drive that altered the economic constitution of the city-states triggered a corollary process of deindustrialization at the opposite ends of the Mediterranean, but the longevity of the process of dissemination of manufacturing into the countryside and secondary cities was not due to it alone. More consequential was the easy availability of wool, the essential raw material for the most popular branch of the textile industry. The animal clip remained within easy reach of all in even the most remote parts of the Inner Sea, thanks to the large number of herds that populated it and that repeatedly and regularly traveled between its lowlands and highlands. With the number of competitors growing without respite and wool obtainable with enviable ease, the competition in woolen textiles lasted for over two centuries, until the latter half of the eighteenth century. Appropriately enough, it was concluded not by the emergence of a few victorious centers that came to dominate the woolen industry, but by the appearance of a vibrant cotton industry on both sides of the English Channel, on the shores of the North Sea.[31]

In sum, the relocation of the textile industry away from the basin's metropolitan centers to its lesser cities and the countryside occurred mainly from the late sixteenth century.[32] The movement was finally reversed from the close of the eighteenth century, with the gradual concentration of manufacturing activities once again in major urban centers.[33] In the interim, however, urban industries of the Mediterranean managed to survive fierce competition and the vagaries of market contractions during the seventeenth-century downturn by perfecting their putting-out systems. Or they tried to render regulations governing guild organizations more flexible than before.[34] With the proliferation in the numbers of producers, the woolen cloth industry became infinitely more varied in terms of quality and price. This stratification created well-delineated niches for a wide range of textiles, leaving the lucrative high-quality cloth industry as the most competitive. Perhaps not ironically, then, places where inferior-quality wool was used to produce lower-quality cloths like the northern regions of Iberia were not as hard hit as the high-quality cloth–producing centers, located mostly to the south of the peninsula, like Toledo, Murcia, Córdoba, Seville, and Cuenca. Similarly, the Plovdiv (Filibe) region of the Ottoman empire, manufacturer of *aba*, rough woolen cloth, flourished in the eighteenth century when its textiles reached far-flung corners of Anatolia.[35] Nonetheless, even in Cuenca, one of the outstanding textile centers of the sixteenth century, which was hit harder than most during the seventeenth-century crisis, one-third of the shrunken labor force was still employed in the woolen industry in the eighteenth, thanks in part to labor mobility that made it possible for most households to diversify their income outside their place of residence.[36]

This resilience was not confined to the woolen industry alone. Silk industry in Bursa, by successfully redistributing its operations in the environs of the city and with growing recourse to domestic supplies of raw silk, went on to employ large numbers of people in the seventeenth and eighteenth centuries; the camelot industry in Ankara thrived thanks to its successful integration of the surrounding villages; and in Valencia, the spinning and weaving of silk textiles was removed to the countryside, with weavers setting up their looms in the eighteenth century in the cities, towns, and villages of the kingdom.[37] What is certain is that the pressures that flowed from the intense competition between 1560 and 1650 subsided eventually. Still, given the level of contraction in economic activity during the seventeenth-century crisis, the survival of manufacturing depended on the continuation of its ruralized forms, as scores of producers went on to serve the highly specialized, stratified, and parcellized markets of the era. Along with the reshuffling of economic activity away from the established textile centers to the countryside was the decline in the number of smaller cities in the Mediterranean during the seventeenth and eighteenth centuries.[38] Notwithstanding the onslaught of the second sixteenth century, the basin recovered some of its manufacturing activities and developed others. By the eighteenth century, Barcelona, Languedoc, Thessaly, Tokat, Aleppo, and Damascus had once again established themselves as textile centers.[39]

The process that started with the age of the Genoese in the 1560s did not, then, come to a close with its end in the 1650s. Instead, the assignment of manufacturing and related activities to cottage industries continued unimpeded.

Given the widespread nature and intensity of competition in the textile industry in the seventeenth and eighteenth centuries, the trend toward ruralization continued unabated until the turn of the nineteenth century. However detrimental the deindustrialization of the established urban centers was to the economic health of the Inner Sea, its percolation into the countryside provided precious *lebensraum,* "providential refuge from poverty," for the denizens of the Mediterranean countryside.[40] In an age when economic activities that previously symbolized the heyday of the Mediterranean were vacating the basin, the widespread dispersal of manufacturing and the resulting demand for industrial crops were more than a welcome addition to the repertoire not only of rural households but also of those who presided over the expansive putting-out networks. Both these factors helped rural households to diversify their crop-mix and weather the trying times during the seventeenth-century crisis and beyond.[41] So, the succor provided by cottage industries to the region's rural households was not of a passing nature.[42] The stimulation of rural manufactures and the demand this generated for animal and cultivated fibers were not the only trends that helped revamp the economic makeup of the Mediterranean. The revision in the spatial redistribution of textile manufacturing itself was complemented by two sets of transformations that offered additional avenues of economic activity in the fields of agriculture and commerce. They too had their origins in the age of the Genoese. And by generating additional sources of revenue to the denizens of the basin, they too helped facilitate the smooth transition from the heyday of the city-states to the waning of the Mediterranean.

Accompanying the revisions in the spatial distribution of the textile industry was the revival in the cultivation of the Inner Sea's tree crops. The westerly and northerly migration of the staples of the Mediterranean trade, as discussed in the previous chapter, left the region bereft of economic activities associated with the heyday of the city-states. The vacuum created by this vegetal egress was precipitately filled in by the advance of vineyards, groves, and gardens.[43] This was a remarkable shift from the days of the city-states, when the basin's center of gravity was located in its coastal plains and valleys. What the march of the tree crops succeeded in doing was to help diversify the crop-mix along the banks of the Inner Sea. The vineyards and orchards that covered it, largely confined to the outskirts of its big cities during the sixteenth century, and the gardens that became part of the landscape at large, not to mention the entrance into the Mediterranean of new crops (e.g., maize, beans), all served to counterbalance harvest deficits deriving from crop shortages.[44] Even the basin's livestock industry, with the share of sedentary sheep on the rise—albeit at a slow pace—or reconstituted on the basis of stall feeding, gave peasant households an additional source of revenue as well as protein.[45] The region's mountains partially re-covered their sylvan resources, given the slowdown in the pace of urbanization and replenishment of the merchant fleets. The smallholding peasantry and sharecroppers found themselves in an environment friendly to their survival on account of the ebbing of commercial pressures during the seventeenth-century crisis. As the reign of cottage industries propped up demand for industrial crops, the return and consolidation of the region's tree crops and the introduction of new crops widened further the opportunities available to petty producers. For one, the channeling of mercantile capital into textile production or land and the subsequent specialization, say, in viticulture or sericulture, raised incomes in certain regions of the Inner Sea.[46] The inclusion of industrial fibers and tree crops in the panoply of economic activities of the region contributed to the orderly transition from the golden days of the age of the city-states to the diminished economic activity of the autumn of the Mediterranean.

Lastly, the aforementioned alterations in the manufacturing and agrarian structures of the Inner Sea entailed parallel adjustments in its commercial edifice. To wit, the ever-widening geographical reach of the rural industries so as to encompass not only Bohemia and Styria, but also Cuenca (in wool) and Ankara (in camelot), called in turn for, and was rendered possible by, the construction of a multilayered and tightly interlocking commodity, credit, and labor flows.[47] Raw materials had to be delivered into the hands of producers who were located at times in faraway corners of the rural world and finished prod-ucts had to be relayed to merchants who regularly frequented towns where fairs of all sizes were held. This development, therefore, involved a proliferation in the number of fairs, for instance, in the region between the Adriatic and the Baltic, where cottage industries were prominent, or in the Ottoman Balkans, whence western-bound raw material was conveyed.[48] The dispersal of manu-facturing called for the services of an extensive web of merchants—local and regional—and *négociants*, canvassing an operational domain that extended from the major port-cities and periodic fairs to producers located in the rural recesses of the basin at large.[49] As commercial webs interlinking producer and consumer markets expanded, the vast commercial space built by these trans-actions fostered a score of merchants (including those of the twin city-states) who carried on with their operations throughout the expanse of the Inner Sea as well as to the north and south of the Brenner Pass.[50] Therefore, in the seven-teenth and eighteenth centuries, the spread of manufacturing and the widening ambit of commercial networks connecting new industrial centers to suppliers of raw materials on the one hand and urban centers to their rural hinterlands on the other radically transformed the commercial architecture of the basin's landscape.[51]

As the region's historically established crop trilogy emerged from oblivion, as manufacturing, once again, became part and parcel of the basin's rural landscapes, and as a wide array of merchants, mostly local, established their presence high and low, the Mediterranean recovered some of the functions that had previously been concentrated in the hands of the traders of the cities of saints. For it was the decline in the city-states' economic might that allowed "lesser" players (i.e., local merchants), "lesser" crops (i.e., tree crops and lesser cereals that were not as remunerative as the exotic crops of the Levant trade), and "lesser" (coarse) cloth to surface, almost by default, so as to undertake an organizing role in the life of the Inner Sea. Despite the long-term and relative economic contraction that became the hallmark of the basin in the seventeenth and eighteenth centuries, there were therefore a number of factors that helped the region to cope with the relative ebb of commercial activity. Nonetheless, the fortunes of the city-states and the Mediterranean were not always diametrically opposed to each other. The region, after all, had recourse to the commercial and financial resources mobilized by its reigning city-states. The former hegemons, with their impressive commercial networks and capital resources, were ready and willing to further these new developments in order to profit from them by changing their theaters and modes of operation.[52]

Since the previous chapter summarized the historical developments in the field of the staples of the Mediterranean trade, this chapter investigates the triple processes that underwrote the restoration of the Mediterranean in the wake of the age of the city-states. The first section sketches the mutation of urban industries from the long sixteenth century: here, the changing spatial redistribution of manufacturing activities between the urban and rural locations across the basin and the significance of cottage industries in contributing to the livelihood of rural households are underscored. The second section examines the ramifications of the advance of tree crops and lesser cereals to fill the void left behind by the migration of sugar to the Atlantic and commercial cereal cultivation to the east of Elbe. The geographical redistribution of economic activities in agriculture as well as in manufacturing across the basin and beyond were not without their consequences. The new commercial networks that came into being from the 1560s gave rise to a new corps of merchants and intermediaries who operated among these widely dispersed locations, which is the subject matter of the third section. This chapter, in short, demonstrates how the basin coped with the end of the era of city-states and the rich trades. The assumption by the region's indigenous crops of a larger economic role, the appearance of new sources of livelihood such as rural manufacturing, and the assumption by the region's lesser merchants of a growing share of its commerce countered the forces of devolution at work. Because of the lack of maritime powers with the kind of commercial and financial might that the city-states possessed and the tenuous hold the new powers held over the Inner Sea, the merchant republics enjoyed greater latitude in coping with, and making most of, the new economic landscape precisely when the transformations in the vegetal and agrarian constitution of the Mediterranean, such as the increase in the share of tree and industrial crops and lesser cereals, provided greater latitude to the denizens of the basin in multiplying their sources of livelihood.

Dispersal of Manufacturing

The Venetian empire was an empire of the waves. At its height, the strength of this empire resided in its mastery over, and orchestration of, flows of merchandise and precious metals. In the fifteenth century, for instance, when the Levant spice trade was thriving under the auspices of Venetian merchants, the annual output of the city's textile industry was a paltry three thousand cloths. Yet the Venetian merchants exported forty-eight thousand cloths a year to the Levant.[53] The re-export of merchandise to the Mamluk and, later, Ottoman lands—the gateways to the spice paradise—was germane to the perpetuation of the rich trades, since bullion shipments covered less than half the value of merchandise dispatched east on board the fleets of the Serenissima. Reportedly, the city of the Doges shipped annually "some 300,000 ducats in cash, as well as 4,000 tons of olive oil, 3,000–4,000 tons of copper, and 300,000 ducats worth of general merchandise" to Alexandria at the end of the fifteenth century. The value of goods dispatched by the merchants of the cities of St. Mark and St. George to the Levant was almost always exceeded by that hauled therefrom by a coefficient of 1:2 to 1:2.5.[54] The merchandise trade, despite the phenomenal expansion it underwent under the stewardship of the Italian city-states, was dwarfed, more so in the fifteenth century, by the mounting volume of spice imports. The imbalance strengthened the enduring fellowship between the bullion and spice trades. After all, "spices went where silver was" on account of the soaring need for silver in the Indian Ocean and Ming China.[55]

Given the occasional decline in the flow of precious metals destined for the twin city-states and the perverse impact this had on their precarious multilateral merchandise balance, it was imperative for these mercantile republics—but more so for Venice after it monopolized the spice trade—to implement a strategy designed to reduce their imports from the south of the Adriatic as well as the north of the Alps by adhering to a strategy of import-substitution.[56] This entailed, first, the nurturing of woolen cloth industry with the express purpose of decreasing reliance on the Flemish (and later Florentine) textiles.[57] Second, it involved the undertaking, or encouragement, of the manufacture of goods imported from the Levant at considerable cost, such as paper, glass, and mirror.[58] It was this two-pronged strategy that prompted the growing role of manufactures in the economic lives of the twin city-states. In both instances, the process that appended manufacturing to the economic repertoire of the mercantile cities unfolded effortlessly owing to the control their merchants jointly exercised over the commerce of raw materials in the Mediterranean. Previously, Venice, even after playing host to woolen and cotton textiles from the twelfth century, had failed to acquire a reputation in the Levant for its cloths. For long, revenues accruing to the Signoria from the city's cloth exports remained minuscule. Raw cotton, on the other hand, accounted for 10 percent of the total value of the city's annual exports to Lombardy.[59] Put simply, manufacturing remained as an economic activity of little import until the hiatus in the spice trade in the early part of the sixteenth century. When the Venetian doges punctiliously reported on the riches of the Republic, they mentioned the stellar returns on the city's maritime trade in glowing terms, yet neither Pruili nor Mocenigo included the city's industrial riches in his list of the Serene Republic's great achievements. For both cities, re-exporting Flemish and Florentine woolen textiles to the Levant markets remained pivotal to their overall commercial enterprise.[60] The situation was reversed only in the latter half of the sixteenth century, when Venetian cloth managed to garner for itself a fame that was only matched by the city's highly esteemed ducat. Thus, a diversification in the array of economic activities in which the city was engaged reached new heights exactly when the Mediterranean world was in the process of losing its glitter as the spice route was diverted away from the Inner Sea from the turn of the sixteenth century to the 1540s or the 1550s.

Indeed, the industrialization wave of the sixteenth century was the Serene Republic's last gasp breathed upon the Mediterranean.[61] Between 1560 and 1650, when the Venetian woolen industry came under challenge by the invasion of lighter woolen cloth from the north, one of the strategies followed by merchants and businessmen to revitalize the city's threatened industries was the relocation of some of their manufacturing operations outside the borders of the Dominant City. The resultant dispersal of manufacturing into the Terraferma and the bypassing of urban guilds remained as one of the defining features of the textile industry not only in Venice—or the quadrilateral formed by Venice,

Genoa, Florence, and Milan—but across the Mediterranean and beyond.[62] What is more, this was not a passing solution to the problems associated with the late sixteenth-century conjuncture but an enduring phenomenon that lasted well into the latter half of the eighteenth century.[63] The intense competition in woolen textiles that came to shape the basin's urban and rural landscapes in the seventeenth and eighteenth centuries reached a conclusion with the reign of cotton textiles on both shores of the English Channel and not as a result of a drastic weeding out of the number of competitors in the woolen industry.

This was not the first time that the areal relocation of textile production either in the *contadi* of the city-states or in the ambient countryside, be it across the Alps, was taking place in accordance with the needs and dictates of the city-states. As touched upon briefly in the preceding chapter, a similar development had already taken place in the thirteenth century in the *barchent*—fustian, a mixture of linen and cotton—industry, relocated in Swabia. It was then that demand for cotton and silk in the Inner Sea shot up owing to the ability of the Italian city-states to choreograph the movement of these raw materials across the basin and the relatively prosperous times of the thirteenth century that were conducive to the introduction of, and experimentation with, expensive garments and clothing.[64] Fortuitously for the nascent cotton and silk industries of the western Mediterranean—all nestled along the shores of the basin, from Montpellier to Genoa, from North Africa to Barcelona, all within easy reach by sea for provisioning in raw materials—the undisputed reign of woolen textiles from the fourteenth to the eighteenth centuries spared them from the fierce competitive pressures of the period.[65] Partly because within the continually widening panoply of woolen cloths, silk and cotton were increasingly employed as supplementary ingredients for manufacturing mixed varieties of textiles rather than as raw materials in their own right.

Strangely enough, then, it was largely the industries of the first wave, silk and cotton, which fared relatively better in the sixteenth century simply because competition in these industries was not as ferocious as in the woolen industry. Competitive pressures building on Levantine cotton and silk textiles did not emanate from similar fabrics manufactured by the Italian city-states but from a different fabric, the light and cheap woolen fabrics from Leiden and London. Demand for Levantine cotton and Caspian silk went on to shape the basin's commercial destiny in the seventeenth and eighteenth centuries. If Dutch and, later, British merchants started to tap into the raw material resources of the Indian Ocean and the Atlantic in the eighteenth century, lesser powers, such as France, came to rely on Levant cotton and silk, providing a steady outlet for these raw materials.[66] In Ottoman lands, the steadiness of demand even during the seventeenth-century downswing encouraged an expansion in the local cultivation of vegetal fibers for export as well as domestic use. The easy availability of industrial crops helped rural industries survive as well as the later expansion of the silk industries of Damascus and Bursa, and the cotton industries of the Balkans, Anatolia, and the Aleppo region in the seventeenth and eighteenth centuries. Equally significant, the process of deindustrialization, which visited the Iberian and Levantine Mediterranean in the 1560–1650 period due to the exportation of industrial crops, lost its potency as soon as the seventeenth-century crisis settled in, and the dispersal of woolen textile manufacturing in the form of cottage industry accelerated across and beyond the Mediterranean.[67]

Though separated by three centuries, these two episodes of devolution of manufacturing from the basin's metropolitan centers to its smaller cities or rural sites were thus intimately interlinked.[68] In the first instance it was the fustian/cotton and silk industries that were decentralized, spread, and, at times, ruralized, whereas in the latter instance, the industry that underwent a similar metamorphosis was predominately the woolen industry. That is to say, whereas the first wave relied on the continual provisioning of fibers imported from, or via, the Levant by the city-states to a wide range of producers, urban as well as rural, the following wave gained strength from the fact that the basic ingredients of the cottage industries—wool, and, in places, linen and hemp—were easily available to all and not subject to frequent supply bottlenecks and price gyrations as was the case with imported fibers.[69] This is why the second wave lasted indeed much longer than the first. Since the timing and tempo of the spread of manufacturing in the Mediterranean was determined by the systemic needs of the leading city-states, the tempo picked up, first, after the 1350s as a result of acute labor shortage in the post-plague period and the profound crisis in the manufacturing sector—especially in Florence after the collapse of the city's mighty companies engaged in the wheat and wool trades. The second wave, on the other hand, started when the spice trade momentarily dried up during the first half of the sixteenth century and accelerated as the Anglo-Flemish textile industry started to dispatch its products to the Mediterranean. So, in both instances, the general contours of change derived from the dictates of the mode of operation of the city-states, Venice in particular. The shift in the main stage of the rich trades from the Mediterranean to the Pontic Sea to the benefit of Genoese traders from the 1250s, and from the Mediterranean to the Indian Ocean, to the benefit of the Portuguese merchants in the first half of the sixteenth century, both encouraged a diversification in the range of economic activities in which Venetian businessmen were engaged, including manufacturing. It is in this context that we examine these two waves as two episodes of the very same movement.

In the twelfth century there were scores of fledgling urban centers engaged in the manufacture of textiles along the shores of the Inner Sea, from Catalonia and Provence (woolens) to the Aegean and Syrian littoral (cotton textiles). Yet from the mid-fourteenth century, concurrent with the reign of transhumance in Iberia, the Balkans, and Anatolia, the conversion of large swathes of land into pasture and the resultant jump in wool trade, especially in the western half of the basin, aggrandized the dominion of woolen cloths significantly.[70] The merchants of the city-states marketed them in the Levant, the land of flax and cotton, where light woolen cloth was in high demand.[71] Yet, however unimpressive the sway of the cotton and silk industries may have been in the western half of the basin, trade in cotton, silk, and English wool was, and continued to be, the essential ingredient of the commercial operations of the city-states. The merchants of the Serenissima and *La Superba* were both engaged, along with the Catalan and Provençal merchants early on, in locating, purchasing, and re-exporting textile fibers, animal and vegetal, across the Mediterranean.

Cotton and silk made their appearance early in the western shores of the Inner Sea, in Sicily, Andalusia, and North Africa. However, not only the quantities produced were modest, but also the quality remained poor when compared with that of Levant cotton and Caspian silk—the latter was transmitted via the Black Sea (and, later, via Aleppo and İzmir). Not surprisingly, wool was more easily available throughout the expanse of the Inner Sea, in the Balkans, Apulia, Sicily, North Africa, Cyprus, and Syria, given the rising star of the golden fleece and the expanding dominion of pastures and grazing lands from the fourteenth century. Witness the shift to sheep husbandry in Majorca, especially after the advent of the Genoese merchants in the western Mediterranean.[72] Beginning in the late thirteenth century, with the proliferation in the number of woolen textile–producing centers such as London, Flanders, Languedoc, and Catalonia, the market for medium and cheap woolens became saturated and the industrial centers in the northern Italian peninsula, in their shift to costlier fabrics, resorted to using English wool.[73] The acquisition of English wool by Florentine companies and, later, by Genoese and Venetian merchants provided the basin's manufacturing centers with their requisite ingredient. When the houses of the Bardi and Peruzzi collapsed and subsequently English wool became scarce, the volume of woolen textile production in Lombardy and Tuscany tumbled and occasioned a disruption in the operations of the Venetian merchants, forcing them to assume a larger role in overseeing manufacturing activities in the peninsula.[74] It was then that the demand for cotton and silk took off. Florence, for instance, "made up for the decline of her woolen industry by manufacturing silk fabrics," exactly when the linen industry in the north was experiencing a similar upturn in its fortunes, as was the *barchent* industry in southern Germany.[75]

Yet, unlike the second wave of industrialization, the reason underlying the northerly spread of the fustian industry from Lombardy did not necessarily spring from the Venetian merchants' need to supply the Levant with textiles in order to trim down the volume of precious metals shipped east. For in the thirteenth century, the spice trade, as depicted in chapter 1, was conducted mostly via the Pontic Sea, not the Levant, and by the Genoese merchants, not the Venetian. The relocation of the rich trades in the Black Sea hastened the Venetian traders' search for new sources of cotton (and sugar), encouraged the transfer of the cultivation of these crops to lands under their control, and reinforced their grip over the cotton trade. Complementing the diversification in the operations of the Venetian merchants was an expansion in their dominion over manufacturing activities at home and in the cities dotting the Po plain. Unlike the Genoese merchants, the liquid funds that the Venetian merchants controlled were not tied up in the spice trade, as was the case before the 1250s. The availability of significant sums of "idle" money in the hands of merchants and the control they established over the flow of raw cotton from the Levant compelled them, first, to reshape the fustian industry at home and, later after the arrival of the Black Death, to market their cotton north of the Brenner Pass.[76] The fact that the flow of raw cotton initially did not generate counterflows of fustian facilitated the northerly cotton trade without causing strife "at home." Moreover, the dissemination of cotton as an industrial crop and its usage in combination with linen and wool served as a boon to the existing industries in the Italian peninsula and elsewhere. The Fugger era reinforced the existing Venetian nexus in that counter-flows of silver from Tyrol and cotton from the Levant via Venice helped propagate the rural, fustian industries in southern Germany, the manufactures of which were eventually shipped to Portugal, Spain, and England.[77] Until the late sixteenth century, the Venetian and Swabian preponderance over the export of fustians, the latter in the hands of the trading companies of the Fuggers of Augsburg or the Kresses of Nuremberg, discouraged the growth of rival industries elsewhere.[78]

The extension, via ruralization, in the reach of the textile industry beyond the Alps was replicated, roughly three centuries later, when the influx of kerseys from the north lacerated urban industries in the region, facilitating the transfer of parts of their textile operations outside the established manufacturing centers of the time. What is more, the impact of this second wave was felt more deeply and widely. Taking into account that woolen cloths were the most widely produced textiles on both sides of the Alps, in Florence as well as in Flanders and Brabant, the impact of the second wave on the economic landscape of the region was more deeply seated and longer lasting than the first—which revolved around cotton and, to a lesser extent, silk. There is no doubt that the cotton industries of the eastern Mediterranean were adversely impacted by the first wave of industrialization along the northern shores of the Inner Sea. Yet it was not the threat of invasion by the fustians that inflicted harm on manufacturing industries in the region. Rather, it was the extrication of cotton from the eastern Mediterranean and the growing usage of cotton grown in the Venetian and Genoese imperial holdings by industries located in the western half of the basin that ended up taking its toll on the textiles of the Levant. The change in the distribution of textile industries across the region was in tune with the proliferation in the number of sites of cotton cultivation, to the detriment of the Levant.[79] Even then, however, the cotton textiles of the Levant were impacted adversely and lightly, because cotton grown in the Atlantic made its way across the ocean in considerable quantities only in the late eighteenth century.[80] Meanwhile, the number of cotton plantations in the eastern Mediterranean grew in the fifteenth century; and later, cotton escaped the plantation template and turned into a cash crop in Anatolia, Syria, and the southeastern Balkans.[81]

Cotton was thus transformed into one of the key commodities in the Levant trade for both Venice and Genoa from early on, and the twin cities remained as its main ports of distribution in the occidental Mediterranean. Concurrent with the expansion of cotton trade across the basin was the development of cotton cloth industries along the western shores of the Inner Sea from the twelfth century as the cultivation of the crop extended in the Aegean region, especially after the 1350s.[82] The region extending from Champagne, southern France, and Catalonia to Pisa, Florence, and the cities that dotted the Po plain came to be peppered with cotton textile centers—of course, alongside those in woolen. The industry experienced a period of phenomenal growth in the thirteenth and fourteenth centuries in particular.[83] Nonetheless, industries in the Levant were not immediately or substantially affected by the growing volume of the cotton textile industry in the Frankish realm. It should be underlined that woolen textiles commanded an insurmountable lead over cotton textiles as far as the volume of goods dispatched to the Levant was concerned.[84] Even then, only certain parts of the eastern Mediterranean, primarily the Mamluk lands, were impacted by it. When Dante, in his *La Divina commedia*, was praising the successes of the Florentine cloth industry, one of the most successful centers of manufacturing that catered to the Levant, his points of reference as fabrics of excellence were still *drappi tartari* and *turchi*—not to mention the carpets and rugs of sheep wool.[85]

In accordance with the new geography of cotton manufacturing, most of the cotton handled by the Venetian merchants was distributed in the textile centers in the Po plain. Some of it was put to use in Venice by its own flourishing cotton industry.[86] That the Venetian cotton industry performed much better than that of Genoa is not all that surprising. Venice had an infinitely better catchment area, from Lajazzo to Acre. What is more, its merchants had a higher stake in the cotton trade, because the momentary transfer of the spice trade from the Levant to the Black Sea compelled these traders to invest more energy and capital in transporting and relocating the cultivation of cotton (and sugar) rather than simply reap the benefits of the spice trade. Witness the fast expansion of cotton cultivation on the shores of the Aegean, where the Venetian merchants threaded and the Venetian colonial enterprise was centered.[87] If the phenomenal expansion in cotton textiles did not generate the goods destined for the eastern Mediterranean markets, this was because the growing cotton exports from the northerly latitudes of the Italian peninsula to Bohemia and beyond permitted merchants engaged in this trade to partake in the draining of Bohemia's silver stocks.

To the south of the Alps, the industrialization wave of the thirteenth century was not simply confined to the field of cotton textiles. The Serenissima, deprived of the revenues that accrued from the spice trade, embarked on an ambitious strategy of import-substitution. As the Venetian merchants' engagement in the Levant deepened in tandem with their dependence on oriental crops, the import-substitution strategy was also carried out in the fields of glass-, paper-, and mirror-making.[88] What disturbed this seemingly orderly transfer of industries to the western Mediterranean and turned the cotton industry into a lightning rod was the velocity of the transfer of the fustian industry to the north of the Brenner Pass as the city-states were hard hit by the tightness in the labor markets on account of the toll taken by the Black Death. The Venetian merchants, emboldened after the 1350s by the return of the rich trades to the Levant, found themselves financially much better off than before. The revival of the spice trade hastened the pace of the manufacturing process in southern Germany: the high profits associated with the pepper and spice trades enabled the Venetian merchants to store growing volumes of cotton at home to be transported north. They remained in charge of distributing Levantine cotton, which was spun and woven across the Alpine passes in southern Germany.[89] In the process, Swabia in general—and Augsburg and Ulm in particular—became primary producers of cotton goods from the fifteenth century, the Fuggers of

Augsburg sending Tyrolean copper and silver to Venice in return for cotton.[90] Some of these textiles found their way into the city-states and their *contadi*, oftentimes causing alarm at home.[91]

As noted, Venice was more deeply ingrained in the cotton trade than its competitors. In addition to cotton that originated in the ports of Beirut and Alexandria, the Venetian merchants were also tapping into cotton supplies from their plantations in Cyprus, Crete, and the Aegean. When, in 1412, Sigismund of Luxemburg, king of Hungary and the newly elected Holy Roman emperor, attempted an economic blockade of Venice to punish the city for taking over Friuli and parts of Dalmatia, he predictably decided to rely on its arch rivals, the Genoese, to keep raw cotton flowing steadily into the hands of those intermediaries in Ulm or Zurich, who would in turn distribute it to producers located in the distant corners of southern Germany. After all, the Genoese, who had rebuilt their emporium in the occidental Mediterranean after the 1350s, had kept their colonies in the Aegean—Chios, first and foremost among them— where cotton was also grown. Yet, the Genoese merchants' feeble presence in the port-cities of the Mamluk realm, coupled with the low quality of cotton grown in Chios (as opposed to that in Crete and Cyprus), rendered it almost impossible for them to supplant the merchants of St. Mark. They may have been well equipped to provide lower grades of cotton to nearby cities, such as Alessandria and Savona, but they were unable to "get cotton to Ulm or Zurich in either adequate quantities or qualities for the fustian-makers in south Germany or Switzerland."[92] The first wave was hence largely determined by the dispersal of cotton and, to a lesser extent, silk manufactures.

That this first wave of industrialization did not have a profound impact on the rest of the Mediterranean was due to the presence and abundance of wool in the Iberian, Italian, and Anatolian peninsulas as well as in North Africa. That the expansion of the textile industry in Catalonia in the fourteenth century was led by woolen cloth despite the long-standing presence of cotton in the region and despite the fact that Marseille was a reliable entrepôt that routed its cotton imports from the Levant to Catalonia demonstrates how limited the extent of cotton manufacturing was.[93] In a similar fashion, despite the progress of the cotton industry in the vicinity of Genoa, it was primarily wool that was carried by its galleys: some of it went to the city's own looms and most of it was mainly transmitted to the Low Countries and Tuscany.[94] Genoa's overwhelming presence in the western Mediterranean after the 1380s and its pivotal role in the distribution of, first, lower-quality Spanish wool and later, high-quality merino wool, gave it an inherent advantage in the procurement of raw wool for its industries, if need be, as did cotton to the Venetian traders. Cotton was not put to use only for textiles, though. Equally, if not more, important for the merchant republics as seagoing powers was the employment of cotton as a strategic asset, in the manufacturing of sails, as had been the case in the Indian Ocean from very early on. In this branch of industry, cotton eventually replaced linen in the Mediterranean as well.[95] Yet, the advance registered by cotton manufactures and the spread of the fustian industry during the thirteenth century were later adversely affected by the easy availability of wool and the popularity of woolen cloths.

The woolen industry, a negligible sector of the Venetian economy since the thirteenth century, experienced an unprecedented vigor during the hiatus in the spice trade in the first half of the sixteenth century. As mentioned above, when the Venetian merchants were re-exporting 48,000 cloths a year to the Levant in the fifteenth century, the city's annual output never exceeded 3,000 cloths. In 1516, the annual output of woolen cloth was around 2,000 pieces; by the 1560s, it had skyrocketed to more than 20,000 cloths a year, cresting at 28,729 pieces in 1602.[96] Venetian textiles established a sterling reputation, particularly in Levantine markets. The onset of the second sixteenth century, then, signaled a long-term change in the structure and spatial distribution of manufacturing across the Mediterranean. From about 1450, the accelerating pace of urbanization in the basin was accompanied not only by a corresponding growth in manufacturing activities but also by the agglomeration and concentration of these activities in sizeable urban centers. Cities like Segovia, Córdoba, Toledo, Salonika, and Bursa experienced rapid growth in manufacturing as the construction of vast, unified imperial economic spaces by the Habsburg and Ottoman dynasties provided new avenues for economic specialization.[97]

With the number of urban centers on the rise and urban demand soaring incessantly, the weight of manufacturing in the economic life of cities increased without a major reversal until the mid-sixteenth century. Thenceforth the industrialization drive hastened because the closing of the Fugger era forced a string of cities—Geneva, Lyon, and Antwerp—to undertake and encourage manufacturing to compensate for their losses in financial intermediation. After the 1550s, Venice was therefore joined by no less than the economic giants of the era past. The second sixteenth century as a result witnessed growing competition between the Italian city-states and the manufacturing centers on both sides of the English Channel—and other industrial centers—over raw materials and markets. To be sure, escalating demands of the manufacturing centers in the north Italian peninsula were felt strongly in the Mediterranean, for rival cities located on the shores of the Atlantic relied for their raw materials on ports of debouchment in northern Iberia.[98]

The manufacturing world of the 1560–1650 period came to be distinguished by the presence of two diverging yet closely interwoven trajectories. In the wealthy quarters of the basin, the sharpening of competition among the major producers located at both ends of the Italian peninsula–Low Countries axis led to the spatial dispersion of manufacturing operations, as a competitive strategy, into the countryside, the *contadi*, and the Terraferma. In the outlier zones, the process of ruralization of manufacturing started later, as in Segovia and Salonika, due to the increasing reliance of the "Italian" and northern merchants on wool imports from the Balkans and Iberia: the inability of artisans to easily obtain raw materials turned ruralization of manufacturing into a survival strategy.[99] What was worse for those who were engaged in the manufacture of textiles in the Mediterranean was that northern merchants marketed light worsted cloth as well as cheap imitations of Venetian and Florentine cloth. Venetian woolen cloth may have set the trend at least until the 1620s, but the trend gained strength with the entrance of the British, Dutch, and, later, French merchants into the Ottoman markets with lighter woolen cloth, the "new draperies."[100]

Thus, on the one hand, industries located in the Italian peninsula and the Flemish lands, despite growing competition, continued to procure the necessary raw materials even from producers in distant lands owing to the economic might and the sweeping compass of their mercantile establishments. On the other hand, the Ottoman and Spanish industrial centers were trying to cope with the inflationary pressures that the industrialization of the city-states was placing on raw material prices. During the second sixteenth century, the woolen and, by association, silk industries of both empires suffered with the diminution in their ability to procure industrial crops as easily and cheaply as before when the industrial fibers found their way to Venice—and France—at an increasing rate.[101] Most urban industries to the east and west of Venice experienced a steady drop in their output. Cloth manufacture in Segovia fell from about 13,000 pieces in 1570–90 to roughly 3,000 in 1700. Silk cloth production in Bursa, after cresting in the 1570s, went on to shrink until the turn of the eighteenth century.[102] Increasing raw material prices set against dropping textile prices arising out of intensifying competition made it increasingly difficult for these producers to survive.[103]

In tune with the increasing pace of expansion in manufacturing across the basin in the second sixteenth century, the prices of raw materials increased swiftly. In the 1550–70 to the 1620–40 period, raw silk prices in Bursa climbed from an average of 73.8 *akçe*s per *lodra* to an average of 290.4 *akçe*s, a nominal increase of 293 percent. Similarly, wool prices followed a steep upward movement: in Segovia they rose from 22.23 *reales* per *arroba* in 1550 to 71 *reales* in 1650, an increase of 219 percent. In the Balkans, wool prices rose 244 percent between 1550 and 1600, whereas English wool prices rose a little over 80 percent, about one-third of the former. The steep rise in prices in the Balkans was due to the fact that the region remained within the immediate operational orbit of Venetian merchants, and wool was not as abundant as in Iberia.[104] Therefore, the rise in the price of wool in the Balkans was more precipitous than that in Segovia: in the former, prices jumped from an index of 100 in 1530 to over 1,000 in 1650, whereas in Segovia—where the market in raw wool was much better organized—they increased from an index of 100 to 615 in the same time period.[105]

The deindustrialization experienced by most urban centers from the mid-sixteenth to the mid-seventeenth centuries set the shape and direction of industrial change in the following centuries. This was because even the demise of industrialization in the Italian peninsula in about the 1650s failed to restore the region's industrial landscape to its former shape and health. By the close of the seventeenth century, Venetian cloth output was down to where it was at the turn of the sixteenth century: a little more than 2,000 pieces, from over 28,000 in 1602. Florentine cloth production also plunged from 30,000 in 1560–80 to approximately 6,000 pieces in 1641–45; and its Milanese sibling did not fare any better either: of the 60 to 70 firms that engaged in cloth production in 1600, with an output of 15,000 pieces, only one was still in business in 1709, sending "a mere 100 pieces to the market."[106] The precipitate fall in the city-states' manufacturing output did not provide succor to the weakened industries of the basin's outlier regions. On the contrary, an intensification of competition, in part due to the easy availability of wool clip, spread textile production even farther into the countryside.[107]

Even the early pioneers and established centers of rural cotton manufacturing were not spared the calamity of the Thirty Years' War and the fall in population and demand. In Augsburg, fustian output dropped from 430,636 pieces in 1612 to 60,500 in 1720. In Nuremberg, too, the number of workshops dropped by one-third in the same period.[108] Cottage industries that produced cloth of low to middle quality proliferated owing to the contraction in markets when urban (textile) industries were taken to task. Rural industries took upon their shoulders a larger part of the basin's manufacturing activities. A petition from the *paraires*, clothiers, in the woolen industry of Barcelona, composed in 1683, related that production was now limited to that of "different towns and villages of the present Principality in such a way that those who previously traded with the *paraires* of the present city today trade today with the *paraires* of the towns and villages."[109] The second wave of ruralization had a profound impact on the destiny of cotton and silk industries in the western Mediterranean, where the difficulty of procuring raw materials was compounded by the intensity of competition in woolen textiles. In this, the demise of Venetian rule played a part. For one, the commercial networks that crisscrossed the textile regions between the Adriatic and the Baltic underwent a transformation from the 1550s that allowed merchants to acquire industrial crops that were conveyed via Frankfurt, Augsburg, and Leipzig rather than Venice.[110] In the process, the share of woolen cloth in the rural industries of the region rose at the expense of the fustians. In a twist of fate, the Anglo-Dutch textiles were now finished and dyed in Nuremberg. This was a significant turn in the evolution of the textile industry across the Alps. The fustian industry was not the only one adversely impacted by the relentless march of the woolen textiles. Fine wool, due to its greater availability, largely replaced silk for furnishing fabrics and particularly for luxury clothing even though demand for silk did not abate during the *beau seizième*.[111]

Given the dominance of woolen textiles in the western Mediterranean, dispersed manufacture remained the norm, as throughout France and Spain.[112] While the manufacture of woolens decreased in Barcelona, it climbed in a number of nearby towns and villages where swift, clear streams and relatively cheap labor and fuel existed. Small towns grew: Tarrasa, Sabadell, Esparaguera, Olesa, Igualada, Monistrol de Montserrat, Castelltersol, and Moya became more important centers for woolens.[113] Spinning of wool in the countryside had become such an essential undertaking by the eighteenth century that when, with the introduction of calico-printing, cotton spinning spilled out from Barcelona, producers of woolens requested the Special Council of Commerce of the city to "oblige the women and girls to spin wool instead of cotton."[114] Woolen manufacture throughout France and the silk industry of Lyon yoked town and country together.[115]

One of the reasons a contraction of manufacturing in Venice did not provide relief, however passing it may have been, to Ottoman artisans was because raw material prices went on to climb relentlessly. Equally important was the ability of the Levant Company—the sole supplier of Levant silk in the "home" market and the beneficiary of high profits that accrued from this trade due to the company's monopoly—to market British woolen cloth in the Levant at prices cheaper than at home. Rising raw silk prices and falling textile prices generated the scissors effect in all its perfection. Between 1621 and 1721, English imports of raw silk from the Levant increased by 275 percent, keeping prices high.[116] The termination of the second sixteenth century in 1650 ended the price inflation, and the consequent fall in prices that followed lasted for nearly a century. The textile industry was not immune to this trend.[117] Woolen industries, high and low, took flight from their urban abode: cutting costs of production when raw material prices were on the rise and cloth prices on the decline was an exacting task, especially if the presence of guilds posed additional restraints on these industries' ability to turn footloose.

Thus the differential trajectories of the 1550–1650 period, with Venetian woolen manufactures enjoying an unprecedented boom on the one hand and those in the Ottoman, Spanish, and French lands bearing the brunt on the other, converged after the 1650s. On the face of escalating competition, the industrial renaissance of Italian city-states was cut short. Under pressure from producers in the Anglo-Flemish lands, the city-states had to face a fate similar to what their Mediterranean brethren had already experienced. The decline in prices that started from the 1650s did not alter the situation drastically, since industrial crops suffered a less dramatic fall in price than grain whereas textile prices plummeted.[118] Prices reached their peak at the turn of the seventeenth century, plateaued, and then held steady for two decades for most goods and went on to decline afterward until the 1750s. The end of the industrial revival in the Mediterranean heartlands did not translate into better times for craftsmen in the Ottoman and Iberian lands.[119] Not only did the continual drop in the prices of light woolen textiles imported by the Dutch and English merchants keep the competitive pressure in place for the local industries but also the hike in demand for silk and cotton by the northern merchants did not let the prices of these raw materials drop significantly. One result of the scissors effect created by price movements of raw materials and finished textiles was the growing import of local supplies: hence the resultant expansion in the culture of mulberry trees, cotton, and hemp.[120]

In a setting, therefore, where raw material prices were falling relatively less than that of finished textiles, ruralization, the favorite stratagem of the second sixteenth century, did not lose momentum.[121] By the 1650s, the flourishing textile centers of the Mediterranean had largely lost the vibrancy they once possessed, at least in their urban form. In terms of manufacturing, Safed, Salonika, and Ávila, for instance, paled beyond recognition.[122] The obverse side of the same coin was, however, that conditions that exacted a heavy toll on urban industries provided a stimulus to the cultivation of industrial crops—and rural manufacturing. Whether these crops were destined for local and regional markets, or crossed imperial borders, was of little import. They commanded better prices than grain, and more land was devoted as a result to the production of cotton, mulberry tree, hemp, and flax. Depressed market conditions that forced rural producers to diversify their economic activities and place growing emphasis on complementary sources of income accruing from cottage industries provided additional momentum to the relocation of manufacturing away from the established centers of the sixteenth century. The impetus originated in the towns, and the cottage industries remained more or less under the control, if not the mercy, of urban merchants.[123]

Easy access to labor desperate for additional income and the availability of industrial crops in nearby locales served to broaden putting-out arrangements, mostly in rural areas, thereby enabling merchants to escape the control of craft associations. Moreover, the pervasiveness of the *verlag* system permitted most merchants (and manufacturers) not to tie up their capital in factories or equipment.[124] These developments in turn compelled rural households to devote more time and energy to the cultivation of industrial fibers, animal and cultivated. As a result, lesser industrial centers and the rural cottage industry that supplanted the textile cities of the previous era were able to cater to the demands of the market.[125] Given the organizational framework woven by merchants and trading companies and the putting-out system, there was, in Languedoc, for instance, "a remarkable continuity of dozens of *fabricant* families, dating back to the early seventeenth century."[126] By the end of the eighteenth century, the production of woolen textiles in the Salonika region had reached its sixteenth-century level. Yet the bulk of the output originated in the countryside surrounding the city rather than in the city itself, as had been the case previously, thanks to the house of Gümüşgerdan, which made its fortune on putting-out arrangements with mountain villages.[127] In Castile, rural production stood well above "that of urban origin." Rural production in Palencia, Segovia, Guadalajara, and Toledo reached 30,291 pieces against the urban total of 14,788 in 1745; "10,437 against 7,688 in 1760; and 33,358 against 20,696 in 1785–95."[128] On the Riviera di Levante, of the 2,064 looms in operation, only 480 were located in Genoa.[129]

Manufacturing ventured into the countryside, out of sheer necessity perhaps, but certainly on account of the organizational capabilities of the merchants, who constructed an impressive *verlag* system, chiefly in response to the changing conditions of the seventeenth century.[130] Not least of all, the high profile assumed by livestock grazing and the presence of nomads advanced the dispersal of woolen industries into the rural recesses of the basin.[131] This was one of the reasons the woolen industry, after its retreat into the countryside, proved to be difficult to extricate from its new setting: the ease with which its provisioning was accomplished gave it an additional advantage. Where pendular migration of flocks presented such an opportunity, the development of woolen industry was almost inevitable: witness the degree of excellence achieved by the mountainous regions in the Balkans or Valencia (where it was referred to as "mountain cloth") in textile manufacturing.[132]

If the fates of the woolen industries of the basin were alike despite the disproportionate dominance of these industries in its western half, the cotton industry in particular followed a different path. In the western Mediterranean, the hold the merchants dealing in the Atlantic trade had on cotton imports brought "complete upheaval" to the practices of textile production and, by extension, the labor market (as in the case of Rouen),[133] because it gave the merchants the liberty to reorganize production as they saw fit from a position of strength. On account of the merchants' control over raw cotton, cotton-spinning and, at times, weaving spread out into villages, buttressing the trend toward ruralization of production. Even when the cotton industry competed with and posed a veritable challenge to the woolen industry, percolation of manufacturing into the countryside continued. Or, as in Barcelona, the development of calico-printing, though diffused in the city's hinterland, also allowed men of money to organize certain parts of the production in manufactory-type units. The steady arrival of American cotton, especially after the Civil War, rather than the capricious sources of Malta, Sicily, and Sardinia, provided the merchants with unimaginable liberties.[134] In the eastern Mediterranean, local production of silk grew on account of fluctuations in the supply of Caspian silk; cotton cultivation and manufactures, too, expanded, especially given the soaring French demand for it.[135] In the period bracketed by 1728/1735 and 1767/1776, while cotton imports into France from the Americas increased by a modest by 14.7 percent—from 1,500 to 22,070 quintals—imports from the Levant soared by 402 percent—from 21,262 to 106,784 quintals. As the share of raw cotton jumped, that of spun cotton fell: from 16,946 quintals in 1700–1702 to 10,805 in 1785–89.[136]

Where cotton production advanced, as in Kilis, Antep, Mardin, Urfa, and Hama, and in the environs of the Anatolian plateau, in Kastamonu and Ankara as well as in the Balkans, cotton textile production picked up.[137] In Tokat in the eighteenth century nearby villages and towns were all integrated into a complex division of labor: only the dyeing and printing of textiles took place within the confines of the city.[138] When Mosul's cotton industry experienced rapid growth in the eighteenth century, merchants certainly controlled the cultivators, spinners, and weavers.[139] The cotton industry, which had a modest share of the market in the sixteenth century, thus came to occupy a discernible position by the turn of the nineteenth, in urban as well as rural locations. The latter, due to its remoteness from the port-cities, remained impervious to the invasion of

British cotton textiles in the nineteenth century.[140] Over the course of the eighteenth century, the production of cotton in Ottoman lands multiplied three-fold, as demand by domestic industries doubled and overseas demand sky-rocketed.[141] Cotton's cash-crop status was sealed.[142]

In the latter half of the eighteenth century, a hike in the prices of raw materials and imported goods encouraged the inhabitants of the Levant to encroach further on their competitors and penetrate into the cotton export markets that catered exclusively to the French. There was an increasing demand from the people of Damascus, who came to smuggle in the fibers that were necessary for their manufacturing. A few decades later, commercial houses in Aleppo and Cairo were especially active in rechanneling cotton supplies toward the interior.[143] Jazzâr's breakup of the French merchants' monopoly and his policy of preference that helped "domestic buyers" must also have contributed to greater local use of cotton.[144] With the egress of manufacturing from larger urban centers, the share of manufacturing-related activities within the revenues of *waqf* holdings was reduced to a minimum during the seventeenth century, remaining at a meager 3.9 percent. That the majority of industrial and commercial space held in perpetuity by holdings that were built in the fifteenth and sixteenth centuries were located in urban centers and that these centers suffered heavily in the following century explains the fall in percentage.[145] The tide turned in the mid-eighteenth century, with the retrenchment of manufacturing to major urban centers.[146]

Similarly, in higher altitudes where mulberry trees were or became abundant, the silk industry prevailed: during the seventeenth century, in fact, sericulture established a respectable presence in the mountainous regions, from the Taurus mountains to Jabal Ansariyya; in the seventeenth century, the vicinity of Latakia, Marqab, al-Bahlûliyya, and their environs were all engaged in "l'arte rurale e la fabbrica delle sete." Silk production in the established—urban—centers of production regained its vigor during eighteenth century. They were joined in by a new silk center, Chios.[147] Silk found a large market in the interior (in Damascus, Aleppo, Hama, Homs, and in certain parts of Anatolia) when its export was brought to a halt during the Napoleonic episode.[148] In 1816, the governor-general of Aleppo, Jalal al-Din Pasha, prohibited by decree the exportation of raw silk from the province, stating that purchases made by French merchants increased prices of these articles to the "prejudice" of the country.[149] Cotton and silk became the "principal branches of commerce."[150] In Valencia, where silk became the "chief fruit of the realm," the enduring stability of the silk industry marked its otherwise volatile manufacturing history. And the Lyon silk industry grew at the expense of the Genoese.[151]

The metamorphosis in the industrial and, relatedly, spatial makeup of the Mediterranean world was in full accord with the transformations in its agrarian structures. For one, the new manufacturing environment worked to the detriment of lands devoted to grains: the culture of industrial crops, sericulture, and livestock husbandry was now at a premium. Moreover, the engagement of rural households in manufacturing helped them to recover some of the losses they suffered from the retreat of cereal agriculture. All in all, the ruralization of manufacturing and the widespread cultivation of industrial crops went hand in hand. In the realm of manufacturing, then, the waning of the Mediterranean was boxed in by two waves of deindustrialization. These waves demarcated, with enviable certainty, a change in the fortunes of the Inner Sea. The first wave was brought about by an invasion of Venetian woolen (and silk) textiles, and the second, that of British cotton goods. In the interim, the dispersal of manufacturing became the order of the day. In terms of longevity, this development mimicked that in the realm of agriculture. As argued in the previous chapter, the assumption by, first, the Baltic and, later, the temperate settlements of the role of the world-economy's granary bracketed the changes in the Mediterranean's agricultural fortunes in the seventeenth through the nineteenth centuries, when the return of the basin's tree crops to its fold occurred.

Restoration of the Triad of the Mediterranean

Unlike in the heyday of the Inner Sea, when the widespread cultivation of oriental crops and the imperialism of wheat privileged the basin's plains and lowlands, the return of the tree crops engendered an elaborate articulation among its lowland, hillside, and mountain agriculture from the sixteenth century.[152] In other words, the blend of wheat, tree crops, and small livestock, a blend that came to be exclusively identified as the triad of the Mediterranean, reestablished its presence throughout the basin starting from the latter half of the sixteenth century. Historically, this was a movement that originated in the Italian and Iberian peninsulas between 1450 and 1560, spread thereafter to the outlying quarters of the basin, and became solidly entrenched in its landscape in the seventeenth and eighteenth centuries. Put paradoxically, the Mediterranean trinity of crops came to blanket the shores of the basin not at its height, but in its waning. As the staples of the age of the city-states, grain, oriental crops, and spices, moved west and north, the staples of the basin, tree crops in particular, returned to their ecological abode. There were two sets of factors at play that fueled this change: economic and ecological. In the case of the former, demand for wine and olive oil picked up along the shores of the North and Baltic seas and the Atlantic in consonance with the rise of Amsterdam and the colonization of the Americas. In the case of the latter, concurrent with the end of the Medieval Optimum at around the mid-sixteenth century, the northerly territorial expansion of vine-growing came to a halt to the benefit of the Inner Sea, the preferred habitat of vineyards. What these two developments signaled was the reversal of a long-term trend that had shaped the Mediterranean roughly from the turn of the millennium with the advent of the global warming trend to the sixteenth century when the Little Ice Age returned with a vengeance as demand for the produce of tree crops turned buoyant.[153]

What had altered the basic traits of the Mediterranean landscape and agriculture from the turn of the millennium to the mid-sixteenth century was, as discussed previously, the westerly migration of crops of the age of the city-states. Broadly speaking, the oriental crops that invaded the Mediterranean from the Indian Ocean sailed west, first within the confines of the expansive Abbasid empire and, later, under the guidance of the twin city-states. And from the late fifteenth century, commercial cereal agriculture inexorably advanced north. Earlier, the indigenous crops of the basin had experienced a redistribution of their sites of cultivation. Viticulture had crawled north from its Mediterranean domicile from the eleventh century under the relatively welcoming conditions of the medieval warming period; the movement may in fact have stretched as far north as the mythical Vinland itself.[154] Not relatedly but in tandem, olive orchards shrank in area and olive oil in significance with the rise of livestock husbandry from the mid-fourteenth century.[155] At the cusp of the sixteenth century, the devolution of the Mediterranean was still under way. Whereas the exodus of the oriental crops and specialized cereal agriculture only gained strength with the rise of the Atlantic and the Baltic, the northerly march of the vineyards, after encountering a veritable setback after the Little Ice Age set in, suffered an eventual reversal, to the benefit of the Mediterranean. With demand for olive oil in the North Sea and the Atlantic on the rise, olive groves also expanded, at times with an impressive speed. The basin, that is, reasserted its ecological jurisdiction over its tree crops from the sixteenth century. Indeed, it was in the eighteenth and early nineteenth centuries that the olive tree was pushed to its maximum limits; it has since retreated to its ecological limits today. To wit, in northern La Mancha, tree crops—olives and vines—occupied almost half of the total land area in the eighteenth century. Today, by contrast, 85 percent of the arable is devoted to cereals.[156] Olive oil, wine, and silk were responsible for the good fortunes of Tuscan Pescia in the late sixteenth and early seventeenth centuries.[157] These two movements then, one of exodus and one of return, were dissimilar not only in direction but in essence. The exodus of the oriental crops and commercial grain cultivation reflected, *strictu sensu,* the dynamics of the world-economy and its continual revision to the benefit of Amsterdam and the United Provinces. The latter, however, marched to the tune of ecological and climatic change, because the northerly invasion of the vineyards owed their successful implantation in the northern perimeter of the basin's ecological belt to the warming trend between 950/1000 and 1250/1300 and, later, to the Medieval Optimum.[158]

These two movements were different not only in character and provenance. The revisions they prompted in the organization of production were also strikingly dissimilar. For one, the westerly movement of oriental crops came to be associated with slavery both within and without the Mediterranean, and the northerly relocation of commercial grain production came to be identified with the second serfdom. Both facilitated, or brought about, specialization and monoculture. On the other, the thriving grapevines and olive groves helped Mediterranean peasant households to diversify their economic activities, mostly in the form of *coltura mista* or *coltura promiscua.*[159] Deeper soils were given over to cereals, strips of limestone to olives, and vines clothed vast expanses.[160] The resurgence of the basin's triad of crops eased the transition from the heyday of city-states to the basin's twilight to its autumn by giving rural producers precious breathing space as did the dispersal of manufacturing. If the vegetal basis of the Inner Sea underwent momentous change from the sixteenth century, three developments helped reshape it in the following two centuries. These triple developments were the return of tree crops; a diminution in the share of large-scale animal husbandry, sheep primarily, and that of wheat in the basin's agrarian makeup; and the reappearance of lesser cereals to make up for the deficit caused by the diminishing import of commercial grain cultivation. These interrelated and concurrent processes endowed the Mediterranean landscape with its by now familiar silhouette.

In part, it was the onset of a colder clime, first, from the turn of the fourteenth century and later from the 1550s, which erected insurmountable barriers to the northerly spread of the vineyards. Viticulture was rekindled, not too surprisingly, at the turn of the millennium, during the medieval warm period.[161] In the thirteenth century, there were over 1,300 vineyards in England, up from the 42 listed in the Domesday Book in the eleventh century. The Cistercians, who became undisputed masters of viticulture, aided the northernbound expansion of the vineyards in Burgundy: between 1000 and 1336, they purchased, and were gifted with, dozens of vineyards, the fruits of which were skillfully drained by the Dutch, Flemish, and Breton merchants, to serve the markets in the southern Netherlands.[162] To the east of Burgundy, it was in

Cracow that wine from Hungary and Moldavia converged.[163] German wines were exported up the Rhine to northern Germany, Scandinavia, and the Baltic. In the fourteenth and fifteenth centuries, when vine-growing was at an ebb along the shores of the Inner Sea, the vintners of Gascony and Gascon wine reached the height of their fame.[164]

The septentrional advance of the grapevine came to a halt or reached its ecological limits in the closing decades of the sixteenth century. Vine plants accordingly vacated some of the recently won territories in order to retreat back to their Mediterranean home. With the onset of the Little Ice Age, vineyards on the lower Seine and Rhineland, in northern France, the Low Countries, northern Germany, Switzerland, lower Hungary, and parts of Austria were abandoned. The process, which started in the years from 1560 to 1600, lasted well past the ages of baroque and enlightenment.[165] The resulting slump in wine production and the hike in its price brought about a switch to beer-drinking north of the Mediterranean ecological zone. With the consolidation of Dutch hegemony, "beer conquered territory even in the south."[166] Yet overall, the geographical balance in viticulture was now tipped toward the Mediterranean, at the expense of the northerly sites of the previous era. Instrumental in this rebirth was the recently found popularity of fortified and specialized wines. What is more, the expansion of vineyards proceeded unhindered in the following centuries on account of the popularity of distilled drinks, brandy, and sherry.[167]

It was not the vineyards alone that experienced a relative growth in the sixteenth century. More land was devoted to the olive groves, less imperialist than vineyards due to the relative narrowness of their "natural" zone of growth. The Reformation and the new ecclesiastical rules governing the consumption of meat and animal fats breathed new life into the olive oil trade from the sixteenth century.[168] Usage of oil spread throughout the Protestant realm, both as a condiment and for cooking, not to mention the needs of the cloth industry, from Antwerp to England.[169] Despite the lasting impact butter had on the haute cuisine of the Mediterranean, the difficulties in preserving and transporting it over long distances rendered it available only to those who were willing to pay premium prices for it.[170] In Italy, for instance, butter was used in the south, simply because the difficulty of obtaining and preserving butter turned it into a favorite of the southern elites who consumed it conspicuously with the purpose of distinguishing themselves from the rest.[171]

As the olive groves and vineyards reestablished their imperium on the hillsides of the basin, in the following centuries the Mediterranean became the almost exclusive habitat of these twin tree crops.[172] In Tendilla, for instance, like the olive-producing Aljarefe region of Seville, the municipal territory was planted in such manner that the region's mountains, slopes, and valleys were densely covered by olive trees; vineyards serendipitously blanketed its rugged terrain.[173] In the eighteenth century, close to 20 percent of the arable was earmarked for vine-growing in Cuenca, and in the Ottoman Morea, 67 percent of the tithe originated in viticulture, as did 51 percent in parts of Macedonia.[174] In large areas of Andalusia and in the Tunisian Sahel, olive tree cultivation even took the form of monoculture.[175] Cereal agriculture, now firmly affixed in the plains between the Atlantic and the Hungarian lowlands, lost the commercial zeal it had acquired in the lowlands of the basin in the fifteenth and sixteenth centuries.

As the splendors of the long sixteenth century lifted the fortunes of the tree crops, due to sustained demand from the fast-growing and demographically colossal imperial capitals, Madrid and Istanbul, and from across the Atlantic, the Mediterranean landscape took on new colors from the seventeenth century with the revitalization of vineyards and olive orchards.[176] Demand for olive oil from the Venetian, Ottoman, and, later, French empires paved the way for the proliferation of olive groves in the Aegean.[177] Prompting similar developments in New Castile and, of course, in Andalusia was the rise in demand from Madrid and New Spain.[178] Bologna and Naples abounded in olives as large and perfect as those of Spain.[179] So did Lombardy, Romagna, and Florence in the seventeenth and early eighteenth centuries. The unwavering character of demand for soap was due to the dispersal of manufacturing from the 1560s, hastening the expansion of the olive groves. In Aleppo, soap-manufacturing became the provenance of larger manufacturing units from the seventeenth century.[180] Despite the predominance of livestock husbandry and animal fats from the fourteenth century, olive oil managed to carve out a lucrative and ever-growing sphere of expansion from the sixteenth century. In short, the olive groves and vineyards, integral constituents of the Mediterranean landscape in the thirteenth century but not after the 1350s, came to reclaim their lot in the basin's landscape.[181] And in the recrudescence of the tree crops, the expansionary spirit of the long sixteenth century certainly played the guiding role.

Henceforth, vineyards that had crawled north, from semitropical to temperate regions during the Little Optimum retreated back to its southern tier.[182] In the following centuries, it was within, and not beyond, the Mediterranean belt that viticulture established a commanding presence: vineyards were more widespread along the shores of the basin than before. Vineyards were also planted on the opposite shores of the Inner Sea, on the North African coast, following the expulsion of the Moriscos from the Iberian peninsula. Near Tunis, grapes were widely grown in the eighteenth century, as was the case around Bizerte and in

Cape Bon, and "raisins, *zebîb*, were sold to distant markets."[183] Between 1630 and 1771, the southern reaches of the Greek peninsula, the Aegean islands, Perigord, and Montpellier devoted more land to viticulture. Patras, for one, thrived and established a commercial presence owing to its famed currants, *uve passe*. In Bas Languedoc, the output of well-situated "maritime vineyards" started to fall in 1645–50 before it collapsed under Colbert, but "the advance of viticulture resumed in the 1730s and 1740s."[184] As a result, vineyards developed primarily on escarpments owing to the clearing of the woodland; later they were pushed out onto the vacant plains. It was in the same century that agriculture in the Ebro region was centered exclusively on vines.[185] In the hill country of Tuscany, or central Italy, the clearing and bonification of new lands from the renaissance through the eighteenth century was tied to the initiative of planting trees and shrubs; the newly terraced slopes in Tuscany were devoted to tree crops, and no less than a hundred thousand new olive trees were planted in the period of reforms. Viticulture thus invaded the Alpine valleys of Provence and Tuscany. The river Po "covered 90,000 perches of vineyards" in Casalmaggiore alone, and 300,000 perches in flooded lowlands from Cremona to Casalmaggiore.[186] In New Castile and Aragón as well, vineyards spread vigorously.[187] The planting of trees festooned with vines happily punctured the cloistered nature of the vineyards and hastened the spread of vines beyond areas of specialized cultivation.[188] Even the merchants of La Rochelle were quite willing to place their capital in vineyards or part-shares in vineyards in the eighteenth century, since they considered that money thus invested could be recovered without too much difficulty and was hence a safe investment.[189] The situation was similar in the Ottoman lands, since shares of vineyards, in ever smaller fractions, changed hands with accelerating velocity. In the seventeenth century, close to 60 percent of the revenues of pious foundations accrued from their vineyards, orchards, and gardens.[190]

The spread of vineyards accelerated during the seventeenth-century crisis, when cereal prices slumped yet wine prices fared relatively better. At times, this posed a serious problem for the provisioning of cities along the Mediterranean coast. Between 1727 and 1731, restrictions were put in place in France on the territorial expansion of the vineyards. They were later repealed in 1759. In Catalonia, farmers complained in 1756 that for many decades, the "vineyards had been eating into the arable land."[191] This was the case in the Ottoman empire: of the revenues accruing to pious foundations from their rural holdings in the seventeenth century, less than one-third derived from cereal production.[192] In seventeenth-century Bursa, for instance, villages received more than 50 percent, oftentimes reaching 70 percent, of their income from orchards and vineyards and not from cereal production.[193] That, time and again, taxes imposed on wine (and tobacco) consumption were revoked, as in 1668, only to be imposed again, in 1688 and 1696, to be revoked shortly, is a testament that it was produced and consumed widely.[194] With cereal prices remaining depressed in comparison to the price of wine, vine-growing experienced a long period of prosperity.

The vine was not alone in commanding the attention of the cultivators. Olive and mulberry trees came to occupy a larger portion of the land under use not only during the second sixteenth century but also later. During the seventeenth century, on the Bolognese and Romagnol hills, in Lombardy, Florence, and the nearby provinces, and even in Rome, orchards of olive covered the hillsides of the Mediterranean more densely than before, and olives were grown abundantly.[195] So olive oil remained an essential part of the diet, and was used for preserving food, especially fish, and for illumination. Yet, changes in global climatic conditions and their impact on the limits of cultivation of the olive tree aside, the growing popularity and the easy availability of animal fats in the carnivorous north during the seventeenth and eighteenth centuries, owing to the switch from cereal to animal husbandry and dairy industries, were not potent enough to change the dietary habits of the denizens of the basin. The Ottoman imperial kitchen, too, which was initially supplied with olive oil from Modon and Coron and in modest quantities, since, here too like the empire's counterparts in the west, butter constituted the main oleaginous ingredient for cooking, later came to consume more olive oil imported from the Aegean and Tunisia.[196] Groves of olives populated the Levantine landscape; the trio of olives, olive oil, and soap asserted themselves in the economic life of the region in the seventeenth century; in Tripoli taxes were paid in olives (and silk and broadcloth) in the seventeenth century. Or, interest payments on loans in and around Damascus were camouflaged behind the curtain of the soap trade in the eighteenth and nineteenth centuries, whereas in the fifteenth century the Mamluks imported olive oil from overseas—Puglia and Tunisia.[197] Surveys of court registers and *waqf* documents indicate that "selling orchards and vineyards was a daily occurrence."[198] Sericulture too established a strong foothold on the upper elevations of the mountainous regions, as in Mount Lebanon, as it advanced to the northern perimeter of the Mediterranean clime, as in Iberian and Italian peninsulas, as discussed in the previous chapter.

The fact that pious foundations that pioneered land reclamation in Ottoman territories during the fifteenth and sixteenth centuries could extract only one-fourth of their revenues from arable production in the eighteenth captures the magnitude of change that took place in the interim period.[199] The growing significance of arboricultural and horticultural goods was thus a trend in tune with the seventeenth-century downturn. As markets contracted, the strategy rural producers followed changed in line with the exigencies of the lean times: producing for an anonymous yet expanding urban market had to leave its place to increasing specialization on the one hand, and growing emphasis on lesser cereals on the other. The first option, specialization, aimed at serving urban markets in high-value agricultural goods—goods coming from orchards, vineyards, and gardens rather than fields. Despite the overall fall in population, urban growth was sustained at times owing to the depopulation of the countryside due to migration. Catering to the urban markets was an option still open to the rural producers.

The rise in the stature of tree crops in the trinity of the Mediterranean accompanied two developments: the transformation of wheat into a humble crop and of animal husbandry, small livestock specifically, into an integral part of smallholding peasantry's scope of activities. Wheat, ousted or eclipsed from the sixteenth century by lucrative tree crops, claimed a less than modest part for itself during the twilight of the Mediterranean. Not before the Continental Blockade, which necessitated the provisioning of the British navy, and, later, the repeal of Corn Laws in Britain in the 1840s did arable production regain salience in the economic life of the basin.[200] Indeed, there was a close correlation between high wheat prices in world markets and the sale of Church and municipal lands in Spain.[201] Between the *beau seizième* and the mid-Victorian boom, grain husbandry was complemented by livestock agriculture especially after the seventeenth century as the local holdings of small livestock increased, as in the Spanish and Anatolian peninsulas. Only in Foggia, the size of the herds remained stable until the end of the eighteenth century.[202]

Wheat, of course, was and has been an honorary member of the Mediterranean trinity. Even in its heyday, however, it remained under pressure from the world of transhumance from the fourteenth century. But the newly gained ability of livestock husbandry to tap into the sylvan resources of the region through transhumance in a systematic manner, that is, the integration of the highlands of the Mediterranean into the seasonal movements of herds, allowed the agrarian and pastoral components of this landscape to inhabit the same economic space without major discord, at least until the mid-fifteenth century. The expansionary spirit of the long sixteenth century, shared by the arable and animal husbandry both, turned this cohabitation into a confrontation between Cain and Abel from the 1450s.[203] What is worse, the economic momentum generated by the influx of the West Indies gold and Potosí silver stimulated wine and olive oil production, multiplying challenges to the predominance of cereals, wheat in particular.[204] A contraction in the share of the arable was a prerequisite for the reconstitution of the time-honored Mediterranean triad that had in fact seen the share of its tree crops diminish between 1350 and 1500 in light of the commanding presence obtained by its two remaining members, wheat and sheep. The steady advance shown by tree crops during the sixteenth century was furthered by the seventeenth-century crisis: a decline in cereal production was hastened, because the deceleration of demographic growth eased the weight of numbers on agrarian resources. As mentioned previously, skyrocketing grain prices between the 1590s and 1620s may have channeled huge investments into land, from the Venetian Terraferma to Provence and to the Lombardian countryside, facilitating the "defection of the bourgeoisie." Complementing this movement in the Ottoman dominions were increasing claims over agricultural surplus.[205] But a passing movement this was, with no lasting revisions of the overall state of Mediterranean agriculture. Nevertheless, until the second half of the eighteenth century and in some places much later, grain production in the basin remained below the levels it had reached previously in the sixteenth century.[206]

One of the reasons bread grains could not fully recover the ground they had lost during the sixteenth century was the expansion of livestock husbandry to the detriment of cereal agriculture. The seventeenth-century crisis sealed this trend by depressing agricultural prices in general and, most important, cereal prices, thereby tipping the balance toward pastoral and non-cereal crops, which fared relatively better.[207] As we will see in chapter 5, mixed-farming and agro-pastoral diversification were both effective assurances not only against lean times during the seventeenth-century crisis but also against the climatic variability of the Little Ice Age. A contraction in arable production allowed more of the land to be devoted to grazing and pasturage. Since livestock products in particular were the least affected by this downward slide in prices, a premium came to be placed on animal husbandry over grain production, relaxing the pressure on peasant holdings. Apart from the favorable terms of trade they enjoyed vis-à-vis bread grains, raw materials of animal origin (e.g., wool, hides, wax, and leather) commanded comparatively lower prices in Ottoman lands than in lands to its west, and it was these products that featured prominently among Ottoman exports during the seventeenth and the early part of the eighteenth centuries.[208]

Therefore, when the Ottoman state, *mîrî*, lands were alienated, they were mostly given over to livestock-raising rather than cereal cultivation. Rüstem Pasha's mastery in circumventing official regulations to encourage grain exports in the 1550s had become a thing of the past by the seventeenth century, as neatly attested to by the spectacular increase in the livestock holdings of the provincial elite. Equally, many large estates of the period allocated most of their lands to livestock-farming, as the literature on *çiftlik*s amply illustrates.[209] Grazing lands did not necessarily take over previously tilled fields, given the relatively high land-labor ratio. In fact, livestock agriculture thrived better in the flora of marshy low-lying lands or untilled plots, the reclamation of which required substantial capital outlays as well as employing or cajoling rural producers to toil under the dreadful conditions of the swampy environments. This was analogous to developments in Spain, where the nobility and the Church appropriated vast areas of land in Andalusia in the seventeenth century, and depopulated the lowlands by devoting them to animal husbandry.[210] Flocks of sheep grazed on heaths and moors, on marshland, forest, and hillsides between Garonne and Bordeaux.[211] And in the Ottoman empire, most of the large estates of the period had a very significant component of livestock-farming. Given the relatively low land-labor ratio, the expansion of grazing lands did not necessarily take place on the debris of tilled lands and of petty producers. By the turn of the eighteenth century, the number of *çiftlik*s engaged in grain production had declined, giving way to animal husbandry. The livestock industry was not exempt from the repercussions of the northward shift in the center of gravity of economic activities.[212] Those who could either afford to buy grain and devote more land to it and those who could manage it with an efficient agricultural base had more of it.

Disputes concerning pastures attest to the cardinal role livestock husbandry played during this period: the provenance of most disputes was not the discontent provoked by the tendency of the arable to expand at the expense of grazing lands. Instead, they reflected increasing and conflicting claims over village commons and pastures. With the expansion in livestock husbandry, inhabitants of these villages found themselves in a defensive posture and had to reassert their established rights over common lands against outsiders, were they entrepreneurs from the neighboring city or members of transhumant populations.[213] This was also the case in Spain, where disputes between local pastoralists and the Mesta colored most of the seventeenth and eighteenth centuries. Overall, from the drastic expansion of the livestock holdings of the Mesta in the Iberian peninsula to the increasing presence of itinerant pastoralism in the Balkan and Anatolian peninsulas, animal husbandry played a major role during the seventeenth-century downturn. Confirmation of this is suggested by the fact that tanneries housed in Aleppo, Cairo, and Tunis, now handling a heavier work load, were located on the outskirts of city limits and by the increasing share of dairy products in popular diets.[214] Sedentary livestock, therefore, increased in number in the mountains of the Mediterranean, and added a gamut of goods offered by the pastoral stocks: meat, leather, wool, and dairy products.[215] It was not exclusively the large-holders who benefited from the expansion of animal husbandry, especially during the taxing times of the seventeenth-century crisis. In Iberia, "the owners of sheep that migrated only short distances or not at all seem to have increased their flocks and gained control over a significant portion of available pasture lands." The local nobility, smallholders, and peasantry in the kingdom of Naples added on to their share of animal stocks, though these trends were later briefly reversed in the eighteenth century. And changes relating to the sheep tax in the Ottoman empire were designed to encourage sheep-raising by the peasantry.[216]

With the shift from arable to pasture during the seventeenth-century crisis, plow land in the low-lying lands of the Mediterranean basin was put down to grass, from Castile, Savoie, Jura, and Bessin to Gruyères, not to mention Bourgogne, Schwerin, and pays d'Auge.[217] The spatial stretch of the movement of turning arable to grazing and the wide compass of rural unrest attest to a Mediterranean-wide process. The factors that emptied out the plains were not particular in character, nor were the attempts to tackle problems of populating the underutilized and unoccupied plains. In Ottoman Anatolia, attempts to sedentarize nomads at the end of the eighteenth century were complemented by the campaign waged in Spain against the Mesta's hold over the countryside, which, it was claimed, privileged husbandry over farming.[218] Both attempts ended equally unsuccessfully. This meant, of course, that lands that were reclaimed during the long sixteenth century were given over to grass and that land reclamation slowed down; it came to a halt in most places or was reversed.

Complementing the return of tree crops and the diminishing significance of wheat and transhumance was a change in the fortunes of lesser cereals to the detriment of high-value crops. With the advent of the seventeenth-century crisis, demand for high-priced cereals such as wheat stagnated or fell with a corresponding fall in demand for these crops.[219] The fall in demand for higher-end cereals was steeper than those for low-priced cereals (such as oats), because the latter were consumed more widely, and not only in times of plenty. The relative stability of demand cushioned these crops against the volatility of price movements. That in most parts of the basin agriculture gave way to animal husbandry was not without a determining influence on the direction of this change: the shift from the cultivation of cereals to increased production of forage crops and animal feed (again, as barley, millet, oats, and maize).[220] This brings us to the third component in the regeneration of the Mediterranean landscape, and involves the appearance of lesser grains along with those of industrial and tree crops discussed above. The import of lesser grains in this period stemmed from the diminished share of wheat in the region's output and diet. Also, when the westerly overland trade picked up in the seventeenth and eighteenth centuries, routes traversing the Anatolian plateau to reach İzmir and the Balkans thrived.[221] The great expansion of overland commerce called for more haulage-power. The corresponding rise in the number of camels, horses, and mules contributed to the concurrent rise in demand for animal feed—oats and barley.[222] Increasing the output of lesser cereals was possible because it was closely connected to the needs of livestock farming, not to mention those who could not afford the more expensive cereals. In light of the contraction in arable and low grain prices, where cereal cultivation dominated, the crops that gained popularity tended to complement rather than supplant bread grains. The popularity of spring crops, buckwheat, and potatoes on both sides of the Channel, the widespread cultivation of maize in the Balkans and of legumes and millet in Anatolia, the march of riziculture in most quarters of the Mediterranean, all provided much-needed relief to the producers by decreasing their dependence on erratic wheat, were they petty producers or *çiftlik*-hands.[223]

The shrinking grain output allowed producers to devise ways to complement the leading winter crops, that is, wheat and barley, in the most likely case that their harvest would fall short of expectations and needs. Summer crops were of no use in making up for the likely deficit, because they were commercial crops. Legumes hence reentered crop rotation on a grand scale. As spring crops, they increased the amount of food available to peasant households, and shielded them from the looming shortage in winter crop harvest. In this, they performed a function similar to the one performed by maize and potatoes elsewhere. When the winter harvest failed, this group of crops kept the producers alive; if that was not the case, then they increased the amount of grain available for marketing since the cultivators marketed high-priced cereals and consumed cheaper ones. The countryside was therefore sown with a wider range of crops than it had been in the sixteenth century. Wheat, of course, was always sown, but in decreasing quantities. Yet, bread grains resumed their expansion in the course of the nineteenth century, this time barley, oats, and rye slowly yielding ground to wheat.[224]

Rice and maize, as we discuss in chapter 5, enjoyed widespread cultivation. In the Levant, rice was mentioned more often in the records of the fifteenth and sixteenth centuries than later, but that was because the Ottoman state tried to organize its cultivation by assigning duties and setting regulations to safeguard its continued presence. It was not mentioned as prominently or as frequently later, not because it lost stature, but because its production was widespread enough not to warrant organization from above. Moreover, growing involvement by nomadic groups made labor easily available, allowing the state to dispense with its duties of supervision.[225] Its cultivation increased in most parts of the Mediterranean, stretching from Valencia via Sicily to Antioch. What is of significance is that rice did march on to northerly locations within the Mediterranean belt. Rice fields in the Italian peninsula shifted definitely from their southerly locations and Sicily to the north, Lombardy.[226]

Finally, in terms of landholding patterns, these changes had two implications. First, where cereal production went unabated under the adverse conditions of the exodus of commercial agriculture and the advent of the Little Ice Age, lordly dominion and coerced labor reigned supreme on the shores of the Mediterranean as attested to by the appearance of rural estates, especially in the Balkans and the Aegean zone.[227] Whether the lands kept under cultivation were devoted to cash crops or not, or whether they exported most of their produce or not, did not matter in terms of the organization of production. Keeping croplands productive required the presence of *çiftlik*-like organization. When this proved impossible, then it was easy to shift to livestock farming, given the abundance of winter grazing fields, which we expand on in the following chapters. Second, the seventeenth-century crisis profited large landholders more than smallholders; the former were better able to withstand the fall in prices, given the size of their holdings. Note should also be taken of the fact that climatic variability, which was an inseparable component of the Little Ice Age, also favored big landholders, for they had a kind of internal insurance against bad harvests by the variety of areas they cultivated, again as we investigate in the following chapters.[228]

The easy availability of raw materials for the manufacturing industry, now dispersed across space, and the supplanting of the rich trades by the "traditional" crops of the Inner Sea, readily available along its shorelines, were accompanied by a change in the nature of economic flows crisscrossing the basin as well as in the dramatis personae who were involved in the commercial operations in the basin. New merchants took advantage of the extended and unregulated commercial space that came into being as a result of the diminishment in the activities of the mighty merchants of the heyday of the Mediterranean, the theme of the following section.

The "Conquering" Local Merchant

As discussed above, the shift away from the Mediterranean to the North Sea inaugurated and consolidated a set of transformations that reshaped the industrial and agrarian edifice of the Inner Sea. Correlatively, commercial networks that interlinked the basin underwent modification, to the detriment of the merchants of the twin city-states, whose august custodianship of the basin was undermined over time in a piecemeal fashion. Previously, the Genoese *nobili vecchi,* in close cooperation with the Castlian throne, under Philip II in particular, had carried the theater of their operations from the Iberian Mediterranean into the Atlantic seaboard. And Venetian merchants, in conjunction with the Sublime Porte, had helped restore the Indian Ocean–Mediterranean nexus— and the Levant trade. The wedding of the operations of the former rival city-states after 1560 entailed two sets of commercial rearrangements. The withdrawal of the Genoese men of money from trade in favor of royal finance prepared the ground for the invasion of the Inner Sea by the northern merchants.[229] Genoese control of American silver concurrently paved the way for the return of the spice trade to its olden route, providing relief to Venetian merchants by resuscitating the Levant trade. Nonetheless, the revival was a mixed blessing for the Serene Republic because by the end of the sixteenth century, the northern merchants had, on account of the withdrawal of their Genoese counterparts from trade, easily made their way, via Livorno, into the Levant— İzmir, Aleppo, and Istanbul.[230] Given that the networks put in place by Venetian merchants were extensive and not easily penetrable, the Protestant merchants did not pose a veritable threat to the Signoria in the Mediterranean spice trade. The marketing of woolen textiles was a different matter, though. Any increase in the share of northern merchants was to the detriment of Venetian fine woolen exports, which had by then become a significant component of the Republic's commercial dealings with the Sublime Porte.[231]

Certainly, the new standing attained by the northern merchants was not confined to the Levant alone. The control the famed merchants of Burgos exercised over the Spanish wool market, which was already weakened by the diversion of exports through Iberia's southeastern coast to the Italian city-states, was further enfeebled by the growing commercial sway of the northern merchants. The golden days of the city of Burgos were thus brought to a rude end, leaving its patricians in the unenviable position of having to continually bargain with the throne about their privileges in the wool trade.[232] The Dutch merchants took control of France's shoreline as well from the North Sea through the Channel to the Atlantic in the seventeenth century.[233] Not coincidentally, the *coup de grâce* to the Serenissima's Levant trade was also delivered by Dutch merchants—by the monopoly they established in the Indian Ocean, to be exact. Once the spice trade was dealt a fatal blow, the Venetian merchants' majestic position in the Levant was finally opened to contestation by the northern powers who were expeditiously conferred "capitulations" by the Sublime Porte:

the British, Dutch, and French merchants solidified their Levantine venture by signing capitulations in 1569, 1581, 1612, and later.[234] The accompanying commercial privileges gave merchants of the north extraterritorial rights—similar in nature to those invested in the merchants of the Signoria—in their commercial undertakings in the Ottoman lands.

In the 1620s, there was a reversal in the fortunes of the both city-states. The *nobili vecchi*'s hold over the finances of Castile began to slip away. The takeover by Portuguese financiers of the position of bankers to the throne, the transformation of Amsterdam into a colossal warehouse, not to mention the periodic bankruptcies suffered by the Castilian throne, worked in concert to loosen the control the Genoese men of money had established over the world-economy's bullion flows. This brought the Serene Republic's dominion over the rich trades to an end. Contributing to the sea change in the Mediterranean was the turn by the Florentines of Lyon to royal finance, after the fairs of the city were replaced by the Piacenza fairs. Establishing themselves in Paris, the members of haute finance contributed to the tilting of the balance at the expense of the Mediterranean.[235]

The Ottoman throne was more than happy to cut its umbilical cord with the Signoria as the Dutch, French, and British merchants progressively established their presence in the Inner Sea, however unsteady this presence turned out to be in the medium run. By fostering competition between powers vying over the Levant trade, the Sublime Porte, thanks to their brewing rivalry, was provided with greater leverage in its dealings in economic matters than before.[236] So were Philip III and the count-duke of Olivares who found out that despite the chronic crises taxing the finances of empire and mounting Genoese losses, there were lenders willing and ready to extend fresh loans to Castile. Nevertheless, the loss of their status of *primii inter pares* did not mean an end to the Venetian and Genoese ventures. After all, the banker-financiers of St. George went on to subscribe to the *asientos,* even though the amounts involved got relatively thinner over time than those extended in the sixteenth century. The Genoese *nobili vecchi* may have been supplanted by Manuel de Paz, Duarte Fernandez, and Jorge de Paz of Portugal, but they carried on their dealings in the empire: silver from Seville and, later, Cádiz continued to find its way to Genoa.[237] The *nobili vecchi* even financed the silk (and woolen) industry in their natal city in order to procure the much-valued white metal after American silver escaped its monopolistic hold.[238] Yet again, the "lure of the sea" beckoned the Genoese from the 1640s.[239] Venetian merchants as well went on to deal in the Mediterranean trade, however diminished.

Concomitant with the demise of the commercial stranglehold of the city-states over maritime trade was the transformation in the composition of the Levant trade in the 1650s as a result of the reorganization of world-economic flows under the aegis of Amsterdam. No longer was it dominated by the spice and pepper trades, which thereafter made up a modest proportion of the Levant trade. Cargoes of coffee, silk, wool, leather, and raw and spun cotton traversed the Inner Sea.[240] Accompanying this revision in the makeup of the Levant trade was an attendant change in the composition of merchants who oversaw and conducted it. The dwindling in the volume of merchandise that flowed from the Indian Ocean to the Mediterranean was confirmed during the first half of the eighteenth century by the forfeiture by the Sublime Porte of the responsibilities in the organization and supervision of the pilgrimage, the moving feast of the earlier centuries, to the hands of the notables of Damascus.[241]

Naturally the array of merchants partaking in the Levant trade widened as the rich trades and the wealthy merchants of the maritime republics lost their commanding position in it. Nonetheless, however vital this trade was to the northern merchants, it never regained the coherence and density it had possessed under the Italian city-states. The Dutch and British merchants were more active in it until the eighteenth century, and the French merchants thereafter. But the Dutch thaler or lion dollar (*esedi guruş*) did not possess the longevity that the Venetian ducat or the Spanish eight-real piece transmitted by the Genoese (*riyal guruş*) did.[242] Nor did the newcomers command the financial might that their predecessors enjoyed. Even though the French traders built an encompassing network in the Levant by means of bills of exchange in the eighteenth century, they still had to partake of the caravan trade along the Ottoman seaboard to procure the necessary financial resources in order to carry on with their operations.[243] Indicative of the inability of the merchants of Marseille to establish a self-sustaining commercial network was their reliance on the dispatchment of growing quantities of silver coins—Maria Theresa thalers minted in Milan—to keep their presence in the Levant.[244] Just as the northern merchants were trying unsuccessfully to weave economic networks of their own, the structure of economic flows in the Mediterranean was undergoing modification, this time to the benefit of local crops and goods. The combination of these developments attributed a larger share of the region's commercial life to its local merchants. In the eighteenth century, for instance, "the conquering Balkan orthodox merchant" was in charge of most of the commercial transactions in the eastern Mediterranean, and the renaissance enjoyed by Barcelona was owed to the local merchants, to those from Catalonia and the Levante. The travails of the Serene Republic in the Levant enabled the trading houses of Marseille to expand east without difficulty.[245]

The sidelining of the merchants of the sibling city-states did not start in the 1650s. Nor was it a smoothly unfolding process. It began in the opening decades of the sixteenth century, and progressed by fits and starts. By the eighteenth century local merchants had solidly entrenched themselves in the commercial networks of the basin. This was, to be sure, a relatively easy achievement in the wake of the erosion of the rich trades. Nevertheless, when British trading houses made their appearance in the region in the nineteenth century, they found it difficult to penetrate into the trading world of the Mediterranean and were forced to take local merchants in their employ. They conducted their business through local intermediaries.[246] And merchants, or *négociants*, from Marseille operated in the Levant through commission agents.[247]

A rise in the number of new merchant groups initially occurred in the first half of the sixteenth century when, during the temporary migration of the spice trade, merchants from the Ottoman lands made their appearance in the Adriatic, reversing the long process of Venetian expansion into the eastern Mediterranean.[248] Instrumental in the westerly movement of Ottoman merchants was the drying-up of the rich trades in the Mediterranean following the circumnavigation of Africa. The disruption of the spice trade proved to be temporary at first; it was limited in duration to the first three or four decades of the sixteenth century.[249] However brief the hiatus was, the return of the rich trades to the Inner Sea did not, as argued above, signal the resumption of the trade patterns of the fifteenth century. Rather, the conduits through which the spices found their way to the Mediterranean multiplied in number, and some of the flows changed venue, from maritime to terrestrial. After all, revisions in the flow of spices reflected structural changes that were taking place within the heart of the world-economy. Antwerp's rise and fall may have been swift, but its place was speedily taken over by Amsterdam. Not only did the imposing economic presence of the United Provinces and the North Sea inevitably redirect economic flows toward Amsterdam, but the Baltic region, with its newfound wealth, could afford to consume greater quantities of meat and, hence, spice.[250] In terms of the supply of spices, the region that stretched from Amsterdam to Danzig was initially dependent on Venice or Antwerp, but its growing prosperity changed the situation. The wealth commanded by the Dutch burgers and the Polish and Hungarian nobility served as a field of force for merchants looking for routes away from the prying eyes of Venetian merchants. Yet this remained a challenging task in the latter half of the sixteenth century, given that the pepper and spice trades plied the waters of the Mediterranean precisely because the shores of the basin, from Seville and Genoa to Cairo, became points of disembarkation for American silver.[251] It was neither the inability of the

Portuguese fleets to hold on to their empire in the Indian Ocean nor even the persistence and ingenuity of the Venetian traders to recoup their share of the Levantine spice trade, but the arrival of American treasure that paved the way for the return of the spice trade to its erstwhile itinerary.

Nonetheless, however forceful the invasion of the Mediterranean by Potosí silver and Spanish *grosso* was in inducing the resumption of the spice trade from the 1550s, it was not powerful enough to reconstitute the world of the late fifteenth century when the Signoria's control over the Levant trade was at its zenith. The recentering of the staple and rich trades along the shores of the North Sea generated a new string of port-cities, from Cádiz and Livorno to İzmir, which served as the outposts and headquarters of the northern merchants in the Inner Sea in their attempts to partake of its rich trades as well as to tap into its resources.[252] It is not surprising that the rise of Amsterdam, with its massive agglomerative power, created similar scenarios at both ends of the basin. In the case of the Iberian peninsula, an abrupt fall in woolen manufactures in Venice, Florence, and Milan in the 1620s meant that Iberian wool exports, which briefly served the Italian city-states, were largely geared to satisfy the needs of northern industries during the seventeenth-century crisis. Dutch merchants managed to penetrate the peninsula's markets and denied local merchants the chance to centralize and control the export of wool on their own terms. Iberia's prime wool market consequently remained "decentralized" for the most part during the seventeenth and eighteenth centuries.[253] The northerly reconfiguration of the economic networks worked to the disadvantage of the Mediterranean, for close to two-thirds of the merchants who operated out of Cádiz, the peninsula's highway to the Americas, were from the north—the northern provinces of the empire—and the denizens of the city.[254] And the Dutch merchants were in charge of the Atlantic coast of France.[255]

If on the Atlantic seaboard the commanding presence of Amsterdam and the United Provinces rendered the Habsburg empire more dependent on the Dutch and, later, the French merchants, in the eastern Mediterranean, where the Venetian presence was deeply rooted, a different scenario unfolded. The changing balance between the North Sea and the Mediterranean inaugurated a wide array of transformations that altered former patterns of trade. By the end of the sixteenth century, the maritime route controlled by Venetian merchants had become one among many.[256] Widely traveled as well were the overland routes that crossed the Fertile Crescent, Anatolia, and the northern Black Sea and reached the westerly centers of consumption and distribution of spices—via Akkerman, Lwów, Kilia and Brašov, Bursa, and Istanbul. Cairo and Alexandria, of course, were still significant nodes of transmission in the Levant trade, and

Venice still had the lion's share. But the appearance of Hormuz as a distribution center of note embarked upon a process of dispersion in the flow of spices: the port city of Bandar Abbas was where the Portuguese eventually unloaded their cargoes of spices, which were then laden onto camels to traverse the Syrian desert to reach Aleppo and therefrom Istanbul. This lengthy voyage was essential to the distribution of spices in Ottoman lands and beyond.[257]

The return of the spice trade in the latter half of the sixteenth century did not restore the old order. The rerouting of some of the trade via Hormuz to Aleppo worked to the detriment of the ports where previously spices had been loaded onto Venetian ships.[258] Aleppo and its port Tripoli captured the pride of place from the second half of the century to the 1620s, when they served as the spice trade's central emporium from which Venetian and other factories as well as Ottoman sea-ports acquired their spices and drugs. Alexandria, which played the main host to the Venetian merchants throughout the fifteenth century, never recovered its former standing. When the spice trade was revived from the 1550s, the Cairo-Alexandria and the Damascus-Beirut routes recaptured some of the trade, but they were overshadowed by the rise of Aleppo and overland routes. A greater number of cities and towns presided over the revitalized overland routes and partook of the riches of the spice trade.[259]

In fine, the overland trade had already started to work to the detriment of Venice's north-south transit trade. The momentum was on the side of terrestrial trade, not only because the higher profile attained by the merchants of the north, the new Christians of Amsterdam, and the *nouveaux riches* of the Baltic grain trade, the magnates of Hungary and the lords of Bohemia and Poland, animated northwestern-bound treks that cut through the Ottoman dominions.[260] More important, the distribution of productive activities in rural (southern) Germany, by increasing the disposable incomes of the inhabitants of these regions, encouraged the consumption of spices. The score of fairs held in the region at the turn of the seventeenth century for the resale of spices did not disappear with the emergence of Amsterdam as the new node of redistribution. Fairs continued to prosper as demand from across the Danube remained strong, and increasingly, was not confined to spices.[261] Amsterdam, as the most significant clearing-house and warehouse of the world-economy, contributed immensely to the thriving of overland routes flowing into—and out of—it. The French merchants who conducted business in Lyon fairs eventually turned into commission agents for the Dutch.[262]

A proliferation in the number of routes that transmitted the western-bound spices and, with them, Indian textiles that were primarily marketed in Ottoman cities, was not the only factor that expanded the realm of operations of the local merchants.[263] The generation of this new mercantile group was abetted by the recentering in the 1550–1650 period of the empires that vaulted over the Mediterranean. After all, the process of economic restructuring engendered a new set of actors who were given the task of catering to the fast-growing and lucrative markets of the new imperial capitals. The capture and rebuilding of Constantinople and the economic *gravitas* it acquired concomitant with the northerly shift of the economic center of the empire, away from Cairo and the Mediterranean to the Aegean and İzmir, gave birth to a new group of merchants who took on the responsibility of provisioning this colossal capital city.[264] In the process, the opposing shores of the Aegean steadily turned into the economic heart of the empire as the Levant trade was transferred northward, "possibly in order to bring it nearer the heart of the empire."[265] The transfer of the capital of the Habsburg empire from Valladolid to Madrid also prompted a new group of traders to cater to the new city, at the expense of those who were in charge of its commerce previously.[266] Lyon, too, whose fortunes the French state tried to bolster for most of the fifteenth and sixteenth centuries, had to bow down to Paris, which was closer to the waves of the Atlantic: Paris flourished on account of "financial capitalism" befitting an imperial capital.[267]

Thus, as Amsterdam started to set the world-economic tempo from the seventeenth century, the economic geography of the Mediterranean was transformed accordingly. Venetian and Florentine businessmen expanded their field of operation to the north across the Brenner Pass to keep up with the increasingly northward-bound flow of economic goods and credit. Understandably, the port-cities that commanded a bigger share of the silver and spice trades were now located on or close to the banks of the Atlantic. All the polities with coastlines on the Mediterranean saw a shift in, and an eventual realignment of, their centers of economic gravity, more in line with the emergence of a Baltic and North Sea–centered world-economy. When the dust settled after the Ottoman-Persian wars, it became apparent that the overland route that surfaced in the latter half of the sixteenth century could sustain voluminous trade as well as underwrite the rise of İzmir in the seventeenth century. Overland trade via Aleppo and central Anatolia, the terminus of which was İzmir, formed the backbone of merchandise flows during the seventeenth and eighteenth centuries.[268] The significance of this should not be underestimated, because the trends that were in place at the end of the sixteenth century were there to stay. After all, the crucial factor, northern dominance, only got stronger, and not weaker. Conversely, Dubrovnik, which had previously made its living from a variety of different shipping trades, some local, others long-distance (among them the Black Sea trade), now abandoned these activities and withdrew into the Adriatic.[269] It was not that the hides and wool of the Balkans had stopped coming into the city, but now they came from the great center at Novi-Bazar, by overland routes, which had replaced the sea routes. Spoleto, likewise, served as an outpost for Venice, because it had excellent access to overland routes delving into the Balkan peninsula.[270] Venetian merchants extended their dominion across the Alps to capture the overland trade. Granted, when the Thirty Years' War turned southern Germany into a war-zone, Venice's overland trade suffered, but did not wane.[271]

Italian merchants were actively engaged in trade in Leipzig, Nuremberg, Frankfurt, Amsterdam, and Hamburg, and visited the cities and fairs of Poland (Lublin, Sandomir, Thorun, Poznan, Gnienzo, Gdansk, and Lwów). Italian goods continued to reach Vienna and particularly Poland, through the busy relay points, such as Cracow. In the early seventeenth century, despite the spectacular losses that big Florentine and Venetian firms suffered, it was north of the Alps, in an area bounded by the routes fanning out from Venice, that "Italian" merchants found compensation for the hard times brought by the economic downturn after 1620.[272] They were "often hated, always suspect, but indispensable," but their growing presence in lands east and north of the Adriatic provided a degree of reprieve from their fall from positions of prominence.[273] In a twist of fate, the queen of the Adriatic, who had reigned over and benefited from maritime trade for centuries on end, was eventually retreating inland and handling overland trade, thereby leaving the Mediterranean open to piracy and the competing seagoing merchants.

Even though the revisions in maritime trade prompted by the spread of "country" (carrying) trade that northern maritime powers practiced, primarily to finance their purchases, changed the structures of the trading world of the Levant, developments in overland trade proved to be more vexing to the established port-cities of the long sixteenth century. Ragusa's demise as a trader in the eighteenth century was analogous to that of Venice, for it too suffered from competition at sea with western caravaneers, but perhaps even more from competition on land with the new network of indigenous Balkan traders and, by extension, trade fairs that rose in the interior during the eighteenth century.[274] The decades following the Habsburg capitulations of 1666 and the expansion of trade with the Ottoman south triggered the settlement of Serbian, Armenian, and Greek communities in commercial centers such as Buda, Vienna, Pressburg, and Leipzig.[275] The development of overland trade and the formation of an expansive trade network expanded the compass of the *verlag* system, which furthered possibilities of rural manufacturing in far-flung parts of the Ottoman realm. It enhanced a trend that was already in operation. So did the increase in the number of beasts of burden; it stimulated demand for animal feed, which was already kindled by the needs of the livestock industry.

Surely, Mediterranean commerce had lost the vibrancy it enjoyed in the sixteenth century. The division of labor underpinning the circulation of goods among its shores had disintegrated almost inexorably, and understandably so. Surely, the return of tree crops opened up new avenues of commercial expansion. And the widespread nature of rural manufacturing that relied on local sources of supply reduced the pace and volume of commerce interweaving its shores. Within the new order, goods leaving Ottoman lands followed overland treks, yet goods flowing into it were more seaborne than not.[276] In fact, even at its nadir, close to 50 tons of silver flowed through the Mediterranean each year to the trading world of Asia. This sum amounted to 50 percent of the total bullion flow to Asia in the seventeenth century, and 25 percent during the eighteenth.[277]

Additionally, the migration of the lucrative crops of the Mediterranean's high age was followed by the return of tree crops—the vine and the olive—as well as the spread of cultivation of industrial crops, which we discussed above in some detail. These agrarian developments complemented the transformations in the field of manufacturing in that they contributed to an extension of commercial networks. Olive oil in Naples, Syria, and Tunisia, viticulture in Provence and Andalusia, currants in Patras, dried fruit in western Anatolia, or silk in Calabria and Mount Lebanon, cotton in Syria, Egypt, and Rumelia not only furnished the producers in the basin with a wider mix of crops, but also contributed to the deepening of commercialization. In catering to these markets which, unlike the spice trade, were not under the monopoly of the Serene Republic, Venetian merchants were joined in and at times outmaneuvered by a score of merchants: Florentine, Spanish, Armenian, Jewish, and Greek, to name a few. Unlike the Genoese, who were equally adept in reaping handsome profits from extending loans to sovereigns as in conducting commercial operations, the Serene Republic's commercial dealings were its life-blood, and here it encountered fierce competition in these expanding markets. In any event, they managed to establish their presence in the prestigious fairs of the seventeenth and eighteenth centuries. By relocating closer to, or on the nodes of, the commercial network that stretched from and through the Levant into Amsterdam, mostly by land, and mostly by routes north of Via Egnatia, they created new avenues of business, as their presence in maritime trade came under pressure because of the contraction in the Mediterranean trade as well as the arrival of a score of new competitors.

Put briefly, two developments helped the reigning city-states to retain their hold over the Mediterranean. The first was the ability of the twin cities to change their activities in accordance with the shift in the economic center of gravity of the Ottoman and Habsburg empires, and the second, the creation of new fields of economic activity as the basin started to recover with the return of its crop trilogy; the resurfacing of manufacturing; and the growing engagement of local merchants, all of which helped the commercial dealings not only of the Venetian but this time also the Genoese merchants. In the case of Venice, its domination of the Mediterranean faded to become a faint memory, yet the basin's maritime trade failed to flourish under the northern powers for two reasons. First, the region had ceased to be the center of the rich trades. As such, the presence of northern powers did not generate a new dynamism in it. Second, the rise of overland trade via the Ottoman empire, with the fairs in Leipzig, Nuremberg, and Frankfurt-am-Main forming its nodes of redistribution, where Venetian merchants still held court, had eclipsed maritime trade. It was not by providence that the fairs in the Balkans burgeoned during the seventeenth and eighteenth centuries, populated by merchants of all sizes and "nations."[278] These fairs came to blanket the region, because the networks put in place were part and parcel of a larger network extending into Amsterdam via the fairs of Leipzig and Nuremberg. As the Ottoman emporium came to be largely drained via İzmir and the Balkans, the Venetian merchants' locus operandi changed accordingly.[279] Unlike in the sixteenth century, neither power had the un-disputed command they previously had over these world-conquering empires, but their umbilical cord was not cut off either. The operations of the city-states were solidly anchored in their respective spheres, the boundaries of which were largely drawn after the war of Chioggia. Genoa remained within the opera-tional realm of the Spanish empire, albeit indirectly, even when it lent money to the French state. It, too, like Venice, shifted its operational terrain in perfect concert with the shift in the economic heart of the Atlantic enterprise, from Iberia to France. Politically, this shift was embodied in the success of Bourbon rule in the eighteenth century. Even when the city's trade revived, the primary destinations their merchants attended were located in the western Mediterra-nean. Venice, on the other hand, remained a mainstay of the eastern Mediterra-nean, of the Ottoman empire, albeit indirectly.

The Genoese never forfeited the strategy of tapping into the silver and gold resources of the Habsburg empire that were transmitted through Seville and Cádiz. The task of provisioning and supplying Spanish America was an enter-prise that, from the beginning, Castile could not, by any conceivable means, handle alone and had to resort to the services of others. This stimulated not only the Dutch economic activities in the north but also that of the maritime provinces of Provence and Catalonia in the south, and opened up new avenues of expansion for the Genoese capital. Toward the end of the seventeenth century, the massive capital resources held by the Genoese were still placed at the employ of—that is, extended as loans to—the Castilian throne, albeit dwindling in volume, but the share of the Papal States, the city of St. Mark as well as the French state had assumed the lead as principal borrowers.[280] As the French involvement in the exploitation of Spanish America, which was considerable from the beginning, gained pace during the eighteenth century with the consolidation of Bourbon rule, it turned into the biggest borrower, outdoing both the Papal States and Venice. Particularly in the eighteenth century, when the task was essentially shifted onto the shoulders of, or was grabbed by, the French state, loans by the Genoese were extended to it in tune with this tectonic change. The growing competition in the Atlantic and the Mediterranean could thus extricate neither the Genoese financiers nor the Venetian merchants from their principal theaters of operation. One could even argue that as the economic hearts of both empires were relocated starting from the seventeenth century, one along the shores of the Aegean and the other in France, the city-states followed that transition, allowing it to make the transition smoothly. How the empires relocated their economic center of gravity determined how the city-states operated. Essentially, the dynamics that interlinked the fortunes of the empires and city-states from early on, which we examined in chapter 1, remained in place, albeit in a revamped form.

As outlined in the preceding pages, the three processes that gained momentum during the age of the Genoese and accelerated further during the 1650–1850 period also engendered an environment that compensated for the departure of the rich trades from the *Mare Nostrum*. As we chronicle in depth in the fourth chapter, the egress of commercial agriculture (grain and sugar) was followed by the rise of the tree crops of the region, not lucrative because available to all; the spread of rural manufacturing raised the need for industrial crops, such as linen and wool, again easily available to all. Surely, both these developments offered greater opportunities for local merchants, as demonstrated by the rise of the "orthodox" Balkan merchants or by the increase in the share of local merchants in coastal trade. The absence of mighty merchant houses from the Inner Sea and, correlatively, the greater participation of local merchants were not strong enough to override the presence of the city-states in the region: the Venetian ducat was as valuable in Ottoman lands in the eighteenth century as it was in the sixteenth, and the Genoese men of money went on to serve, without serious competition, as financiers of the states of the region. Both played crucial parts in olive oil, grain, cotton, and silk trade.[281] This did not mean that the city-states managed to retain their former power, nor did it mean that the part they played should be overemphasized. But the fact that they were not rudely routed gave them the luxury to make the transition to their lesser, diminished position in a leisurely fashion. The final act of the city-states, now deprived of the imperial connection that they lacked and profited from in the fifteenth and sixteenth centuries, was to create a territorial state of which they were a part. Braudel asserts that the forging of the kingdom of Italy in 1861 was owed to the Genoese.

To recapitulate, this chapter, following up on the framework that was laid out in the first chapter, traced the manifestations of the waning of the Mediterranean. The general trends that encompassed the Mediterranean held sway in Ottoman lands: as elsewhere, the contraction in arable production was complemented by the rise of tree crops and sheep-farming. Ruralization of production helped disseminate manufacturing into rural areas, and kindled the cultivation of industrial crops. The development of overland trade increased demand for lesser cereals as animal feed, kept the putting-out system in functioning order, and encouraged the transport of goods that were not as bulky and low-value as grain: the products of the livestock industry (wool, leather, and hides) found wider circulation within and beyond Ottoman lands. In tandem, these developments took a heavy toll on cereal agriculture by encouraging animal-farming, industrial crops, and lesser cereals—as happened in other corners of the Mediterranean world.

That the plains and cereal agriculture suffered serious losses in the seventeenth and eighteenth centuries goes without saying. Still, this does not explain how the basin's agricultural landscape was reconstituted, this time at the expense of the arable. If tree-crop plantations multiplied and summer grazing ground for sheep-raising mushroomed on the hills of the Mediterranean, a transformation of large scale that necessitated the relocation of the economic heart of the region in upper altitudes, the movement has to be explained with reference to the ability of its rural infrastructure to be able to accommodate such encompassing changes. This is what the second part tackles.

PART II

OF MALARIAL PLAINS AND ARBOREAL HILLS

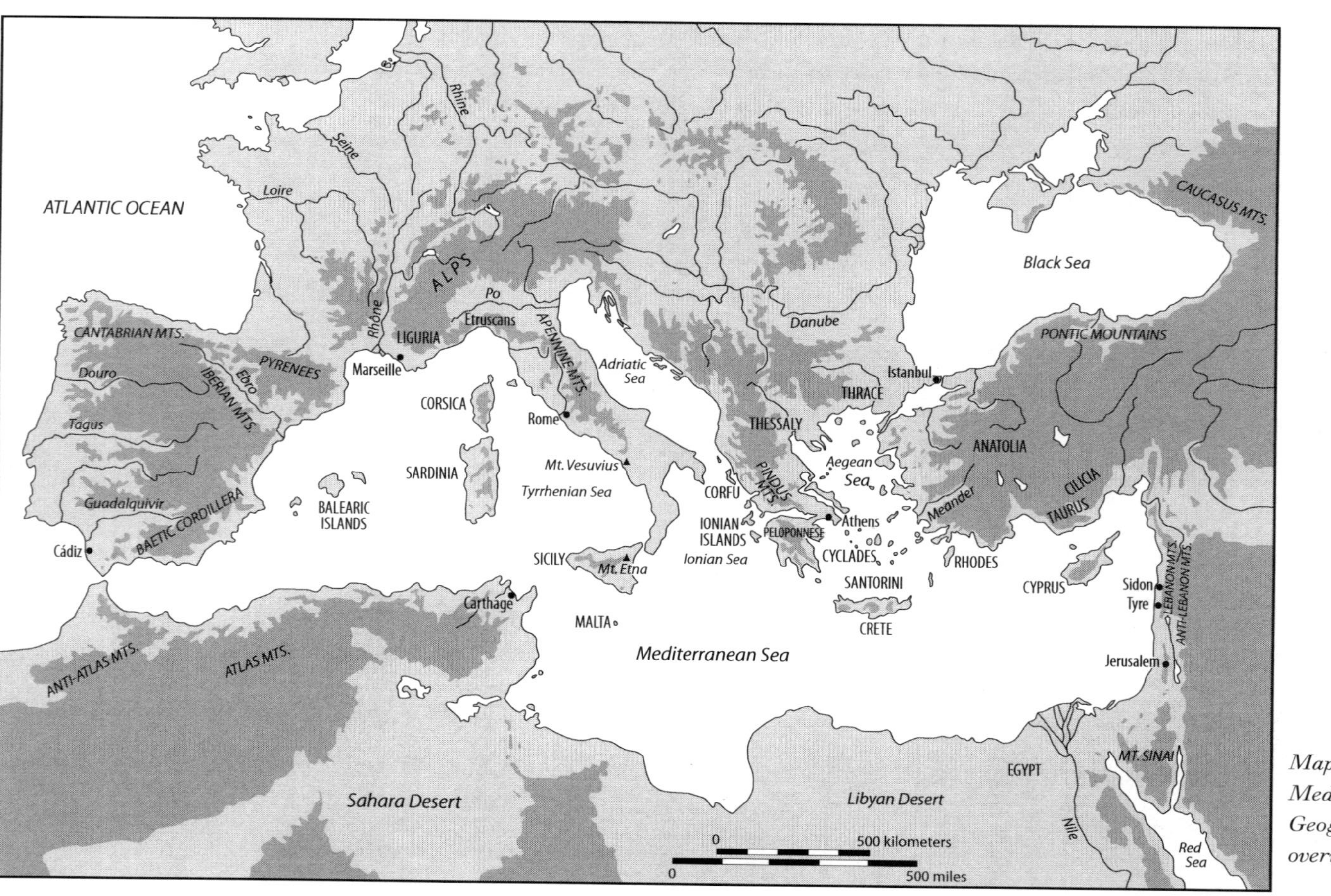

Map 2. The Mediterranean: Geographical overview

CHAPTER FOUR

Reversal in the Fortunes of the Plains

The second sixteenth century commenced in the 1560s. By then, the relatively short-lived medieval warm period had come to an end and the Little Ice Age had promptly resumed after the brief hiatus of the *beau seizième*. The consequences of the resumption of the Little Ice Age had become distressingly apparent in the closing decades of the century.[1] In 1590, the duke of Tuscany was forced to travel to Danzig in search of grain, since the alluvial plains of Maremma, the granary of Tuscany, had been flooded. The following year was marked by the arrival in the Mediterranean of Baltic grain, a sea change in the provisioning of the Italian city-states and in the nature of the Levant trade. In fact, grand dukes of Tuscany, from Cosimo on, had tried to subdue and divert the waters that frequently submerged the Maremmas—obviously without much success.[2] The inundation of Florence's breadbasket in 1590 was neither the first nor the last. A series of floods by the river Arno had already caused havoc in the region: the "wrath of God" flood that struck the city of St. Giovanni in 1547 without warning was followed by the floods of 1557 and 1589.[3] What is more, the situation did not get any better in the following centuries. Indeed, between the turn of the sixteenth and the turn of nineteenth centuries, heightened fluvial activity in the Arno valley deposited large volumes of sediment at the mouth of the river at the rate of 5.15 million cubic meters per annum. In the latter half of the twentieth century, by contrast, the volume of transported sediment was much diminished and had tapered off to 1.91 million cubic meters per year.[4] Stated differently, increased wetness, unseasonable rains, floods, soil erosion, and silting that distinguished the Little Ice Age together led to a reversion of the formerly cultivated lands to nature, and hence reinforced the contraction in commercial agriculture, prompted by the rise of *pax Neerlandica*, by forcing producers to vacate the lowlands—the signposts of the Mediterranean landscape in its autumn.

Predictably, serial flooding was not confined to Tuscany. The Po valley was "interrupted for long tracts between one city and another by great extents of heath, marshland, and waste, not to speak of woods and fens."[5] In the sixteenth century, the fields were flooded around Mantova and Brescia. Nor was the fluvial activity limited to the Italian peninsula. From the riparian valleys drained by the river Meander in the Aegean to the banks of the Guadalquivir in Iberia, most quarters of the Mediterranean were repeatedly visited by inundations in the latter half of the sixteenth century.[6] Indeed one of the basic features of the Little Ice Age was an episodic increase in climatic variability—not only in precipitation and deluge but also in drought.[7] Often the change in climatic fortunes prompted frequent and sporadic flooding in the summer and autumn months: on the lowermost Rhône, for example, summer flooding originated in the Swiss and French Alps, and autumn inundation was brought about by heavy rains in the Cévennes and Durance. The Loire was capable of submerging the whole valley between Roanne and Orléans under water.[8] In some alpine communities, sounding the bells at times of "bad weather" became a specialized calling in the seventeenth century, worthy of regular pay.[9] The piedmont regions of the basin's mountainous range suffered the most from intermittent inundations, from the plain of Mitidja in the south to the Pontine marshes in the north. More often than not, the severity of the inundations was magnified by, or resulted from, the denudation of the region's mountainsides.

The deforestation wave of the first sixteenth century was largely occasioned by the fast pace of land reclamation on the one hand, and urbanization on the other. Growing demand for land and timber that caused the carving up of the woodlands played a critical part in accelerating deforestation in the relatively warm decades of the 1450–1550 period.[10] The hills and sloping surfaces of the Mediterranean—barely able to absorb runaway water even when generously blanketed with trees because of the calcareous nature of its soil—were in consequence surcharged with water. What is worse, the extension of cultivation to unsuitable slopes aggravated the environmental consequences of heavy rains.[11] Increased hydrological activity precipitated soil erosion and gullying, altered river channels, and facilitated the formation of flood-plains and deltas at the lower altitudes owing to the flatness of coastal plains. Increased precipitation and the denudation of the mountains exacted a heavy toll on the low-lying plains of the basin by rendering these landscapes vulnerable to the assaults of run-off water and frequent flooding. In the process, it turned some of the previously cultivated terrain into badlands such as the *calanche* of the Italian peninsula and the *malas tierras* of the Iberian peninsula, and increased the extent of waste lands in the Anatolian peninsula and the Levant.[12]

In certain parts of the basin, the extent of deforestation became manifest in the latter half of the sixteenth century judging by the difficulties encountered by shipbuilders in procurement of timber.[13] A "wood famine" brought about by the disappearance of woodlands forced the navies and shipbuilders, who were in search of new supplies of timber, to travel farther and farther away from the Mediterranean. Venetian shipbuilders, for instance, given the size of the city's naval and commercial fleets, surely figured prominently among those who led the pursuit for fresh supplies of oak following the depletion of resources in the Veneto and the Adriatic.[14] The Spanish shipbuilding industry, too, could no longer rely on Castilian forests alone in the latter half of the sixteenth century; instead, it started mining the sylvan resources of the peninsula's northern latitudes, as attested to by the copiousness of royal decrees issued by Philip II to ensure the protection of timber supplies in the Cantabrian and Basque provinces of the empire.[15] This was hardly surprising, for in the span of four centuries following the establishment of the Mesta, much of the northern Meseta and Extramadura had been "cleared" of forests.[16] Later, access to the verdant sylvans of the Americas facilitated the replenishment of the kingdom's stocks of timber, which were then used for Atlantic-bound vessels: the launching of shipbuilding in Havana and Cartagena further eased the pressure exerted by the wood famine that struck the Mediterranean. In the Levant, the Ottoman navy, unlike the Mamluk fleets, which made use of timber imported from the Gulf of Antalya (Adalia), ordinarily relied on the forests of the Black Sea and the Sea of Marmara rather than of the Mediterranean. Neither Evliyâ Çelebi nor Katip Çelebi mentioned the lumber trade in the mid-seventeenth century. And the navy drew on the forests of Karaman in central Anatolia for its operations in the Red Sea and the Persian Gulf.[17]

In sum, the menace of deforestation loomed large over the Inner Sea in the late sixteenth century, and not solely because of the needs of the shipbuilding industry. Rather, the scarcity of oak timber had become general throughout the basin due to a variety of reasons. Along the North African shores of the western Mediterranean, for example, what occasioned the clearing of woodlands was the flourishing sugar industry, which consumed large amounts of fuel. The attendant shortage of timber hastened, in part, the relocation of sugarcane cultivation to new, insular sites of production where sylvan resources were bounteous and, at times, innocent of human presence.[18] More important, the dizzying pace of population growth between 1450 and 1560 and the accelerating velocity of land reclamation and the parallel industrialization of the city-states that accompanied it eventually inflicted a heavy ecological price on the basin's landscape. Given the rapid growth in the number of souls inhabiting the shores of the Mediterranean, it is only to be expected that the expansion of herding, as in Iberia and Puglia, and the sedentarization of the incoming pastoral populations, as in Anatolia and the Balkans, intensified the pressures exerted on the basin's environmental resources. In the western Mediterranean, the preponderance of herding in the economic life of Iberia unleashed a dynamic that was extremely damaging to the tree cover in the peninsula, as acknowledged by contemporaries themselves. Devouring brush fires set by shepherds to hasten the growth of pasture grasses, especially in light of the phenomenal rise in the size of migrant herds, voracious herbivores, under the Mesta (and in the kingdom of Naples under the Dogana), often led to a thinning of the forest cover and forced the Cortez from the first half of the sixteenth century to repeatedly take measures to reverse the erosion of the tree cover.[19]

In the eastern Mediterranean, the gradual settlement of nomadic populations opened new fields to tillage, at times at the expense of woodlands. Often herding was not replaced by, but went on to complement, the agricultural pursuits of the newly sedentarized populations. This agro-pastoral vocation was more pronounced in certain parts of the Ottoman empire—along the Aegean shores of Anatolia and Thrace, and in the eastern Balkans in particular—than in others.[20] In the eastern Aegean, for instance, where the key commercial nodes of the Venetian and Genoese (and later Dutch) maritime empires—namely, Phocea, Çeşme, İzmir, and Aya Solug (Ephesus)—were located and served as outlets for the Denizli-Aydın-Manisa region (ancient Lydia and Karia), sedentarization of formerly pastoral peoples took place swiftly.[21] The pace of settlement prompted the dual processes of deforestation and soil erosion. Here, the depletion of the sylvan resources did not derive from the intensity of shipbuilding activity. As mentioned above, for the construction of its naval fleets, the Sublime Porte sourced timber from the Balkans and the shores of the Black Sea.[22] Demand generated by the replacement of galleys by galleons in the seventeenth century deepened the navy's dependence on timber from the Pontic mountain range.[23] Instead, deforestation was due to the rapid pace of reclamation that took place from the 1450s, once the turmoil generated in western Anatolia by Timur's relentless advance subsided and the resurgence of local rulers was cut short by their subjugation to the house of Osman. In the Aegean coast, most of the reclamation and settlement activity was concentrated along the banks of two major rivers, the Gediz and the Menderes, which irrigated the region's lush lands. These rivers naturally constituted the ecological backbone of rural settlements, since they boxed in the fertile valleys and the rich river basins of the region.

After a prolonged period of settlement that had lasted for over two centuries, the region started to exhibit signs of ecological exhaustion by the 1570s.[24] In Manisa province, for instance, an increase in the arable lagged dismally behind population growth. While land under tillage expanded at a modest rate of 9 percent from 1531 to 1575, the population multiplied by 39 percent. The situation was similar in Hamid province, where a demographic growth of 81 percent between 1522 and 1568 far outstripped the expansion in the arable, which remained at a modest 22 percent.[25] In Lâzıkıyye/Denizli, those holding a full- or half-farmstead decreased significantly.[26] With the number of hands and mouths multiplying at a pace incommensurate with the pace of reclamation, the assarting of trees to prepare the ground for the sowing of crops picked up pace, leading to denudations of the forests and, by extension, to frequent flooding. The heightened fluvial activity created havoc in the collection of taxes because rural producers, after they were visited by big floods, were forced to periodically relocate their fields away from the water-sodden river banks, "forcing tax collectors to draw and redraw administrative boundaries."[27] Given the aquatic nature of the flood-plains, it was understandable that those who migrated to the region later were insistent on taking up residence on the hillsides and on altitudes closer to the "oak-growing range" than the malarious lowlands.[28] By all accounts streams and river plains were not desirable locations for the pastoral peoples for settling down, since the rivers "could dry up in the long summer drought and often flooded in rainy season."[29] In passing, it needs to be mentioned that the aquatic nature of the new landscape was not without its rewards. In both Manisa and Aydın, the cultivation of aquatic crops and rice paddies in particular expanded their ambit to such a degree that it was said that rice would yield most generously regardless of wherever and whenever it was sown.[30] The number of water mills used to dehusk rice multiplied as did that of the buffalo, natural toilers of heavy and sodden lands. They were mostly imported, most probably from Siroz, and in part bred in the region.[31]

But this was only part, and perhaps a not so significant part, of the story. More important, there was a steep rise in the number of nomadic groups, which steadily expanded their sedentary and pastoral activities along the banks of the Menderes and the Gediz rivers.[32] In Aya Solug, for instance, two vineyards that belonged to the Azize Hatun foundation had already been turned into summer pasture by 1583.[33] Naturally, the intensification in the assarting of trees by new settlers occasioned a reduction in the radius of pasturage as a result of the decline in the geographical mobility of nomadic, itinerant populations. The settlement of these sizeable pastoral peoples and the resultant intensification of the grazing needs of their herds inevitably diminished the carrying capacity of land that was set aside for feeding smaller numbers of herds. As the wood and grass cover grew thinner, the pace of erosion and alluviation speeded up. Subsequently, swamps grew at the river deltas opening up to the Aegean Sea. Malaria and fever promptly followed suit in due course. The town of Ephesus / Aya Solug suffered the consequences of the developments along the Menderes River valley since the town was built on land where the river swelled into a great estuary.[34] With the silting-in that followed and the inevitable appearance of malaria, the town's population dropped at a time when the Aegean was slowly but surely emerging as a new commercial zone in concert with the arrival of northern merchants.[35] Karaburun and İzmir, spared of fluvial and riverine misfortunes, burgeoned as port-cities precisely when Ephesus was struck by adversity. The loss of the Genoese entrepôt of Chios to the Ottomans in 1566 provided additional fuel to commercial vibrancy along the Aegean shores of Asia Minor. Despite the economic efflorescence of İzmir and its environs, however, the nearby town of Aya Solug, sapped of its populations and pestered by the ongoing alluviation and visitations of malaria, helplessly witnessed its commercial functions vanish owing to its fever-infested air.[36]

Initially, the building of a *han*, inn, and the holding of a weekly market on the mainland in a site near Aya Solug with the purpose of relocating commercial functions previously performed by the old town failed to successfully transfer these activities to the newly chosen site. The basic infrastructural amenities of a new commercial town were properly put in place only when Öküz Mehmed Pasha, the grand vizier of Ahmed I and the future governor of Aleppo, was given proprietary rights over the site between 1614 and 1616: within the town walls was founded a *han* with shops, including those that marketed high-priced imports, textiles, a coffee house, and a bakery. Later, the port authority and the garrison, both of which were initially located in Aya Solug, were transferred to the new location. This was followed in most likelihood by the relocation of some, if not most, of the remaining inhabitants of the old town. In 1667, when the new settlement had a population of three thousand, the deserted Aya Solug contained merely sixteen tax-paying households.[37] The new town, in its embryonic form, was known under a plethora of names, as is often the case with many newly established settlements. To the merchants who frequented it, it was appropriately known as Scala Nova, Neapolis, or Echelle Neufue. Officially, the new settlement was known as the Kaza of Aniya: its name was derived from the castle and the port of Byzantine Anaia. Yet more meaningfully than the first two, at least for our purposes here, to the locals it was known by the common appellation of Kuşadası, "Island of the Birds," and fortuitously, not of mosquitoes. Befitting its name, the new city was heavily protected against malaria by a stretch of high, wooded mountains: it had, quite simply, a healthy, salubrious climate. The town's promontory jutting out in the sea was designed to protect its harbor from the perils of silting. On the other hand, the old town, now reverted back to nature, as if it were, was taken over by the elements of such landscapes: it was populated by deer, wild boars, and jackals.[38]

Surely, the transportation into the Mediterranean of earth from farms and the mud that accumulated when rivers cut new channels through the ancient flats was hardly a novel development. The ecological pulse of the planet had, and has, always made itself felt in one form or another. In fact, the takeover of the plains of the Inner Sea by malaria itself was part and parcel of a historically established cycle, if not a pattern.[39] It occurred, at least in some parts of the Mediterranean, with some regularity, mostly following periods of intensive reclamation or deforestation or both. It was such an integral part of life on the shores of the Inner Sea that the denizens of the basin were fully acquainted with the marshy environment that came into being as a result and knew how to exploit it fully: the marshlands, the reeds, and the wetlands provided the denizens of the region with new, additional sources of livelihood. So, the formation of pools of stagnant water or wetlands did not necessarily turn the plains into a "marginal" environment that remained unutilized. They were used, among others, for seasonal pasturage or salt-making, as in the case of Ebro valley.[40] The reappearance, albeit with unpredictable regularity, of such a landscape offered different avenues of exploitation than simply putting these plains under the plow, as we discuss in the following chapter.

Given the awesome sway and power of the Little Ice Age, the alluviation and silting-in of the Aegean shores of Anatolia was surely not a singular event. The process of sedimentation exposed the fragility of the ecological situation in the latter half of the sixteenth century not only—and not even primarily—in the outlier zones of the Mediterranean, but equally in its central zone, namely, the septentrional regions of the Italian peninsula. Torcello, for one, an outlying island in the Veneto, was afflicted by a similar malady at the turn of the sixteenth century, for it fell victim to the progressive silting of the northern lagoon by the waters of Brenta. By the second half of the century, flooding and erosion had become endemic, with Constanziaca and Ammiana reverting to marshland and Torcello and Mazzorbo condemned to an undistinguished existence, thanks to the "programmatic neglect" shown by the Venetian authorities, who were very well conversant in hydrological matters.[41] In Valpolicella, in north-central Italy, the advent of erosion had become such a force to be reckoned with that it became one of the principal factors affecting decision-making processes of rural producers when choosing the crops to be planted.[42] The Iberian peninsula or Languedoc was not immune to floods either. In the southeastern coast of Mediterranean Iberia, the period between 1580 and 1630 was marked by repeated floodings, with the year 1617 earning the notoriety of being the "deluge year." The lower Rhône was often visited by periodic inundations in the latter half of the sixteenth century and floods reached their high point in the 1590s. In fact, at the mouth of Bras de Fer, the shore advanced by 50 meters per annum between 1587 and 1711 when erosion in the Rhône basin was much higher than in the preceding and succeeding centuries. The case of Castellò d'Empùries tells a similar story: the town, which stood at the mouth of the Muga in the early eighteenth century, is now 3 kilometers inland, separated from the sea by floodplains, wetlands, and lagoons.[43] It was no surprise, then, that marshy lagoons alone claimed an area perhaps as extensive as 6,500 square kilometers along the shores of the Mediterranean.[44]

After the 1550s, binary processes of deforestation and erosion gained considerable pace with decades. The early phase of the Little Ice Age, which stretched from the 1300s to the 1450s, was marked not only by famines. Village desertions and migrations—*völkerwanderungen*—had become part and parcel of the social life of the Mediterranean countryside.[45] It was characterized by an atmosphere of warfare and large-scale mobilization, both in the Levant, owing to the westerly expansion of the Mongol empire, and in the western Mediterranean in the 1350–1450 period, owing to the Hundred Years' War. Ironically enough, in both instances, the withdrawal of cultivation generated a fallowing effect.[46] This allowed the basin's low landscapes to recover from the maladies brought on by the fast and sustained pace of economic expansion since the turn of the millennium.[47] If vast tracts of land were perforce put to fallow during this period of widespread warfare, the same of course cannot be said of the hills and mountains, which offered refuge to those who fled the lowlands. Nonetheless, the embryonic state of long-distance transhumance, developed only in the twelfth century or so, when coupled with the state of the basin's hillsides and mountains, then "merely a poorer version of the realities found in the plains," did not prompt the kind of decimation later inflicted on the arboreal resources of the basin by the spread of livestock husbandry.[48]

During the Medieval Optimum, the processes of land reclamation and deforestation returned with a vengeance and the full ramifications of this ecological turnaround became evident when the Little Ice Age resumed in the 1550s.[49] Notwithstanding changes in its intensity, this era of climatic variability, dotted with droughts as well as wetter and humid conditions, lasted, without a lengthy reversal, until the latter half of the nineteenth century.[50] The plains hemming the Mediterranean's mountainous range lost a significant portion of their populations, since the low landscapes, maritime and inland, turned swampy and hence malarial. This was more so the case in the Mediterranean littoral where, due to the proximity of the sea, the water table was close to the surface and usually there was "too little gradient to allow adequate drainage by gravity or natural flow."[51] But the state in which the low-lying plains found themselves was in no way an exclusive property of the coastal plains. The inland plains were subject to the same set of forces as well, albeit under different conditions, as attested by the condition of the Cappadocian, Thessalian, and Lombardian plains.[52]

The basin endured the initial vexations of the phenomena associated with the Little Ice Age, at least for a while—until the mid-seventeenth century at the latest. From then on, the forfeiture and desertion of the low-lying landscapes proceeded without recess, especially when the few remaining signs, however feeble, of the Ottoman wheat exports died out more or less in the 1650s.[53] With the completion of the migration of oriental crops from the Mediterranean after the firm anchoring of sugar production in the Atlantic in the 1650s and with the pivotal position assumed by the Baltic grain trade in world-economic flows, the commercial environment of the fifteenth and sixteenth centuries was considerably punctured. The seventeenth-century crisis that followed the deluge-ridden late sixteenth century encouraged livestock agriculture and the diversification of agricultural crops: during the 1620/1650–1730/1750 period, the terms of trade favored industrial crops and the lesser cereals over bread grains. Neither the expansion of livestock farming nor the introduction of new crops was particularly beneficial to the economic livelihood of the plains. Instead, the resurgence of lesser crops and livestock farming left their stamp on the higher altitudes of the basin. Given the scale and longevity of the relentless march of the ominous duo of swamps and malaria to the detriment of sown lands, the process of the abandonment of the arable was not arrested all at once. Far from it. Intimations of a reversal in the fortunes of the lowlands emerged at the end of the seventeenth-century downturn: it was then the onset of recovery, tentative though it may have been, became evident. It was not until the late eighteenth and mostly the latter half of the nineteenth centuries that these malarial low-lands were brought under tillage, thanks to the draining of stagnant pools of water.[54]

The coastal plains along the Mediterranean shores of the Anatolian peninsula therefore remained "vacant" and were put to use by nomadic populations in their winter cycle of transhumance.[55] In other corners of the Inner Sea, the situation was not all that different. The Tavoliere of Foggia and parts of Valencia, to wit, presented a similar view to the traveler, not to mention Provence, where the advance of vineyards into the lowlands during the eighteenth century was due to the sparseness of habitation in the plains. In the Maremma of Siena, the arable shrank by 33 percent in the course of the seventeenth century.[56] The Iberian peninsula was also thin in population: it had a population density of 15 persons per square kilometer in 1712–17. The ratio went up only in the nineteenth century: by 1833, it had reached 24, still unimpressive in comparison to more densely settled regions of the Mediterranean.[57] The popularity of sites on slightly higher altitudes, or in locations where silting and sedimentation did not pose an imminent threat like the low-lying basins, peaked in the seventeenth and eighteenth centuries.[58] Judging by travelers' accounts, the coastal plains, which previously hosted commercial cultivation or were put to use as arable during the expansionary period between 1450 and 1650, were largely abandoned in the following two centuries: the depopulated lands yawned "like gaping wounds."[59] The lowlands of Cilicia, previously home to the relatively long-lasting and economically viable Dulgadir principality as well as to Lesser Armenia, which was crucial to the operations of the merchants of the city-states, were converted into grazing fields or reclaimed by the wild until the opening decades of the nineteenth century.[60] Evliyâ Çelebi attributed the sparseness of population in the region to the presence of fever and malaria: even the sparrows, he asserted, that flew over these insalubrious lands did not have the time and the opportunity to develop immunity to malaria and were as a result plagued by it. In the Aegean, only a tiny portion of the plain of Manisa was under tillage at the turn of the eighteenth century.[61] Owing to the shift in the center of trade to the Aegean shores of the empire in the seventeenth and eighteenth centuries, the Levantine littoral was sparsely populated. When Napoleon's army brought Acre under siege, the principal enemies it had to struggle against were the malaria-ridden marshlands and the perilous *via maris* on the coast. The southern shores of Anatolia, the Gulf of Alexandretta in particular, were no exception. The plain of Antioch was given over to riziculture as its highlands started to thrive, thanks to the spread of tobacco cultivation.[62] At the turn of the nineteenth century, the *Mare Internum* presented a landscape that differed sharply from that at the turn of the seventeenth: the basin's forbidding coastal plains were closely watched over by the ambient hills and mountainsides where the economic heart of the region was now located. After all, *cultura promiscua,* which became the gold standard of Mediterranean agriculture from the seventeenth century, as discussed in the previous chapters, was rendered possible by the simultaneous exploitation of all the gradients of the basin's mountainous landscape, from its lowlands and hills to its uplands.[63] This was a wide-reaching change that affected most regions and altitudes of the Mediterranean.

Therefore, the ecological change that started to unfold from the 1560s hastened the return of the formerly cultivated lands to nature, and neatly magnified the impact of the economic change precipitated by the rise of *pax Neerlandica.* For Amsterdam's centrality and its dependence on the Baltic for its grain lifted the commercial pressures on the bread lands of the Mediterranean. The implications of this retreat which, broadly speaking, resulted in the growing significance of horticulture, tree crops, and aquatic crops (such as rice) as well as crops dependent on irrigation (such as cotton), did not come about as a result of one single factor, but was the outcome of a variety of factors at play, some of which had a longer lease on life than others.[64] The first section of this chapter therefore charts these temporalities, from the longest-spanning, the Little Ice Age, to the shortest in duration, the seventeenth-century crisis, via the intermediate cycle of hegemony that was bracketed by the rise of *pax Neerlandica* on the one hand and by that of *pax Britannica* on the other. Since these triple processes worked together and in a complementary and self-reinforcing fashion, the transformation of the Mediterranean landscape proved to be equally long-lived and entailed slow-moving yet profound changes in the way the lands of the Inner Sea were put to use. The second section accordingly traces how the emptying of the lowlands was followed as a result by the gradual ascent of the settlements that blanketed the lowlands to higher altitudes, where neither malaria nor silting would pose a threat. The impact of malaria was reduced somewhat by the temporary cultivation of the lowlands, since the inhabitants of villages located on the hillsides descended onto the lowlands to sow them, only to leave these lands at once to retreat to their new environment. And this increased the number of claims on lands that were cultivated on a temporary basis. As the ferocity of the Little Ice Age started to subside, demands of the newly forming *pax Britannica* prompted a new round of agrarian commercialization and breathed new life into the basin. The third section therefore depicts how the change in the conjuncture demanded a gradual descent of rural settlements to lowlands, and the outcome was, once again, an attempt at either expanding the agricultural basis of the basin by planting cotton or turning these lands into bread lands. The region had come full circle, and starting from the 1870s, greater emphasis was placed on drainage, which complemented the picture. In all corners of the basin, millions of hectares of new land were drained, reclaimed, and opened up to cultivation, tilting the balance once again to the benefit of the lowlands.

Three Lives of the Inner Sea

In the 1450s, the long period of economic downturn and the political turmoil that accompanied it came to a halt, and agricultural production, which had stagnated at first and contracted later, started once again to show signs of life. After all, two hundred years of struggle between the houses of Aragón and Anjou over the two Sicilies had finally been decided in favor of the former. The confrontation between the houses of Osman and Paleologus over the control of the Golden Horde and the Black Sea was concluded in 1453, also in favor of the former. The signing of the Peace of Lodi in 1454 completed this picture by establishing a modus vivendi among the warring city-states, be it only for less than half a century. In the western Mediterranean, the Hundred Years' War came to an end and the onslaught of the plague turned into a lingering memory. In the Levant, the turbulence created by the consolidation of the Mongolian realm—transcontinental in scale—gradually ebbed away, leaving its place to smaller, fragmented imperial polities. At both ends of the basin, therefore, the tasks faced by peasants and lords alike in the mid-fifteenth century were the settlement and exploitation of deserted and virgin lands, the (re)building of (old) villages, and the freeing of fields of the encroaching brush and weeds, signs of the fallowing effect of the period between 1250/1300 and 1450.[65] The resumption of economic growth and the concomitant restoration of political order heralded the end of the fifteenth-century crisis and opened up new avenues for economic expansion. Yet among the factors that facilitated and framed the economic upswing from the mid-fifteenth century, albeit in a surreptitious fashion, was the hiatus in the Little Ice Age. Earlier, its advent was signaled by the onset of famines that plagued, first and foremost, the shores of the North Sea from 1315 to 1322. It is worth noting that the merchants who traveled south in a desperate search for grain returned home empty-handed, since the Mediterranean, itself short of bread grains, did not have any on offer, even for sky-high prices.[66] Turbulent though the arrival of the Little Ice Age was, its first phase came to an end not before too long. The beginning of the Medieval Optimum in the mid-fifteenth century played a pivotal role in the phenomenal expansion between 1450 and 1550, since the warming trend momentarily reversed the climatic turnaround ushered in at the end of the thirteenth century.

The period that stretched from the mid-fifteenth century to the politically tumultuous late sixteenth was an era not only of economic growth but also of extensive land reclamation and colonization, as discussed in the second chapter. The expansionary élan of the period was unmistakable despite the fact that the rate of growth in agricultural production started to trail behind that of demographic growth roughly from the third quarter of the century. There were definite signs across the Inner Sea that the times of plenty were coming to an end: unseasonable rains, droughts, increased rainfall, landslides, and soil erosion started to make themselves felt with annoying regularity. The shift in the ecological fortunes of the basin was noticeable, especially when compared with the mostly dry years of the 1450–1550 period.[67] The most spectacular manifestation of this reversal of fortune took place in the Black Sea, which froze in 1620; the last time a similar deep freeze had occurred in the Pontic Sea was in 1232.[68] Less dramatic but more consequential to the inhabitants of the basin was the damage inflicted on its lowlands by the increasing wetness in the latter half of the sixteenth century in the form of unyielding erosion.

Even earlier, Leonardo da Vinci had been alarmed by the undoing of mountains by "rains and rivers" in Tuscany.[69] By mid-century, the land given over to the plow to meet the needs of growing numbers of people was already extended to cover the sloping surfaces of the basin, which were not the most suitable terrain for agriculture. Generally speaking, the tilling of the slopes, which required the felling of trees and the clearing of the woodland cover, aggravated the environmental consequences of the "heavy rains" of the late sixteenth century by hastening the tempo of areal erosion.[70] What helped cushion the impact of the deterioration in ecological conditions on the economic fortunes of the Inner Sea, even if for a relatively brief while, was the advent of the century of the Genoese exactly when the Little Ice Age was returning. As discussed in previous chapters, the resumption of eastern-bound precious metal flows and the resurgence of the Levant trade on the one hand, and the vibrancy of the Atlantic trade on the other, invested the Mediterranean with an economic vigor that was denied to it when the spice trade had shifted away from the Inner Sea.

Though the impact of the Medieval Optimum on the expansionary élan of the sixteenth century is well documented, the same cannot be said of the Little Ice Age, even at its height—in the seventeenth and eighteenth centuries. Documentation that captures constraints placed on agricultural producers and landholders by greater climatic variability is rather sporadic and circumstantial. But then again, the historiographical imbalance in favor of the sixteenth century is in no way confined to the region's ecological history alone. It is easier, of course, to chart the trajectory of the Mediterranean in its heyday than in its twilight. Befitting its world-economic eminence, the Mediterranean was then well-lit owing to the luminosity of its reigning city-states: the trails the merchants of these cities left behind in their missives as well as the dispatches detailing their exploits and travails fully capture the density, scope, and sway of their operations.[71] The golden age of the basin was also well-lit by the massive documentation amassed by the imperial polities at the zenith of their power. The most comprehensive cadastral surveys that the Ottoman bureaucrats compiled offer a detailed cross-section of the agrarian structures of the empire in the sixteenth century, not exclusively but mostly in the latter half of it. Philip II's famous *Relaciones topográficas* too provide us with a picture of the state of his territorial possessions in Iberia in between 1575 and 1580.[72] The relatively well-documented world of the golden age of the Mediterranean was already coming to a close by then, ushering in the relatively dim world of the seventeenth and eighteenth centuries when destinies previously intertwined by the Inner Sea seemed to be drifting apart.

This relative dearth or absence of detailed documentation captures the fact that the imperial governments' ability to effectively administer their realms had started to wane at the conclusion of the sixteenth century. The travails of imperial bureaucracies and the waning of the city-states that presided over the Greater Mediterranean, as argued in part I, were both symptoms of the shift in the center of gravity of the world-economy away from the Inner Sea. To a certain extent, the demise of the cities of St. Mark and St. George and their attempts to hold on to their eroding power shaped the region's destiny during the seventeenth and eighteenth centuries, providing in the process a degree of unity and coherence to its economic life. Be that as it may, the northerly migration of specialized grain cultivation to Pomerania and Poland, the westerly and Atlantic-bound migration of sugar production, and the diversion of the "rich" trades away from the Mediterranean could be neither contained nor reversed. The economic devolution that followed attenuated the power of the city-states.[73]

Given that the empires built by Venice and Genoa were aquatic in nature and their commercial operations interwove the basin's port-cities (and their hinterland) more readily than the rest, it is almost a foregone conclusion that the primary sites of the previous order would be adversely and disproportionately affected by the end of the age of the city-states. Almost surgically, it was the coastal regions and the maritime plains of the Inner Sea that suffered the aftereffects of this devolution the most. For one, when the rise of the Baltic grain trade arrested the commercialization of arable agriculture in the Inner Sea, it was the littoral regions of the basin that were hit by the contraction of lands devoted to cereal culture even though the grain trade went on in a reduced fashion. Similarly, the footloose character of sugarcane and cotton cultivation deprived the region's valleys and maritime plains previously devoted to these lucrative crops of their main livelihood, reducing further the extent of commercial cultivation in locations accessible by sea. And the diversion of the spice trade, by denying the Mediterranean the high profits delivered by the rich trades that underwrote the basin's commercial buoyancy, reduced the degree of interconnectedness among the shores of the basin by lowering the volume of merchandise trade. The resulting configuration of these three forces led to a contraction, first, in the degree of commercialization within and across the basin as emphasized above and, second and more important for our purposes here, in the arable devoted to the cultivation of cash crops.[74] The emptying of the maritime lowlands and the attendant decline in the numbers of people who inhabited these lands served as the markers of the Mediterranean landscape until the latter half of the nineteenth century.

The Inner Sea's loss of ground vis-à-vis the Baltic and the North Sea was certainly among the key factors that set the main tenor of change and determined the tempo of the withdrawal of cultivation from the lowlands. Still, however pivotal the redistribution of world-economic activities were in the hollowing of the Mediterranean, the spatial shift in the center of gravity does not fully account for how the retreat from the lowlands, a process with a deep historical breadth, endured for so long despite repeated efforts by landlords and others to repopulate and reclaim the vacated lands, mostly during the eighteenth century. Indeed, claims over the low landscapes multiplied in number precisely during the second half of the Little Ice Age, when withdrawal from these lands was at its height: pastoral groups who used these vacated lands as pasture in winter; rural producers who either periodically tilled them when and if conditions allowed or made use of the resources offered by the wetlands, which had taken over the untended fields, for securing a livelihood or additional income; landlords, lay and ecclesiastical, desirous of extending their estates and putting these idle lands in use when conditions, ecological and economic, became ripe.[75]

This did not mean that the multiplicity, and competing nature, of the claims prevented the exploitation of these lands by most, if not all, of the claimant parties as long as demand for cereal and commercial crops sagged.[76] Only when demand for commercial crops recovered from the latter half of the eighteenth century and agricultural production was given a momentous and lasting boost by the repeal of the Corn Laws in Britain did it become apparent that competing claims over land that came into being during the lethargic seventeenth and eighteenth centuries could not be readily accommodated within the existing set of landholding and usufruct arrangements. The widespread sale of state-owned lands in the wake of the Napoleonic era and the passage of new land codes helped crystallize rights over land—generally classified as communal, waste, and paludal—and mostly in favor of those who were willing to pay for, and reclaim, them. Settlement of claims over abandoned lands sealed the movement that favored the bringing of the basin's lowlands under cultivation.[77]

It is to be expected that such a large-scale, long-term movement owed its existence not simply to the migration of certain economic activities away from the Inner Sea but to a wide array of historical developments that ranged from the demise of the city-states and the exodus of commercial crops to the ecological pulse of the basin. All in all, the processes that framed the withdrawal of cultivation and settlements from, and their return approximately two centuries later to, the low-lying plains were threefold. Of differing duration and nature, these three processes framed the retreat from the lowlands in different ways. The first of the triple processes, the longest in span, was framed by the advent and conclusion of the Little Ice Age.[78] It commenced much earlier than the other two, in the late thirteenth century. In this long-term shift toward increased climatic variability that lasted until the mid-nineteenth century, there was a relatively lengthy hiatus that spanned roughly from the 1450s to the 1550s, the Medieval or the Little Optimum.[79] The main climatic tendencies of the Little Ice Age were either reversed, or in more instances lost their severity, during the Medieval Optimum. Yet, the period of relative warming was not a simple break in the climatic and ecological trends of the era. By contrast, the tempo of urbanization and land reclamation, which accelerated during the economic expansion and favorable climatic conditions of the 1450–1550 period, rendered the basin more vulnerable to the depredations of the Little Ice Age upon its return. The extent of deforestation that took place in the interim as a result of the soaring demand for timber for construction, industrial production, and shipbuilding was more substantial than before.

At any rate, the period of cooling, deluge, and drought, all damaging to agriculture, stretched, with their ups and downs, until the 1850s—if not until the 1870s.[80] At the height of the Little Ice Age, from 1550 to 1700, "temperatures were 1.2 to 1.4 degrees Celsius below those of the Medieval Warm period."[81] At times, temperatures returned to their sixteenth-century levels; at other times flooding lost its ferocity. But the climatic variability of the period and the atmosphere of uncertainty it fostered with regard to agricultural production remained in place, and discouraged attempts to keep the low landscapes under cultivation.[82] The neglected state of low-lying plains was thus one of the reasons there was a retreat from the arable, since the aquatic environment that the Little Ice Age fashioned engendered insalubrious conditions on the shores of the Mediterranean. In the eighteenth century, there were attempts by imperial authorities in Spain to control or restrict the expansion of wet rice cultivation for "fear of infestations associated with the marshes."[83] The Cilician and the Pamphylian plains on the shores of the Mediterranean Anatolia were bereft of permanent human settlements. When inhabitants of the encircling mountains and hills descended on to the plains at harvest time, it was as if they had ventured into the netherworld of festering malaria, and as "dead men tilling," they were not expected to return in full health, if at all.[84]

Even when climatic conditions got intermittently better, it was not easy to undo the damage already inflicted on the lowlands by the expansion of the swamps: draining low-lying lands required financial resources beyond the means of most petty producers. In the case of landlords even, investing in drainage to create new land during the seventeenth-century crisis when agricultural prices remained stagnant at best was not the rewarding strategy it later became in the nineteenth century. This is why at the height of the Little Ice Age when the lower altitudes of the basin were kept sown, the exploitation of the land took place principally under the organizational and administrative umbrella of the all-encompassing *latifondi* and *çiftlik*s, which were capable of cajoling and mobilizing labor but reluctant all the same to invest in land bonification. Otherwise, these "idle" lands were set aside as grazing fields for large numbers of livestock, as in Puglia, or money was invested in it for the "forage revolution" of the eighteenth century, as in Venice.[85] Conversely, the smooth gradients of the mountainous range where small landholders reigned were devoted to polyculture: intensive mixed agriculture of olives, vines, and fruits.[86] As we discuss below, it was the disappearance of this climatic variability from the mid-nineteenth century, magnified by the soaring British demand for lowland crops from the 1840s, which reversed the trend. The Sisyphean task undertaken by the central states, draining companies, and would-be large-estate owners in the latter half of the nineteenth century to reclaim the vacated lands was rendered easier by the retreat of the glaciers and the end of the Little Ice Age.

Initially compounding and later healing some of the afflictions wrought by the Little Ice Age was a second process, not as lengthy as the climatic and ecological process briefly sketched above but long-lasting nevertheless. It was of a different provenance, for it pertained to the cycles of hegemony of the world-system, and the division of labor that came about as the new hegemon recast the spatial redistribution of economic activities across the basin almost *ab novo*. The temporal span of this process was bracketed by two cycles of hegemony: it framed the retreat of cultivation from the low-lying lands of the Inner Sea, and spanned approximately from the 1590s to the 1830s. It began in the 1590s with the commencement of Amsterdam's (and the United Province's) ascent to the apex of the capitalist world-system.[87] Even though the United Provinces' high age came to a close in the 1670s, the fact that the Anglo-Dutch rivalry that followed in its wake took place away from the Mediterranean left the basin's world-economic standing unchanged. This state of affairs remained in place until the 1830s—that is, until the United Kingdom's gradual ascent to hegemonic position.

Seen from the vantage point of the Mediterranean, the first instance of hegemonic rise and the subsequent revision in the global division of labor found its symbolic expression in the arrival, in the 1590s, of the northern ships in the Inner Sea: these ships transported not only Baltic grain but also Atlantic sugar.[88] Britain's rise to hegemony from the 1830s was symbolically announced by the arrival in the basin of the steamship—to be followed by a rise in demand for wheat and cotton by Manchester and London.[89] Mounting requests for these dual crops turned the low-lying fields, largely utilized by herds and intermittently tilled by rural producers, into lands with potentially valuable assets, if properly drained. The task was undertaken, first, by enterprising landlords and, later, by the statal agencies during the interwar period. If, earlier, the draining companies were successful in rehabilitating sodden lands in the relatively wealthy parts of the basin, such as Venice and Provence, it was primarily the central states that assumed the responsibility of drainage in the late nineteenth and twentieth centuries.[90] The turnaround in the fortunes of the low landscapes then came about thanks to the launching by the states of the region of large-scale drainage projects. The previously vaporous lands of the low-lying plains were subsequently brought back to life in a systematic fashion, from the Mitidja plain and the Po plain to the Guadalquivir valley and the Konya plain.[91] In any event, the basin's low-lying landscapes were brought under the plow by and by and given over to agriculture. When seen from this vantage point, it can be stated that the hollowing of the coastal regions of the Inner Sea was brought about by the abatement in the centrality of the Mediterranean pursuant to the consolidation of *pax Neerlandica*. The abandonment of the lowlands reversed course with the onset of *pax Britannica* which, due to its mounting demand for cotton and wheat, stimulated the settlement and reclamation of the basin's coastal plains. In sum, two cycles of hegemony bracketed the period of retreat from the lowlands and the decreasing import of cereal husbandry.

Nested within these two drawn-out processes, which exacted a toll on the basin's cereal culture, was the seventeenth-century crisis. In duration, it was in fact the shortest of all three: it stretched from 1620/1650 to 1730/1750.[92] The impact of the crisis on the basin's economic restructuring was straightforward enough. The economic downturn, since it was accompanied by a decline or slowdown in demographic growth, acted as a brake on arable production, since demand for bread grains shrank.[93] That the number of hands to be put to work in the fields declined correspondingly only compounded the trend: those branches of agriculture that did not require intensive labor survived the crisis easier. It was this new equilibrium that energized livestock husbandry, since it was significantly less labor intensive than cereal agriculture.[94] However short-lived the crisis was in comparison to the other two processes, like them, it helped tip the balance in the Mediterranean against cereal husbandry, to the benefit of livestock agriculture and industrial crops. In sum, then, the combination of these three intertwined processes—different in nature, duration, and provenance—acted to create an environment that was not beneficial to the expansion of bread lands and, in particular, low-lying plains from the late sixteenth century to the late nineteenth.

Of the three process that framed this great agrarian cycle, it was the Little Ice Age that had the longest historical span. The first phase of the Little Ice Age, as if it were, spanned from the turn of the thirteenth to the mid-fifteenth, and in some places the turn of the sixteenth, centuries. Its advent at the turn of the thirteenth century was heralded by the onset of unusual cold, early frosts, and deadly famines that collectively caused villages to be abandoned, sometimes permanently, in distant corners of the basin, from Castile to Asia Minor. Surely, the impact of the Little Ice Age was far from uniform.[95] But overall, its arrival hastened the reversion of lands, which were recovered from waste and woodland since the turn of the millennium, to fallow and pasture and the desertion of settlements built on marginal lands. Lands that remained under the plow were the least prone to floods given the increased tempo of precipitation from the late thirteenth century.[96] On both sides of the English Channel, the poor harvest of 1314 was followed by two years of heavy precipitation, which caused catastrophic floods and pushed the price of wheat to unprecedentedly high levels. In 1316, persistent cold and wet conditions did not leave enough time for cereal crops to ripen and "the price of a measure of wheat rose from 5 to 40 shillings; and at Louvain, in seven months, from 5 to 16 *livres.*"[97] The situation did not get any better in the following years, hence the dreary sobriquet the "great famine," which captures the widespread hunger, deluges, social turmoil, and epidemic diseases that reigned during the 1315–22 period.

The period of cooling and wetness was not confined to the north of the Alps.[98] Northern merchants who traveled south to procure wheat could not find cereal there in part because of policies designed to keep available supplies at home.[99] The reason behind the introduction of such measures was that the south— Languedoc, for instance—itself was desolated by famine and dearth for a total of twenty years between 1302 and 1348. Not as catastrophic as the great famine, which devastated both shores of the English Channel, harvest failures, torrential rains, and a succession of bad winters still had an impact on the economic well-being of the region.[100] In the 1430s in particular, an abnormal amount of cyclonic activity in parts of the Mediterranean caused "more rainfall than is now normal," but not extremes of temperature.[101] Because of problems with drainage, marshlands, and malaria, economic life flourished away from the vapors of the lowlands. From the southernmost corner of Tuscany to the mountains of Romagna, villages on the higher slopes of the mountains and hills encasing the Inner Sea came to provide haven for larger numbers of people in the first half of the fifteenth century.[102] In Pistoia, the population of the mountains showed a healthy and steady increase over the years 1244 to 1427 as the Black Death and the crisis of the fourteenth century were plaguing the lowlands. Knowing the terrain and its perilous nature, villagers entrenched in the 200 to 500 meter range did not even plow the slopes to minimize the impact of erosion; they fertilized it with night soil instead.[103] This, however, was a luxury that most peasants did not have: in Byzantium, slopes were generally given over to pasturage which shrank as cultivated area increased. In fact, the farming of the slopes pushed back the forest until the fourteenth century.[104] The situation was not all that different in Iberia, where settlement and reclamation expanded the arable with speed: there were four major floods between 1250 and 1333. Frequent droughts that often lasted for more than a year and fairly frequent floods and torrential rains marked Valencian agriculture in the fourteenth and fifteenth centuries, as confirmed by the increasing frequency of prayers for rain.[105]

Surely, the period was one of climatic variability, characterized by heightened fluvial activity but also by summer rain and occasional droughts.[106] Droughts held sway over the eastern latitudes of the basin, from Asia Minor to Russia. In Russia, there were twenty-two drought years in the fourteenth and fifteenth centuries—eleven in each century—and an equal number of floods. In Asia Minor there were no recorded famines in the twelfth century, but the peninsula was scourged in the 1300–1333 period by droughts and invasions of locusts.[107] Judging by the rural settlement patterns in different parts of the Ottoman empire in the fourteenth and fifteenth centuries, it is more than likely that there was increased precipitation, just as in Russia. It was in this period that alluviation was at work in Bithynia, the birthplace of the Ottoman empire, as well as in Macedonia.[108] In Bithynia, most of the rural settlements were located on the floors of its alluvial valleys and, more important, away from river banks, secure from the perils of flooding and malaria. The narrow Pontic coast of Asia Minor, with its steep gradients, was also largely devoid of permanent sizeable settlements, and remained so until the eighteenth century.[109]

Following a widespread abandonment of low-lying lands, pastures and meadows took over from cereal fields; farms specialized in the cultivation of fodder multiplied; and the raising of livestock spread with impressive speed, as confirmed by the rise in meat trade as well as in meat consumption. From the fourteenth century to the 1550s, meat became available to large numbers of people as a result of a retreat from agriculture. In these carnivorous times, meat consumption increased in urban and rural areas, from Tuscany to Tours, from Germany to Sicily.[110] In Ottoman Edirne (Adrianople), the portions of meat served in the soup kitchens of the imperial foundations were considerably higher in the mid-fifteenth century than the allotments of the seventeenth century.[111] On account of the growing dominion of herds of sheep (and goats) across the high- and lowlands of the basin, mutton consumption too increased, more so in the eastern Mediterranean than in the west, where pork reigned. By curbing demand for the arboreal resources of the basin, the economic downswing that started in 1250–1300 period placed the high landscapes largely at the employ of small livestock. Later, the grievous times of the Black Death did not eliminate but further reduced the demand for the sylvan resources of the ambient mountains.[112] The low landscapes of the basin where the population density had increased steadily from the turn of the millennium suddenly found themselves drained of people.[113] At the end of the fifteenth century in the vicinity of Béziers, the number of vineyards to every one hundred fields had declined from forty in 1353 to six in 1407; the town ditches of Montpellier, previously home to vines, orchards, and gardens, were now home to "briars and thorns, snakes and lizards."[114] Yet, the hiatus in the pace of settlement and the halt in land reclamation was relatively brief. When economic and ecological conditions turned favorable from the 1450s, the return to the land was precipitous, unlike in the eighteenth and nineteenth centuries, as we discuss below. It needs to be noted that ecological changes that the Medieval Optimum prompted in terms of mounting demand for timber and land would have its ramifications later in the following centuries.

The resumption of the Little Ice Age in the 1560s exposed the region to the environmental vagaries of the steady and breakneck economic growth in the 1450–1550 period. Given the relative brevity of the period when the uplands attracted populations from the low landscapes from the onset of the Black Death to the 1450s, and given the absence of newly established claims over these forfeited lands, a return to the plains took place rather swiftly.[115] The exception was Tuscany, where Florentine patricians managed to partition large portions of it among themselves.[116] The resettlement of the plains and the lowlands when the slopes were being tilled more intensively than before and when the woodlands were under attack put the producers in the low landscapes in a highly risky situation after the 1560s. An increase in precipitation associated with the Little Ice Age rendered it more likely that the lowlands would turn into marshlands and swamps. The retreat from the lands reclaimed during the Medieval Optimum played a part in the emptying of the coastal and inland plains throughout the basin, reducing the surface area devoted to grains in the seventeenth and eighteenth centuries. Increased fluvial activity that surfaced thrice in all its intensity during the course of the Little Ice Age—c. 1570–1638, 1760–1800, and 1830–70[117]—pushed the limits of settlement higher. This upward movement restored the settlement pattern most readily identified with the Mediterranean prior to the colonization of its lowlands, where ambient hills and mountains of the basin were dotted with castles and hill towns overlooking the valleys and plains.

That the plains were evacuated from the turn of the seventeenth century and remained so—devoid of "permanent" settlements—turned out to be extremely beneficial for aspiring landholders in the eighteenth and nineteenth centuries.[118] When world-economic growth picked up in the 1730s, lands vacated previously were put to use by landlords, be it intermittently, to introduce the large-scale cultivation of cash crops by employing laborers desperate enough to work in inhospitable environments. Given the lacustrine and insalubrious nature of the lowlands, attracting voluntary laborers for performing onerous tasks in such risky environments was certainly not easy. Nonetheless, short-term migratory movements from the surrounding hilly and mountainous regions provided the lowlands with a much needed reserve army of labor. Towns of the upper valleys in the Biferno region, for instance, furnished Apulia with harvest laborers as did the mountain dwellers in Lombardy in the low-lying marshy regions of the Durazzo plains.[119] Here and there, parts of the lowlands came under commercial cultivation, but these lands were put to full and effective use only during the latter half of the nineteenth century, when a new round of reclamation and colonization of the lowlands started in earnest.

The process lasted well into the second half of the twentieth century. In the Ottoman case, the introduction in 1858 of the new Land Code, a code designed to institute allodial rights over *mîrî* land, precipitated the pace of the seizure of such lands. Appropriating wastelands, *mawât*, had the additional advantage of advancing claims over large tracts of land without necessarily alienating the peasantry who had enjoyed rights of usufruct over the *mîrî* land until then.[120] Following the dissolution of the Mesta in 1836, the abolishment of señorial privileges, and the sale of Church lands, farming also picked up pace in Spain.[121] Large agrarian units flourished in the Tuscan, Pontine, and Po marshes.[122] The liquidation in the 1820s and 1830s of "feudal" arrangements in the kingdom of Naples—hence in Sardinia and Sicily, the breadbaskets of the Mediterranean before the sixteenth century—reflected the trend that was at work along the shores of the basin. So for the most part of the seventeenth and eighteenth centuries, the low landscapes of the Inner Sea failed to retain their status as the basin's favorite sites of cultivation due partly to their swampy and malarious nature.[123]

The second process was bracketed, not incidentally, by the cycles of hegemony of the world-system. The ascent of a new hegemonic power was premised on the refashioning of the global division of labor to meet its needs and to establish its hold over the system. The restructuring of the world-economy, first, in the 1590s and, later, in the 1830s, enveloped the second temporality. The first revision in the structuring of world-economic flows occurred in harmony with the construction of *pax Neerlandica* and the second, with that of *pax Britannica*. From the viewpoint of the Mediterranean, this took the form of the arrival, first, of northern ships during the grain crisis of 1590 and, later, of the steamship in the 1830s. The arrival of the northern merchants confirmed the decline in the ability of the city-states to regiment and choreograph economic flows crisscrossing the Inner Sea. After all, the Amsterdam grain market overtook that in Antwerp in value in 1544, and a separate Corn Exchange was already in place in Amsterdam by the beginning of the seventeenth century.[124] Besides the grain trade, there were other signs of an increasing northern presence in the Inner Sea: one was the rise of Livorno, the base from which the northern merchants launched their first assault on the city-states' hold over the basin; İzmir and Aleppo, gateways to the silk trade; and Cádiz, the nerve center of their commercial and financial operations across the Atlantic. Surely, the Mediterranean, as a wealthy region, never lost the affectations of the northern merchants, but it never commanded the kind of gravitas the Baltic or the Dutch Indies possessed. However successful the anchoring of the northerners in the Venetian millpond was, the commercial networks they put in place never attained the degree of sophistication and density achieved under the city-states. As discussed in chapter 2, the absence of lasting financial and commercial arrangements was eloquently captured by the popularity of "country trade" along the shores of the Levant: the unwavering need by the northern powers to procure goods or bullion, in return for their shipping services, to pay for their imports testified to the contraction in the basin of the level of commercialization in the wake of the demise of the city-states.[125]

Yet, it is the latter-day apparition of these two waves of northern invasion that commands the literature: the force invested in the arrival of the steamship designates the seventeenth and eighteenth centuries as a period when processes framing the unity of the Mediterranean region at its height had fast become a thing of the past and the resuscitation of this unity was contingent on the advent of *pax Britannica*. Not unexpectedly, the long period of "delinking" between these two invasions corresponds to the periodization that is in currency in the existing historiography of the Mediterranean, according to which the seventeenth and eighteenth centuries serve merely to bridge the golden age of the sixteenth century and the reform-ridden nineteenth.[126] That, at the end of the eighteenth century, the average size of northern ships calling at Ottoman ports was barely over 200 tons and that of the "Turkish and Greek" ships only 125 tons is called upon to testify to the advance nineteenth-century shipping and transportation registered after the 1830s.[127] Most ships built by the Signoria, however, usually carried in the environ of 300 tons of burden; the Genoese carracks and the Venetian ships laden with grain carried up to 1,000 to 2,000 tons.[128] Swift though the steamships may have been, they managed to surpass the 1,000-ton mark only in 1874. The relative brevity and the fairly low volume of maritime trade conducted by the northerners in the coming two centuries is one of the reasons this interim period does not receive the attention it deserves.

The two periods that bracketed the waning of the Mediterranean, the first prior to the northern invasion of the 1590s and the second following the arrival of the steamship in the 1830s, were both periods of strong agricultural growth. As such, the long sixteenth century expansion and the mid-Victorian boom both witnessed the reclamation of vast stretches of land and an increase in the region's agricultural output and were closely related to the cycles of hegemony outlined above. The two periods of agricultural growth followed revisions in the axial division of labor under the aegis of the newly ascending powers. As a result, in both cases, vast expanses of cereal lands were brought within the orbit of the world-economy under the aegis of the ascending hegemon. The former wave under *pax Neerlandica* opened up the east of the Elbe to cultivation to serve the needs of Amsterdam and other urban centers along the shores of the

North Sea and the Atlantic. And the latter wave under *pax Britannica*, after the repeal of the Corn Laws in 1846, not only brought about the incorporation of the vast Russian plains, their produce channeled into world markets through Odessa, but more important gave a big momentum to the opening up of temperate settlements to cultivation.[129]

In the former case, the ability of the nobility to utilize coerced labor—under the second serfdom—was a result of the region's growing incorporation into the world-economy. The facility with which the produce of this region found its way to the Baltic to be unloaded into the vast storage houses in Amsterdam and in Bergen, and the ability of Dutch merchants to access this source were of paramount significance in turning the region into a basic supplier of grain. To their south, most of the producers lost their allodial rights on land to become sharecroppers.[130] Equally significant was the inflationary wave of the latter half of the sixteenth century, which hit the Mediterranean more fiercely than the Baltic; the price differential that allowed the former to respond to the growing needs of the world-economy started to decline from the 1560s. Attempts to perpetuate the impetus created during the wheat boom of the mid-century faded under the pressure of the inflationary environment of the post-1560 period, which pushed prices up, including those of agricultural goods, along the Mediterranean coast, rendering the Baltic grain relatively cheaper.[131]

When the inflationary surge started to subside in the 1620s, the grain trade had already become an integral part—in fact, the "mother trade"—of *pax Neerlandica*. From then onward, it was the Baltic trade that came to be associated, almost exclusively and at least until the 1670s, with cereal production.[132] For producers or landholders in Ottoman lands, the inflationary surge, accompanied by soaring demand due to sustained urban growth and exorbitant prices, reinforced the commercialization drive of the sixteenth century. When grain prices started to decline after the dizzying heights of the early seventeenth century, the Ottoman Mediterranean had ceased to occupy the position it did in the mid-sixteenth century. Despite the fact that the price of wheat in Istanbul, after reaching its highest point in 1606 at 80 *akçe*s a *kile*, fell and fluctuated between 30 and 40 *akçe*s for the rest of the century, the Baltic had already been transformed by the Dutch merchants into their granary.[133] That the world-economy was in a period of recession, and that this created a fiercely competitive environment in which prices and populations were both declining in tandem, made things worse for producers in the Mediterranean. Hence, the economic momentum shaping the basin during the long sixteenth century ended prematurely. Producers in Pomerania, Poland, and Silesia were exempted from the immediate impact of the onset of the downturn because of the ongoing vibrancy of the United Provinces market, a luxury the Mediterranean producers did not have. From the 1620s, the evacuation of the plains reconstituted small peasantry on the hilly slopes, on the highlands overlooking the plains, and in mountainous regions. From then on, putting the plains under cultivation entailed either adhering to sharecropping arrangements or resorting to outright coercion. This was a significant reversal from the trends of the sixteenth century, when agricultural production and lands sown to cereals expanded considerably: the density of rural settlements along the shores of the Inner Sea was definitely higher at the end of the sixteenth century than it was in the mid-nineteenth.[134]

The second invasion of the Mediterranean by the northerners in the 1830s was as pivotal in the transformation of the region's rural landscape as was the first. As mentioned, the abolition of the Corn Laws signaled a reshaping of the world-economy's division of labor. And, as was the case in the late sixteenth century, this was a time of agricultural expansion. Large-scale land reclamation or the opening of new lands to cultivation led the way: it started during the eighteenth century with the settlement of Siberia and extended into the newly settled vast temperate lands in the latter half of the nineteenth century. This was also a time in the Ottoman lands when the start of steady growth in agricultural production, complemented by the institution of the new Land Code, served as a stimulus to gradually bringing maritime and inland plains under the plow. Once again, reminiscent of the initial *çiftlik* formation in the closing decades of the sixteenth century, there was a discernible attempt to form larger agricultural estates.[135] Yet the flooding of the world markets by wheat that originated in the temperate world cut the momentum short from the turn of the twentieth century. With the impetus given by the Victorian boom, agricultural production mounted first, and then stalled, very much like it had in the mid-sixteenth century. The expansionary thrust of the 1850–1950 period was reminiscent of the 1550–1650 period. Both were cut short by a reversal in agrarian prices: the great wave of "enclosures" that closed the sixteenth century lost its momentum during the seventeenth-century crisis with the reversal in agrarian prices. The second wave slowed down considerably when terms of trade turned against agricultural goods from the turn of the twentieth century.[136]

And the last, and the shortest, temporality that created forces similar in nature and operated throughout the Mediterranean and beyond was the world-economy's secular downturn, namely, the seventeenth-century crisis.[137] The stagnation that followed an expansionary sixteenth century and a fall in population growth tipped the balance, as discussed at length below, in favor of lesser grains, industrial crops, and livestock farming. Most of the lands previously earmarked for cereals were, as a result, given over to pastureland. This had the effect of intensifying the agrarian trends put in place by the waning of the Mediterranean.[138]

To conclude, the autumn of the Mediterranean was comprised of not one, but three processes. It started before the onset of the seventeenth century. The return of the Little Ice Age and its ecological repercussions were felt throughout the basin starting from the 1560s. That was also when the ships of the Atlantic ventured into the Mediterranean, yet this first invasion proved to be brief; they returned with greater force in the 1590s. Neither the arrival of northern steam-ships in the 1830s, nor the end of the seventeenth-century crisis were, in and by themselves, forceful enough to terminate the long-term abandonment of the plains. The Little Ice Age, which started showing signs of reversal in the 1840s, came to a close in the 1870s, completing the great cycle. The global warming that followed provided the ecological component of the availability and ac-cessibility of the plains due to lessened wetness and humidity. This great cycle, which presided over the withdrawal from the lowlands, bracketed a sea change in the basin's agrarian, industrial, and commercial constitution.

Singly and together, these three processes had fundamental parts to play in the reshaping of the Mediterranean during the seventeenth and eighteenth centuries. Their overall and cumulative effect was twofold. Economically, all three processes contributed to the contraction of arable agriculture, as outlined and depicted in part I. Ecologically, they induced a shift away from the low landscapes: the center of gravity of the basin's economic life, which had been concentrated in its lowlands since the turn of the millennium, was now located elsewhere, in ambient heights, the subject matter of part II. The processes that prompted a decline in commercial arable production and the plight of permanent cultivation and habitation away from the low landscapes were two-pronged, and the two complemented each other in a remarkable fashion. That attempts to increase land supply by drainage were abandoned due to increased wetness and erosion generated two different but closely related outcomes. On the one hand, new, or previously deserted, fields on higher altitudes were brought under cultivation. At the turn of the seventeenth century, peasants in Languedoc reclaimed the old *agri deserti* and areas of secondary growth at an accelerated rate. They also reduced the area of forest through the clearance of assarts on the land across the Rhône, as place-names associated with burning testify.[139] On the other hand, lowlands largely emptied out as a result of sedi-mentation and frequent inundation were put to some use. Where and when the battle against the marshes was lost and the drainage of lands was deemed to be unprofitable, villagers were allowed to use shallow marshes for pasture and other agricultural activities.[140] The first of these double processes—that is, the ascent of rural settlements—is the topic of the following chapter. The discussion that follows here dwells upon the retreat of permanent settlements from the lowlands and, relatedly, the momentary return of these lands back to the forces of nature.

Away from the Unhealthy Vapors of the Lowlands

Although the coastal plains that enveloped the Inner Sea were modest in size, and cultivation in the inland plains was not intensive, the desertion of the lowlands nonetheless altered the agrarian geography of the Mediterranean basin. The region's bread lands were almost exclusively confined to the better-drained soils of the valley basins and the floors of intermontane depressions, save for the low-lying plains that were frequently repossessed by the miasmic and noxious vapors of the lowlands and by the anopheles mosquitoes.[141] Surely, the gentle, more rounded, deep-soiled slopes of the hill regions of the Mediterranean were equally good cereal lands, but they were of limited spread. For most of the period under consideration, shortages of wheat output loomed large and were "of depressing regularity."[142] Grain and commercial agriculture may have been partially evicted from the plains, but it survived at higher altitudes, oftentimes in the form of winter wheat.[143] Pivotal in the relocation of arable agriculture away from the lowlands were the changes in the ecological makeup of the basin after the 1560s.

That the lowlands of the Mediterranean were perilously prone to frequent and intermittent flooding from the latter half of the sixteenth century largely shaped the destiny of the Inner Sea in the seventeenth and eighteenth centuries. Granted, the glaciers advanced, ice floes got denser, and winters became more severe. But the Little Ice Age was not a period of uninterrupted humidity and wetness. Rather, the fluvial activity and cold spells it unleashed were concentrated episodically, and the intensity of these phenomena waxed and waned over time.[144] Still, increased climatic variability played a crucial part in the desertion of low-lying lands. To contain the gullying waters of the rivers that were routinely transformed into foaming torrents within a few hours, in the seventeenth century the Ottomans built in the Balkans "high hog-backed bridges without piles, to give as a free passage as possible to sudden rises in the water level."[145] In fact, most of the repair and reconstruction of the empire's stone bridges, which were continually weakened by high waters and flooding, took place precisely between the seventeenth and nineteenth centuries.[146] The situation was not all that different in Spain at the end of the seventeenth century from Córdoba to New Castile, where some of the bridges were destroyed by floods. The bridge at the entrance gate of Córdoba, built by the Moriscos, was destroyed by a seasonal flood in the 1680s. The Spaniards built another bridge on top of the old one, with seventeen arches to accommodate the increased volume of riverine flow.[147] An area of 140,000 hectares in the basin of the lower Guadalquivir was taken over by the *marismas*. The Italian peninsula was dotted by landscapes of marshes, wetlands, and rice fields.[148]

Thus, the episodic and recurrent flooding affiliated with the Little Ice Age created an environment conducive to the march of marshlands, much to the detriment of lands won from the wild—or "invented"—during the long period of reclamation that started at the turn of the millennium and went on until the late sixteenth century, even though the process was momentarily reversed during the fifteenth-century crisis.[149] Given the dizzying pace of reclamation and settlement after the 1450s, low-lying lands that hemmed the mountains of the Mediterranean were fully exposed to the calamities that areal erosion could inflict on them owing to increased wetness, including the takeover of the arduously built arable by paludal and stagnant waters. In the process, the lowlands lost their appeal as preferred sites of settlement and cultivation from the early part of the seventeenth century. Conversely, the hillsides and higher altitudes found favor among the former residents of the low landscapes as well as the newcomers, and this stayed so well into the late nineteenth century, if not the twentieth. In the interim, the population of the lowlands declined, to the benefit of ambient hills and higher ground.[150] With the low-lying lands menaced by inundation and economic slowdown in progress, what struck travelers in the Roman and Iberian lowlands was the emptiness of the plains. The inland low landscapes of Corsica, Sardinia, and Cyprus were desolate as well. The Ottoman countryside was not immune to this ecological trend. That two-thirds of today's villages and nine-tenths of the cultivated parts of inner Anatolia date back only to the latter half of the nineteenth century is a perfect testament to how scarcely populated the peninsula was before then.[151]

The threat of looming inundation may have been present for most of the Little Ice Age, but the intensity of deluges associated with it was subject to ebb and flow. The threat took on a menacing tone from the 1570s: a great number of coastal and inland plains populating the basin turned amphibious.[152] Flooding turned into a perennial problem in Andalusia and in Lower Languedoc (not to mention the sown fields of Maremma in Tuscany) at the close of the sixteenth century.[153] Because of malarial conditions, some of the plains could only be exploited periodically. Almost habitually, the coastal plains were deserted at dusk by those who inhabited them during the daytime for hilltop villages:

agricultural hands who tilled them and shepherds who tended their herds in these lowlands left them for the safety of higher ground, withdrawing from the realm of the fen-lands and the mosquito. The thin settlement network of the Pamphylian and Cilician plains of the Ottoman Mediterranean in the sixteenth century was erected by nomadic or transhumant groups who, having built their villages on the surrounding heights, brought certain parts of the maritime plains under temporary cultivation and used the rest as winter grazing. The coastal plain in Cilicia was also sparsely populated by settlements that contained less than a hundred souls at the most whereas villages on the hilly environs housed at least three hundred. In both cases, vast swampy and malarial stretches posed a serious threat to the nomads who tilled the plains on a temporary basis as they did to Napoleon who, in his campaign to capture Acre, was forced to move his army on dangerous hilly terrain through a long defile rather than cross the marshy *via maris* that cut through the plain of Sharon.[154] In Algeria and Morocco, it was malaria that decimated the French settlers, and some villages had to be colonized several times over in the Mitidja, the "graveyard of colonization." The British and French armies suffered tremendously in Macedonia in the Great War, despite the protection extended by quinine.[155] In the period between 1887 and 1920, over 2 million suffered from malaria in Italy.[156]

Even though the first wave of heavy flooding that engulfed the Mediterranean took place from the 1570s to the 1630s, the loss of productive land to the wild was not initially regarded as a grave problem, at least until the 1650s.[157] This was because, as discussed previously, agricultural production, which had kept up with population growth for over a century until the 1560s, started to trail behind, resulting in the relative scarcity of bread grains. And the upward pressure this deficit placed on cereal prices survived well into the mid-seventeenth century. Along with soaring grain prices, farming gained a new esteem among would-be capitalists and moneyed urbanites as the return of money to land picked up in pace, not only in the northern Italian peninsula but also in Catalonia.[158] In the eastern Mediterranean, the initial impulse provided by mounting urban demand was later compounded by the demise of the Levant trade from the 1620s as a result of which part of the wealth accumulated from previous commercial undertakings was channeled to land: the Cairene merchants invested in the production of flax and sugarcane.[159] Those who had already established their claim over agricultural surplus tried to consolidate and increase their share of it: members of the Ottoman military-administrative cadre in Aleppo and Damascus, for instance, strengthened their hold over the countryside by advancing loans to the peasants, buying *mîrî* land, and recruiting sharecroppers.[160]

In Valencia, lands left behind by the Moriscos after 1609 were appropriated by the nobility, not immediately but eventually. Given the insatiable financial needs of the Habsburg throne, nobles for the most part invested in royal finance rather than in land. All the same, wealthy citizens of Madrid, Toledo, Burgos, and Valladolid were investing in nearby villages during the last quarter of the sixteenth century. Urban investment in rural property was turning into a strategy preferred by greater numbers of people.[161] The consolidation of land holdings on the one hand, and the increasing purchasing power of the big land engrosser on the other, led to the retreat of the smallholder in Languedoc.[162] That is, significant sums of money moved into the countryside. In the case of the Italian city-states, money flowed into their *contadi*—and in the case of Venice, into the Terraferma—on account of the wealth these illustrious cities commanded and the habitual crop shortages they experienced.[163] The upward movement of agricultural prices and the return of money to land, at least in the affluent parts of the basin, encouraged attempts at controlling and reversing the spread of marshlands through drainage. Yet, the steep rise in grain prices, which attracted money to land, slowed down and ended in the 1650s at the latest. Grain prices remained depressed for over a century.[164] The hold of the nobility, lay and ecclesiastical, and the wealthy over land got stronger in the following decades on account of plummeting agricultural prices. But by the same token, however, their land holdings could not be organized as *grundherrschaft*-like units precisely because of depressed market conditions during the seventeenth-century crisis.[165] The sordid state of the countryside and its corollary, the attendant growth in the share of urban populations, were acknowledged in the Ottoman empire by the reliance of the new taxation system introduced at the turn of the seventeenth century on *avârız*, "extraordinary taxes," designed to tap into urban wealth instead of on agricultural production, as was the case previously. These developments arrested the momentum of sedentarization, and nomadic life once again became more appealing.[166] During the eighteenth century, when the imperial throne routinely resorted to tax-farming to raise money, it was mostly urban stations and trades that captured the attention of bidders, not the countryside. In the Iberian peninsula, the widespread practice of the sale of towns to raise revenue for the imperial throne eloquently anticipated the developments in Ottoman lands, in spirit as well as in practice.[167]

From the second sixteenth century, the return of swamps and desertion of land therefore became a region-wide phenomenon. And when organized attempts were made to reverse the advance of marshlands, they were invariably located in the opulent zones of the Mediterranean such as the Italian peninsula, and not in Andalusia or Cilicia. The task required great capital outlays and infinite expense, which the financially strapped imperial bureaucracies and enterprising nobles could ill-afford. When grain prices were at their height at the turn of the seventeenth century, the Roman countryside hence presented a landscape that stood apart from the rest of the region, with the possible exception of Lower Languedoc, where marshy areas near Arles, Narbonne, and Fréjus were drained, albeit on a limited scale.[168] As long as money was pouring into agriculture, reclaiming lands from marshlands remained a priority. The fall in grain prices slowed down, if not reversed, the return of wealthy classes to farming, and with it attempts to reclaim land waned. Drainage and land improvement in the fluvial environment of the Little Ice Age, surely costly for the petty producers, seemed costly for the landlords and nobility as well. Even in Italy, efforts by pontifical and urban governments and capitalists to reclaim land in the 1650s could not have been more ill-timed. Investment in land was initiated at a time—the late renaissance—when in the valleys and coastal zones of the Mediterranean, low-lying marshy landscapes were reappearing, and worse, spreading.[169]

In fact, the draining of Lower Languedoc was discontinued in the 1660s.[170] As attempts to drain swamplands tapered off in the Roman campagna, the Mediterranean lowlands presented a much more uniform façade after the 1650s, as its plains were laid to siege by the marshlands and vegetation that invaded any open ground in its wake and slowly devoured even grazing lands. Malaria-bearing anopheline mosquitoes swarmed the coasts and pushed back the limits of human habitation. The oppressive presence of fever—*ısıtma*—was singled out by Evliyâ Çelebi as the root cause of the sparseness of population on the banks of Ottoman Mediterranean.[171] In Ottoman Palestine, six hundred out of seven hundred villages were located in the mountains in the opening decades of the nineteenth century, not in the valleys or plains afflicted by malaria. The Cilician plains were no more than badly drained, fever-ridden, and thinly populated lands.[172] In the Iberian peninsula as well, malaria, which established itself in the lowlands and rice fields in the late seventeenth and eighteenth centuries, had replaced all other diseases as the principal affliction of the realm.[173] In the foothills of the Cévennes and the Massif Central, there had been some thirty epidemics of malaria, some lasting five to seven years between 1398 and 1792.[174] The withdrawal of agriculture and the parting of permanent rural life, in turn, helped extend the realm of the "wild"—that of the swamps, the reeds, and of course, the much-maligned sovereign of this natural milieu, the malarial mosquito.[175]

When travelers in the seventeenth century visited the Mediterranean countryside, they were struck, even in the relatively better attended Roman Campaign, by the landscape welcoming them, a landscape rolling as far away as the eye could see yet populated by shepherds and flocks of sheep, and "occasional farms or cornfields, isolated and set at a long distance from any other."[176] To the west of the Agro Romano stood the great marsh, or *albufera,* of Valencia, fed by the rivers Turia and Júcar. The marshes of the Ebro river delta and Lower Provence commanded an imposing presence, and only a tiny portion of their sodden lands was brought back to cultivation before the nineteenth century, in the case of the former not before the 1850s.[177] In the east, the situation was not any different: vast tracts of brush and heath continued to cover the countryside along the route between Gallipoli and Adrianople.[178] According to the merchants and consuls who were stationed in Alexandretta, the port-city of the Cilician maritime plain and one of Aleppo's outposts on the Mediterranean, marshlands that occupied the lowlands made the air so "heavy and unbearable" that they thought they were in imminent mortal danger.[179] The marshes, in this instance, tested the capabilities of beasts of burden, horses and camels mostly. For they were laden with merchandise and forced to cross this inhospitable territory on their way to and from the port.[180] The local populace was able to evade the menace of pending fever and malaria only by moving up to the refreshing altitudes of the surrounding mountain range in the summer months when the air was at its heaviest in the plains.[181] Throughout the width and breadth of the Mediterranean world, the seventeenth and eighteenth centuries consequently witnessed a drastic reduction in commercial cereal agriculture in concert with the emptying of the lowlands.

Put differently, the waning of the Mediterranean entailed not only a shift of commercial agriculture away from the Inner Sea to the Baltic and the Atlantic but also a transferal of the basin's center of economic activity from its lowlands to its hillsides and mountains. Gone were the times when the Mediterranean itself, from Sicily and Cyprus to the Levant and North Africa, was the site of vibrant commercial cultivation. The changed landscape was starting to make itself felt by the mid-seventeenth century, when Evliyâ Çelebi toured the Ottoman realm. The thinly settled, low-lying plains still silhouetted the coastal regions of the Inner Sea at the close of the eighteenth century when Volney journeyed east and when Swinburne traveled the Spanish and Italian peninsulas.[182] When compared with the heyday of the city-states, the change in the basin's landscape could not have been more striking. When ibn Battûta sailed along the shores of the Mamluk empire and Ottoman and *beylik* Anatolia in the mid-fourteenth century, the eastern Mediterranean he saw was copiously dotted by port-cities, prominent and minor, from Payas to Alaiyye in the north to Tripoli in the south.[183] Despite the heavy toll successive waves of crusader attacks and the Black Death had inflicted on the region, and despite the appearance during the crusades of malaria on its shores,[184] the coastal regions of the basin were livelier than in the seventeenth century. The change in landscape did not mean, of course, that the lowlands were completely deserted. Unlike the plague, malaria did not usually kill its victims, but sapped their energy and vitality.

Owing to the retreat of cultivation, the Mediterranean plains—inland and coastal—remained home to nature for most of the seventeenth through the mid-nineteenth centuries. As such, in the eyes of settled populations the lowlands were infested not only by mosquitoes but also by serfs and outcasts, shepherds and wanderers, those with very little agricultural occupation and those who made a living from the marshlands.[185] Travelers who visited the region were struck by the fact that most villages were built on high ground, a long way from low-lying territories.[186] In other words, evacuation of the plains had populated the highlands, and most rural settlements chose a site midway between crops in the valley (or the plains) and the forests of the mountainsides. The plains of the Mediterranean remained thinly populated until the nineteenth century. If the wetter clime of the closing decades of the sixteenth and the opening decades of the seventeenth centuries placed the lowlands of the basin under fluvial siege, then the fall in agricultural production in concordance with the slowdown in demographic growth sealed the trend.

The contraction of cereal husbandry was further strengthened during the seventeenth-century crisis, which reinforced the conversion of the arable into pasture. Just witness the slowdown in the tempo of sedentarization in the Ottoman lands and the pastoral *redux* in Iberia and Puglia. The basin consequently suffered from frequent shortages. Scarcity of corn was menacing Catalonia as late as the closing decades of the eighteenth century when Swinburne visited the region. At the time, the principality of Barcelona could not produce "above five months provision. Without the importation from America, Sicily and the north of Europe, it would run the risk of being famished."[187] This was when the absence of habitation and large quantities of uncultivated land was the order of the day, from Valencia to Andalusia, where the depopulated wastes were of "vast extent."[188] It was only from the end of the eighteenth century that the Mediterranean shores of Iberia were woven into a new commercial web, this time around the centrality of the port of Barcelona.[189] Provence did not fare any better. In 1796, the lower Rhône region of Provence had "grain sufficient for only five months," and this was not an exceptional year but more representative of a longer, debilitating trend.[190] That cereal cultivation and trade in Ottoman lands lost the momentum they possessed prior to the seventeenth century can be gleaned from the correspondence regarding the *waqf* villages. What they attest to is an increase in the number of complaints about the inability of peasants to pay their dues on time and in full as well as the draconian demands of money-lenders who profited from this inability.[191] The situation was similar in Languedoc, where debt-ridden tax-payers abandoned their holdings and "decamped." In the second half of the seventeenth century, the amount of grain handed over by tenants to the granaries of landlords was down by 20 to 50 percent. Under such demanding conditions where debts and arrears accumulated pitilessly on account of falling prices and economic contraction, there was an exodus from land, reaching its apogee between 1690 and 1720. The exodus continued until the 1740s, and in the process, the neglected arable turned to "waste."[192] The sharecropping arrangements of the seventeenth century, so widespread in southern France (and the Italian peninsula), eventually gave way to the formation of large estates precisely because neither the landholders nor the sharecroppers were content with the returns. The weakening of sharecropping arrangements gained pace in the eighteenth century. In certain parts of the Ottoman empire, land surface measurements were replaced from the seventeenth century by seed capacity measurement, a telling indication of increasing availability of land as a result of forfeiture and extensive agriculture.[193]

The Italian peninsula did not present a considerably different façade.[194] In the seventeenth and eighteenth centuries, the peninsula was not peppered, as was the case in its heyday, by landscapes of orderly rotation and fallow, but was speckled by scores of pastures, thickets, fens, and marshes punctured by a few open fields. These open fields owed their presence not to regular, systematic tilling but to temporary clearings, because breeding had by then taken over as the principal economic activity supplanting or complementing agriculture. More crucially, this kind of landscape extended in the seventeenth century not through the "Roman Agro, the Corsi of Calabria, the Tavoliere of Puglia, and the Maremmas, but also through vast Sicilian, Neapolitan, and Sardinian fiefs."[195] In most regions of the peninsula, the system of temporary clearings came to prevail over systems of fallow and biennial rotation. In other parts of the peninsula, when the rural economy exhibited resilience, it was mostly due to the expansion of vineyards and mulberry plantations, as in Lombardy, or to the proliferation of chestnut trees on higher altitudes.[196]

So, a retreat from the lowlands and the contraction of commercial grain production along the banks of the Inner Sea took place in unison. Even in places where commercial grain cultivation did not retreat, at least initially, in the face of the march of tree crops, grain shortages or failures that stemmed from climatic variability placed restrictions on production: Sicily, for instance, exported close to 40,000 tons of wheat a year for most of the seventeenth century, but the years 1605–8, 1617, 1622, 1628–29, 1635, 1641, 1648–51, and 1660 were all marked by crop shortages.[197] Overall, then, the Australia and Canada of the Mediterranean in the sixteenth century suffered the consequences of climatic fluctuations in the seventeenth century. Even here, though, eventually more land was devoted to mulberry and olive trees. Accordingly, the grain trade conducted through the port of Marseille oscillated between a paltry 1,200 and 1,600 tons in the mid-seventeenth century.[198] Although the volume of grain dispatched from one province of the Ottoman empire to another in case of crop failure reached at times 10,000 tons (as in the case of the shipment from Hama to Aleppo in the taxing year of 1757),[199] the amounts exported from its ports remained modest. In 1709, the tax-farmer in charge of the customs of Aleppo had 2,400 tons (20,000 *charges*) of wheat in storage, and the Pasha of Urfa 3,600 tons (30,000 *charges*), both of which were shipped to France to alleviate the miseries inflicted by the famine of that year.[200] In the eighteenth century, approximately 6,000 tons of wheat were shipped from the Ottoman lands to the French ports; the amount reached "an exceptional average of over 200,000 *charges* (24,000 tons) in 1750–1754."[201] And in the western Mediterranean, 12,000 tons (100,000 *charges*) of wheat were delivered per annum from the port of Constantine to the Compagnie d'Afrique.[202] Tunisia, too, exported grain for most part of the eighteenth century, albeit sporadically, but even the modest sums its plains supplied shrank drastically in the first half of the nineteenth.[203] After the early seventeenth century, Cyprus, which had been a grain surplus zone, no longer offered such surpluses. Consequently, provisionalist policies became more common than not. In the urban centers of the basin, real wages fell as grain prices skyrocketed, forcing municipal and imperial authorities to enact measures to ensure the steady provisioning of urban populations by stabilizing the violent gyrations in grain prices.[204] The erratic and episodic nature of the grain trade as well as the relative thinness of its volume remained in place.

It was not coincidental that as grain production in the Inner Sea diminished, the regions that took over the task of provisioning Istanbul—namely, the coastal plains circumscribing the Black Sea—were all located in the uppermost septentrional latitudes of the climatic and ecological limits of the Mediterranean.[205] The northerly shift in the location of the principal sites of cereal husbandry that supplied Istanbul paralleled the northerly shift in the location of the breadbaskets of the world-economy during the seventeenth and eighteenth centuries: the movement that was inaugurated by the transformation of Baltic grain into the mother trade of the Dutch empire was sustained by the assumption of this function by Sweden, England, and England's American colonies.[206] The maritime and inland plains of the Pontic Sea were not the only regions that profited from the retreat of cereal husbandry in the Mediterranean. Nilotic Egypt, though of the basin geographically and economically, was not necessarily so ecologically and climatically. As such, it was outside the boundaries of the Mediterranean clime, as were the northern reaches of the Euxine Sea. Both, however, went on to produce considerable quantities of grain that were shipped to Istanbul—and Mecca, in the case of Egypt, at least until the turn of the eighteenth century.[207] Moldavia alone shipped 28,000 tons (350,000 hectoliters) of wheat annually to Istanbul, which consumed close to 130,000 tons of wheat a year in the seventeenth century. In the nineteenth century, more than 60 percent of the capital's bread grains originated in the Balkan provinces.[208] As discussed in the previous chapter, the shift had started earlier: the Aegean shores of the empire came to supplant the established port-cities of the sixteenth century, with İzmir and Salonika gaining prominence.[209] Therefore, deprived of most of the economic activities that underwrote its prosperity, the coastal plains of the Levant remained sparsely populated. Meanwhile, Egypt and the Scythian coast, lands that were major districts of grain production under the Roman empire, reassumed their position. Conversely, no district of the Mediterranean "proper" remained host to large-scale commercial grain cultivation. That the downturn in cereal culture occurred within the Mediterranean clime suggests that forces other than, and in addition to, those of the world-economy were at play. These forces were ecological in nature.

Grain husbandry may have suffered a double blow due to the waning of the Mediterranean and the advent of the Little Ice Age. But the latter, with its wetter clime, assisted the expansion of rye cultivation in the Iberian and Italian peninsulas, rice and maize cultivation in various corners of the basin, all of which deflated the significance of wheat.[210] After all, the shortened growing season and increased precipitation associated with the Little Ice Age compelled producers to switch to crops that could cope with the vagaries of the new age. If the advance of the marshlands was exacting on winter crops and cereals, it was not necessarily so for other crops. It helped establish an environment, which, if utilized to its fullest extent, could facilitate the spread of aquatic crops such as rice, or summer crops such as cotton that depended on irrigation, as we examine in chapter 5. Indeed, the organizational capabilities of the Ottoman imperial state previously mobilized to sustain riziculture were no longer needed. Rice cultivation, which provided a staple of immense importance to the army in its long campaigns, had become enough of a staple so as not to warrant imperial supervision.[211] The gradual and unyielding advance in cotton cultivation and the consolidation of the sway of livestock husbandry, too, owed their luster to the new ecological order.

Malaria- or fever-ridden, subject to periodic incursions by floods, and re-claimed by the "wild" from the seventeenth century, the plains ceased to serve as the frontier and forefront of territorial expansion and settlement. When and if tilled, the low landscapes were put under the reign of the buffalo or large numbers of cattle, given the lacustrine nature of these lands. The Pontine marshes were a refuge for herds of wild buffalo, in the midst of wildlife.[212] In the eighteenth century, six plowings were advised for the heavier soils of the low-lands of eastern Languedoc whereas four sufficed on lighter land.[213] In Ottoman lands, for example, the spread of large agrarian estates, *çiftlik*s, a creation of the marshy lowlands, served as a stimulus to the diffusion of the use of water buffalo.[214] Or, large numbers of plow oxen were put to work, "no less than twenty-four ploughs at work in the same field, each drawn by a pair of oxen," on the rich plains of Andalusia.[215] Not surprisingly, some if not most of these large estates were devoted to livestock-rearing,[216] given the extending range of land that could be utilized as pastures and grazing land. More often than not, they were inhabited by temporary dwellings or were sparsely populated.

In the Veneto, too, it was livestock agriculture that attracted most of the urban investments to satisfy the growing demand for meat and dairy products. And here too, the phenomenon of building villas not only assumed greater significance in the seventeenth and eighteenth centuries: 332 new villas were built in the seventeenth century and 403 in the eighteenth, up from 84 built in the fifteenth. But also, by subjugating the surrounding countryside to their needs, they "removed vital ingredients" from it, and contributed to the "further disaggregation of the degraded landscape," now a site of "temporary clearings and nomadic pasturage."[217] In this light, neither the agricultural contraction that brought the reclamation process to a halt, nor the shift that favored non-cereal crops and led to a severe contraction in cereal production, were par-ticularly Ottoman phenomena. Nor was the rise of livestock farming—and livestock-related—goods.[218] During the seventeenth and eighteenth centuries, it was the trade in livestock goods that replaced the cereals: hides, wool, wax, and leather crossed the borders more often than cereals.

Seen from the vantage point of ecological change in the Mediterranean, two points need underlining. First, these two centuries, *mutatis mutandis*, favored large landowners or landholders in the low-lying lands, for large estates were better equipped to withstand the fall in prices, given the size of their holdings. By extending the spread of their holdings, large landholders reduced their exposure to adverse market conditions by passing risks on to producers through various arrangements. Or they spread risks emanating from the volatile times of the demise of the Mediterranean by obtaining lifetime tax-farming. Climatic variability of the seventeenth and eighteenth centuries favored big landholders. Since landowners and landholders engrossed—or tightened their control over—land, mechanisms of "internal insurance" they built against bad harvests were refined by the variety of areas they either cultivated or farmed-out. The producers lost some of the liberties they enjoyed during the sixteenth century, forced to switch from the freehold arrangements of the *beau seizième* to the straitjacket of sharecropping; some were demoted from their petty-producing status to the status of hired hands who were put to work on larger estates, and still others lost the certitudes of the prebendal, *timar*, system to face the vagaries of the tax-farming—*iltizâm*—system. These changes did reflect the lean times of the seventeenth-century downturn faced by all producers in the Inner Sea.

The Ottoman *çiftlik*s, however, should not be seen as poor replicas of the "second serfdom," but as a manifestation of the basic ecological and economic traits of the Mediterranean landscape during the seventeenth and eighteenth centuries. Unlike manorial estates in east of the Elbe, the *locus classicus* of commercial grain production in the seventeenth century, *çiftlik*s in Ottoman lands were limited in number and ranged from 30 to 500 hectares in size.[219] Moreover, since these estates were ordinarily built on *mawât*, wasteland, their formation and consolidation was rarely accompanied by the alienation of the direct producers. With the retreat of agriculture during the Little Ice Age, when the *mîrî* lands were being alienated, the lands that went on to be exploited were more naturally devoted to livestock-raising than to cereal cultivation. In matters agricultural, then, developments along the Aegean shores of the Ottoman empire reflected trends commonly shared by the other quarters of the Mediterranean in their growing emphasis on non-cereal agriculture. As the perfect embodiment of these trends, the *çiftlik*s did not represent an anomaly, but in fact conformed to the norm.[220] (Moreover, transformations outside the realm of livestock and cereal husbandry can better be captured not within the purview of the *çiftlik* system but from the vantage point of the *waqf*, mortmain, and *tamlîk*, quiritary, systems.)[221] The *latifondo* economy in Italy, too, vaunted a wide array of crops and products, but a large part of its resources was devoted to a combination of raising cereals and sheep.[222]

Seen from a bird's eye-view, however momentous the rise in agricultural production during the sixteenth century was, the expansion was comparatively less impressive than what preceded and followed it. Agricultural expansion of the sixteenth century helped recover ground lost during the fourteenth-century crisis.[223] The recovery did not boost agricultural output as remarkably as had been the case between 1000 and 1250, when colonization of vast expanses of land was realized by inventing lands through land clearance, or *débroussaillement*, assarting forests, and draining marshes.[224] A similar development took place during the late eighteenth and nineteenth centuries with the opening, first, of the Russian plains and, later, of the temperate regions. And it was during these two periods of exceptional growth that large companies dealing with and specializing in grain trade emerged on the historical scene: witness the rise of the Bardi and Peruzzi who built their fortunes on the shoulders of the phenomenal expansion of the thirteenth century; Cargill and Continental followed in their footsteps to build their empires from the late nineteenth century and are still with us.[225] There was a long-term contraction in the Mediterranean basin's agricultural output, and the levels of agricultural output that were attained before the advent of the plague in the mid-fourteenth century were not reached much less surpassed until the agricultural "take-off" of the late eighteenth or early nineteenth centuries, a contraction that lasted more than four centuries.

Second, a similar agrarian organization came into being to the north and east of the Alps, from Poland, the new epicenter of commercial agriculture, to the Balkans, now largely deprived of its role of provisioning the Italian city-states. The reason behind similarities in organizational forms can be reduced to an ecological necessity, keeping lowlands under cultivation, rather than to the demands of the world-economy. In other words, arguing that Ottoman *çiftlik*s were organized differently than their counterparts elsewhere in that their animal farming was as significant as their farming operations obscures the similarities in the ecological setting and its implications on the organization of production. The wealth that fell under the hands of the nobles in the east of the Elbe provided solace from ecological distress. *Çiftlik*-owners, on the other hand, did not have to resort to such funds. On account of ecological change, livestock agriculture became and remained a remunerative economic activity. "Wasteland" suited to large holdings was therefore plentiful and easily accessible. If the lingering presence of *çiftlik*s cannot be tethered to the presence of export markets, the differential impact of the Little Ice Age in undermining wheat cultivation has to be entertained. By increasing wetness, it aided rye cultivation.

The ecological undercurrents or implications of these changes are examined in the following chapter.

Overall, then, it was not the geographical relocation of the agrarian heart of the world-economic order away from the Inner Sea alone that left its deep imprint on the degree and nature of commercial agriculture on its shores. There was an attendant transformation in the basin's landscape concomitant with the change in its ecological makeup. As discussed in the second chapter, prior to the turn of the seventeenth century and the Little Ice Age, the preferred sites of commercial agriculture in the basin were and remained its plains. From the seventeenth century on, *a contrario*, it was the strategic hills, fluvial terraces, rising ground, foothills of mountains, and higher elevations that came to be favored locations for settlement and tillage.[226] From the plains of Esdraelon and Akkar on the eastern littoral of the Mediterranean to those of Çukurova and Antalya on the Mediterranean shores and Salonika on the northern shores of the Aegean, from the continental plains of Konya to those surrounding Mount Erciyes and Jabal Nablus or Mount Ansariyya, the lower altitudes of the Ottoman landscape were rarely inhabited on a permanent basis.[227] Too, the lowest points of the basin were largely evacuated: the plain of Murcia, the mouth of the Danube, the inland plains of Cyprus, and the Amik plain of Antioch.

The advent of Baltic grain into the Mediterranean, the gradual rise of second serfdom in lands devoted to producing grain, and the dampening effect of Baltic grain on cereal prices lent support to the process of abandoning the plains as well as cereal husbandry. From then on, the practice of keeping low-lying plains under cultivation took on two distinct and contrasting forms. These lands were tilled on a temporary basis by pastoral peoples who used them as their winter pastures or by the inhabitants of surrounding villages who had established satellite settlements therein. This went on during the eighteenth century, but with the end of the seventeenth-century crisis between 1730 and 1750, attempts and struggles to establish rights over these abandoned and "underutilized" lands multiplied. Sizeable agricultural holdings came into being through the acquisition of land, as in the case of Languedoc, or through the consolidation of claims and rights of usufruct and property over such lands by lay or temporal bodies, as in the Spanish and Ottoman lands: these sizeable holdings were scarcely turned into capitalist estates. The march of riziculture in Italy in the seventeenth and eighteenth centuries, or the widespread cultivation of maize and tobacco in the Balkans, took place under the watchful eyes and the organizational ability of the nobility or large estate-holders.[228]

If not farmed, the abandoned lowlands were mostly given over to pasture.

Often, the earmarking of lowlands for the use of herds took place in a de facto rather than a de jure fashion. Given the multiplicity of oftentimes conflicting claims over lands that were not planted with cereals for long periods of time by occasional tillers, grazers, and the inhabitants of nearby villages, there were attempts by some claimants to enclose such lands for their exclusive use or turn them into "enfeoffed" property.[229] The easy availability of such lands led to the encouragement given by the Ottoman state to ruling class members to "reclaim" waste lands: this allowed the placement of abandoned, *mawât,* lands under single ownership with the express purpose of turning them into large estates.[230] Grazing, under the direction of influential organizations such as the Dogano and the Mesta, expanded at the expense of other economic activities: hence, the tightening of these organizations' hold over such lands, especially during the eighteenth century, referred in the literature as "refeudalization."[231] In other words, the forfeiture of the low-lying lands of the basin by their previous tillers did not mean that claims of usufruct or quiritary rights over them abated accordingly. To the contrary, these erratically exploited lands put to work by multiple parties not only provided an additional safety valve for populations who were stricken ecologically and economically, by giving them an additional breathing space, but also constituted fertile ground for contestation among those who utilized them.

Reversal in the Fortunes of Low-Lying Lands

That the low landscapes, dotted by marshy plains, were afflicted by rapid erosion, floods, and malaria did not necessarily mean that attempts to improve the sodden lands of the plains were either lacking or rare. In fact, as discussed previously, the *beau seizième* was a time of reclamation and settlement of new lands, but the sixteenth-century recovery and the Medieval Optimum that accompanied it both lacked longevity. Under the intimidating shadow of the Little Ice Age, trying to fully recoup losses from increased erosion and flooding in the post-1350 period proved to be a Herculean task for most, with the predictable exception of the Italian city-states. Given its brevity, the long sixteenth-century recovery was overshadowed not only by the previous wave of sustained and prolonged reclamation, which brought large swathes of low-lying plains under tillage from 1000 to 1350, but also by the following wave that started in the 1750s and extended well into the latter half of the twentieth century.[232] Nonetheless, relatively short-lived though the recovery of the Medieval Optimum may have been, the pace at which the colonization and improvement of new lands progressed during the long sixteenth century was still impressive. An excellent indication of this economic revival was the high settlement density reached in the low-lying parts of the Ottoman empire in the late sixteenth century. The number and population of villages and hamlets that blanketed the Konya plain in inner Anatolia and the coastal and inland plains around Damascus, as recorded in sixteenth-century imperial land surveys and tax registers, were only slightly lower than their namesakes in the mid-twentieth century.[233]

Between the sixteenth and the nineteenth centuries, the expansion of the arable slowed down at first, and came to a halt later, and sizeable stretches of the Mediterranean plains reverted back to the wild. Even when growth resumed during the eighteenth century and even when the urban bourgeoisie and local notables-cum-landowners made strides in bringing new lands under cultivation, the pace of reclamation remained anything but impressive: in Lower Provence, a total of 3,000 hectares of land was drained between the end of the sixteenth century and the Revolution: this amounted to 3 percent of the arable in Arles. The drainage of the marshes of Petit-Poitou failed to substantially increase land supply: several thousand hectares were added to the land under tillage.[234] The ratio of cultivated land in France, which had dropped to 36 percent of the whole by the end of the Wars of Religion, remained so for a considerable period of time.[235] In fact, land transfer that took place in the seventeenth and eighteenth centuries due to the reclassification of marshlands as common lands was so vast that the crown had to ban it, because the de facto transfer of land was eroding its tax base.[236] The situation was not decidedly different in Mediterranean Spain (Valencia, Murcia, Aragón, Albacete), where the amount of newly plowed land increased by a little over 400 hectares in the seventeenth century, and in Valdichiana, where 2,500 hectares of land along the main channel of the lower Chiana was restored to regular cultivation by the mid-eighteenth century.[237] Stated differently, even when there was an expansion in the region's arable, the gains were infinitesimal. This held true for the Ottoman empire as well: land converted to regular cropping rose by 35 percent over a span of a century and a half—between 1700 and 1850—as opposed to 103 percent worldwide. Crucially, part of the expansion took place at the expense of forests. As settlements moved onto higher elevations, forests shrank by over 10 percent due to the felling of trees and thinning of the wood cover not only in the Levant and North Africa but also in the northwestern Mediterranean. French forests diminished in extent in the eighteenth century. The rate at which the forests were assarted along the shores of the Inner Sea was considerably higher than the world average of 4 percent, as we discuss in the following chapter.[238] Meanwhile, the area covered by grasslands and pasture declined imperceptibly, by 0.4 percent: the world

average was 0.3 percent.[239] The Ottoman countryside was still scantily populated at the close of the eighteenth century. The Rishwân tribe, for example,
which spent summers on the Anatolian plateau, east of Ankara, changed its
winter pasturage from the east of Aleppo to its west: its wintering quarters
shifted from plains around the salt lake of Jabbûl to the Amik plain, without any
hindrance. One of the first tasks set by Ibrahim Pasha during his occupation of
Syria in the 1830s was to settle Egyptian immigrants in the swampy valley of
Baisan.[240] In France, too, over the centuries, lords who found drainage to be an
unprofitable undertaking allowed their villagers to use shallow marshes for
pasture and other agricultural activities.[241]

Rural settlements that crowded the region's low landscapes in the long
sixteenth century thus got sparser after the 1650s. The tendency toward the
emptying of the low-lying lands was not reversed when economic conditions
improved roughly from the mid-eighteenth century on account of the multiplicity and longevity of the processes that subjugated the region to the fluvial
offensive and climatic variability of the Little Ice Age. The tendential decrease
in the density of rural settlements remained stubbornly in place until the latter
half of the nineteenth century, though attempts at draining marshes started in
the 1750s. Still, that the lower landscapes of the basin remained under siege by
ecological and climatic forces did not mean that they were left to the pestilential
mosquito alone. There were two reasons for this. The first was the momentary
return of capital to land. In the eastern Mediterranean, for instance, fortunes
accumulated in the Levant trade were partly invested in land and in commercial agriculture at least until the mid-seventeenth century, if not later.[242] The
foray of money into agriculture cleared the way for wealthy classes to establish
their presence in the countryside without much resistance, since the inhabitants
of these sites, if not already relocated elsewhere, were more than happy to
consent to any tenurial arrangement that would allow them to hold on to their
land. Genoese merchants bought up estates in the kingdom of Naples; large
seigniorial villas mushroomed in the Venetian countryside. The eighteenth
century was "one of the golden ages of château-building" in France.[243] During
the ages of enlightenment and reason, large agricultural estates remained as the
key fixtures of the Mediterranean landscape. The phenomenon was not confined to lay estates alone. Ecclesiastical estates followed the same trend. The
Church managed to recover the property it lost during the civil war in France;
in the Italian peninsula in the seventeenth century, close to one-third of all
landed property was classified as ecclesiastical mortmain. In the Ottoman empire too, in the eighteenth century the revenues accruing to the pious foundations were in the range of one-third of the state revenue.[244] The plains were

thus speckled by large-holdings, controlled by temporal and spiritual bodies, when demand for cotton and wheat picked up pace after the conclusion of the seventeenth-century crisis. Most of these holdings, then, on account of the forfeiture of lands after the 1650s, were located on lands that were considered to be waste, paludal, or common, such as the coastal plains of the Inner Sea as well as its interior basins along routes of transport. Thessaly, Epirus, Macedonia, Thrace, the Maritsa valley, Danubian Bulgaria, the Kossovo-Metohija basin, the coastal plains of Albania, and western Asia Minor were the primary sites of large estate formation in the Ottoman empire, as we discussed above.[245]

Under these conditions, it was the landowners who in effect took over the responsibility of combating the forces of nature—the advance of marshlands, strictly speaking. Capital outlay needed for the fight was the landowners' burden. When and if their attempts at rehabilitating these lands failed, as they often did, or when they found them to be costly, as they often did, they resorted to force, a method that was never absent from the *çiftlik*s in the Ottoman empire. More often than not, the employment of force to recruit and subordinate labor was a good gauge of how wretched the conditions in the lowlands were. A core group of sharecroppers fixed in the lowlands and a mobile labor force of seasonal and temporary hands who were imported from the mountains to carry out the labor-intensive parts of the production process helped keep agriculture in the plains alive. If utilizing force was one way of keeping lands under the plow, so was paying laborers high wages to compensate for toiling in the disease-ridden environment of the plains. This was the case in Maremma.[246] Yet, this option was open only to a few. Otherwise, landholders who remained anchored in the plains had to find ways to tempt (through sharecropping) or coax people from the uplands to populate the lowlands in their possession.[247]

The second reason the low-lying lands remained in use was a growing adherence by the denizens of the basin to strategies designed to accommodate climatic variability and deterioration of land—both of which reflected the changed rural landscape. For one, access to, and exploitation of, land was governed by a complex set of rights on account of the shifting nature of cultivation and grazing. There were many towns and villages in Spanish Iberia and the Ottoman empire, more so in the case of the former, where sown lands were periodically reallocated among peasants so as not to anchor one's fortunes in the same plot of land but to give them the chance to cultivate plots of differing quality and location.[248] Pastures and woodlands were held in common. Cultivating lands at different altitudes offered a similar guarantee against the unpredictability of natural forces. Shifting cultivation practiced by pastoral populations provided the producers with spatial flexibility as well. In addition, labor mobility, especially in the case of Spain, enabled the rural producers to diversify their sources of income as insurance against bad times. The producers, that is, were not powerless against exacting ecological forces.

Where the countryside was not as strongly gripped by the powerful—as in Andalusia, where the majority of land was held by the *señoríos,* or as in the Italian city-states—but was held in lien by the municipal corporations, communal rearrangements allowed the usage of these lands by periodically reallocating them. Two-thirds of the land was in the form of "locally controlled pasture, common, forest, or waste."[249] Wastelands, *baldíos,* were "vast tracts of poor-quality scrub and marsh used extensively for pastures and gathering." Uncultivated tracts, *realengas,* were also used for "grazing and collecting."[250] Reallocation of land, as we will see, allowed the producers to redefine the limits of the arable on a continual basis in line with the new ecological forces. In seventeenth-century Provence, it was the pastures located on higher elevations, *terres gastes,* which were turned into cereal lands. Ironically, the clearing of the forests gave the waters of the Rhône more freedom to invade and further sodden the low landscapes.[251] As the clearing of the forests placed bears and wolfs under peril, the lowlands were given over to wild boars, foxes, and antelopes. When inhabited, these lands were covered, as in the case of Konya, by huts of simple trellis-work of sticks and reeds, suspended on four poles and elevated from the soggy ground.[252]

Not surprisingly, a relatively bright spot in this ecological setting was Venice and its Terraferma where, between 1500 and 1800, close to 150,000 hectares of land was brought under cultivation through drainage. After all, the Italian school of hydraulics was famed for perfecting techniques of landfill in the plain. By diverting the muddy waters of the flood to lowlands that were flooded or easily flooded, their technique attempted to make the lowlands—where alluvial sediment of the earth carried by the water was deposited—progressively rise in level.[253] Yet even in Venice, most of the advance took place before the mid-seventeenth century when the Mediterranean was losing its resplendent glare. Until then, land reclamation through drainage and irrigation followed relative price movements, and relative scarcity of land led to the draining of marshes and the construction of irrigation canals. The onset of the seventeenth-century crisis, which depressed cereal prices, and the newfound centrality of Baltic grain to the functioning of the world-economy brought this dynamic to an end. Only in the United Provinces, befitting a hegemon, were forces of nature tamed successfully: the amount of land drained from the sixteenth to the nineteenth centuries reached 280,000 hectares. Not only that, but Cornelius Meijer and the master hydrologist Cruquius, equipped with the idea of drainage windmills, were making tours along the shores of the Mediterranean, in Provence in particular, advising urban governments and landowners on matters related to draining marshlands and fenlands, though the scope of their operations remained relatively modest.[254] It was also then that the most advanced drainage technology of the time was put to use by Humphrey Bradley, who was granted the newly created office of *maistre des digues* by Henry IV in 1599, for the reclamation of vast amounts of land—in the west near Nantes, in Normandy near Mont St. Michel, on the northeastern coast, and along the Mediterranean near Arles.[255] Again, the results of this wave of reclamation were all but impressive. Otherwise, where the power of wealth was not a factor, most were powerless against the colossal and unrelenting forces of the wetter and colder climate. At the turn of the nineteenth century, Aleppine merchants were trying to bring 600 hectares of land under cultivation by draining the swamps on the coast of Antioch, with little success.[256]

The commercial and ecological conditions of the seventeenth and eighteenth centuries therefore inescapably reversed the trends of the long sixteenth century. Most rural settlements found themselves perched, once again, in the hills overlooking the Inner Sea and the plains, thanks to the introduction and improvement of strategies that allowed peasants and sharecroppers to keep the imperiled lands under cultivation, such as periodic redistribution of land and temporary clearings. The slowdown in the pace of land bonification clearly manifested the limits of ecological recovery that took place during the *beau seizième*. Impressive though the pace of reclamation was, the fresh outbreak of malaria and the fight against the marshes from the late sixteenth century decreased the effectiveness of land improvement.[257] By the mid-seventeenth century, marshlands had secured their advance in the French countryside. Past the 1650s and until the Revolution in 1789, however, little was accomplished in the French Mediterranean by way of bonification.[258] It was not coincidental that the only irrigation canal built in Provence between 1350 and 1750 was constructed in the 1550s, just before the start of the Wars of Religion.[259] What was ironic was of course that just when malaria was building its imperium along the banks of the Inner Sea, the cure for it was also making its debut in the basin owing to the efforts of the Jesuits. The Holy City itself was malarious and in need of a cure for fevers. The congress that elected the Jesuit vicar-general every three years provided the procurators who represented the Peruvian Jesuits to take with them, from San Pablo to Rome, new supplies of the febrifuge bark, otherwise known as cinchona, because the plant was used for curing chills and hence to treat the disease. In the mid-seventeenth century, trunks filled with the bark were transported to Rome via Spain. Its distribution unfortunately remained confined to the curia rather than circulating more widely.[260] And why quinine was effective in reducing the severity of the symptoms of malaria remained a mystery until the turn of the twentieth century.[261]

The return of the Little Ice Age therefore cut short the recovery of the sixteenth century. Despite the fact that the age of the Genoese reactivated the Mediterranean economy and momentarily reduced the impact of growing climatic variability, economic recuperation of the basin was cut short, more so at the opposing ends of the Inner Sea. However fast-paced the process of reclamation was, the gains of the 1450–1560 period were not as impressive as that of the 1000–1250/1300 period, when not only was the process more prolonged, but also the pace of colonization and settlement was much faster. In the eastern Mediterranean, for instance, the sedentarization of large nomadic populations that moved west was largely completed by the mid-fourteenth century. So was the takeover, colonization, and settlement of new lands in the Iberian peninsula and France (and elsewhere).[262] Drainage and irrigation of the lowlands were essential to the success of the invention of new lands. The building and upkeep of networks of drainage channels and the maintenance of the economic health of flood plains were owed to the efforts either of individual landowners, as was the case in the lower Rhône region, or urban municipalities, as was in the case in Pistoia, Tuscany.[263] In most places, the movement was led by abbeys: they played an important role in draining as in the pays d'Auge in Normandy as well as a large part of southeastern Provence before 1400. More generally, Benedictine and Cistercian abbeys founded on marshy terrain reclaimed vast amounts of land from the twelfth to the fourteenth centuries by crisscrossing them with a network of *dugali* (larger common drainage ditches).[264] The crucial role that water—its control and distribution—plays in most *waqf* operations, which acted as the forerunners of colonization from the Levant to North Africa and Valencia, points to a similar process in the eastern and southern shores of the Mediterranean.[265] Where abbeys and religious foundations did not play as prominent a role, protection against flooding was provided by local men who were placed in charge of performing the tasks of inspecting and repairing dykes and ditches and of manning them whenever flood threatened.[266] So, hydraulic systematization allowed landholders to sow their plots without fear of inundation.[267] When the drainage and irrigation channels were neglected, and if the mountains were too quickly deforested, altering the conditions of the flow of the streams, as was the case after the mid-fourteenth century, conditions in the lowlands worsened accordingly.

If the arrival of the Black Death in the 1340s and the shortage of hands that ensued rendered the upkeep of this network almost impossible, the contemporaneous growth in the number of herds traversing the Mediterranean lowlands and mountains on a regular basis must have played a role in the reshaping of the region's landscape. The newcomers to the basin's upper elevations may have made things worse by opening up new lands by felling trees as their sheep and goats were thinning out the *maquis* cover, thereby compounding the volume of run-off water pouring into the lowlands. This of course reduced the ability of the soil to absorb water. Judging by the thinness of silt deposition on its shores, Mediterranean Iberia seems to have been exempt from the malarial menace before the thirteenth century, before large-scale transhumance became popular.[268] In Tuscany too, there was no evidence of transhumance before the thirteenth century, and in Byzantium before the twelfth.[269] In effect, it was the tilling of the slopes in the Byzantium that eventually paved the way for the systematic practice of summer pasturage and long-distance pastoralism from the eleventh or twelfth century.[270] In the Italian peninsula as well, no major feature of the mountains sharply distinguished them from the plains until the twelfth century or so.[271] Granted, the mounting need by urban centers for the resources of the highlands (e.g., timber, mining) opened the uplands to intensified human activity and settlement.[272] Yet, the concomitant arrival of large-scale transhumance and the turnabout in climatic conditions played a significant part in the ecological transformation of the Mediterranean.

As the threat of erosion reappeared with the onset of the Little Ice Age, the situation changed: great floods inundated the Ebro valley repeatedly (in 1325, 1329, 1380, 1448, 1466, 1488, and 1518). The alpine streams, which remained "well-behaved" until the mid-fourteenth century, turned torrential and the Rhône floods started to visit the foothills with greater frequency and ferocity, and changed the balance to the benefit of the uplands.[273] The crisis of the fourteenth century and the arrival of the Black Death, which undermined the economic well-being of the plains, by reducing demand for draught animals (and timber) did not inflict as heavy a toll on the uplands as it did in the lower altitudes of the basin.[274] Urban investment in land in certain regions of the basin gained pace (in the richer zones of the Italian peninsula, such as in Florence, where rural investment shot up after the Black Death), but overall, the desertions of the era, the sharp decline in population, and the frequency of intense flooding hollowed out the lower river valleys and the plains.[275] The clearing of mountains, however, did not cease because of population influx from the lowlands.[276] In the Lower Rhône region, the work of draining the marshes started only in the fifteenth century, and uplands overlooking the low-lying plains remained the preferred sites of settlement. After 1350, local authorities seem to have been little concerned with either improving the quality of already drained fields or extending the cultivated area by draining marshes.[277] Livestock husbandry went on to recast the landscape until the latter half of the eighteenth century, maybe not too incidentally to the beginning of a new wave of reclamation.

The gradual and timid return to land that followed the end of the seventeenth-century crisis was dissimilar in nature to that which followed the fifteenth-century crisis. In the case of the latter, the end of the economic downturn in the 1450s was accompanied by a precipitous return. After the 1750s, however, it was the urban patriciate who "embarked upon the conquest of the country-side."[278] In spirit, the tax-farming system that gained eminence in the Ottoman dominions—long-term in duration in the seventeenth century and turned into life-term later—was a variation in this path of rural transformation. Whether they were administered by landowners or landholders, these estates inhabited similar ecological environments and operated under similar constraints. Curiously enough, even the gothic architecture of the modest mansions and watch-towers in the estates of the Ottoman notables on the Aegean Sea coast replicated the "Italian" style, owing to the migration of construction workers from the former colonies of Genoa. Symbolically, the replication of the gothic style on the opposing shores of the Aegean bespoke to a deeper similarity that emerged from the manner in which the lowlands were put to use.[279]

The next round of colonization in the basin's lowlands took place at the height of the *pax Britannica*, when agricultural production gained new momentum worldwide.[280] With the start of mid-eighteenth-century expansion appeared the prospect of extending the arable, first, for planting cotton, the demand for which soared, and then, for cultivating wheat, from the 1840s. This prospect involved a turnabout, because it entailed turning pastures and lands laid to waste during the seventeenth and eighteenth centuries back into crop-lands and, thereby, the draining of lowlands. Under *pax Britannica*, world wheat production more than doubled, from 916 million bushels in 1831–40 to 2.1 billion in 1881–87.[281] Starting timidly in the nineteenth century, the process assumed greater force at the turn of the twentieth, when the plains of the Ebro, Cilicia, and Thessaly, just to mention three, were opened up to permanent settlement. Others had to wait until after World War II.[282] From this point of view, the Mediterranean landscape at the end of the nineteenth century had more in common with that of the late sixteenth century than even that of the early nineteenth. Improved security, drainage of lowlands and plains, eradication of malaria, and the demands of the *pax Britannica* all contributed to this reversal of fortune.

The tempo of reclamation hastened in the latter half of the nineteenth century: the land that was brought under cultivation in the Ottoman empire in sixty years, from 1860 to 1920, exceeded the new additions to the arable registered in the previous century and a half, from 1700 to 1850, by a million hectares.[283] The movement gained further momentum during the twentieth century. In Syria, an enormous amount of land that was infrequently used in the past was brought into regular cultivation from 1850 to 1950, and hundreds of places developed from hamlets to sizeable villages. Excluding the Jazira, about 2.5 million hectares of new land were plowed up and about 2,000 villages were established on this newly won land; the figures for Transjordan were 40,000 hectares and 300 villages.[284] During the same time period, a total of 2.8 million hectares of marshlands was reclaimed in Italy in the course of a century or so after the unification. More impressive, by 1950, land reclamation had been applied in some form or other to 7 million hectares.[285] In Spain, the arable increased by 165,000 hectares (20,000 hectares of which was along the banks of the Mediterranean), and in France the share of the arable jumped from 36 percent of the total in 1600 to 60 percent by 1890.[286]

The ebbing of the Little Ice Age altered the environmental and ecological components of the order that laid down the foundations of the ascent onto the mountainous zones. In tandem with the revisions in the world-economy's axial division of labor, it unleashed two sets of transformations, closely related and complementary, which modified the region's agrarian order from the second half of the nineteenth century, an order familiar to us today. The first of these transformations was the start of the re-colonization of the plains, due in no less part to global warming, and the resumption of urban growth, both of which abetted the resurgence of the imperialism of wheat. Surely, King Wheat was the principal agent which both made and hastened the pace of reclamation of the plains. In 1845, for one, wheat was an export crop of no significance in Ottoman lands. Yet, by 1850, it had reached the fourth rank, and by 1855 the second. And the empire's volume of wheat production more than doubled, from 20 million bushels in 1831–40 to 40 to 50 million in 1888. By the end of the century, wheat had come to reclaim the primacy it had enjoyed during the long sixteenth century.[288] The central Anatolian plateau, which was an important yet second-rank producer of grain, turned itself into the leading producer of the region in the twentieth century. And in the piedmont regions of the eastern Taurus range (Antioch, Kilis, Antep, Urfa), the plains turned productive and were devoted to wheat. Comparatively diminutive in size, fallow fields became a rarity in these plains.[289] The villages were still small in size, though: scattered over the fertile and extensive plain of Kilis was a population of 50,000 that resided in 640 villages.[290]

An important factor that gave a big boost to the settlement of the plains was the passing of new land codes in the Ottoman empire in 1858; in France following the Revolution; and in Spain and Italy during the Napoleonic era. In the mid-nineteenth century, when the Ottoman central authorities were striving hard to settle the nomads by force in order to hasten the turning of pastureland into cornfields, the Land Code designed to frame the new wave of colonization functioned as a perfect complement to it. That the low-lying lands were largely unoccupied allowed the new code to be implemented without displacing rural producers. Claims on these lands, if legitimate, were honored; if unclaimed, they were put up for sale by the central government. Such vast expanses of land were acquired with the hopes of reclaiming them, an expensive task open to the wealthy few. The Sursuq family, for instance, who extended their holdings after the new Land Code, had come into possession in 1872 of lands in excess of 20,000 hectares (200,000 *dönüm*s).[291] Unlike Rumelia, where most of the *çiftlik*s were built in the plains in the seventeenth and eighteenth centuries, the Anatolian and Syrian plains, which were set aside for temporary cultivation and grazing, remained largely uninhabited, very much like *mawât*, wasteland.[292] In Spain too, "approximately 10 million hectares of church and municipal land" were sold between 1836 and 1900.[293] The revolution in France permanently clarified authority and property over land and water. Claims to marshland were transferred to villages, while claims to water and rights of way were preempted by the state. The law also prescribed simple rules for drainage, and empowered the government to force reclamation if landowners could not agree to carry out projects by themselves.[294]

The second set of transformations was closely related to the first. With more land in the plains brought under the plow or under the control of those who became landowners with the passage of the 1858 Land Code, the space of maneuver available to livestock-holders shrank notably. From then on, most regions were "forced to choose a single major economic activity."[295] Since the mid-nineteenth century, a number of more favored regions have ceased to grow the full range of Mediterranean crops and specialized in one or two cash crops.[296] The import of these two sets of transformations on the Ottoman countryside cannot be emphasized enough: the landscape that envelops the region today, to a large extent, came into being during the 1850–1950 period.[297]

At any rate, when land improvement was shouldered by smallholders and in a piecemeal fashion, the amount of land reclaimed remained modest. This was the case in Albufera lagoon in Valencia: reclamation of lands in the Ribera Baja allowed 6,163 hectares of rice to be sown between 1796 and 1920. As a result of better drainage, incidences of malaria in the Ribera Alta, which had previously alarmed the officials and caused the imposition of limits on the extension of cultivation, declined from the 1860s.[298] In Spain between 1912 and 1947, some 900 hectares of the Albufera of Valencia were reclaimed illegally by piecemeal reclamation, particularly around the edges of lakes and lagoons.[299] Ambitious schemes were started, not many for the first time but more successfully than before. Many of the lands reclaimed in the region since the nineteenth century have been developed by large commercial companies or the states: in Italy, the reclamation of marshlands was almost entirely confined to the Po plain in the 1861–1915 period: 330,000 hectares were drained. So were the Pontine marshes (75,000 hectares), the Maccarese plain (41,200 hectares), and the Tirso in Sardinia (126,500 hectares).[300] In Emilia, the area of low-lying marshland decreased from 89,000 hectares in 1870 to 78,000 in 1906, 60,000 in 1925, and 56,000 in 1950.[301] Only recently, toward 1900, was the Mitidja, behind Algiers, finally claimed for cultivation; marshes in the plain of Salonika in 1922. It was on the eve of World War II that work was finished on the draining of the Ebro delta and the Pontine marshes.[302] Under the watch of the drainage companies and the newly independent states, the momentum of colonization and bonification picked up considerably, well into the latter half of the twentieth century. Unlike the recovery of the sixteenth century, the draining of paludal lands in the nineteenth century turned, once again as was the case in the 1000–1350 period, into a prolonged and sustained process of land improvement.

The withdrawal of commercial cultivation from the low-lying, marshy plains and the resurging popularity of hill- and mountainside cultivation and settlement were what distinguished the seventeenth- and eighteenth-century Mediterranean from its fifteenth- and sixteenth-century predecessor. During the course of the nineteenth century, but mostly gaining velocity from the 1850s, the low landscapes of the Inner Sea were steadily yet inexorably re-colonized. The tempo of recovery reached a crescendo when upland populations started to descend into the lowlands, to relieve the demographic overshoot that was taxing the higher altitudes of the basin. In order to depict the symbiotic relationship between the lowlands and the uplands, the following chapter chronicles how the world of the hills, previously given a big boost by increased climatic variability, erosion, and flooding, came to be subjugated to, and reshaped by, the colonization of the lowlands.

CHAPTER FIVE

New World of the Hills

The second sixteenth century commenced in the 1560s. It was then—with the onset, most prominently, of the Wars of Religion—that the upheavals and traumas that violently shook the empires in the Mediterranean until the mid-seventeenth century commenced. The principal factors that triggered this wave of social and political turbulence were twofold: on the one hand, an intensification in the pressure on land occasioned by the sustained demographic growth during the long sixteenth century and, on the other, the inflationary upsurge generated by the arrival of American silver in the Inner Sea during the age of the Genoese.[1] The rebellions that razed settlements across the Ottoman realm from the 1590s to the 1650s, albeit with periods of quiescence in the interim; the revolts in Habsburg Catalonia, Naples, and Sicily in the 1640s and later; and, of course, the Wars of Religion in France brought forth, or contributed to, the desertion of lands under cultivation. Deepening the atmosphere of insecurity in the countryside was an escalation in taxes imposed predominately on agricultural producers to provide succor to the financially strapped imperial bureaucracies of the era. Conjointly these factors prompted an egress from rural areas.[2] The demographic outflow caused tears and lacerations in the rural fabric of the Mediterranean countryside. Correlatively, in the urban centers of the basin, real wages fell as grain prices skyrocketed due to inflationary pressures, forcing municipal and imperial authorities, from Genoa and Venice to Andalusia and

Istanbul, to enact measures to ensure steady provisioning of urban populations by stabilizing the violent gyrations in grain prices. So erratic was the grain harvest that in 1615 "people in the streets in Madrid, money in hand, were begging for bread."[3] Provence is said to have evaded peasant insurrections during the conflict-ridden seventeenth century primarily because grain, imported via Marseille, was always available.[4]

In the face of great fluctuations in grain output and trade that marked Mediterranean agriculture in the latter half of the sixteenth century and the growing magnitude of rural-to-urban migratory movements, measures introduced by imperial and urban bodies to cope with the drop in food supplies ranged from denying newly arriving rural migrants and the poor the rights extended to urban citizens in terms of food to encouraging or forcing uprooted peasants to return to their villages.[5] Additionally, mandating or cajoling nomadic populations to take up sedentary agrarian life was a policy often proclaimed by the Sublime Porte, even though such measures usually came to naught.[6] Deeply rooted in the social fabric of the Inner Sea, the afflictions that struck the Mediterranean countryside were thus not aleatory but systemic; as such, they were not easy to cure with short-term measures. Ominously then, the social and political tumult of the late sixteenth and the first half of the seventeenth centuries brought down the number of rural hands due to the displacement of people in the countryside at a time when the number of mouths to feed in the major and secondary cities was still multiplying.[7]

Accompanying and deepening the social turmoil that fueled waves of displacement and migration in the countryside was a concomitant weakening in centripetal forces. The process was initially set in motion by the inflationary pressures of the late sixteenth century, which unfavorably affected the finances and the capabilities of the central governments.[8] At times the vacuum created by the enfeeblement in the political and military might of imperial bureaucracies was promptly filled by governors, rebellious or not, whose urgent need to solidify an autonomous economic base vis-à-vis the weakened central seat of power demanded that they seek new sources of income and, thus, new avenues of remunerative commercial activity. Fakhr al-Din Maan, the governor of the subprovinces of Beirut and Sidon, was a case in point. The prominence gained by silk exports in Mount Lebanon's economic life in the opening decades of the seventeenth century and the governor's desire to expand the scale of silk production under his jurisdiction led him to encourage the migration of Maronite populations from the lowlands north of Mount Lebanon to the highlands of the Shuf mountains and Kisrawan.[9] In charge of two major gateways to the Mediterranean and in search of independence from the Sublime Porte, Fakhr al-Din

Maan also established close diplomatic relations with the dukes of Florence—who were intent on increasing their raw silk importation—albeit briefly and without consequence.[10] His political aspirations may not have borne fruit in his lifetime, but from an economic point of view, his legacy proved to be a lasting one, since sericulture later became deeply ingrained in the agricultural fabric of Mount Lebanon. Serendipitously for those engaged in silk production, recurrent problems in the westerly shipment of Caspian silk via Aleppo animated the local sericulture industry in the latter half of the seventeenth century. Thenceforth, the produce of the region was not only dispatched overseas, but also marketed in situ to satisfy the needs of local spinners and weavers.[11] The political travails of the seventeenth and eighteenth centuries that imperiled the countryside, especially the low-lying regions of the basin, thus resulted, among other consequences, in the relocation of the region's economic center of gravity to its hillsides, where tree crops thrived and small livestock ranged. Similar forces were at work in other parts of the empire, prompting widespread cultivation of mulberry trees for silk cloth manufacturing.[12] The re-anchoring of rural settlements on higher altitudes flowed from, and neatly complemented, the return of low-lying landscapes to the mosquitoes and the marshes, which we outlined in the previous chapter. How the inverse process of ascent that carried rural settlements from lower to higher altitudes unfolded is, in turn, the subject matter of this chapter.

Growing reliance on the produce of the highlands, tree crops in this instance, by the inhabitants of the region evinced a shift in the agrarian composition of the basin, because roughly from the turn of the millennium to the late sixteenth century it was the coastal plains of the Levant that provided the triumphant city-states with valuable commercial crops. Progressively from the seventeenth century, the hillsides and higher altitudes of the Mediterranean came to host the region's principal commercial crops—tree crops. What happened in Mount Lebanon was not a development specific to the Levant, but part of a larger movement that reconfigured the vast and multifarious landscapes of the basin at large. Along the Mediterranean shores of France, for example, the expanding dominion of the mulberry tree, along with that of olive and chestnut trees (and vines), transported the economic heart of the region to the relatively higher elevations of ambient hills and mountainsides.[13] And for most of the seventeenth and eighteenth centuries, silk and silk products (in the form of black taffeta and ribbon exports) remained the mainstay of the economy of the Lyon region, wine that of Marseille, and wool (and cloth) that of the Languedoc region: in 1750, two-thirds of France's exports came from its southern tier, outstripping the northern provinces.[14]

The changing balance in the Spanish empire between its center located on the Mesetas and its periphery at the expense of the former generated a similar economic dynamic.[15] The provinces that surrounded New Castile, all land-bound, suffered from the sprawling and overpowering dominion established over them by the new Habsburg capital, Madrid.[16] By contrast, the littoral provinces of the empire, like Catalonia and Valencia, escaped the demands of the capital city, and remained subject to the rhythms of a larger commercial world even when they were loosely integrated into the networks that extended into the Atlantic.[17] The mulberry tree and the vine played a crucial part in the economic buoyancy of the maritime regions of the Iberian peninsula, as they did in the Levant and Mediterranean France. Surely, the expulsion of the Moriscos, who were chiefly settled in the rough and mountainous parts of Valencia and were conversant with sericulture, opened the way for the arrival of, and the takeover of vacated lands by, the incoming Catholic families from both sides of the Pyrenees. And it was owing to the lucrative business offered by the mulberry tree and the vine—at the expense of sugarcane and, less significantly, rice—that the newly arriving populations found it relatively easy to make a living from the lands abandoned by the Moriscos. The sudden removal of the workforce severely curtailed the output of the sugar plantations and rice fields at the turn of the seventeenth century; riziculture, however, reemerged in the eighteenth, as we discuss shortly.[18] Of the lucrative branches of economic activity that the new settlers took over, sericulture and viticulture proved to be the most popular: the output of the former was dispatched to the textile industries in Seville and Toledo, and that of the latter was destined for transatlantic trade.[19] Extensive plantations of vines, olives, and citrus trees adorned the southern tier of the Italian peninsula as well. In Tuscany too, every act of land clearing, improvement, and systematization was primarily tied until the nineteenth century to the "initiative of planting trees and shrubs."[20]

Yet the all-accommodating markets of the prosperous long sixteenth century became a faint memory with the onset of the seventeenth-century crisis, which induced a contraction in the volume of economic activity. Worse, the wet and cold years that came to outnumber years of drought from the last quarter of the sixteenth century inflicted untold damage in Valencia on the mulberry trees, since the river Júcar "burst its banks several years in succession, laying waste some of the best mulberry in the kingdom" in the 1620s.[21] Rural producers thus concentrated their efforts on the exploitation of dry hills and steppes—and not the *huerta*, the rich, irrigated plain, in part because bringing marshlands under control entailed the employment of oxen and, by extension, availability of pasture lands. Yet grazing land was always in short supply because of the intensive nature of cultivation in the *huerta,* as was the case in Alcira. The scarcity of arable consequently turned the hillsides into principal sites of cultivation. Luckily for the Valencians, a decline in demand for silk from northern Italian cities and from the manufacturing centers in Castile was compensated, in some measure, by a rise in demand from the surviving rural textile industries of peninsular Iberia.[22] Mulberry trees as a result went on to populate the landscape: in Alberuqie, they occupied 22 percent of the arable in 1667, with another 11 percent partly given over to them; between one-fifth and one-quarter of the *huerta* in Simat de Valldigna was planted with mulberry trees, and another one-third partly so.[23] Later, wider commercial avenues opened up by Barcelona's rising star in the transatlantic trade helped the region to weather the storms of the eighteenth century by exporting wine, citrus, and rice.[24] With the rise in demand for cereals, rice resurfaced in the eighteenth century as an indispensable staple, and occupied higher and well-drained lands in the coastal region. In the Júcar valley, for instance, rice harvest came to surpass that of wheat in the eighteenth century.[25] That is, not only was most of the rough and mountainous range previously inhabited by the Moriscos eventually taken over by the new denizens of the region; the newcomers too managed to fully tap into the variant potentialities of the Mediterranean ecology, from the rice fields in the lowlands to the orchards and vineyards on the hills.

Given the scale and longevity of the transformation in the ecological and demographic makeup of the Inner Sea to the benefit of hillsides, it can easily be surmised that what drove large numbers of people to relocate their settlements to higher elevations was not political or social turbulence in the lowlands alone, and that reasons for it were naturally manifold.[26] The magnitude and long duration of the upward relocation of rural settlements points to the presence of deeply seated factors. The return of money to the land was one such factor. Certainly, this was a process more pronounced in the affluent regions of the basin, where fortunes made in long-distance commerce and in manufactures, especially in the second half of the sixteenth century, were later diverted to land. In Liguria, for instance, it was land hunger that forced the dispossessed to relocate their settlements in higher altitudes. The stranglehold established by the Genoese nobility over the Ligurian lowlands thus obliged the populations of this region to take refuge on higher ground.[27] Unlike the earlier wave of colonization of the highlands in the thirteenth and fourteenth centuries, when the need for montane resources triggered the incursions of the peoples of the lowlands, from the sixteenth century it was the availability of land suitable for simple subsistence that appealed to the new colonizers. This was attested to by the precipitate extension of the cultivation of chestnut trees. In higher altitudes, chestnut substituted for the bread grains of the lowlands:[28] it was pounded into flour for black bread, and chestnut—and, no longer grain—porridge filled the stomachs of the poor.[29] Popular before the assault of the Black Death, the chestnut tree, which had lost the preeminent role it performed in determining the contours of the Ligurian landscape until the sixteenth century, resurfaced with relish and zest.[30] The same pattern was evident in Languedoc, especially in its Cévennes region—the mountainous zone to the west of the Rhône, reputed as the Châtaigneraie (the "chestnut zone")—and in Corsica.[31] In any event, the retreat of the smallholder in Languedoc following the sale of Church property in 1569, and the acquisition of vast tracts of bread lands in Corsica by the Genoese merchants created a hunger for land in the low-lying plains of both localities, hastening the migration of the dispossessed to ambient and distant heights where there was land to be had, however toilsome was the task of clearing it for cultivation.[32]

An additional factor that tipped the balance against the low landscapes in the relatively affluent parts of the Mediterranean was the spread of an atmosphere of heightened insecurity along the shorelines of the sea as a result of generalized piracy. The dissolution of the maritime order posed a threat to the safety of commercially valuable ports and their neighboring coasts as the Signoria's powers of policing its *Mare Nostrum* weakened. Proliferation of acts of piracy contributed to the depopulation of the littoral zones of the basin.[33] Last but not least, the ferocity of the wave of pestilence that struck the basin from the 1650s only made matters worse by accelerating migratory movements mostly toward higher elevations, where the air was salubrious and remoteness from plague-ridden centers of settlement offered a precious shield.[34] The specter of plague lingered in the Levant until the 1840s.[35] The abandonment of the lowlands in favor of hilly slopes and mountainsides was thus a phenomenon that was induced by a wide array of factors, interrelated, self-reinforcing, and accumulative. It was also circum-Mediterranean in scale, as demonstrated by the cases of Valencia, Jabal al-Shuf, Jabal Kisrawan, and the Cévennes region.

Compounding the disorder in, and egress from, the lowlands due to political tumult, piracy, and pestilence was the onset of the Little Ice Age, as outlined in the previous chapter. Though heightened climatic variability disproportionately plagued the low landscapes, it too added to the growing attraction of the uplands by hastening the ascent of rural settlements onto higher, better-drained lands, hills, and mountainsides that stood at a safe distance from the malarial low-lying lands and, by association, feverous maladies affiliated with such environments. Moreover, the revival in tree-crop cultivation in the Mediterranean from the 1560s, as chronicled in part I, was facilitated by, and in turn solidified, the new spatial redistribution of economic activities. After all, the basin's hill-sides were perfect hosts to tree crops; its sheltering mountains provided sanctuary for the livestock in the form of summer pastures; its high valleys were used as arable; and the low-lying plains, wrapped in the welcoming warmth of the coast, served as winter pastures and, whenever possible, as bread lands.[36] It was not surprising that the number of hill settlements increased relative to those in the open country in the seventeenth and eighteenth centuries. Nor was the newly gained popularity of cereal cultivation on higher altitudes, especially of winter wheat, surprising.[37] The withdrawal of commercial cultivation away from the low-lying, marshy plains, and the re-centering of economic life around hill- and mountainside cultivation and habitation, in most places in the form of dispersed settlements, were what distinguished the seventeenth- and eighteenth-century Mediterranean from its fifteenth- and sixteenth-century predecessor.

Demonstrating the historical trajectory of higher landscapes was the village of Aín, roughly 50 kilometers north of Valencia and situated in the heart of the mountainous range of Sierra de Espadán, which crests at 1,083 meters.[38] After the Moriscos were expelled in 1609, the village was taken over by newly arriving Castilian- and Catalan-speaking peoples. During the course of the seventeenth century, population growth in Aín remained sluggish: it increased on average by 0.4 percent per annum. It needs to be pointed out that the exploitation of the Sierra on which the Morisco villages were etched had already decreased in intensity between the revolt of the Alpajurras in 1568–70 and the final act of expulsion of the Moriscos from the realm in 1609 owing to the decline in population in the decades after the revolt. The population of the village, after a feeble recovery from the lows it had reached in the 1570s, fell further after 1609, since the newcomers were substantially fewer in number than the departing Moriscos.[39]

Most of the land that became vacant as a result of the exodus thus initially reverted to nature, or was put to good use by transhumant herds from Aragón. The manure provided by migrant sheep herds and the resultant fertility of the arable, however diminished in acreage, and the abundance of irrigated land where yields were higher allowed the new inhabitants of the village to find sustenance from agricultural pursuits, perhaps without much toil. Wheat that was grown for the consumption of local inhabitants was, as we observed above, winter wheat. The drop in the number of inhabitants and in the intensity of cultivation during the lethargic seventeenth century accordingly abetted the region's recovery from the ailment of thinning sylvan resources that took place, *grosso modo*, from the eleventh century, particularly in the fifteenth and sixteenth centuries. With the fall in demographic numbers, previously cultivated lands and pastures reverted to woodland. The forests, destroyed in the fifteenth century, consequently "regenerated, soils stabilized, and stream channels were in equilibrium."[40]

The situation changed in the eighteenth century: the population, for one, trebled in the 1700–1787 period.[41] In concert with this demographic turnabout, pressures on the Sierra's ecological resources mounted. At first, slopes surrounding the village were brought under tillage with the aid of dry-farming techniques. The forests on the lower slopes receded, however; in their stead, tree crops—olive groves, figs, and vineyards—were planted. By the end of the eighteenth century, the orchards and vineyards had taken over from the former oak woodlands.[42] In the span of less than a century, Aín was transformed from a modest host to transhumant herds to a "typically" Mediterranean village, dotted with vineyards, orchards, and cereal fields sown to wheat. Local goats and sheep, though outnumbered by one to three by itinerant herds, complemented the Mediterranean triad. At the turn of the nineteenth century, wheat fields and olive groves populated most of the arable, and their products, commercially speaking, were of roughly equal value. Lesser in importance but still commercially significant were maize, figs, vegetables, and wine exports. Following these two categories at a respectful distance were carobs and raisins.

Economically and socially, however, the village reached a crisis point at the beginning of the nineteenth century with a sharp hike in the number of landless and an unrelenting pressure on the Sierra's natural resources. The intensification of agriculture provided momentary breathing room to the inhabitants of Aín: olive and cork production increased tremendously; maize was displaced by summer wheat and vegetables. The region took advantage of the phylloxera that hit the French wine industry in the 1860s, and expanded its vineyards.[43] Later, during the opening decades of the twentieth century, more and more land was devoted to cork oak planting in order to supply the expanding village cork industry. Even so, these new economic activities could not relieve the village of the strains generated by the rise in demographic numbers. With a population density of 159 persons per square kilometer of cultivable land, the village reached the height of its population in the 1880s, even though in most parts of the Sierra, the demographic climax was attained earlier, in the 1850s. Thenceforth, demographic decline set in and the population density fell: in 1950, it was down to 90.6, a drop of over 40 percent.[44] What precipitated the fall in population was the growing opportunity for migration.

From the second half of the nineteenth century, the revitalization of economic life, first, along the coast, and the gradual re-colonization of maritime and, later, inland plains to plant cotton or sow wheat—two crops for which demand was burgeoning as we underscored in the previous chapter—created new sources of revenue for the inhabitants of "congested" highlands. Furthermore, demand for industrial hands and service workers in the thriving port-cities and urban centers of the Inner Sea enhanced the attractiveness of the plains. The inhabitants of Aín, too, responded to the beckoning call of Barcelona. The ecological overshoot that resulted from a fall in the carrying capacity of land due to population growth compelled migrants to take up residence in the lower altitudes of the Sierra. A similar scenario, with significant variations, was unfolding in far-flung parts of the Mediterranean littoral.[45] Surely, crops that gave sustenance to the rural settlements on higher altitudes were varied: vine (and bark) in the case of Aín; chestnut trees in the Apennines region; silk in Mount Lebanon; and tobacco in Jabal Ansariyya. And very much like in Aín, the Apennines, Mount Lebanon, and Jabal 'Alawi started to experience population loss from the latter half of the nineteenth century, when the tempo of economic activity quickened in the lowlands and in the plains. In Jabal 'Alawi, tobacco cultivation, the *sahilî* or coastal variety, which provided a relatively comfortable existence to the inhabitants of the region for most of the seventeenth and eighteenth centuries, failed to accommodate rapid demographic growth with similar ease.[46]

Migration to ambient lowlands hastened by the extension of cotton cultivation to meet the demands of the French and English textile industries accelerated over time, especially during the Crimean War and the American Civil War when the Atlantic cotton and the Pontic grain trades came to a standstill, prompting commercial cultivation in the Mediterranean.[47] When the economic climate in the low-lying lands improved due to the jump in demand for cotton, the magnetic pull of high altitudes unavoidably lost force. The building of railroads, by boosting demand for timber and hands, both of which were plentifully abundant on the mountainous zones of the basin, had its share in the denudation and depopulation of the uplands. The inland plains were brought within the ambit of commercial agriculture which, in turn, enhanced downstream migration.[48] The rise and fall of Aín, in short, symbolized the trajectory of many an upland village which, after the better days of the seventeenth and eighteenth centuries, was forced in the nineteenth century to yield to ecological forces at home and to the lure of employment down below.[49] The descent of the inhabitants of the highlands and hill villages to cities located on the coast or in the interior to be employed in large agricultural estates or to provide workers for the manufacturing and service sectors of nearby urban centers was only one way of dealing with the strictures placed on the realm of human action by environmental factors.

In this scenario, the flux and reflux of the uplands was not the result of one major factor, but a confluence of a set of variables, which ranged from pestilence and piracy to the depopulation of the lowlands, not to mention epidemics of malaria in marshy zones. Similar movements of ascent of rural settlements had surely taken place before the late sixteenth century. In this instance, though, there was one new factor that considerably hastened the colonization of the uplands. This was the onset of a new round of vegetal dissemination in the wake of the Columbian exchange.[50] Historically, the Mediterranean basin witnessed two major waves of crop diffusion since the eighth century or so. The first originated in the Indian Ocean, and involved the dissemination and acculturation of tropical crops, the political and commercial repercussions of which we examined in part I.[51] One of the distinguishing features of the first wave was that the crops introduced during this westerly vegetal migration thrived best in the warm coastal plains and valleys with ample supplies of water for irrigation. The acclimatization and commercial cultivation of these exotic crops breathed new life into the littoral plains for centuries to come, as confirmed by the thalassic order put in place by the twin city-states in the Greater Mediterranean, an order that privileged maritime plains, coastal zones, and port-cities. Sailors traveled from one seaside inn to another, dining in one and supping in the next.[52] This aquatic world, which placed coastal "fields of the sea" at its center, involved a terrestrial spatial order. After all, it was the economic and political efflorescence of coastal areas that indirectly prompted the slow renaissance of the mountains. The economic momentum set off by the spread of commercially remunerative crops subjugated the mountain range to the demands of the aquatic world by assigning it the task of providing coastal areas with much-needed laborers and montane resources.[53] With the relatively brief exception of the 1350–1450 downswing, when the pace of economic growth slowed down and the mountainous sanctuary came to host growing numbers of people who fled from the Black Death and land enclosers, it was the coastal zones and littoral plains that set the pace of change along the shores of the Inner Sea. This wave lasted from the ninth to the late sixteenth centuries, as did the primacy of the lowlands.[54]

Hastened in the closing decades of the sixteenth century, the exodus of oriental crops from the Mediterranean shattered this economic and spatial order, since the transfer of main sites of sugar and cotton cultivation to the Atlantic Mediterranean largely deprived the lowlands of their economic dynamism.[55] Not incidentally, it was exactly at this juncture that the second wave of crop diffusion, which broadcast the crops of the Columbian exchange, began to penetrate the basin, in the process enriching the repertoire of crops available to rural producers. It modified the floral landscape of the Mediterranean. In essence, vegetal transformations caused by the second wave were quite unlike those caused by the first, because most of the crops that crossed the Atlantic survived at higher—as well as lower—altitudes. True to their provenance, tobacco, maize, and beans survived easily on upper elevations.[56] If the imprint left behind by this round of vegetal dissemination was not as noticeable as the first, it was simply because none of the food crops were successful from a commercial point of view. Rather they were effortlessly incorporated into the diets of the region's poor. The popularity of American food crops as the gruel or meat of the poor therefore reflected the dwindling fortunes of the basin's peasantry. More important, the arrival of these famine relief crops was taking place concurrently with the withdrawal of agriculture from the low-lying plains. As such, they contributed to the upward movement of the post-sixteenth century by providing the denizens of the region with cereal and animal feed (in the form of maize), commercial crops (in the form of tobacco), legumes (in the form of beans), and a wide range of garden vegetables. Most of the newly introduced crops were integrated into the crop-rotation cycle in place in the Mediterranean, and came to be planted in the spring and summer. In the process, they became integral to the region's diet, from the Veneto to the shores of the Nile.[57] With the waning of economic dynamism diffused by the coastal zones owing to the gradual dissolution of the order fashioned under the heavy hand of the Serenissima and the commercial slowdown that ensued, these new crops, along with itinerant herds and tree crops, helped secure subsistence at higher altitudes.

Spatially speaking, then, these two waves and the agrarian orders they engendered differed considerably. During the westerly migration of tropical crops, the demands of the coastal zones, the privileged sites of acclimatization, determined the shape and form of the basin's landscape. Not only the valuable lands of coastal zones but also rights to common lands, including the woodlands, were steadily appropriated by the wealthy and the powerful headquartered in cities along the coastal zone. Given the economic benefits derived from such control, it was not curious that even the majestic mountain range that wrapped the coastal plains was, *sensu lato,* collaterally affected by the economic and political efflorescence along the Mediterranean littoral. What altered this spatial order was a twofold change. The first was the oft-repeated migration of oriental crops to the Indies. Befitting patterns of the world-economy, the affixation of sugar production in the Caribbean hit first and foremost the coastal zones of basin, in the Levant as well as in Sicily and Iberia—the subject matter of part I. The second change pertained to the domain of ecology: the ultimate surrender from the 1560s of low-lying lands to the triumphal march of the swamps and malaria.[58] American food crops as the agents of the second wave of dissemination were, due to their vegetal constitution, not bound to take up residence in the imperiled lowlands but higher up, away from the feverous vapors of the lowlands—the subject matter of the previous chapter.

If the first wave prepared the ground for, and was in turn strengthened by, the high age of city-states, the second wave witnessed the waning of the city-states and, by association, the Mediterranean. After all, it was during the autumnal period of the age of the city-states when the Inner Sea could no longer host and prosper from the cultivation and trade of exotic crops that the region recovered its olden signal tree crops.[59] The vine and the olive tree had lost their commercial vibrancy to a large extent under the domination of the city of St. Mark—because of exotic crops and goods from which it derived its wealth due to its ability to monopolize these "scarce" resources.[60] From the seventeenth century, the renaissance in the cultivation of trees and vines occasioned almost spontaneously a transformation in the character of settlements that blanketed the Mediterranean relief. Perched and hill settlements reasserted themselves now that the supremacy of the coastal zones down below was no longer unchallenged, and the long-extant rapport between the lowlands and the mountains was fractured. Blessed by the arrival of American food crops and the newfound popularity of tree crops, hill and upland villages, hamlets, and scattered farmsteads came to dominate the basin's silhouette, gradually fashioning the Mediterranean landscape we are familiar with today. Equally germane in the formation of this rural landscape was the availability of lands in the lower-lying altitudes of the coastal range, which were now reclaimed by "nature." Unlike the homologous episode of the 1300–1450 period, after which a precipitous return to land took place, this proved difficult on account of the conquest of lowlands by the advancing marshes or by the urban patriciate and notables.[61] Given the difficulties associated with the draining and reclamation of marshlands until the "conquest of water,"[62] and conflictual claims over such lands, removal to lower ground by hill villages and dispersion of settlements proceeded unhindered, "frequently producing double villages or leading to more scattered settlement in the neighboring plains and valleys."[63]

How life in the Mediterranean uplands changed between the sixteenth and the eighteenth centuries, a process that eloquently complemented the withdrawal of cultivation and settlement from the lowlands, is the focus of this chapter. Of the three factors that helped sustain the ascent of the economic heart of the region to higher ground, the resurgence of lesser grains and the spread of tree crops were already discussed in previous chapters. Thus the first section of this chapter emphasizes the introduction of American food crops in the wake of the departure of oriental crops. It first explores the dissemination of these food crops that primarily benefited inhabitants of the uplands. As to be expected, the cultivation of these new crops was not strictly confined to the upper elevations of the basin: maize, for example, allowed large-holders in the Balkans to expand cultivation into low-lying lands preferably drained of stagnant water, and to stave off incursions of marauding bands or occasional intrusions of unruly and displaced mountaineers. But they made their reputation as crops resistant to the inclement conditions of higher altitudes.[64] As "garden crops," they facilitated the survival of many rural households, befitting the lean times of the seventeenth and eighteenth centuries.[65] It is also important to note that in concert with the spread of American food crops, the resurgent popularity of tree crops and lesser grains constituted the essential elements of the polycultural makeup of agriculture along the Mediterranean littoral. From the famous *coltura mista* of the Italian peninsula to Mount Lebanon, where vineyards were sown with wheat, intercropping became the governing principle of the new agrarian order. Mixed-cropping was not the only strategy employed by peasant households to establish deeper roots in the hilly and mountainous environments of the basin. As the second section explains, peasants adopted different strategies to cope with these changes: diversifying their sources of household income through labor mobility, especially in the Iberian peninsula, and diversifying the ecological environments they cultivated, prominent in Ottoman Anatolia, and by utilizing hamlets as agents in the colonization of vacant lands, as in the south of the Eu-Geneva line in France.[66] And finally, the third section demonstrates the limits of growth prompted by the ascent of settlements, and the toll exacted on highland environments by the jump in population from the late sixteenth century—as attested to by the case of Aín. That the tempo of economic growth was picking up in the low-lying landscapes exactly when the uplands were experiencing a slowdown turned out to be providential. Neatly complementing the draining of the lowlands, as highlighted in the previous chapter, the downward current of proletarians descending from the mountains assisted in the re-colonization and settlement of the plains: it thus relieved the former of its demographic strains, and populated the latter, neatly putting on display the complementarity of the two. Coming to a close was not the heyday of the mountains of the Mediterranean alone but also the Little Ice Age and the great agrarian cycle itself.

Vegetal and Ecological Basis of the
Mediterranean after the Age of the Genoese

As the heyday of city-states was coming to a close and the Little Ice Age was wreaking havoc in the lowlands beginning from the latter half of the sixteenth century, two developments loomed large in the basin. The first emanated, as sketched in chapter 4, from the desertion of lowlands that were re-conquered by the noxious miasmas of malaria: hence the consequent contraction of the arable led to a growing need for substitute cereals. The second development originated in the growing dominion of tree crops, as outlined in chapter 3, and reflected the economic prosperity of the age of the Genoese: hence the reemergence of tree crops and the ongoing popularity of livestock husbandry stimulated and eased the relocation of the economic life of the Mediterranean onto higher altitudes. These two processes were closely related: the abandonment of low-lying lands entailed a shift in the basin's center of economic life toward higher ground. On account of the contraction in the volume of cereal production in the relatively confined fields of the uplands, these two parallel developments boosted the demand for staples that could be successfully sown at high altitudes. This was not because sylvan resources could not in principle afford food security, but the felling of trees and human plenitude in the seventeenth and eighteenth centuries rendered the situation precarious.[67] The reemergence of lesser cereals helped redress the shortage in grains to a certain extent. Crucially, a factor that gained prominence in the age of the Genoese—the arrival of American food crops—satisfied both demands at once: maize and the potato, both of which could easily be grown at higher elevations, provided much needed substitute cereals. If maize replaced the costly wheat flour in the Veneto, as in *polenta*, in Genoa, it was the potato that replaced hard wheat flour, as in *tròfie*.[68] By the nineteenth century, in parts of Mount Lebanon, millet had come to be grown only when the maize harvest failed to satisfy expectations.[69] The potato served a similar function in the north: grain consumption in Flanders alone fell from 758 grams per person per day to 475 grams as potatoes replaced about 40 percent of cereal consumption.[70] Not merely this, as crops that could be used as animal feed, maize and the potato both helped livestock husbandry. As horticultural crops, they facilitated mixed-cropping: witness the marriage of trees to vines, the union between arboriculture and horticulture, or the association of the cultivation of grain with that of vines. Put differently, the American food crops satisfied most of the demands of the new ecological order.

The growing import of the new food crops became evident in Venice during the big famine of 1590. As discussed in chapter 2, northern grain arrived without delay. But so did an unfamiliar grain called *grano turco* or *sorgo turco*—maize, to be precise. It was also known, among others, as *bled d'Espagne*, which confirmed that the new crop had by the end of the sixteenth century become part and parcel of the vegetal makeup of the Mediterranean, from its eastern latitudes to the Levant. Introduced into the Veneto in the late 1530s or thereabouts, in most likelihood from the port-cities of Atlantic Iberia or, as the name of the new crops suggests, the shores of the Levant, maize cultivation subsequently spread to the entire Terraferma.[71] Therefrom it was imported into Venice in considerable quantities in order to alleviate pressing crop shortages. Principally if not exclusively, the urban poor and the peasantry of the Veneto evaded death by hunger thanks to "sweet" or "golden" polenta prepared with maize when wheat flour became scarce and dear. To ease the pressure on shrinking wheat supplies, the Venetian Grains Office, in a change of long-standing policy, authorized any mixture of grains to be baked and sold.[72] Not only in the 1590s but time and again in the following centuries, the triumphant march of Indian corn helped those in the countryside to survive the exacting times of famine and scarcity. Initially, the new crop was also grown in regions near Venice. Later, however, it became a permanent denizen of the Terraferma, but not of the Triumphant City. It was absent from the tables of the rich, and its classification by the agronomists of the period primarily as animal fodder captured its lowly, plebeian character.[73] A crop which, in its natal lands, was consumed roasted, boiled, whole, grated, or mashed came to replace local cereals and beans to form the essential ingredient of *polenta* (or *millasse* in southern France), and when given a chance, replaced most minor cereals.[74]

That is, just as the cultivation of sugar and, to a certain extent, cotton were being relocated in the plantations in the Atlantic to the detriment of the Mediterranean, American food crops were gradually and furtively making their way into the Inner Sea. Jointly, these two sets of vegetal transformations, one of egress that culminated in the mid-seventeenth century with the anchoring of sugar production in the Atlantic and one of diffusion that commenced in the first half of the sixteenth century, attested to the economic devolution of the basin. In sharp contradistinction to the crops of the Indian Ocean provenance that were lucrative, hence "noble" or "patrician," and thrived across the semi-tropical latitudes of the Mediterranean, the crops that arrived from the Atlantic were "proletarian" or "plebeian" in nature, especially since they started out as horticultural crops, available in the vegetable gardens of most rural households long before they ventured out into the fields.[75] Symbolically, the exodus of lucrative plantation crops, which was followed by the advent of maize, "poor man's cereal," and beans, "poor man's meat," reflected the reversal in the fortunes of the Inner Sea.[76] What is of cardinal significance from our point of view is that whereas the newly arriving oriental crops blessed the maritime plains from the eighth or ninth century to the sixteenth, it was the crops of the Columbian exchange that fittingly helped some of the inhabitants of the basin to escape from the scourges of crop shortages and famines. In accordance with the declining significance of wheat on the one hand and the upward ascent of rural settlements on the other, the spread of American food crops, as substitute cereals, helped buttress the transfer of rural settlements away from the infestious lowlands to the ambient highlands.

It was evident that in the wake of the gradual and steady withdrawal of cereal cultivation from the plains of the Inner Sea starting from the late sixteenth century, wheat had to be replaced by, or supplemented with, alternate bread crops. After all, most bread grains did not survive, or were not equally prolific, on upper elevations, not only for want of level ground but also due to the harshness of winters. Of tougher constitution than wheat and more resilient to frost and lack of sunshine, barley fared better on higher altitudes, yet it was not always the crop best fit to serve the needs of most peasant households. Unlike wheat, it was not an indispensable ingredient of peasant diet; it was cultivated as fodder, for horses primarily and not for sheep and goats that most mountain and hillside peasants owned and tended.[77] Rather, three crops gained popularity as substitute cereals in the Mediterranean in the seventeenth and eighteenth centuries: maize of the Columbian exchange; millet and rye of the Old World; and rice of the aquatic world.[78] Maize, for one, could mount to dizzying heights with ease, and inhabit altitudes other crops feared to tread; millet grew on altitudes much higher, and could withstand tougher climatic conditions, than wheat and barley; and rye fared well at high altitudes. Normally due to insufficient moisture, rye was unsuitable for the Mediterranean.[79] Arguably, the changed climatic conditions of the Little Ice Age proved beneficial to its wider cultivation in the Italian and Iberian peninsulas.[80] Rice, by contrast, made use of the aquatic environment of the lowlands and valley floors. Already popular in the eastern Mediterranean and known and grown in the Italian and Iberian peninsulas where it was introduced earlier, its cultivation was resumed from the late sixteenth century. It was also introduced into some new areas.[81] Though initially classified as an exotic crop, rice, in a twist of fate, had by the eighteenth century taken on connotations of a "poor man's fare." In some parts of the basin, it became a credible alternative to existent grains.[82] During the waning of the Mediterranean, when crop shortages and hunger held sway, it was these three sets of crops that solidified their hold on the shores of the Inner Sea. Mention also needs to be made of the fact that the potato, which, like maize, served as an alternate crop and livestock feed, gained popularity in the basin in the nineteenth century, though in a limited fashion.[83]

The diffusion of American food crops as cereal substitutes across the Mediterranean was anything but even or uniform. There were two distinctly discernible modes of acclimatization: even though the potato could be grown on lands unfit for—and more important, at altitudes inhospitable to—most other crops, its territorial expanse came to be initially limited to the vast plains beyond the northern limits of the olive tree, from the Baltic to the Black Sea, whereas maize came to be grown fairly extensively in the Mediterranean as well as in the Balkans.[84] The spread of maize and, to a lesser extent, the potato proved to be highly rewarding in the long run, since, in terms of yields, both cultigens had an indisputable edge over wheat, barley, and lesser cereals. Along the shores of the Inner Sea, yields from cereals were low in the fifteenth and sixteenth centuries, four- to fivefold being the norm. Shareholding arrangements the Ottoman state had with the *ortakçı*, sharecroppers, which stipulated the amount of seed to be sown as well as how the harvest was going to be apportioned between it and the producers, indicate yields ranging between four- and sixfold.[85] These figures paled in comparison with yields on maize and potato which reached, at least in theory, close to eighty- to a hundredfold at times. Owing to their abundant yields and promiscuity, both crops provided calories relatively inexpensively when compared with their well-established low-yielding competitors.[86]

The introduction of American food crops as brand-new vegetal additions to the basin's flora was of significance for three reasons. First, these crops thrived on upper—as well as lower—elevations, and were resistant to the inclement conditions at high altitudes.[87] Due to its ability to effortlessly ensure subsistence, maize, for instance, enabled the mountain populations of the Balkans to survive without being exposed to the malarial environments and the Ottoman armies in charge of the lowlands.[88] Despite the threat of pellagra, the phenomenal success registered by the spread of maize across the Mediterranean by the nineteenth century, from the Straits of Gibraltar to the Black Sea, was owed in large part to its ability to take root in lands deemed to be waste because of their sandiness, altitude, or aridity.[89] Maize received a warm reception in the humid environments of the Pontic mountains, Mount Lebanon, as well as in the Veneto and the Midi; tobacco established a commanding presence in the Levant as a cash crop where competition from the plantations in the Atlantic was not as fierce as it was in the western quarters of the Greater Mediterranean. Even though maize made its early appearance in the mid-sixteenth century, it was not until the mid-seventeenth century that it was marketed as a commercial crop, as in

Béziers. Evidently, it was the dire times of the seventeenth-century slowdown that encouraged experimentation with new crops. Maize was firmly in place in the Anti-Lebanon range in the eighteenth century.[90] Tobacco, entrenched in the hills of the Levant, became a successful commercial crop.

The status of American food crops changed depending on the location where they were sown and harvested. As they climbed up to higher elevations, though, as a rule of thumb, they were more likely than not to be plebeian, only because they could be planted as garden crops and were exempt from taxation, a privilege extended to new crops in most parts of the Inner Sea.[91] And if they were sown in lower elevations, then their "plantation crop" status surfaced in full force, making it more likely that they would be planted in lands drained of stagnant water, and staved off from the incursions of marauding bands or the occasional intrusions of unruly and displaced mountaineers. Limited though this development was, it needs to be mentioned that there was immense peasant resistance to the transformation of maize into a field crop by aspiring landlords, a resistance that often fueled rebellions against those who wanted to move these crops outside the confines of their horticultural abode.[92] This was seen as a not so subtle preparatory measure for the transformation of the cultivation of maize into a single-crop system. Moreover, in those parts of the basin where landlords and nobility had monopolistic rights over the milling of cereals, the new crops thrived with a vengeance since they escaped the familiar traps of lordly rent-extraction through milling.[93]

Second, the arrival of new crops helped reestablish a balance that was previously in wide currency in the region.[94] A momentum was given to the diffusion of the crops of the Columbian exchange in the Mediterranean countryside due in part to their ability to complement existing cultivation patterns. Most of the crops that made their way into the region were of a similar constitution with that of the flora that was in place. As a result, the American food crops widened the range of supplementary crops by adding new varieties to the region's vegetal and alimentary repository: the broad bean, for instance, popular in the region, came to be complemented by a wide range of beans. String beans and lima beans joined the ranks of the chief products of Spain in the seventeenth century, and were also grown in the fields around Aleppo and Jerusalem even in the sixteenth century. The same was valid for squash, which joined the ranks of the cucumber family.[95] Both leading food crops had the ability to graft themselves into the existing rural landscape with relative ease, without major readjustments in the existent cultivation systems.

In its Mediterranean-bound journey, maize was accompanied by two other American cultivars, beans and squash. This new triad found a receptive environment, partly because it complemented the famous trinity of wheat, olives, and grapes. The American triad neither competed with nor supplanted these crops. It simply complemented them. While wheat and other bread grains were planted in autumn and winter, the new crops were planted in the spring.[96] Neither was there an incompatibility in harvesting, as wheat was harvested in June, grapes and olives in the autumn, and the American crops at the end of the winter. Given that the cultivation of American food crops remained limited in compass and that their widespread adoption started from the latter half of the seventeenth century, what was more germane in the upward-bound movement of rural settlements was the resurgence of spring crops, mostly legumes, and their incorporation into crop-rotation patterns.[97] Most of the summer crops grown in the region were cash crops: in case of shortage in the output of winter cereals, they were not of much assistance in making up for the deficit in grains. Lesser crops, millet first and foremost, and legumes (i.e., lentils, peas, and beans) helped to fill the vacuum created by the contraction in wheat cultivation.[98] The adoption of the legume family enriched the variety of food intake at the best of times; more significantly, it guaranteed rural households' survival at the worst of times, that is, when winter crops failed. The newfound prominence of lesser grains was unavoidable, given the region's geographic and climatic peculiarities: the predominant mountainous relief; thin soils of weathered limestone; the paucity of alluvial valley lands and coastal plains; the practical restriction of grain crops to the winter or rainy season; the elimination of summer grain crops to compensate for scant harvests from the autumn sowing; the unreliability of the fall and spring rains on which the success of the crops depended, especially in the eastern basin.

Third, and relatedly, not only could they be grown and sown in a manner complementary to the existing mix of crops, but they also helped fashion new crop-rotation patterns. With the inclusion of potatoes in the rotation, fallow land was put to use rather than left idle. With the inclusion of maize, the number of summer crops increased to the producer's benefit because most of the summer crops that were planted in the region were raw materials and not alimentary products. In short, the adoption of maize involved less agrarian reorganization and almost no capital investment. It adapted itself to diverse climates, soils, and altitudes, as long as there was sufficient moisture.[99] And since maize used soils and seasons that had not been fully utilized by the peasants of the region, it eased the transition to a system of continuous rotation. In fact, in northern Italy, the transition from biennial rotation of grain and forage to continuous rotation occurred owing to the popularity of maize, which functioned as a replenishing crop.[100] Shortage in winter harvests could thus have been made up for much faster and comfortably. In this sense, the new food crops functioned just as the introduction of spring grains did on both sides of the English Channel: they both increased overall cereal output; diversification in grain prices provided protection against climatic variability.[101] In Palestine, for instance, maize facilitated biennial cropping by complementing sorghum and sesame as summer crops; wheat, barley, and lentils were the main winter crops.[102] Complementing the capability of American crops to serve as staples to fill in the vacuum left behind by the contraction in wheat cultivation and their ability to flourish in the new ecological setting of the Mediterranean, at its higher altitudes as well as in the fluvial environments of the basin, was their versatility: they could be, and were, used as animal feed, if need be, in the face of the expanding dominion of animal husbandry well into the first half of the nineteenth century.[103] What is more, with bigger numbers of herds roaming the hills of the Mediterranean and in need of fodder, maize provided, besides human food, green fodder, available in summer, the season of scarcity in the lowlands. Both provided cheap fodder for livestock as well as cheap subsistence food for rural households, thereby increasing the share of marketable surplus. The peasant and his herds consumed the unsavory potato and maize, and saved the valuable wheat for marketing purposes. Hence they abetted commercialization: it was thanks to maize that Toulouse specialized in the grain trade, and Egypt and the Balkans in large-scale cotton production.[104] In the Toulouse region, an overwhelming 96 percent of the wheat and over 99 percent of the rye grown were exported.[105] In other parts of the region, peasants, who subsisted in plots of maize, raised crops and livestock for the market or as rent destined to landlords.[106] Belonging to a wide array of plant families, crops originating in the Americas did not demand similar ecological and climatological environments. The acclimatization of the American "vegetal portmanteau," *pace* Crosby—potatoes, maize, tomatoes, zucchini, squash, lima beans, haricot beans, red and green peppers—proceeded smoothly. The process was exceptionally successful in that when the tomato, as naturalized in the Old World, made its way back into the Americas, it was considered a new crop, deserving a new name, rather than a variation or offspring of the one that had left its shores three centuries before.[107]

Thus, unlike the prior wave of dissemination of agricultural crops, which emanated from the Indian Ocean and introduced the Mediterranean world to commercial crops such as sugar and cotton, the diffusion of American food crops was less noticeable and more elusive than that of their predecessors. One reason for this was that crops disseminated by the Columbian exchange were, as remarked above, mostly plebeian in nature and averse to large-scale production.

As crops originating in the highlands of the Americas and hence accustomed to inclement weather conditions at higher altitudes, they ensured subsistence for peasant households, high and low. This was in stark contrast to the patrician character of the crops dispersed by the previous wave, which were highly commercial; necessitated considerable outlays of capital; were not averse to plantations as their production site; and more important, thrived in the warmer climates of the coastal plains. Commercial cultivation of such crops was instrumental in bringing the plains under tillage early on.[108] New plebeian crops, experimented with first as garden crops, were hidden from the gaze of travelers more easily than field crops. They did not alter existing production patterns but complemented them; as such their impact was not duly appreciated.[109] So it is quite difficult to map with certitude the inroads made by these crops into the farming world of the Mediterranean. Nor is much known about the extent of these initially horticultural crops.

What is certain, however, is that these crops not only found a receptive environment in which to flourish, but some were disseminated farther with the Ottoman empire's westward expansion. For the advance of the empire came to be associated with an extension in the cultivation of rice in the fifteenth century; of maize, sesame, and peppers—paprika in particular—in the sixteenth century; and of tobacco and coffee in the seventeenth. What is certain, also, is that in the Mediterranean basin, some of the crops that originated in the Americas as a result came to be identified, in name and in provenance, with their Levantine hosts: maize was referred to as *grano turco,* or "Saracen millet," and squash as "Turkish cucumber" or "zucco of Syria" (hence, zucchini).[110] As if to resist their absence from historical records, the American crops entered the region's cuisine with the force to assert their staying power: the newly arriving bell-peppers, tomatoes, and squash were now served stuffed, in the process replacing that precious vegetable of the Old World: cucumber. Their presence was not confined to the field of dietary regimes; the new crops penetrated into the realm of the symbolic and the confessional, shaping anew the life-worlds of the denizens of the empire. For the inhabitants of Jabal Ansariyya, the 'Alawis, for instance, tomato and pumpkin, both of the Americas, became forbidden foods.[111] For such an interdiction to be issued, the inhabitants of the mountain must have been familiar with these crops from very early on. Yet, almost from the seventeenth century onward, the very same region, along with a score of other localities in different corners of the empire, owed its livelihood to a great extent to tobacco cultivation, another American crop.[112]

And despite all the similarities between maize and the potato, the diffusion of these crops into the Old World took different itineraries. A mostly northerly path was taken by the potato, which traveled from northern France to Poland and beyond, and complemented cereal culture. From the Caucasus to the plain of Nablus, smallholders cultivated it as a guarantee against possible crop shortages. Along the shores of the Inner Sea, it came to inhabit the mountains of Spain, Greece, Italy, and Morocco before the nineteenth century. And a meridional route was taken by maize, a crop hosted primarily but not exclusively by mountainous regions. From Haut-Languedoc and Gascony to the Basque region and northern Italy, from the Balkans and the Black Sea coast to Mount Lebanon and the Romanian plains, where humidity was not lacking, despite hesitant beginnings, it entered the repertoire of small producers and plantations alike, at first as a fodder crop.[113] Where it became a preserve of small producers, it is difficult to trace its development, but where it was turned into a field crop, then it became a telltale sign of commercialization, because it served as a staple for the poor.[114]

"Pilot" family gardens and orchards hosted these crops first: during their growth cycle, in the hot and dry summers of the Mediterranean, irrigation was more likely in small plots of land, and also in the hilly parts of the region. The cultivation of maize was possible only in humid zones like northern Italy, in the provinces of Venice and Lombardy, or the valleys of the region, in Mount Lebanon and the Black Sea coast.[115] In the northern parts of Mount Lebanon, in particular, maize became an indispensable element of the agrarian landscape. At the turn of the nineteenth century, barley and "white maize," that is, millet, were grown only when the maize harvest proved to be insufficient. In times of dearth and famine, many were "reduced to such indigence as to subsist on Dhourra or Indian corn" alone. The situation was largely unchanged at the end of the century.[116] The advantage maize and legumes had over the potato stemmed from the fact that the region was more than familiar with beans, broad beans to be exact, which were not completely unknown and used for soil enrichment because of the nitrogen the beans released into the earth. By the late seventeenth century, maize and, later, beans had become firmly integrated into peasant farming and polyculture as an alternative staple cereal, particularly in the damper districts of the shores of the Mediterranean and Atlantic Iberia.[117]

A crop that followed a trajectory different from that of the American food crops was tobacco. As a commercial crop, it found a warmer reception in the Levant than in the western Mediterranean due to the latter's proximity to the tobacco plantations on the Atlantic. Its advent in the eastern Mediterranean dates back to the turn of the seventeenth century. The pace at which it established a footing in the region must have been swift, because the time-span extending from its initial arrival in Istanbul in 1591–92 to the establishment of its commercial presence by the 1620s was less than a mere three decades.[118] In Lefke, close to Bilecik, a tax was already being levied on tobacco in 1661–62.[119] The growth of its cultivation during the course of the century must have been sufficiently alarming to some: a series of bans were imposed on its production and consumption. Notwithstanding the bans and a total prohibition imposed on its cultivation in the 1683–97 period, the measures put in place fell significantly short of their aim of eradication. Imposition of a luxury tax on its use was opted for instead.[120]

The arrival of tobacco, or at least certain varieties of it, followed at times a long and circuitous route. The Yucatan type, called *başı-bağlı* or *Persotchian*, and widely cultivated in the Balkans, for instance, reached Latakia not via the Mediterranean but via the Indian Ocean: on its route to the Persian Gulf and the Ottoman lands, tobacco traveled through Japan, China, Java, and India, where it was experimented with and grown from the beginning of the seventeenth century. Complementing its maritime trek with the well-known terrestrial routes of the time that fanned out from the Gulf, a route revised and popularized by the Portuguese presence in the Indian Ocean, it thenceforth followed a northwesterly route as the Ottoman empire itself became a significant consumer of oriental goods in the latter half of the sixteenth century rather than merely transmit them west. Still, to some, this trajectory is not a certainty, for they acknowledge the presence in the aroma of the Persian variety a Chinese and Tibetan timbre: this connotes a terrestrial transmission, whereas the Indian provenance implies a maritime diffusion.[121]

Again, following the usual transportation and distribution networks built by the silk merchants of the time, canvassing the empire from east to west, from Syria and, in particular, Latakia where its production took hold, it found its way, first, to the Black Sea Coast (Bafra, Samsun, and Trabzon), and thereafter, to Greece and Macedonia. This Persian type is the most widely diffused and cultivated type in Macedonia, and in particular in Yanina (Epirus) and Albania. In fact, in Macedonia, some villages in Drama were inhabited by people of Persian descent (the name "Persotchian" given to a village two hours outside of Drama); it appears that those who mastered the art of tobacco-growing migrated first to establish its cultivation elsewhere. Villages in the districts of Edirne, Montenegro, and Herzegovina appear to have been recipients of such emigrant flows, tobacco known under the name Shiraz. The cultivation of the crop in Syria was not confined to Latakia, however, even though the region came to be identified with a certain type of tobacco: it was also grown in an area stretching from Sidon to northern Syria, which turned into a major exporter of it. If the northern latitudes of the region came to specialize in the *sahilî*, coastal, variety, its southern tier grew the *barranî*, inland, variety. Another route the crop followed was the usual Mediterranean route, and in line with the rise of ports of the Aegean as the spice trade resumed during the latter half of the sixteenth century. Tobaccos of either Peruvian or Brazilian origin were grown in Macedonia and western Anatolia: in Xanthe (Karasu/Nestos); in Kavala (Drama, Kavala, Seres, and Kırlıkova); and in İzmir (Ayasoluk, Tira, Ademis).[122] There are other varieties, but their field of diffusion gravitated to the very same localities that hosted those of the Persian and Mediterranean kind. Salonika and Edirne, Beirut and the environs of Damascus all became producers of differing significance and of different varieties. Given its great flexibility in adapting itself to the most varied climes and soils, the compass of its cultivation is hardly surprising. In accordance with the commercial nature and widespread cultivation of tobacco, a policy of revenue generation by the imperial treasury, cautious at first, was introduced in the 1690s. This culminated in the encouragement of tobacco cultivation in the latter part of the eighteenth century.[123] Revenues accruing from it were diligently tracked and subsequently taxed by the central state: duties levied on tobacco constituted one of the biggest sources of revenue for the Porte originating in the province of Aleppo. At the end of the eighteenth century, it led all other revenue farms. Whereas the revenue derived from 20 to 25 villages brought in 116,000 piasters in total, tobacco alone fetched 60,000 piasters, followed by coffee with 20,000 piasters.[124] Tobacco, then, provided the peasants of higher landscapes with a valuable cash crop. Furthermore, its trade and distribution remained in the hands of merchants who were resident in or nearby the actual areas of tobacco cultivation and not large merchants from Istanbul or other major market centers.[125] The movement away from the low-lying lands hence found strong support from the dissemination of American food crops. Maize in Mount Lebanon and the Pontic Mountains, tobacco in Mount Ansariyya, the valleys of the Meander, the hills of Thessaly, and the Kurdish mountains found refuge in and thrived on higher altitudes as well. This was in stark contrast with what happened with the crops that migrated from the Indian Ocean. Tobacco, as a successful cash crop, offered an avenue that was not available to other American food crops, for most excelled in providing sustenance on lands deemed to be of little use.

Of course, the advent of American food crops, in and by themselves, may have had a limited impact on the overall structure of Mediterranean agriculture. But the impact of their spread was inflated by parallel and contemporaneous developments that all helped popularize mixed cultivation, which was the best insurance against climatic variability. After all, the new food crops, singly or severally, fitted comfortably into the existing cultivation patterns. So not only did the range of crops at the service of peasant households proliferate, it also became easier to effectuate a transition to continuous, or relatively intensive modes of, cultivation owing to the versatility of American vegetal portmanteau. That the new crops were initially acclimatized as garden crops reinforced the horticultural infrastructure of the countryside, which was already paved by the distension in the dominion of tree crops. In addition, the diversification in the range of crops brought on by altitudinal diversity allowed for intensive methods of land use. The resurgence of lesser cereals, now fortified by the advent of maize, potatoes, and legumes, came to fill the void created by the contraction in wheat production.[126] Cultivators could make better use of their land holdings, because a small proportion was dedicated to cereal husbandry: since land under fallow shrank in sympathy with the arable, more land was available for tilling. At the time, most of the lands in Anatolia were tilled after being left for fallow for a year or two, if not more.[127] In France, the elimination of fallow was one of the centerpieces of reform advocated by the agronomists in the eighteenth and nineteenth centuries.[128]

The diversification in the composition of crops had a serendipitous ecological outcome. Intercropping became an effective mechanism in the struggle against erosion: a good mixture of grain farming with arboriculture was an excellent prevention against it, for polyculture provided the ground exposed to erosion with some cover and for longer periods of time. Furthermore, mixed agriculture helped the land retain moisture.[129] The upward movement of rural settlements and the rise in the number of populous villages, when coupled with the scarcity of level lands to be devoted to grain cultivation, compelled producers to devise a solution that worked to the benefit of the environment. This was of course the spreading practice of intercropping, *coltura promiscua.* Mixing tree crops with grain sown between rows of mulberry or olive trees, or grain sown between rows of trees festooned with vine, made excellent use of limited resources while increasing agricultural output.[130] The interculture of trees and grain crops, as native to the Mediterranean climate as its ancient triad of crops, returned in full vigor. In Mount Lebanon, vineyards were not cultivated with spades, but with oxen, "for they were planted with straight rows of trees far enough one from another."[131] In the higher precipices of the Pontic range, trees were festooned with vines.[132] Moreover, harvest failure in one of the crops did not automatically lead to famine or financial ruin, for it was accompanied by a host of other crops.

The popularity of intercropping in the Ottoman realm is confirmed by the frequency of references made to it in documents concerning the pious foundations, *waqf*s. In the inventories of the rural assets of these foundations, it becomes apparent that orchards and vineyards always contained items necessary for grain production: seed, implements, and a storage space for these agricultural items and for the harvest were always provided for by the foundation itself. What is more, the *responsa* collections from the seventeenth and eighteenth centuries contain scores of inquiries about changes in the status of land as a result of intercropping.[133] Mixed-cropping posed a serious problem to the Ottoman Land Code in that the clarity expressed in the fifteenth- and sixteenth-century codes corresponded to the rural landscape of the times where the arable was clearly demarcated and separated from private property. This no longer was the case in the following centuries because of the changed crop-mix and landscape. In principle, any land that was plowed was *mîrî* and subject to tithe, whereas orchards, vineyards, and gardens were private property. Therefore, the opportunity for producers to claim *mîrî* lands as their own transpired when intercropping became an integral part of the landscape, as abundantly evident in court records. Intercropping gave the producers the occasion to change the definition of their lands with the change in usage. Even though arable production continued along with the cultivation of tree crops, as long as the land was classified as an orchard or a vineyard, it was exempted from regular tithe and peasant taxes. Those lands that were not classified as *mîrî* paid fixed tax-revenues.[134] The widespread usage of intercropping may have had an environmental role to play, but if the cultivators opted for it, this was because of the titular promise it held for them.

The introduction of American food crops and the expansion in garden agriculture, and the momentum gathered by mixed-cropping as a result of the expanding imperium of tree crops and lesser cereals, not only reflected the new spatial redistribution of economic activities to the benefit of hilly regions, but also responded to the climatic variability of the Little Ice Age. Switching to (cereal) crops able to withstand the harshness of colder and wetter winters or flourish in a shorter span of time was one of the options as attested to by the growing stature of rye and millet at both ends of the Inner Sea.[135] The Italian peninsula experienced a similar development wherein producers employed different strategies to make up for the changing climatic conditions and the new impositions and problems this climatic shift had on agricultural production. In Verona, for example, the effects of "land-erosion and flash flooding" had a devastating impact on many communes during the seventeenth century so as to render the exploitation of land extremely vexing. In response to these circumstances, rural producers, rather than rely on a single crop, diversified their crop-mix, opting for mixed-cropping (i.e., cereals, grapes, and mulberry trees)[136] During the baroque era, land consolidation in the region slowed down. Scattering parcels over a variety of soil types was another strategy that the rural producers used to minimize the effects of climatic variability. Sharecropping lost the pervasiveness it commanded previously and the popularity it enjoyed in the eyes of landholders, and made way for the renting of land for fixed annual payments. In the higher elevations of the region, olive groves doubled as pasture lands and hay fields, a manifestation of the fact that these altitudes were becoming host to settled populations and herds rather than transhumant herds and timber suppliers.[137] In Liguria, too, the second period of colonization of the highlands took place from the second half of the sixteenth and the seventeenth centuries, mostly for subsistence, given that the acreage in the lowlands was shrinking due to ecological reasons as well as the growing sway of urbanites over rural land.[138]

Millet (and panic), which required both heat and moisture, too, found ideal conditions in the irrigated fields of the Nile valley and the Cilician lowlands.[139] Yet, that millet came to command a highly visible presence after the sixteenth century attests to increased humidity associated with the Little Ice Age. One of the indications of its popularity, though an indirect one, is that wherever maize appeared, the name given to it by the peasants invariably associated it with millet, as if the two belonged to the same crop family.[140] Millet was mentioned rarely in the sixteenth century, but judging by the travelogues we have at hand, it made great headway after that century.[141] That crops which demanded moisture, such as millet and rye, came to occupy a commanding position in the region was owed, in no small measure, to the Little Ice Age. *Boza*, made out of millet and enriched by opium, became an important drink and an intoxicant, as wine was in other parts of the empire. Legumes came to decorate, in abundance, the sacred dish of the Sufi *tekke*s: *aşure*. The dish was a latecomer to the *tekke* culinary culture. Its slowly growing and revered presence as a phenomenon of the post-fifteenth century attests to the growing importance of legumes.[142]

Overall, the significance of the introduction and percolation of the crops of the Columbian exchange and the floral transformations they inaugurated were twofold. First, maize, millet, and rye emerged as credible candidates to supplement or, at times, supplant King Wheat. As we will see in the final section of this chapter, rice, as a crop of the wet lowlands, also joined the group of substitute cereals. The deficit that stemmed from the contraction of croplands in the low landscapes was thus covered owing to the growing cultivation of new crops and the resurgence of lesser crops. Second, the garden crop quality of the American food crops was an excellent complement to the expanding imperium of tree crops. Growing emphasis on horticulture and arboriculture and perfection of mixed cultivation contributed to the well-being of mountain and hillside villages. The diversification in the polycultural makeup of the vegetal composition of the Mediterranean helped it withstand better the tribulations of the Little Ice Age. More important from our viewpoint, both these developments concerned rural settlements at the higher altitudes of the Mediterranean mountains. And both aided, in some form, the ascent of villages and hamlets onto higher ground, allowing them to keep a healthy distance away from malarial marshlands. It is the ascent of rural settlements that accelerated in the seventeenth and eighteenth centuries that the following section depicts.

Perched between the Mountains and the Plains

As we depicted in detail in part I, the low landscapes of the Mediterranean were brought under tillage and settled densely during the age of the city-states. Of the diverse ecological zones of the Inner Sea, it was, for the most part, the coastal zones and maritime plains that set the tempo of change, and determined the pace of the flux and reflux of the hills and mountains by the demands they placed on the higher altitudes of the basin for most part of its history.[143] Even though the low-lying plains of which the region was abundantly short played a paramount part in the economic and ecological order of the day, it is the impressive mountainous skeleton that commands the region's silhouette. The Mediterranean, after all, is nearly ringed around by mountains that frame and shelter it. These towering mountains provide an ocular unity to the region's landscape, and transport "the imagination from Antioch to Jerusalem" and from the Alps to Mount Lebanon with enviable ease and speed.[144] Only in the east of the coastal hills of the Atlas Mountains is this unity fractured by the sheltering sky. The mountains have had a sprawling and overpowering presence throughout the width and breadth of the basin. Overlooking narrow, discontinuous, and relatively confined plains as well as wide plateaus, the Mediterranean uplands, dotted and blanketed by woodlands, pastures, villages, and hamlets, have occupied a prominent place in the region's economic history.[145]

Thus far, however, we have chronicled the withdrawal of commercial agriculture from the low-lying marshy landscapes and the economic devolution of the littoral zones of the Inner Sea, and examined how the undoing of mountains by rains and rivers contributed to this outcome. The vantage point of our analysis has consistently been that of the low landscapes. The previous chapter stressed that the transition from the long sixteenth century when the coastal plains constituted main theaters of colonization and settlement to one where the hills and hillsides took on the task of accommodating growing numbers of inhabitants on their slopes occurred gradually. And as erosion and sedimentation accelerated on account of the fast-paced expansion of the long sixteenth century and the return of the Little Ice Age, the low-lying plains lost favor to the benefit of ambient higher ground. That is, rather than take it for granted that the predominance of hill towns and settlements have historically been one of the defining features of the Mediterranean landscape, we constituted that dominance historically, by tracing the transformations in the lowlands.[146]

The full implications of the scale of misfortunes that primarily hit the plains—cereal lands—from the 1560s can be appreciated in light of the orographic features of the basin: only 15 percent of Provence and 11 percent of Spain lie below 300 meters above sea level; so does less than one-quarter of the Italian peninsula.[147] Laden with vertiginous relief and bereft of sizeable plains, the basin is not generously endowed with level, cereal lands.[148] To be sure, the region's topography allows cereal cultivation in its countless and oftentimes fertile valleys, intermontane basins, and plateaus. But the sheer presence of the impressive mass of mountains perforce reduce the *ager*, surface available as arable land. To wit, the Anatolian peninsula rises to 1,162 meters above sea level: over one-fourth of its land mass lies above the 1,500-meter relief mark, and 60 percent of its surface area above 1,000 meters. Similarly, 56 percent of Spain's surface area lies between 440 and 1,130 meters.[149] When the sparseness of extensive plains was coupled by scarcity of labor, it is not surprising that an average of 10 percent of the arable in most provinces of the Ottoman empire was put to agrarian use in the nineteenth century: on the brink of the Great War, the area sown to crops in Ottoman Anatolia was only 6.3 million hectares—12 million at the most if an allowance is made for land left for fallow, assuming that land was sown every other year.[150] Even in the 1930s, of the 77 million hectares of the surface area of the Turkish Republic, 11 million, a mere 14 percent, was sown to crops.[151] The altitudinal range offered by the basin's mountains and hills overlooking its narrow coastal plains and the more sizeable high plains discouraged monoculture especially when the low landscapes and valley floors at times became less easily accessible. Elevational distribution of economic activities in the basin can thus be said to have dictated the popularity of polyculture.

Scarcity of level land may be seen not merely as constituting a constraint that the orographic attributes of the Mediterranean imposed on its inhabitants, but as a stimulus for diversifying the mix of crops they cultivated. This landscape, owing to the ecological variety it housed, which extended from the warm plains of the coast to the snow-capped summits of the mountains, was also an ideal setting for an integrated network of villages, reserve fields, and temporary settlements that took up residence in different altitudes of it, especially on its hillsides and valleys.[152] This altitudinal variety eventually led to the formation of a pattern of agglomerated settlements whereby different reaches of the landscape were put to use in a complementary fashion.[153] In Montpellier basin, for instance, village communities were aligned in long strips just as in the eastern Mediterranean, in Palestine, to concomitantly exploit the plains and the *garrigue*.[154] Spatially dispersed though these settlements were, they did not stand in isolation. The versatility and flexibility of this tightly knit network became manifest when some of the ecological zones that marked the Mediterranean landscape became at times less welcoming or inaccessible to its inhabitants. Wherever they were settled, most rural households could easily refashion their economic activities so as to invest more time and energy in one or two ecological zones without completely vacating the others. So withdrawal from the plains did not curtail access to the resources of these landscapes; it was possible to engage in temporary cultivation, drain manageable sizes of the marshes to extract salt or to "invent" new land in it, or use the wetlands for poaching or hunting game. Conversely, when rural settlements were disproportionately located on the hills, tilling the highlands, using the pastures for grazing, or exploiting the resources of their woodlands were, and remained, part and parcel of the means of livelihood available to rural populations. Alternatively, downstream migration from alpine regions to low-lying districts provided laborers necessary for working the fields, as the slopes of Massif Central served as a reservoir of labor destined for the Iberian peninsula in the seventeenth and eighteenth centuries.[155]

In brief, the primacy of mountainous landscapes has been all but timeless, or congenital. The density of populations and herds that inhabited higher ground, and the nature and intensity of the exploitation of montane resources, have been subject to ebbs and flows over time, and the rhythm of change has largely been set by developments in the plains and hills fencing in the mountains.[156] The destinies of these ecological zones remained tightly intertwined as well, at times in a complementary fashion and at others adversarial. As we discussed in the previous chapter, the lowlands of the basin, after being reclaimed and tilled, almost routinely fell victim to the unhealthy vapors of the swamps and fens, and rural life took refuge where mountain and plain met—above the fever-inducing zone, away from great extents of "heath, marshland, and waste."[157] In this wave of upstream migration, there were three new developments that gave the ascent of rural settlements from the sixteenth century its *differentia specifica*. These developments abetted the denizens of the region to establish a firmer hold on higher altitudes.

The first development was the rise and consolidation of transhumance, which had emerged as a systemic practice from the fourteenth century. It made possible the institution of regular contacts between varying altitudes of the landscape, and gained considerable momentum in the long sixteenth century. The golden age of herding economy in the Iberian peninsula and the Ottoman empire extended from the 1450s to the 1580s, and in the kingdom of Naples from the 1470s to the 1610s, with a hiatus in the first half of the sixteenth century.[158] Surely, developments in differing elevations of the Mediterranean mountains were interrelated from early on. Judging by the developments in the Byzantine empire, in its Anatolian and Balkan peninsulas, one of the original reasons for widespread adherence to the practice of transhumance was the tilling of slopes and the inexorable conversion of pastures into fields in the 1000–1250/1300 period.[159] But competition over higher ground accelerated considerably in the long sixteenth century as demand for wool (and meat) and cereals soared, fanning the clash between the sown and the grassland. This pushed the upper limits of summer grazing or the arable due to mounting land hunger in the lowlands. The second and related development was the clearing of land at higher altitudes primarily for settlement and tillage as well as lush pastures, and not just for the exploitation of sylvan and metallurgical resources as was the case before. The number of hill and mountain settlements, seasonal as well as permanent, multiplied. The establishment of regular connections between various altitudes of the basin became easier owing to the formation of well-knit networks woven by livestock migration, but also by labor and information flows. The return of money to land at the end of the sixteenth century threatened communal property and village rights, and land hunger at lower altitudes hastened the colonization of highlands, principally for subsistence.[160] The state of the plains during the Little Ice Age turned the upward movement into a process with a long lifespan.[161] And finally, the introduction of American crops, such as maize, tobacco, and potatoes, facilitated survival on higher elevations—as well as on lower landscapes. That these crops started out in "kitchen gardens" in their new host environments before they ventured out into the fields eased their acculturation in higher elevations, primarily because they served as substitute cereals.[162] Unlike the crops of Indian Ocean provenance, which flourished in the semitropical climates of the littoral regions in the early part of the second millennium, crops of American provenance, arriving from the sixteenth century, thrived easily at high, as well as low, altitudes, and expanded the range of options available to rural households to feed themselves as well as the poultry and livestock they possessed.[163]

Along with the newfound popularity of tree crops and lesser cereals, which was sketched in the third chapter, the aforementioned three developments rendered the wave of upland colonization after the 1560s different in nature than the previous wave which, partly due to the momentum given by the acclimatization of tropical crops along the maritime plains and coastal valleys of the basin, was largely centered around its lowlands. This process impacted the mountains only indirectly. When at the turn of the millennium the "slow renaissance of the mountains" started, it was the reclamation of the lowlands, the growth of cities, and the revival of maritime trade that launched the intensive exploitation of the highlands.[164] The steady and long-term economic growth during the expansionary period from 1000 to 1250/1300 generated a need in the lowlands and the hilly parts of the region for the products of the woodlands. When the colonizers of this era started to mine the arboreal resources of the uplands, they had no need for haylofts and stables. This of course changed after the sixteenth century.[165] Early on, what the colonizers were after was timber, which remained in high demand, thanks to the momentum given to shipbuilding by the crusaders who headed to the Holy Land. Need for timber did not subside when the crusades lost momentum. Yet the need to replenish the fleets of the affluent and seagoing city-states kept demand for timber robust as these cities built and consolidated their aquatic empires.[166]

The restructuring of the Mediterranean trade from the 1350s, as we have seen in chapter 1, also prompted the mercantile city-states to broaden their economic base by including in it the manufacture of wares previously imported from the Levant. The import-substitution strategy pursued to reach this end boosted demand for lumber and metals, plentifully abundant in the uplands. Besides, the golden age of the city-states generated a rampant demand for stone for upgrading their urban infrastructure. Endowed with rich mountain resources, Pistoia, for instance, burgeoned in consequence and in an impressive fashion in the fourteenth and fifteenth centuries.[167] Meadows, too, attracted newcomers. With the contraction in rights to common lands after the thirteenth century, not only the rights of pasturing but also that of gleaning stubble for feeding livestock came to elude the poor peasantry. Initially, parts of the mountainous zone were beyond the ambit of enclosers. Yet, by the fifteenth century, rights to forest use were also well established in most parts of the north-central peninsula.[168] In Tuscany, zones that previously remained beyond the reach of the long hand of Florentine nobility were subsequently brought under the dominion of the Tuscan men of money. The uplands were no longer a refuge for the poor, now that rights over the resources of woodlands, from trees to stubble, were earmarked more carefully. The growing grasp of sylvan resources by the moneyed was best demonstrated by the domestification of swine in peasant farms instead of in the woodlands.[169] "Enclosed" and market-responsive, mountain villages were hence at times economically better off than their counterparts in the plains, and attempts to tax these populations triggered peasant insurrections in the fifteenth century that tested the determination of the Florentine ruling class.[170] What transpired in the plains and hills of north-central peninsular Italy thus determined what happened in ambient heights.

To the west of the Italian peninsula and on both sides of the Pillars of Hercules, the situation was similar. In the ninth and tenth centuries, the acclimatization of new crops, sugarcane, rice, and cotton, all of which thrived in the coastal plains, breathed new life into the low landscapes.[171] The intensive garden agriculture that the Arab populations practiced was largely based in the sun-scorched plains. Both these processes of acclimatization added to the impulse provided by the advent of new crops, and encouraged lowland-bound labor movements from the mountains.[172] In the Byzantine lands as well, at least in the Greek peninsula, the period from the ninth century was marked for all practical purposes by the withdrawal of trees and the appearance of plants associated with agriculture and deforestation. The trend was reversed, as elsewhere, in the fourteenth century with the extension of "arboreal formations and dense plant cover, resulting in the spread of wilderness" and of pigs.[173] Overall, the tempo of change in the highlands was set in a manner to respond to the needs of agricultural revival in the plains and the urban renaissance taking place along the shores of the Inner Sea. The breakneck pace of demographic growth and the land hunger of the long sixteenth century reversed the return of the forest, which had been attained during the fourteenth century as a result of the Black Death and the lengthy wars of the period that reduced population.[174]

The second factor that differentiated the earlier wave of colonization from the latter was a change in the scale of transhumant migration and the change in the nature of upland settlements. From the late fourteenth century, the steady rise in the number of itinerant sheep started to modify the economic composition of the Mediterranean. The networks that the ever peripatetic herds wove among the different altitudes of the region's mountains altered its landscape in a dramatic fashion. Before the appearance of transhumance as a systematic form of animal husbandry and as an integral part of the Mediterranean economic life, the mountains were "poorer versions" of the plains and lowlands.[175] The advent of Turkish nomadic populations from the east created in the former Byzantine territories an environment similar to that in the western Mediterranean. The new migratory movement that transhumance entailed transformed mountain pastures into an integral part of the landscape, which was now crisscrossed by herds and peoples on a regular basis, weaving a tight network between the lowlands and the highlands and among a score of way-stations, from the vaporous lowlands, to the cool, verdant summer pastures of the high landscapes or from the freezing temperatures and snow-covered pastures of the mountains to the welcoming climates of the plains.[176] For one, these migratory movements oversaw the formation of a communication network that interwove different parts of the region. The case of the Cathars in the thirteenth century eloquently speaks to the fact that this network could be put to "subversive" use in escaping the orthodoxy of the Church, which was headquartered in the lowlands.[177]

Crucially, the extension in the spatial stretch of transhumant populations and the subsequent exploration and exploitation of different elevations of the basin's mountains gave the populations inhabiting them a new range of choices. As this continual, seasonal movement increased the ability of the region's populations to tap into the resources of disparate ecological settings, in the process, spatial flexibility, as if it were, was built into the life-worlds not only of the nomads and transhumant populations but of the denizens of the basin. In the space of a couple of days of travel, it was possible to put different ecological environments to use. The availability at different altitudes of the region of a wide range of activities, which generated manifold sources of livelihood and income, helped peasants diversify their economic repertoire. The parish of Santa Maria Morello was located below 500 meters, "but its fields, pastures, and woodlands extended up to the slopes which peaked at 934 meters."[178] There were, in fact, many parishes that were located at similar altitudes, but their lands extended up mountain slopes to peaks as high as 1,000 meters. Access to variegated heights of the mountains of the Mediterranean gained a new timbre with the addition of transhumant migrations into its fold. Mulberry trees, after all, fixed the band of settlement at about 800 to 1,400 meters; olive trees climbed up to 1,100 meters, and vines, 1,700 meters.[179]

Spatially, this ecological feature of the Mediterranean landscape manifested itself in the waxing and waning of satellite settlements during periods of agrarian contraction or regeneration. Subsidiary settlements, *gastinae*, occupied the hills of the Levant at the time of the crusades, and, similarly, *chôrion* speckled the Byzantine landscape, perpetually changing its status back and forth to *kômè*, village "proper."[180] Yet, due to their fleeting existence, these impermanent settlements are usually seen as remains of villages "proper" rather than as full-fledged components of the Mediterranean landscape. Given the climatic and edaphic character of the region, it was oftentimes imperative to "abandon a site in order to move to another one nearby," and the deserted lands were later "re-occupied or re-exploited."[181] Temporary cultivation of land and the flexibility of rural settlements were mechanisms that allowed peasants and nomads to fully exploit and take advantage of the potentialities of their ecological habitat as well as providing a guarantee against calamities, natural or otherwise.

In other words, temporary settlements and satellite villages were, and remained, an indispensable component of the Mediterranean landscape regardless of the economic fortunes of the basin. In times of robust growth, these adjunct settlements were transformed into villages, and in times of depression and turmoil, they were the first sites to be vacated. They served as signposts of rebirth or decay. The *grange, bastide,* or *mas* in lower Provence eloquently captures the ambidextrous nature of these rural units. All these terms denote a property around a farmstead, oftentimes with the owner's house alongside, and they capture two contradictory processes: dismembering of village territory, but also extension into, and colonization of, marginal land.[182] *Mazra'a* functioned the same way in Ottoman lands: these fields proliferated in times of agrarian upswing, served as tentacles of established villages, and abetted the colonization of virgin lands.[183] In times of land forfeiture and outmigration, villages were demoted to *mazra'a* status. In short, the latticework of temporary and permanent settlements was firmly embedded in the rural infrastructures of the Inner Sea. Notwithstanding their omnipresence in the Mediterranean, adjunct settlements took on additional import from the late sixteenth century by providing hill villages access to diverse ecological and altitudinal environments.[184]

In the Ottoman lands, for instance, there already was an extensive network of rural settlements early in the sixteenth century: in 1530, the total number of villages in the province of Karaman and Rûm stood at 6,447, and that of *mazra'as* at 3,759.[185] During the course of the sixteenth century, as agricultural expansion set in solidly responding to the vertiginous pace of urbanization and the buoyancy of economic growth, new land was brought under tillage. In addition to the establishment of new villages *ab ovo,* turning *mazra'as* into *ab novo* villages and inhabiting them on a permanent basis gained popular currency in expanding the arable.[186] Tribal confederations, owing to the sizeable labor force they commanded and to their organizational skills, proved to be equally important agencies in the process of colonization. Of the 629 villages in the subprovince of Yozgad, 96 had names containing the suffix, *kışlak,* winter pasture (or *kışla,* its abbreviated form). Obviously, tribal confederations that used the lower altitudes of the Anatolian steppe to graze their flocks in winters eventually settled down. Turning pastures and temporary settlements located in climatically more hospitable lower altitudes into villages and arable was common enough so as to constitute a pattern.[187] Since these lands were already planted, however intermittently, farming them on a continual basis required little additional toil. As villages became more populous, they extended their tentacles into the arable in their proximity in the form of new *mazra'as,* increasing the amount of land available for planting. More often than not, peasants from neighboring villages colonized these new fields conjointly. Equally common was the farming-out of these fields to individuals who, in most likeli-hood, farmed them out in turn or brought them under cultivation through sharecropping.[188]

By the mid-sixteenth century, the expansive Ottoman realm had come to encompass more than 550,000 enumerated villages—or rural settlements, to be precise.[189] At the time, most of these settlements were inhabited on a relatively permanent basis, and some temporarily. During the course of the following three centuries, however, the share of permanent settlements, which had swiftly increased between the 1450s and the 1590s, started to diminish, in absolute and in relative terms, and remained low, a point raised repeatedly by western trav-elers as well as reform-minded bureaucrats. The number of villages dotting the realm is said to have decreased at least by one-fourth by the first half of the seventeenth century, eventually plummeting down to "75,000 in all" by the turn of the nineteenth.[190] In Old Castile, too, in 1800 there were 22,318 isolated churches, where presumably hamlets or villages once stood.[191] Exaggerated and historically inaccurate though the conjectures given by the travelers about the extent of the exodus may have been, what is of import from our viewpoint is not the exact magnitude of this change, but its overall direction—and the repercus-sions this depopulation, or *wüstungen* if you will, had on production and rural settlement patterns. Not incidentally, the void left behind by the sharp decline in the number of villages inhabiting the plains and the flat-lands was filled by impermanent settlements. By the mid-nineteenth century, these "satellite" vil-lages had become deeply entrenched in the Ottoman countryside, high and low. Slowly yet relentlessly, uninhabited or sparsely inhabited reserve fields popu-lated deserted lowlands as much as the hilly and mountainous terrain overlook-ing it.[192]

If the lowlands through which the main trunk and travel routes connecting disparate parts of the empire had fallen into disuse as hastily as most travel accounts would have us believe, it would not be imprudent to assume that some of the inhabitants of these rural units, rather than merely join those who swarmed into urban settlements, were displaced onto higher altitudes where inhabitants felt safer against the incursions of the *celâlî* bands roving through the countryside. Settled, transhumant or nomadic, these populations on the move colonized in effect new altitudes towering over the low-lying lands, which by then had turned into vast stretches of land cultivated haphazardly, if at all. Understandably, most travelogues dwelt on the decline in the number of vil-lages without accounting for the parallel growth of those of unfixed settle-ments, precisely because they lingered in the plains and could chart the travails of its villages better than the seemingly erratically tended fields of its surroundings. When the two developments are analyzed in unison, however, as an inter-related phenomenon rather than as disparate and independent processes, then a different picture emerges.[193] In this picture, the shifting balance between permanent and temporary settlements provides the setting within which the countryside was reshaped as settlements advanced or retreated. Abandonment of significant portions of the extant arable in the plains was indeed accompanied by the concomitant colonization of new lands, notwithstanding the overall contraction in the arable.

These temporary settlements served different functions, and existed within the interstices of an agro-pastoral world: *mazra'a* (reserve fields), *kom* (ranch for animal breeding), *oba* (grazing area), *divan* (hamlet), and *ağıl* (sheepfold) represented elementary ways of taking over cultivable or pasture land.[194] The possibility of a relatively easy transition from settled to temporary settlements and back was not a preserve of the Ottoman countryside alone, but was a geographic attribute it shared with the rest of the mountainous Mediterranean. Whether the reserve fields on the precipitous slopes of the Mediterranean hills were temporarily inhabited by transhumant groups, or they were cultivated by nearby villagers whenever the occasion called for, these temporary settlements enabled the Ottoman *reaya* to increase its gross product.[195] Yet, in Ottoman imperial Land Codes and registers, an overwhelming majority of which date back to the fifteenth and sixteenth centuries, villages, *karyes*, loom larger than any other rural unit populating the countryside, since they were permanent settlements, permanent at least in the medium run: as the embodiments of the *çift-hane, iugum-caput,* system, they formed the principal components of the rural order.[196]

Yet, the same documents equally reveal that the number of temporary settlements, *mazra'as*, seldom fell below one-third of that of villages; in certain places, they in fact outnumbered permanent settlements. The number of *mazra'as* was two or three times greater than the number of villages. As a result, by 1800, about half the Anatolian population depended on various types of periodic settlements.[197] To most, this is not particularly surprising, because it took place in the seventeenth and eighteenth centuries, and by then, the central bureaucracy's ability to patrol the countryside had been drastically diminished, and the rural order underlying settled agriculture had fractured beyond repair. Likewise, in France, the 1891 census listed 36,144 communes (towns, burgs, and villages, large and small), but 491,800 "hamlets, villages, and sections of communes." Scattered populations were attached to certain communes. The units of concentrated population thus had on average thirteen outlying settlements around them.[198]

It should be emphasized that before the animation of economic life in the maritime and inland plains, hamlet-like dispersed settlements and adjunct villages that flourished on the uplands from the late sixteenth to the eighteenth centuries had widened the operational orbit of rural households and communities into ecological environments of variant altitude and composition.[199] The proliferation of hilltop villages and settlements throughout the basin was mirrored in the dispersion of rural settlements in the Italian peninsula as a result, partly, of the popularity of the *mezzadria* system; in Provence there was an increase in the numbers of *mas* and *bastides*, dispersed settlements that helped colonize marginal land. In Spain, on hilly terrain, the abundance of springs and the importance of tree crops favored a long-established dispersion, aided by the careful use of terraces: *alquerías* (isolated farms), *quintas* (vineyards), and *barracas* (rice fields), as did the *mazra's* in the Ottoman lands.[200] Functionally, the establishment of satellite or adjunct villages in the lower altitudes and in the plains expanded the operational terrain of the mother village perched on the hillside. Also, high labor mobility allowed denizens of the basin to diversify their sources of income.[201] The availability of expansive *saltus* lands on the declivities and higher altitudes of the mountainous zone and the wide range of use they could be put to, from livestock farming to arboriculture and from forestry to viticulture, allowed a diversified agricultural base. Both strategies, diversifying ecological environments and sources of household income, were especially efficacious during the taxing times of the seventeenth-century crisis, but also throughout the eighteenth century.

The spatial order described above could be preserved by geographic mobility, and thrived on short-distance and short-term mobility: pendular migration, that is, by settled populations among different altitudinal ranges. During harvest time at lower altitudes, which of course was always earlier than higher elevations, peasants spent more time on the threshing floor, and slept at night in makeshift habitations rather than travel back to their home "village." In Spain, for instance, village territory enclosed lands that stretched from dry upper elevations "across the first plain and then the irrigated *huerta* to the coast."[202] The panoply of lands the villagers put to use ranged from dry *terra alta* and well-drained *terra rodina*, to the heavier soil of the plains, *terra archila*, and the *terra chopet* of the flood plains. In the hilly regions of Palestine, too, village lands extended in long strips to the east, and these lands were cultivated when conditions allowed. During the plowing and harvesting seasons, the villagers slept out in the fields until the season's work was completed rather than return to the village every night.[203] Given this spatial mobility, most villages located in mountainous zones came to be composed of two living quarters, if not more:

complementing the mother village was the offshoot village at lower altitudes where cereal cultivation took place. Formation of upper and lower villages mirrored the process best: village names that were preceded by *"zir-, bâlâ-, yukarı-,* and *aşağı,* or in the Balkans *donli-/dolne-/dolnje* and *gorni-/gorne-/ gornje"* and *kato/pano* abounded. In the Levant, the suffixes *al-fawqa/al-tahta* and *al-kabira/al-saghira* captured the same process.[204] The process that produced double villages led to dispersed settlement in neighboring plains and valleys in France as well.[205]

Most mountain villages had as much cultivated land outside the village, dispersed over wide stretches.[206] And the kind of mobility this occasioned, within the confines of one village or more, became so much part of the land-scape that it was even sanctioned by the legists of the time, much to the chagrin of timariots in whose interest it was to tie the *reaya* to the land. Peasants wandered "from one village to another," and found "everywhere large tracts to cultivate."[207] In Castile and elsewhere, too, there was high degree of spatial mobility: villagers migrated from place to place, temporarily and permanently. That most of the peasants in Iberia were alienated from their land and had been turned into hired hands contributed to this mobility.[208] Land would be worked at almost any distance from the village in the Greek peninsula as well.[209] It was not unusual in the Ottoman empire that villagers had holdings in more than one village. This was certainly the case in eighteenth-century Damascus, but, judging from endowment deeds, it appears to have been a widespread phenomenon.[210] This was in contrast to the orderly sixteenth century, when the movements of the *reaya* were monitored more closely. Hence, by the eighteenth century, tilling, tending, and buying land in different villages had become a common occurrence. The imposition of charges on land sales in fact provoked the revolt in Kisrawan in 1858 and led to a confrontation between lords and the peasants who were buying land from them.[211] In terms of tax-farming operations, too, geographic mobility created conflicting claims, especially in the collection of extraordinary taxes and dues. The family of migratory movements— seasonal, pendular, short- and long-distance—was more diverse in nature after the sixteenth century than before.[212] To say that the plains were left unattended and unmanned did not necessarily mean that they were not put to use or remained untilled. It meant that the mobility that allowed the upward-bound movement of settlement units, as examined in the previous section, was complemented by the lowland-bound mobility of labor. The ways in which the lowlands were brought under cultivation or utilized differed, depending on the modality of interaction between the plains and the mountains, which was different in each instance. In Rumelia, where landholders had to attract laborers to work at their estates, the most common strategy was to offer sharecropping arrangements, which spread to many parts of the region—Macedonia, Thrace, Bulgaria, and Serbia—between 1600 and 1800, to reach its height in the nineteenth century, when the plains were back in favor. To render the lowlands attractive, remunerating farm laborers in whole or in part with assignments of small plots of land was a common practice in the *latifundium* system. The coexistence of great estates with sharecropping was a feature common to most places where wastelands were brought under the plow.[213]

The rarity of vast plains thus left its impress on the agrarian fabric of the region. Settlements habitually advanced and retreated on the outer fringes of land occupation. And the transition was not necessarily from full-fledged to deserted villages or vice versa. The lines that separated villages, farmsteads, and offshoot settlements were neither thick nor definitive. The interplay between them as well as their ability to metamorphose into numerous forms were the essence of their mobility and flexibility. The fluidity embedded in this mode of rural settlement was well served by the geographical setting of the Mediterranean. Laden with vertiginous relief and short of sizeable plains, this topography did not endow the region's landscape generously.[214] The basin was short of vast stretches of land suitable for large-scale cereal culture in contradistinction to the Atlantic and Elbian plains of the north.

In light of the steep rise in the number of peasants earning their livelihood on higher altitudes, the scarcity of level and easily cultivable land was a restrictive factor.[215] Village arable met the demands of its inhabitants with great difficulty. This, in response, prompted the building of new *mazra'a*s and offshoot fields in level fields and the plains where temporary grain cultivation was possible, and on different altitudes of the mountain range where the ecological diversity was translated into a diversified crop-mix. Also, the vital role played by the cultivation of labor-intensive crops demanded close supervision of outlying gardens, orchards, and vineyards.[216] Upland villages were forced to mobilize all the resources available to them in order to guarantee their subsistence and survival: engage in polyculture, export labor, and find a safe cereal. Utilizing the Mediterranean relief to cultivate fields located in different parts of it in order to minimize exposure to risk and diversifying the crop-mix were equally effective in countering the fluctuations in harvest yields. Periodic redistribution of land among cultivators was an equally popular measure in reducing chances of perpetual misfortune. Shifting agriculture was another expression of this mobility, since it was based on the burning of undergrowth so as to clear land for cereal cultivation for a year or two, and the ashes acted as a fertilizer. In this bush-fallow agriculture, land was subsequently abandoned for as many years to recover its fertility.[217]

Until the mid-nineteenth century these uninhabited plains, maritime and inland, were cultivated by the residents of villages located on bordering hills and mountains.[218] Settled solidly on higher altitudes yet somehow strained by the scarcity of resources these highlands offered, hill villages extended their "tentacles," thereby establishing offshoot settlements in the plains that they inhabited for only a small part of the year. This was because the workforce from parent villages would descend to the plains, first, to sow the fields, and later, to harvest the crop. Over the course of the nineteenth century, as the plains became more hospitable for extended stay, the to-and-fro movement from parent villages outlived its usefulness. At much the same time as population growth resumed and need for cereals was on the rise, the descent from the mountains, now threatened by ecological overshoot, gained speed. Consequently, new villages sprouted up in the lower-lying lands during the latter half of the nineteenth century and cultivation in the plains ceased being ephemeral.

Above we outlined the factors that tipped the balance between the highlands and lowlands to the benefit of the former. The ascent of rural settlements accompanied by the "vulgarization" of intensive methods of land use, and the ease with which self-subsistence was attained thanks to the advent of American food crops paved the ground for hill and mountain villages to thrive within the confines of the resource base available to them. As is often the case, however, success did not breed success. Growing pressure on montane resources due to demographic growth on the one hand, and the gradual recovery of the lowlands due to the commercial impetus given by *pax Britannica* on the other, undermined the economic and spatial order of the lethargic seventeenth and eighteenth centuries. In light of these dual developments, the staying power of upland settlements weakened. Progressively, they were subject to the vagaries of limited availability of level land and, more germane to our analysis, denudation of the mountains and resultant erosion, among others, on account of demographic growth. The descent of rural settlements and recuperation of the coastal and riverine plains acted as catalysts of change, putting an end to the renaissance of hills and mountains, as we discuss in the following section.

Descent of Rural Settlements

During the seventeenth and eighteenth centuries, then, the uplands of the Mediterranean, with the partial help of the crops of the Columbian exchange but, more important, due to the increasing intensity of mixed cultivation, shouldered the responsibility of accommodating the region's variegated agriculture on its hills, slopes, valleys, and intermontane basins.[219] This became more pronounced once the emporia of oriental crops and spices built by the northern Italian city-states along the coastlines of the Inner Sea was gradually fractured. The relatively empty and disease-ridden state of the plains, maritime and riverine, which compelled rural households to center their operations on higher ground, presented them with the chance to plant alluvial lowlands on a temporary basis, if need be. The expansive ecological complex that came into being from the late sixteenth century therefore extended from warm winter pastures that hemmed the skirts of the region's mountain range to its alpine pastures, a territorial complex that bestowed the inhabitants of the region with short- and long-distance mobility. This altitudinal redistribution of economic activities that favored hillsides and uplands was conducive to the development of horticulture, now boosted by the introduction of American food crops; arboriculture, now rejuvenated by the spread of tree crops; and livestock husbandry.[220] As a result, there were intimations of a revival of the "traditional" Mediterranean ecological order associated with transhumant flocks on the *garrigue* lands, wheat and irrigated crops in the plains and valley floors, olives and vines on the stony slopes. Not only that, the increasing ability of local populations to access hydraulic energy due to intensified riverine activity assisted the spread of rural crafts in the hilly and mountainous parts of the region: witness, among others, the dispersion of textile dyeing and cloth washing.[221] Unlike in the long sixteenth century, the absence of vibrant demand from, and full-time employment in, the lowlands during the seventeenth-century crisis and after rendered it imperative for upland communities to diversify their sources of livelihood and income. So, on account of the sodden and malarial state of the plains, the uplands of the Inner Sea were forced to rely on their own resources more than usual in the seventeenth and eighteenth centuries. Certainly, commercial agricultural cultivation in the plains was still a possibility, but since the scale of such production was limited, it failed to draw migrants from nearby mountains. That coercive measures had to be deployed more often than not in order to attract and keep migrants in place attested to the unattractive nature of the wet lowlands, however remunerative the task at hand was.

If the heyday of the uplands was in part derived from the misfortunes of the low landscapes, then it follows that when signs of life were to reappear in the latter, that would have immediate repercussions on the economic life of the highlands. This was indeed what happened during the latter half of the nineteenth century. Once draining marshlands gained momentum, followed by the offering of titular titles to abandoned or waste lands, which mostly belonged to the states, to the highest bidders, the reversal of fortune set in. People of the highlands found themselves, once again, engulfed in downstream migration, seasonal and, more germane to our discussion, long-term. The reclamation of vast tracts of land in the coastal and inland plains by and large registered the end of the grand agrarian cycle that started with the return of the Little Ice Age in the 1560s. The return of wheat, as King Corn, signaled the beginnings of the end. By the end of the nineteenth century, wheat along with cotton had regained their status as one of the most distinguished inhabitants of the basin.[222]

In between the attempts to reclaim the plains in the sixteenth century, mostly in the relatively affluent parts of the Mediterranean, and the massive drainage projects of the late nineteenth across the basin, people of the highlands had little exposure to the limited opportunities offered by the low landscapes. Petty producers experimented with new crops, as exhibited by the spread of maize and tobacco (and the potato) in the humid zones of the region. Where humidity was lacking, they increased the share of legumes, now fortified by the arrival of American legumes, in crop rotation, and relied on them as foodstuffs; added arboriculture to their arsenal of economic activities; and diversified their crop composition.[223] Moreover, given the almost serendipitous availability of abundant and verdant grazing pursuant to the evacuation of the lowlands, livestock agriculture expanded as much by default as by choice. The resuscitation of legumes and the extension of stock-breeding was an integral part of the producers' strategies to protect themselves against harvest failures, and make up for the deficit caused by the contraction in cereal production. It needs to be underlined though that however damaging to the economic health of the lowlands the contraction in arable production was, the period from the 1550s to the 1870s also witnessed a spread of aquatic and irrigated crops such as cotton and rice, both of which opened up the possibility of bringing relatively lower elevations of the basin under tillage, be it temporarily. The growing significance, and the spread of cultivation, of cotton almost exclusively in the eastern Mediterranean, and rice in the western Mediterranean, particularly in Spain, in the eighteenth century once again brought level and extensive lowlands within the realm of operations of some peasant households, but in more likelihood of large landholders. Cotton plantations, as inhabitants of the lowlands, served as points of attractions for downstream labor.

Cotton production in the Balkans increased threefold in the eighteenth century, and mostly in the lacustrine environments of the lowlands, understandably because of the irrigated nature of the crop.[224] Since these lands were usually under the control of the notables of nearby towns and villages, hands hired to work in these inhospitable conditions suffered the consequences of exposure to malarial infestations.[225] As a crop of the lowlands, which were badly in need of rehabilitation, cotton demanded the organizational capabilities of large estates, local notables, and tribal confederations, especially if its cultivation was commercial and large-scale—which was mostly the case. In the Balkans, its natural habitat was the *çiftlik*s. It was these landholders who pioneered and enforced the planting of cotton, or cash crops in general. One of the options available to those who inhabited the mountains was employment in the plains, as sharecroppers, *yarımcı*, or waged laborers, *aylakçı*, under the direction and watchful eyes of landholders.[226] At the employ of *çiftlik*-owners, the conditions in which migrants from mountain villages found themselves in the marshy and malaria-ridden low-lying lands of the plains, home to large estates, were abject. After all, the migrants had to toil in the "marshes between Lárisa and Volos, the muddy banks of Lake Jezero, or the wet river-valleys of the region."[227] The unforgiving nature of the plains necessitated the coaxing of labor by mercenary forces, *kırjalı*s and *hayduk*s, who were assigned the task of enforcing the rule of the landholders to keep the migrants in place, in case discontent with working conditions could not be contained.

But these developments, even though largely confined to the *çiftlik*s located near the Aegean, Adriatic, and Black Sea shores and even though it was these agricultural estates that took the lead in introducing the cultivation and extending the domain of cash crops, were not unparalleled.[228] On the easternmost shores of the Mediterranean, the task of mobilizing labor in the critical stages of cotton cultivation was performed by provincial notables or tribal confederations. The extension of cotton cultivation in Palestine under al-Jazzâr and Zâhir al-'Umar testifies to this.[229] One of the first tasks set by Ibrahim Pasha during his occupation of Syria in the 1830s was to settle immigrants from Egypt in the swampy valley of Baisan.[230] The perpetually mobile tribes that shuttled between the mountains, valleys, and plains served a similar function by providing the necessary labor, when needed, in the planting stage or during the harvest. This pattern was later successfully reproduced in the Cilician lowlands at the end of the nineteenth century.[231] Cotton's availability increased its share in, and animated, rural industries, as attested to by the number of cloth-washing centers and dye-houses in the registers of pious foundations.[232] It came to occupy a position similar to that of wool in the western Mediterranean. The health of rural industries depended on it, for it became the industrial crop of the Levant par excellence. If wool reigned over the highlands, cotton then ruled the plains. And the presence of putting-out arrangements functioned as a reliable outlet.

Riziculture, too, like cotton cultivation, necessitated irrigation, and spread in the lacustrine marshes, wet river valleys, and piedmont regions.[233] Initially, rice spread in limited areas in the fifteenth and sixteenth centuries. It reached the plains of Lombardy and the ill-drained districts of the Po valley in the early sixteenth century. By the end of the century, there was a steady supply of the new cereal to the city of Genoa, and it was being grown on a commercial basis in the eastern parts of the Po lowlands. The Veneto, Emilia, Piedmont, and the region around Mantova were intimately familiar with it. To the west, it had spread along the Ligurian coast to reach Nice by the end of the century. Rice was grown wherever there were natural fens, or places that were flooded artificially. Yet, since riziculture involved the landscape of marshes and fens, rice cultivators had to live with the vicissitudes of such environments: malarial infestations affiliated not only with the marshes but increasingly attributed to the new paddies.[234] Such worries may have ruled the day during the *beau seizième*, but not in the lean centuries that followed. The decline it suffered in reputation proved to be fleeting, though, for rice regained popularity in Spain, not as a creature of the wet lowlands but as a high-yielding crop, in the closing decades of the seventeenth and the eighteenth centuries. Despite attempts by imperial authorities to control or limit the expansion of rice planting and despite mud and fever all year round, it managed to come into vogue "as an alternative to traditional grains."[235] After all, it was the best way of putting marshy coast-lands and river-lands to work. The outstanding point about rice was that, very much like maize, it contributed to the supply of staple, not luxury, foods. Despite all its travails, it complemented the region's agricultural economy at its weakest point.[236] Between the sixteenth and eighteenth centuries, in fact, the spread of cultivation of rice in the northern regions of the Italian peninsula occurred exclusively through stable rice fields established in expanding marshlands.[237] In Ottoman lands, rice cultivation, labor-intensive in character, was tended on a part-time basis by *göçer evler*, migratory populations who commuted between their summer and winter pastures or by itinerant day-laborers.[238] There, too, the work was performed under unhealthy conditions, but it lasted only a few weeks. Rice may not have been "of fine quality," but it was "abundant in the low-lands of Anatolia."[239] Like cotton, "the development of riziculture represented less the interests of peasant farmers than those of land-owners with considerable investment potential. The reshaping of existing fields and the extension of irrigation and drainage channels were costly and demanded both skills and coordination of efforts not easily available to all."[240]

In short, a core group of sharecroppers fixed in the lowlands and a mobile labor force of seasonal and temporary hands who were imported from the mountains to carry out labor-intensive parts of the production process helped keep agriculture in the plains alive. The employment of force often to recruit and subordinate labor was a good gauge of how miserable the conditions in the lowlands were. If utilizing force was one way of keeping lands under the plow, so was paying laborers high wages to compensate for toiling in the disease-ridden environment of the plains. This was the case in Maremma.[241] Yet, this option was open only to a few. Otherwise, the landholders remained anchored in the plains, and had to find ways to tempt or coax people of higher altitudes to populate their fields in the lowlands.[242] But overall, access to the plains by the people of surrounding mountains and hills was limited, and when it was available, it took the form of sharecropping, for it allowed a periodic adaptation of cash or in-kind payments to the volatile market conditions.

Yet, given the sluggishness of agricultural growth during the seventeenth-century crisis and the high population density on the mountainous regions of Rumelia, the share of the migrants who found employment in the lowlands was relatively low. Two options were available to the inhabitants of the mountains. One, of course, was migration. And migrate they did in large numbers and worked in distant parts of the Mediterranean.[243] The development of cottage industries provided another option.[244] The sophisticated nature of trade networks canvassing the region, as discussed in the second chapter, well-endowed with an extensive *verlag* system, furthered the possibilities of rural manufacturing, especially when coupled with the growing role of sheep-grazing on the higher altitudes of the region.[245]

In any event, despite the fluctuations in demand for cotton and growing rice exports from Spain, the presence of these two crops and, as we mentioned previously, of vineyards that had taken over abandoned plains, the magnetic field created by these cash crops was not forceful enough to turn the cultivation of low landscapes into a permanent undertaking. In view of this, the uplands remained the main theaters of agro-sylvan and pastoral expansion. Yet the ecological transformations of the 1560s–1870s period could not be easily contained due to the growing need to put the hilly slopes to work, as was the case in Aín. The congregation of growing numbers of livestock, hamlets, farmsteads, and peasants in relatively confined hilly terrain may have bred success in accommodating growing numbers of people, but only in the short run. In the final analysis, it was unavoidable that there would be growing strains in the ecological and economic makeup of these regions due to increased demographic density, which put the resources of the highlands to a serious test, again, as was the case in Aín. It was a certitude that the erosive influence of land clearings, the tilling of slopes, and the intensive exploitation of montane resources would act as a check against the congestion of highlands. It was, after all, the cultivation of hillsides and uplands—and not grazing per se—which precipitated erosion by exposing bare soil to occasional heavy storms.[246] In the short run, the need to struggle with, and prevent, erosion that resulted from accelerated pace of agrarian exploitation was made easier thanks to the widespread popularity of intercropping. After all, mixture of grain farming with arboriculture was a good prevention against erosion.[247] So was terracing, which changed form and picked up pace during the eighteenth century.[248] But it was not sufficient to counter the growing problem of erosion.

The ascent of rural settlements toward agronomically attractive altitudes of the region's highlands was hence not exempt from ecological constraints. Extensive exploitation of the uplands, the felling of trees to assart land, and the thinning out of the maquis cover by sheep and goats compounded the volume of run-off water drenching the lowlands, reducing the ability of the soil to absorb water. During the seventeenth and eighteenth centuries, active forest clearance took place in the alpine valleys of Provence and Italy, notably for viticulture.[249] Not surprisingly, there were strong signs of land erosion during the second half of the seventeenth century and the eighteenth, and sloping fields suffered the most from it.[250] In Sardinia, forest exploitation and land clearance, which could be associated with seventeenth- and eighteenth-century land hunger, caused the deposition of rough sediments on the Gulf of Oristano. Exceptional levels of erosion were, by implication, a result of climatic wetness of the Little Ice Age.[251] Since the eighteenth century, population pressure in the mountain valleys had encouraged sporadic enclosure of small clearings in the common pastures and active clearance of woodland.[252] The same observations held true for the Ottoman empire as well. Between 1700 and 1850, land converted to regular cropping rose by 35 percent, as opposed to 103 percent worldwide. As settlements moved onto higher altitudes, woodlands shrank by over 10 percent due to assartion and deforestation, a figure that matched that of the western Mediterranean (a figure much higher than the 4 percent world average). Meanwhile, the area covered by grasslands and pasture declined imperceptibly, by 0.4 percent: the world average was 0.3 percent.[253] Even when at the end of the eighteenth century, fodder crops were grown during the fallow year, a system that slowed down erosion, this was a development that took place primarily on both sides of the Channel rather than in the Mediterranean.[254]

A telling index of deforestation was the appearance, in the lower altitudes of the region, of animals that left their sylvan habitat more often than previously. Along with widespread flight and famine, what compounded the destructiveness of the period was the appearance of wild animals, ravaging settled villages, at least in the piedmontese regions of Erzurum, Kars, and Erivan. The fauna that invaded the lower altitudes of the mountainous regions had not been an integral part of the countryside prior to the 1590s.[255] A very similar set of forces was at play in the Balkans, from 1606 to 1609 in particular, where a fear of wolves grew, following reports of the frequent forays of wolves into villages as well as isolated hamlets.[256] This was a telltale sign of inclement weather in the abode of wild animals on higher altitudes. Due to the severity of winters, increased precipitation in the form of snow, and due to the fact that snow stayed on the ground longer than before, wild animals had to descend farther below their usual hunting grounds in order to fetch food. It became a regular occurrence after the 1600s. In fact, wolf-hunting was "an index of the health of the countryside, and even of the towns."[257] Over time, the number of tigers and leopards that descended onto the settled zones of the hills and valleys, and even low-lying lands multiplied; so did the frequency of such incursions. Not only did wild animals, tigers, leopards, lions, and bears become more visible than before in localities near settled zones, they also appeared in major cities: the appearance of a "panther" that descended down to the very center of Damascus in 1677 was such an event.[258] It was eventually killed, its hide proudly presented to the governor. In the next century or two, such incidences became commonplace.[259] But the descent of wildlife onto lower altitudes on the one hand, and the settlement in large numbers of the population on higher altitudes on the other, took a very heavy toll on the fauna during the seventeenth and eighteenth centuries. By the end of the nineteenth century, tigers and lions had become almost extinct on the mountains of the eastern Mediterranean. In fact, a tiger was sighted for the last time in 1853 near Ayas, and the leopard killed outside of İzmir in 1843 was the last recorded one.[260] The number of wolves that showed up in and around Franche-Comte was at its height in the closing decades of the eighteenth century, and it was not until 1850 or so that wolves died out.[261]

Besides the intensification of agriculture and destruction of forests due to the expansion of livestock husbandry or agriculture, the arrival of railroads with skyrocketing demand for timber delivered the coup de grâce. It resulted in slope denudation and valley alluviation. Soil erosion between 1760 and 1880 was speedy, due to the resumption of urbanization, consolidation of fields, shifts to monocultures, upland deforestation, and herding.[262] Deforestation was more brisk on the slopes facing the inland plains, where declivities were not as steep as the coastal strip, so it embraced human settlements better, in part because less steep declivities reduced the danger of violent flooding. At the same time, this hospitality opened up the way for deforestation. Once covered with forests as dense as those on its opposing declivities, albeit poorer in species, they suffered the most. Timber trade partly denuded the mountains: timber from Cyprus, the

Taurus mountains, and Mount Lebanon met the needs of the region.[263] The arrival of the railroad deepened the economic malaise afflicting the mountains. On the one hand, it accelerated the pace of drainage the railroad companies initiated in return for future claims on agricultural surplus; and on the other, the need for timber denuded the mountains, depriving them of their resources.[264] The eastern Mediterranean had more forests at the beginning of the nineteenth century than at the end of Ottoman rule.[265] All in all, the carrying capacity of the mountains was seriously tested and strained by the increase in the number of inhabitants, human and animal, and by the utilization of more of its resources as its higher altitudes remained under snow longer.[266] What made the matters worse for the tree crops of the basin was the arrival of diseases in the latter half of the nineteenth century: the vineyards were hit by phylloxera in the 1880s; the silk worm disease, the pebrine and muscardine, was later followed by the wartime destruction of mulberry trees; the disease that hit chestnut trees damaged arboriculture seriously, making the attraction of the lowlands, or emigration, evident.[267]

The waning of the Little Ice Age, concurrent with the travails of arboriculture, altered the environmental and ecological parameters of the order that prepared the groundwork for the retreat onto the mountainous zones. The revisions in the world-economy's axial division of labor precipitated two sets of transformations, intimately related, which reshaped the region's landscape from the latter half of the nineteenth century, a landscape that is still with us today. The first of these transformations was the onset of the resettlement of the plains, due in no less part to the draining of the swamplands in earnest, and the revival of urban growth, both of which aided the resumption of the imperialism of wheat. No doubt, wheat was the agent, and the catalyst, which hastened the pace of clearing of the swamps and fen-lands. Wheat, which as late as 1845 held no more than eighteenth rank on the list of Ottoman exports to Great Britain, had reached the fourth rank by 1850, and the second by 1855. The empire's wheat production soared from 20 million bushels in 1831–40 to 50 million in 1888.[268] By the century's end, it had already reclaimed the supremacy it enjoyed in the sixteenth century.[269] The rise of temperate settlements from the 1890s altered the picture tremendously. The emergence on world-economic stage of Baltic grain, from the 1590s, and the emergence of the temperate zones of the British empire, from the 1890s, as the breadbaskets of the world-economy created analogous developments across the Mediterranean, but with one difference. The momentum of drainage did not come to an end with shrinking export markets, but went on past 1945.

That the plains were ostensibly bereft of permanent settlements allowed the imperial or statal authorities to designate them as part of the *mîrî* land, as wasteland. It was these lands, seemingly vacant yet put to use intermittently by nomads, nearby villagers or others, where claims of usufruct or ownership overlapped and remained under dispute: in the western Mediterranean they were liquidated by the modernizing states from the 1830s, because the multiplicity of claims, it was argued, prevented their use in line with the principles of economies of scale. In the eastern Mediterranean, these were the lands that were alienated first with the passing of the Land Code of 1858.

In Rumelia, for instance, the pace of *çiftlik* formation increased, attracting new hands. The number of *çiftlik*s in the Gazi Evrenos *waqf* increased from 10 percent of its agricultural holdings in 1702–4 to over 50 percent at the end of the nineteenth century, absorbing some of the migration from the mountains. There is evidence that laborers used to come to the plain of Salonika from as far as the Pindus mountains, despite the pending malaria threat around the lake of Yenice-i Vardar and in the lower course of the river Vardar.[270] In 1839–40, the Kızıl Deli Sultan *zâviye* in the subprovince of Dimetoka owned 180 hectares (1,800 *dönüm*s) of arable fields, but only 0.3 hectares (3 *dönüm*s) of pasture land.[271] Stock-raising was giving way to cereal cultivation.

This could not have come at a more propitious time, because the balance established between the vacant yet sown plains and the populous highlands had started to exhibit signs of exhaustion by the latter half of the eighteenth century. Outmigration from the mountains accelerated in the latter half of the eighteenth century. From then on, the mountains of the Mediterranean took on the duty of tirelessly exporting labor. This was due to the success registered by the mountains in giving refuge and providing livelihood for continuously growing numbers of people. Put briefly, the hilly terrain was stretched to its limits, and the first signs of this was the acceleration of migratory movements to lower altitudes. In the Balkans the *çiftlik*s, however abominable the working conditions were, provided an outlet for a wide stream of migrants. The movement gained strength in the nineteenth century, when such opportunities arose in other parts of the empire with the acquisition or appropriation of lands after the Land Code.

The transformation of lands hitherto lain "waste" into if not private property then something akin to it accelerated the settlement and tillage of low-lying lands without alienating large numbers of rural producers. The hilly and mountainous regions of the empire remained largely exempt from the changes the Land Code inaugurated or sanctioned in the plains. As labor exporters, they felt its indirect repercussions because, in most parts of the empire, land on the higher altitudes was for all intents and purposes private property.[272] Even the failings of the law, namely, its inability to distribute title deeds to each and every producer and pastoralist, which abetted the concentration of vast stretches of land under the hands of tribal sheikhs or rural notables, accelerated the attack on virgin and waste lands. After all, the slow toil and the patient work of putting these waste, at times barren, lands under the plow, and the process of draining could be more successfully carried out by those who were in command of substantial pecuniary means or a sizeable army of laborers. Such an undertaking was beyond the means and quixotic efforts of smallholders.

The developments in the Mediterranean basin followed the same economic rhythms. Of the general movements shaping it, the movement away from the plains from the sixteenth century onward, on its shores in particular but not necessarily beyond the olive trees, was of cardinal import. So were the migratory movements, albeit mostly temporary, which gained pace during the seventeenth and eighteenth centuries. The two phenomena were closely interrelated. As concentration on higher altitudes increased in disproportion to the resources at the command of these mountainous regions, households with permanent settlements on these higher grounds had to export labor power to extend their reach beyond their immediate and not so remunerative surroundings with an eye to increase the flow of additional income. The population movements in the alpine regions of the Italian peninsula as well as those originating in the Pyrenees, French or Spanish, into the valleys or the plains surrounding it were movements with an impressive spatial stretch. Migrants from and across the Alps re-colonized abandoned lands in Provence under the *métayage* contracts of *facherie*.[273] This was a two-pronged movement. On the one hand as permanent settlements, or the core of the household operations, shifted up into higher territory to stay clear of malarial environments and plagues, the households felt on the other hand the need to stretch its network wider from such commanding yet somehow restrictive environment into the plains as hands for the sowing or the harvest, or as skilled hands into the urban centers, nearby or far away. Therefore, geographical mobility was an integral part of this new landscape. It was not only transhumant groups that continued to freely circulate between the lowlands and the high mountain ranges, but also the human stock had to search for complementary sources of income for some in lower altitudes and for others in distant lands on a long-term and at times permanent basis.

That the task was not easy can be surmised from the results obtained by the turn of the twentieth century. Despite the presence of conditions extremely favorable to the opening of the lowlands to cultivation, the inroads the bonification measures made into virgin territory remained limited. What accelerated it later was the arrival, at the turn of the century, of railways crisscrossing the plains.[274] The turning of uncultivated lands into proper fields needed laborers as much as moneyed urbanites or notables as the experience in Ottoman Balkans testifies. And there were two sources that came to provide the labor force necessary for the endeavor. Although labor migrations were concomitant, they were of differing provenance. The first was provided by the influx of migrants flocking into the empire from its far-flung provinces as these provinces changed hands.[275] And the second was provided by emigrants descending onto lower altitudes from the mountainous zones towering over the empire.

It was in the latter half of the nineteenth century that descent from the mountains started in earnest. Prior to the 1850s, there were, of course, waves of outmigration, but they were and remained local in character and small in scale. The inhabitants of Mount Ansariyya, the 'Alawis, who survived on tobacco cultivation for long, started to spread out to the surrounding areas, the plains in particular: the plain of Akkar in the south, the plain of Latakia in the west, and toward al-Ghab in the east. The movement gained momentum in the 1860s, when a score of villages were created. This essentially peaceful advance was limited to the left bank of the Orontes, but all the eastern part of al-Ghab was settled by 'Alawi populations. In fact, the 'Alawi colonization of the region from the steppes east of Homs and Hama and southwest of Aleppo was accomplished by the *agha*s of Hama as much as by migrants. To the north, the 'Alawis had by the end of the nineteenth century partially colonized the Cilician plain where, in the vicinity of Tarsus and Adana, they numbered eighty thousand.[276]

The İsmailis undertook a similar migration, starting from the fortress atop the 'Alawite range. From the middle of the nineteenth century, a sizeable colony came into being in the region of Salamiyya, to the east of Hama and at the edge of the Syrian desert. In a similar fashion, the Tahtacıs, who had sojourned on the higher altitudes of the Taurus range, descended onto lower altitudes. Adana and Tarsus on the Cilician plain attracted migrants from neighboring higher-lying regions: Diyarbakır, Harput, Bitlis, and elsewhere.[277] More timely still was the influx of large numbers of immigrants, *muhacir*s, from the northern latitudes of the empire, destined, in the second half of the eighteenth century, to Rumelia, and throughout the nineteenth, to Anatolia and beyond. The influx of migrants went uninterrupted for most of the nineteenth century. The movement originated in the Crimea in the 1780s, and the first wave of migration headed for Rumelia. The number of migrants reaching Anatolia increased over time, only to be compounded by those coming from Rumelia as hopes of cessation from the empire rose and nationalist mobilization became the order of the day. From the 1820s onward, migrants came from the Crimea and the Caucuses, and from 1878 onward from the Balkans. By the turn of the twentieth century, there were close to 4 million immigrants in Anatolia alone. They constituted one-third of its population in 1914.[278]

Although labor exports from mountainous regions did not start in earnest until the nineteenth century, when their resources proved insufficient to feed multiplying numbers of inhabitants, the mountains exhibited a capacity to export labor, even in the eighteenth century. This was an early sign of the overshoot that awaited the mountains: Mount Ansariyya experienced chronic overpopulation, and sent waves of migrants north, to Antioch; east, to Hama, Homs, and Aleppo; south, to the plain of Akkar; and west, to the maritime plains and Latakia. Maronites from Jobbet-Bisharret and Batroun moved south, to Kisrawan, the Druze territories, Bekaa, and the maritime plains, to Batroun and Jbayl. Jabal 'Amil was populated by Shi'ites who were displaced from Mount Lebanon by the expansion of the Maronites, and the Shi'ites were displaced toward the Bekaa. Latakia thrived during the eighteenth century, because it was able to attract and rely on short- and long-term migration.[279] The readjustment of the rural settlement pattern to the new ecology and the flexibility built in this pattern of which we spoke in the section above were contingent on labor mobility. That is why it remained unexceptionally high throughout the period under examination.

If the settling of the plains was not initiated by the new landowners, then it was initiated by pastoral groups—in which case it was done in an authoritative and collective manner. This explains in part the popularity of *mushâ'*, historically confined to grain-producing zones and the low-lying lands, at times classified as "salty."[280] For one, it facilitated the colonization of new agricultural lands by minimizing vicissitudes associated with it for individual producers by periodically apportioning land or rights over land, because the arrangements were or seem to have largely been the property of lands devoted to the dry-farming of cereals.[281] Land was also held in common and subject to redistribution in Ula, Çukurova, Antalya, Esdraelon, Konya, Kars, Söke, east of Aleppo, and Jordan.[282] This process of colonization was given an additional thrust by the flexibility of mortmain holdings: long-term or perpetual lease of agricultural lands, *hikr*, allowed foundations to rehabilitate their lands by relying on the initiatives of those who were willing to invest in such lands. At times, this turned out to be a method of appropriating *waqf* lands.[283]

The influx of new hands into the cereal lands did not alter the general equilibrium between land and labor. Labor was still scarce in proportion to the amount of land sown to crops. The balance was not tilted in favor of capital, which meant that an arrangement (such as sharecropping) that would allure the newcomers had to be fashioned. A relatively thinly populated countryside was the stage for the changes that took root during the Victorian boom. The political unrest that shook Jabal Hawran at the end of the century was aimed at abolishing the custom according to which the shaykhs received one-third of the crop.[284] The 'Alawis who descended onto the plains worked as sharecroppers, *murâbi*'s; they were entitled to one-fourth of the proceeds of their labor.[285] The need to reclaim wide tracts of land, albeit on a small scale, and open them to tillage increased the demand for rural hands and hastened the outflow of labor from mountainous regions. Overall, reclaiming vast stretches of land put to work with a relatively limited number of hands encouraged sharecropping arrangements. Reclamation of these lands, as the case in the opening of east of Elbe to cultivation during the sixteenth century, brought with it new relations of production, as was the case with the second serfdom. Even though huge expanses of land were brought under the plow, and either the would-be land-lords, like the Sursuqs in Palestine, invested considerable sums of money or tribal sheikhs owing to their command of huge pools of labor undertook an organized enterprise, the process did not result in the formation of large estates. Those who mobilized resources not easily available to an ordinary producer eventually had to give latitude to the newcomers, and solidified their position as rent-collectors than landlords. This strengthened the social fabric of Ottoman agriculture by replicating the land tenure dominant in the hilly and moun-tainous regions of the empire.

Finally, the ascent of rural settlements redefined the theater of contestation between the desert and the sown. The rather swift expansion of the arable during the long sixteenth century was at the expense of lands put to use by pastoral and transhumant farmers. The conversion of grasslands into corn fields in most corners of the empire was inevitably accompanied by a series of conflicts arising from transhumant populations' incursion into ever-expanding lands sown to grains. This was to be expected given the fact that these pastoral populations traveled long distances between their summer and winter quar-ters, and increasing segments of the land they used as their winter quarters were located in lower altitudes of warmer temperatures and were now devoted to grains.

This settling of incoming migrants in government-appointed localities, stra-tegically chosen, hastened the colonization of the plains. The massive relocation of migrants and the settlement of nomadic populations went hand in hand. As settled populations came to occupy land previously used for grazing, nomads' field of maneuver shrank. The incoming migrants were strategically placed in sparsely settled regions: in western and central Anatolia, where the gentle slopes of the mountainous range were already occupied, it was the low-lying lands and the center of the basins, difficult to till and more difficult to drain, which were given over to the migrants. The second popular location for the migrants were the terraces overlooking rich alluvial valleys that were already inhabited by nomads: these terraces were rocky with thin topsoil.[286]

The outmigration that originated in the high-lying regions was, again, a Mediterranean-wide phenomenon, for the factors that pushed populations out had a very strong ecological component that transcended political boundaries. In France, the migrants came from "the Massif Central, the Alps, the Pyrenees, the Jura and certain regions on the edge of the Paris basin—in short from areas which, taken together with others nearby, could still today be called 'the poorer France.' "[287] The migrations were not only local; the high region of Auvergne and of course the Pyrenees sent migrants to Spain from very early times. The mountainous zones of Italy too were, or became, large exporters of labor.[288] Most of the paving men and diggers of aqueducts in Istanbul were of Albanian provenance, and most of those who worked in the Hijaz railroad were from mountainous Italy, Greece, and Montenegro.[289]

The central government tried to accelerate the pace of sedentarization by organizing military campaigns against strong nomadic confederations of the time to force them to settle, in the Cilician plains, in eastern Anatolia and on the northern fringes of Syria. When the conditions were not ripe, however forceful the imperial army and threat of punishment was, these campaigns proved to be futile, as was the case at the end of the eighteenth century. The panorama in the mid-nineteenth century was different, however. The passing of the Land Code, the rise of a group of landowners, and the reclamation of the plains altered it substantially. If for nothing, the contraction in winter pastures in maritime and inland plains limited their migratory movements, which, in turn, made them more receptive to settlement. Moreover, they were more sedentarized at the end of the eighteenth century than before, because temporary agricultural fields were tilled by pastoral groups anyhow. Consequently, settling pastoral groups was an easier task than it had been before.[290]

The second set of transformations was closely related to the first. The increase in wheat cultivation and the arable devoted to it brought back the same constraints on animal grazing as existed during the sixteenth-century expansion. With more land in the plains brought under the plow or under the control of those who became landowners with the passage of the 1858 Land Code, the space of maneuver available to livestock holders shrank notably. From then on, most regions were "forced to choose a single major economic activity."[291] This was a dramatic about-turn from the polyculture of the seventeenth and eighteenth centuries. Since the mid-nineteenth century, a number of more favored regions ceased to grow the full range of Mediterranean crops and specialized in one or two cash crops.[292] For example, the monoculture of vines in Bas Languedoc and parts of Roussillion, and the intensive forms of commercialized polyculture in Province and Roussillion were introduced.[293] The easy availability of Pontic grain from the late eighteenth century was more than instrumental in this.[294]

In fine, the descent of the inhabitants of highland and hill villages to cities located on the coast or in the interior to be employed on large agricultural estates or to work in the nearby cities' manufacturing and service sectors was only one way of dealing with the strictures placed on the realm of human action by environmental factors. In other cases, in fact, a development reminiscent of what happened in the sixteenth century reoccurred: villages previously located "midway between the crops in the valley and the forests of the mountainsides" or "between the vines on the upper hillsides and the cereals down in the valley" started to "slide downhill," at times they even "snowballed."[295] The hamlet-like dispersed settlements of the mountains that flourished from the late sixteenth to the eighteenth centuries finally had to give in to the power of attraction of the lowlands. Whenever conditions permitted, villagers built new dwellings, at times temporary and at times permanent, farther down the slope.[296] These examples attest to the fact that the new world of the hills had lost its attractiveness by the mid-nineteenth century at the latest.

The gradual yet inexorable return of the center of economic life to the lowlands of the Mediterranean was triggered by soaring demand for cotton and wheat from Manchester and London at the height of the mid-Victorian boom. Drainage projects that brought vast areas of marshland under the plow by extending croplands in the maritime and riverine plains in effect signaled the end of the heyday of hills and mountains. The massive downward migration of mountain peoples reflected the strains that had built up in the higher altitudes since the late sixteenth century. Arguably, the success shown by the fresh round of drainage and reclamation of the lowlands was in part due to the conclusion of the humid Little Ice Age. That the end of the prominence of the hills and mountains and that of the Little Ice Age were concurrent was not incidental, for the two were closely connected. Also coming to a close was the great agrarian cycle that stretched from the 1550s to the 1870s and temporally framed this geohistorical movement.

Conclusion

The Mediterranean between the
Leek-Green Sea and the Green Sea

The Mediterranean was all but a timeless entity with invariable ecological and agrarian features. Some of the characteristics we attribute to it today as "traditional" or "olden" are products of relatively recent developments, the historical pedigree of which is not lost in the mists of time. The Inner Sea may of late present a picture in which vineyards, silver-green olive trees, and golden-yellow cereal fields, along with herds of sheep and goats ranging across the *garrigue* lands, populate its landscapes. But this picture was subject to frequent modification and revision. If Herodotus, Lucien Febvre conjectured, could wander once again in the mid-twentieth century along the shores of his beloved wine-dark sea, he certainly would be unable to recognize the landscape of the Aegean basin he knew so intimately, and he would be astonished when he saw the flora that is today assumed to be quintessentially Mediterranean.[1] In part, this was because starting with the eighth century or thereabouts, two waves of vegetal migration substantially enriched the floral composition and reshaped the ecological layout of the Inner Sea. As we explored in the previous pages, the first wave of crop migration emanated in the Indian Ocean, and the latter, in the Americas. Indeed, Febvre brought to notice the cypresses of Persia as a symbol of the occidental migration of flora from the Indian Ocean, and the cactus plants as a symbol of the Columbian exchange that diffused the flora of the Americas across the oceans. Transformations in the composition of the Mediterranean landscape were not limited to the vegetal and ecological alone. Crucially, they determined the economic and political contours of the age of city-states, from its dawn to its eclipse.

Indeed, it was the economic might of the twin city-states that underwrote and hastened the westerly migration and acclimatization of oriental crops. The rise and consolidation of the Venetian maritime empire, for one, became contingent on the valorization of exotic, "scarce," and lucrative goods. This occurred at the expense of the region's "indigenous"—tree—crops that were easily available to all. Not surprisingly, only after the Genoese merchants managed to perfect an organizational setting for large-scale cultivation of oriental crops under the new skies and within the fold of the "plantation complex" in the Atlantic did the lucrative crops lose their luster first but not exclusively in the Venetian Levant.[2] It was at this juncture, when the economic devolution of the basin was setting in under the ephemeral glow of the age of the Genoese, that the tree crops of the Mediterranean resurfaced. Only after the weight of the rich trades was lifted off the shoulders of the Inner Sea and the economic order imposed by the Serenissima came undone was the region able to resume the features it possessed prior to the reign of the city-states.[3] The return of the basin's "natural" flora did not simply revive the landscape of the older times, however. From the sixteenth century, the indigenous crops routinely sown by the peasants were enriched by the spread of American food crops that were dispersed during the second wave of worldwide vegetal migration. These food crops provided the region's poor with new staples: they entered not only the crop-rotation cycle by supplanting low-yielding crops, but also the diets of the denizens of the basin by replacing market-worthy cereals.[4] The differential nature of these two sets of crops, lucrative in the case of the former and plebeian in that of the second, fittingly captured the waxing and waning in the fortunes of the Inner Sea. In sum, the two phenomena that reshaped the landscape—crop migration and dissemination on the one hand, and the making and unmaking of the Mediterranean on the other—were closely interrelated.

For one, the consolidation of the rule of the city-states and the diffusion of lucrative crops were two sides of the same coin. For the perpetual rivalry between the Venetian and Genoese men of money reordered the economic life of the basin, and it was then that sugar and cotton were elevated to the status of "golden crops." Establishing control over the cultivation of oriental crops, preferably in dominions under the city-states' direct control, and overseeing the rich trades were prerequisites for ruling the waves of the Inner Sea. The city-states marked their territory and perfected their modus operandi accordingly. No doubt, the western-bound movement of oriental goods at times changed venue not necessarily owing to the efforts or the might of the city-states' merchants who were minor players in the western outposts of the emporium over which the Mongol empire held sway. It was mostly due to the fluctuations in the ability of the empire of the steppes to effectively patrol and secure commodity flows under its imperium that fueled changes in conduit. During the thirteenth and fourteenth centuries, the principal commercial nexus shifted from the southerly maritime circuit, which pitted Baghdad against Cairo for the privilege of serving as the hub of this commercial world, to the continental land routes, the termini of which were located on the shores of the Pontic Sea.[5] From the mid-fourteenth century, the southerly maritime route regained favor after the Mongol empire began to weaken. If the terrestrial circuit blessed the fortunes of the city of St. George, which was in charge of the Euxine Sea, the Indian Ocean route favored the city of St. Mark, which remained in charge of the Levant. Changes in the volume and provenance of the spice trade further contributed to the import of tropical crops.

As we emphasized in previous chapters, these crops thrived in the semitropical climates of the coastal zones, and designated the maritime plains and valleys along the Mediterranean littoral as well as the "liquid plains"[6] that the city-states reclaimed for their own benefit as their main theaters of operation. The spread of tropical crops thus reshaped the ecological order of the basin in favor of its lowlands. Initially, greater precipitation and widespread use of irrigation in the eastern shores of the basin contributed to the warm welcome the tropical crops received in the Levant.[7] The plains and irrigated coastal valleys were transformed to accommodate the commercial cultivation of the new flora. To wit, the coastal plains in the Levant as well as in Sicily, Cyprus, Crete, and the westerly latitudes of North Africa, among others, were devoted to the cultivation of oriental crops. The advance of these exotic crops inescapably determined the destiny of the basin's popular crops. But this was not the only reason for the diminishing import of tree crops.[8] Climatic fluctuations were equally responsible. From the ninth or the tenth century with the onset of the period of global warming, vineyards easily advanced north, and the center of *grand commerce* in wine was located mainly in the viticultural regions along the Atlantic banks of Gascony. The arrival of the Little Ice Age may have slowed down the pace of northerly advance of vineyards, but it did not halt it all at once. The northern-bound movement was reversed in earnest from the latter half of the sixteenth century.[9] It was then, when new vineyards were included in the supply chain, that the Mediterranean varietals became an integral part of this trade. Additionally, the progressive conversion of some of the arable into pastureland from the fourteenth century had an adverse impact on agriculture by encouraging the conversion of cereal-growing lands into pastures. And the reign of livestock husbandry lasted well into the nineteenth century. The erosion caused by the thinning of plant and vegetation cover had dire consequences in the long run in the formation of the malarial lowlands.[10]

Climatic variability associated with the return of the Little Ice Age increased the frequency of droughts and torrential rain, taxing bread crop harvests, leading to crop shortages and, worse, to famines. Equally crucial, of course, was the abandoning of many villages. The desertion was a large-scale long-term movement that stretched from the shores of the English Channel to the banks of the Mediterranean and, despite ebbs and flows in the tempo of forfeiture of cereal lands, persisted until the latter half of the eighteenth century.[11] Finally, the trade in olive oil also ebbed once animal fats gained popularity concomitant with the rise of transhumance after the fourteenth century; in the eastern Mediterranean, the advent of Turkish tribes, with their huge livestock holdings, created a similar situation. Not only did the wine and olive oil trades and the commercial cultivation of tree crops subside in significance then, but sugar and cotton as plantation crops came to occupy the center stage. Olive plantations in Malta were felled in the fifteenth century under the Aragonese rule to make room for the growing of cotton.[12]

Pivotally, after the relocation of the Genoese empire in the western Mediterranean, one of the strategies the merchants of St. George employed was to re-create the *stato do mar* the Venetian businessmen and colonialists carved out for themselves in the Levant. This involved finding new sites of cultivation for the tropical crops that were by then acclimatized in the eastern Mediterranean, in Cyprus, Crete, and elsewhere. The opening of the "Isles of Paradise" offered land which, unlike the exhausted plains and valleys of the Levant, provided more than ample room for the cultivation of these crops, which were extremely taxing on soil. The perpetual need for new land for cultivating sugarcane and new energy sources for manufacturing sugar rendered the industry footloose. That the newly reclaimed islands along the banks of the Atlantic were densely wooded (such as the Madeiras) was a luxury that producers in the Levant sorely lacked. Energy needs of the sugar industry in particular were well served by the presence of virgin forests.[13] Finally, easier availability of slaves firmly secured the advantageous position of the western Mediterranean. If the Venetian rule transformed its off-shore territorial holdings into "free export zones," then the Genoese did the same, but in a different set of off-shore sites in the Greater Mediterranean and the Atlantic, which the geographer al-Mas'udi referred to as the "Green Sea."[14] Put differently, the diffusion of crops that entered the Inner Sea and reshaped its landscape generated later a perverse effect by continuing its westerly journey to the islands peppering the Green Sea. The westerly journey of exotic crops and their growing availability to Genoese merchants undermined the economic and political heyday of the Serenissima. During the age of the city-states, the Mediterranean landscape may have become a medal on which the effigy of the Signoria was struck, but the impress left by this effigy was removed by the exodus of oriental crops and the gradual restoration of the basin's vegetal makeup. Ecological change went hand in hand with economic change.

If the advent of plantation-friendly oriental crops placed the locus of commercial activity along the coastal plains and in the islands of the basin and was thereby highly visible, the dispersion of the crops of the second wave proved to be elusive at best. Unlike the crops of the Indian Ocean, which blessed the city-states with rich trades, the food crops that originated in the Americas were conspicuously absent from the commercial realm. Indeed, tobacco was the only crop that established a noticeable commercial presence, and even then only in the eastern Mediterranean. In the western Mediterranean, the transformation of tobacco into a plantation crop in the Americas denied it the warm reception it received in the Levant.[15] All the same, the contribution of American food crops to commercialization was immense. The increasing cultivation and consumption of maize and the potato, two principal cereal substitutes, by the peasantry boosted the amount of agricultural produce thrown into commercial circulation. The second wave of crop migration failed to provide new commercial crops for the inhabitants of the basin, but by offering alternate cereals, it abetted the marketing of crops that fetched higher prices than lowly maize or potatoes. Rural producers were able to survive on cheap cereal substitutes, which permitted them to market a larger share of their agricultural produce. Witness the popularity of maize in one of the most significant cotton exporters of the nineteenth century, Egypt, or wheat-exporting Romania in the eighteenth century.[16] The potato grew prolifically in the wide plains in north of the olive-tree line, mostly in lands that were originally earmarked for sowing rye, the cultivation of which was threatened by the increased wetness during the Little Ice Age. More often than not the new crops thrived at higher altitudes and, as poor men's crops, they were immediately integrated into the crop-rotation patterns of the region. They paved the way for continuous crop rotation, and were rapidly transformed into humble denizens of the vegetal makeup of the region.[17] Furthermore, these substitute cereals strengthened sharecropping arrangements, in that most of the rents were handed over to the landlord in the form of wheat, and the peasants were allowed to retain most of the maize they grew as subsistence crops.

American food crops arrived at a serendipitous moment in the history of the Mediterranean—during the age of the Genoese. The abandonment of lowlands to the malarial mosquito and the ascent of rural settlements to higher altitudes where cereal lands were in short supply required the cultivation of crops that would flourish easily in less clement climates than in the lowlands and provide inhabitants of the uplands with substitute cereals. As we explored in the previous chapter, maize and the potato performed this function, and legumes, specifically beans, which complemented the native leguminious family of the region, served as meat for the poor. Not only were the new crops prolific; their incorporation into the existing crop rotation facilitated a transition from biennial to continuous rotation. Put paradoxically, as the flora and fauna of the Old World, weeds and sheep in particular, were reshaping the landscape of the Americas to the detriment of indigenous crops and animals, American food crops were reviving the agrarian landscape of the Old World.[18] If the former wave came to be identified with the city-states' awesome commercial might in abetting the spread of tropical crops within and without the Mediterranean, the latter wave could not have been more different in its outcome. Rather than serve the mighty and widen the range of commodities comprising rich trades, it led to a betterment in the lot of peasant households. Even when they were cultivated on large-estate lands, as in the *çiftlik*s in the Balkans or the *latifundia*, it was in order to provide cheap subsistence crops for hired hands rather than as commercial crops destined for the market.[19]

Modifications launched by these waves of crop dissemination pertained, but were not confined, to the realm of the flora. Continual revisions in the vegetal makeup of the basin had distinct spatial and economic implications. Natives of varying ecological zones, the new crops preferred sites and belts of cultivation of differing characteristics as their new habitats. Tropical crops from the Indian Ocean, the "Leek-Green Sea" as Ptolemy called it,[20] thrived best in the coastal zones of the Inner Sea. The devotion of the basin's low landscapes to the cultivation, and later the egress, of tropical crops shaped the course of Mediterranean history during the first half of the second millennium. In sharp contradistinction to the first wave, the American food crops were not as discriminating: even though they provided higher yields in diverse climatic conditions, they flourished best where there was enough humidity or precipitation for the crops to ripen. So in the Mediterranean, they fared better mainly but not exclusively in mountain valleys. These two waves privileged two vastly different topographies. During the first phase, which lasted until the late sixteenth century, it was the low-lying lands of the basin that constituted the economic heart of the Inner Sea, preparing the way for the reign of city-states. During the second wave, however, the higher altitudes of the Inner Sea provided sanctuary to those who had escaped the expanding marshlands below, and the displaced populations were aided in the ascent of their rural settlements owing to the versatility and prolificacy of American food crops.[21] Deeply ingrained in the landscape as one of its "natural" inhabitants, these crops did not turn errant, unlike the crops of the previous wave. The spatial compass of the crops of the second wave spread quickly, standing therefore in stark contrast to that of the commercially lucrative crops of the first wave: the former enlivened the plains, and the latter the uplands.

What is more, the two groups of crops were grown under dissimilar organizational arrangements: the first group was prone to large-scale cultivation, preferably in a plantation-type setting, whereas the latter group was much more versatile and was principally grown as horticultural crops, at least for a long while. The latter were friendly to small peasants. This was the principal reason these crops were so easily incorporated into the crop-rotation patterns and daily diets of the region's inhabitants. Sugar, after all, served as the navigator in the settlement and exploitation of "sugar islands" in the Atlantic. It made and unmade the fortunes of city-states. Maize, on the other hand, may have replaced other cereals in polenta, but it also invited pellagra. It made and unmade the fortunes of rural households. If the former underwrote the heyday of the city-states from the fourteenth to the sixteenth centuries, the latter simply served as "famine relief crops" in the seventeenth and eighteenth centuries.[22]

In sum, the diffusion of two sets of crops defined two Mediterraneans, and the second wave lasted until the third quarter of the nineteenth century, when the great agrarian cycle plotted in the previous chapters came to a close, albeit a temporary one, since the end of this cycle was, by definition, the beginning of a new one. During the course of this great agrarian cycle, two developments served to reshape the Mediterranean. First, the dissolution of the age of city-states lifted the commercial siege laid by the Venetian and Genoese merchants and financiers. With the egress of commercial crops, tree crops resurfaced; cereal-growing and small livestock breeding was reduced in size, losing their status as monoculture and exclusive specialization, and becoming progressively part and parcel of smallholding rural households. Second, the dispersion of the crops of the Columbian exchange, as cultigens supportive of mixed cultivation and horticulture, hastened the return of the tree crops of the region and supplanted cereals where conditions did not favor arable cultivation. Economic transformation went hand in hand with ecological transformation, and the Mediterranean regained some of its former characteristics to constitute what we presently see as its *differentia specifica*.

The factors that drove and constituted the agrarian cycle were manifold. The erosion in the hegemonic might of the city-states, the return of the Little Ice Age, and the vegetal migrations of the Columbian exchange, among others, enhanced the trends that colored the afterglow and waning of the Mediterranean. Reflecting the multilayered structure of the agrarian cycle, its conclusion, too, signaled not only the closing of the Little Ice Age, but also the easing of the ecological bind that vexed the region's mountainous zones. The end of the cycle also heralded the onset of *pax Britannica*, the relaunching of enormous drainage projects, and the transformation of the eradication of malaria into a tool of empire-building with the help of quinine which, ironically, was available from the latter half of the sixteenth century when the Mediterranean lowlands needed it the most.[23] Belatedly, the Jesuit bark, now mostly in the hands of Protestant missionaries, scientists, and colonial administrators, served to turn the inhospitable environments of would-be colonies, which normally served as graveyards of colonists, into habitable lands for the settlers from, and local administrators of, the so-called great powers.[24] That the idea of building a statue of a giant mosquito as the principal historical actor who fended off agents of colonization was entertained in the days of decolonization in Africa in the 1960s speaks to the import of malaria in the colonization of the low landscapes.[25]

Put paradoxically, the great agrarian cycle extended from the introduction of *chincona* into the Mediterranean in the 1560s, when its use remained limited, to its widespread use as quinine in the 1870s, when imperial powers employed it in their colonies in the tropical world.[26] That is to say, the feverish maladies that pestered the inhabitants of the low-lying Mediterranean posed similar problems elsewhere. The reclamation of the low-lying lands, particularly the plains, brought the second cycle to a close. The specialization in the eastern Mediterranean in tree crops was assisted by the arrival of Black Sea grain, as was the case prior to the fifteenth century, and in the western Mediterranean by the arrival of grain from the British Isles and British America. Thanks to easy availability of cereals, greater emphasis could now be placed on horticulture, cultivation of citrus trees, and bananas, which once again, invaded the low-lying lands as commercial crops. The long sixteenth century and the period starting with the latter half of the nineteenth century therefore proved to be similar in nature: in both instances the lowlands were put to full use, unlike the interim period, which divided the two when the plains were largely deserted. The interim period left its impress on the rural landscape by ameliorating the chances of small peasantry to survive through the depressed agricultural conditions of the seventeenth and eighteenth centuries. Small peasant holdings, in peril at the end of the sixteenth century, survived thanks to the reversal in conditions beginning in the late sixteenth century.

Climatic variability, for one, which distinguished the period spanning from the late sixteenth century to the third quarter of the nineteenth century, reversed the subdivision of land under the human plenitude of the long sixteenth century. Glacial though its pace may have been, land concentration surely progressed during the seventeenth-century crisis. When market opportunities diminished during the economic slowdown, landlords took possession of new lands, but the process did not take the form of enclosures as happened on the insular side of the Channel. Nor did it take the form of renewed peasant serfdom, as was the case in the cereal-growing lands that stretched from the Baltic to the Black Sea. The rural depression that hastened land engrossment at the same time prevented landlords from turning their control over land into direct subjugation of the peasantry. Sharecropping, instead, became the norm, from southern France via Italy to the Balkans. And these crop-sharing arrangements were strengthened by the introduction of crops, most of which were horticultural in nature, that maintained a balance between the harvest handed over to the landlords and the crops the peasants were allowed to keep. As the small peasantry was allowed a precarious subsistence, the landlords found recompense in the sustenance of their grasp over the countryside.[27] Now embedded in extensive household networks called the "Mediterranean family," the peasantry managed to survive without major losses and, with the exception of few regions, remained in place even in the latter half of the nineteenth century, mostly under sharecropping arrangements. The consolidation of Church and *waqf* lands did not necessarily work to the detriment of the peasantry. One factor that encouraged the popularity of sharecropping arrangements was the change in the land-labor ratio to the benefit of the latter. What endowed labor with a greater bargaining power was the loss of population in the countryside and the desertion of previously cultivated lands, which jointly forced landowners to offer better terms to those who were already tilling, or were willing to till, the soil. If the desertion of villages and the arable caused frequent changes in the status of land, the erratic tilling that ensued made it more difficult to establish solid rights over it, as claims over them, mostly conflicting with each other, proliferated.

In fine, between the mid-sixteenth and the mid-nineteenth centuries, the three economic processes depicted in part I offered precocious latitude to the city-states that reigned from the fourteenth to the seventeenth centuries in times of diminishing fortunes due to the devolution of the Inner Sea. The inability of newly ascending maritime powers that rode the waves of the Mediterranean to establish a solid and encompassing economic scaffolding gave the sibling cities longer lease on life than customarily allowed. Relatedly, the three ecological processes depicted in part II underwrote the transformations in the composition of activities, which in turn offered peasant households a more diversified base of livelihood than previously. The change in the floral and vegetal composition of the basin, when accompanied by the retreat to the hills, enabled the inhabitants of the basin to survive fairly well despite the general reversal of fortune.

The completion of the great agrarian cycle in the third quarter of the nineteenth century introduced a new set of factors that altered the landscape of the Inner Sea. The reclamation of the low-lying plains and the human plenitude have so far been in the spirit of those unleashed during the long sixteenth century. The new round of settlement of the basin's coastal and inland plains, which started with the drainage of stagnant waters beginning in the late nineteenth century, gained further momentum in the twentieth. Ironically enough, this new wave too is creating a panorama akin to that of the age of the Genoese. Once again, as in the 1550–1650 period when lands previously devoted to monocrop, commercial cultivation were eventually deserted and the region's staple crops took refuge on the hillsides and mountains of the Mediterranean, today the coastal plains, this time colonized by the tourism industry, are forcing the crops of the lowlands to seek shelter elsewhere. The cycle has come full circle.

Introduction

1. Braudel, *Mediterranean*, 1: 14.

2. A good example of this paradigm is Ashtor, *Social and economic history*, 300–331.

3. See, for instance, the articles on how the Mediterranean was split along religious lines and would not be reunited again after 1500: Balard, "A Christian Mediterranean," and Green, "Resurgent Islam."

4. Flynn and Giráldez, "Silk for silver," 52–68.

5. Matthee, *Politics of trade*, ch. 3; Ülker, "XVII. ve XVIII. yüzyıllarda ipek," 336.

6. Klein, *Mesta*, 105. The "classical age" of the Ottoman empire too ended at around 1600: İnalcık, "Ottoman state."

7. Braudel, *Civilization and capitalism*, 3: 170; Morineau, *Incroyable gazettes;* García-Arenal and Wiegers, *Man of three worlds*, ch. 3.

8. Badoer, the Venetian ambassador in Spain, reported in 1557 that the Republic of Genoa "lends credits to all and waits on everyone": Kamen, *Empire*, 54, 63, 69; Stein and Stein, *Silver, trade,* 11–12.

9. The long odyssey of precious metals from the Americas to the Mediterranean was elegantly summed up by the seventeenth-century Spanish poet Francisco de Quevedo: "Born in the Indies, passing away in Spain, buried in Genoa, sir Money is a powerful king." Lopez, "Market expansion," 462.

10. Glahn, *Fountain of fortune*, 146; Kołodziejczyk, "Export of silver," 105–15.

11. Özbaran, *Yemen'den Basra'ya;* İnalcık, "Ottoman state," 319–40.

12. Godinho, "Le repli vénitien," 283–300.

13. Israel, *Dutch primacy,* 610–19; on this northerly movement: Brunelle, "Immigration, assimilation," 203–30; Braudel, *Identity*, 2: 565–76.

14. On how the Castilian merchants of the Antwerp–Medino del Campo–Lyon circuit were sidelined by the Genoese: Martín, *Lettres marchandes;* also Vilar, *History of gold,* 88, 162.

15. Wee, "Anvers," 1073; Lees and Hohenberg, "Urban decline," 439–61; Thijs, "Technologie et productivité," 353–58.

16. Gioffrè, *Gênes;* Arrighi, *Long twentieth century*, 125, 131–32.

17. Gascon, *Grand commerce*, 1: 211–22; Gascon, "Quelques aspects," 45–64.

18. Gentil da Silva, *Banque et crédit,* 1: 711–16; Braudel, *Identity*, 1: 197–98.

19. Kirk, *Genoa*, 88–89.

20. Gökbilgin, "Venedik," 119–220; Valensi, *Birth*, 9–21.

21. Goffman, *Ottoman empire*, 137–64; Lane, *Venice*, 285–94.

22. Braudel, *Mediterranean*, 1: 616–17.

23. See the articles in Hamilton et al., *Friends and rivals;* Ülker, "XVII. Yüzyılın ikinci yarısında," 261–320.

24. Braudel, "Le siècle des Génois," 446.

25. Faroqhi, "Crisis and change," 423–24.

26. Kévonian, "Marchands arméniens," 199–244; Arbel, *Trading nations*, chs. 2, 8.

27. Greene, "Beyond the northern invasion," 46–48; Filippini, "Les nations à Livourne," 581–94; Sombart, *Jews*, 22–27.

28. On changes in the granting of *asientos* in favor of new Christian financiers from Lisbon: Boyajian, *Portuguese bankers;* Devèze, *L'Espagne*, 188–89.

29. Braudel, *Civilization and capitalism*, 2: 338; Kirk, "Little country," 407–21.

30. Braudel, *Civilization and capitalism*, 3: 172; Kirk, *Genoa*, 176–80.

31. Wee, "Structural changes," 14–33.

32. Braudel, "Le siècle des Génois," 437–42.

33. Braudel, *Mediterranean*, 1: 14.

34. "The *coup de grâce* was not delivered by Vasco de Gama but by the Dutch, well over a century later": Day, "Levant trade," 808.

35. Cf. Cipolla, "Decline of Italy," 178–87; Elliott, "Decline of Spain," 52–75; Issawi, *An economic history*, 3–13.

36. Braudel, *Civilization and capitalism*, 3: 326–30; Pyne, *Vestal fire*, 106.

37. Elliott, *Imperial Spain*, 333–41; Murphey, *Ottoman warfare*, 3–6.

38. Braudel, *Out of Italy*, 204–6.

39. Issawi, *An economic history*, 1–16.

40. For a critique of theses centered on this "discontinuity": Ringrose, *Spain, Europe*, ch. 1; İslamoğlu and Keyder, "Agenda for Ottoman history."

41. Providentially for Ottoman history, the world-economy as outlined in broad strokes by Fernand Braudel is centered on the Mediterranean, and it is through the optical field of this world that his account of the momentous changes of the long sixteenth century is filtered. Owing almost exclusively to Ö. L. Barkan's work, Ottoman historiography was initially receptive to Braudel's investment of the Mediterranean world with an organizing force and to his relocation of the empire's history within this world's spatial and temporal coordinates. His holistic interpretation failed to command sustained interest despite the original impact it had. Much sooner than later, the Mediterranean-centered world-economy ceased to function as a frame of reference for Ottoman history: İnalcık, "Impact of the *Annales* school."

42. Samsonowicz, "Relations commerciales," 537–45.

43. On the intensity and commercial nature of agricultural production on the eastern shores of the Mediterranean: Yusuf, *Economic survey*, 53; Planhol, *Les fondements*, 76; on the exodus of sugar from the Mediterranean: Mintz, *Sweetness and power*, 30–31; on the changing balance between the North Sea and the Mediterranean in terms of commercial grain production: Pelizzon, "Grain flour," 160–63.

44. Güçer, "Onaltıncı yüzyılın"; Delano Smith, *Western Mediterranean Europe*, 195.

45. Tielhof, '*Mother of all trades,*' chs. 2–3; Vries and Woude, *First modern economy*, 322–29.

46. Craeybeckx, *Un grand commerce.*

47. Carter, *Trade and urban development*, 304.

48. Gutmann, *Toward the modern economy.*

49. Braudel, *Out of Italy*, 17.

50. Heers, *Gênes*, 229.

51. Crouzet-Pavan, *Venice triumphant*, chs. 2–3.

52. Houtte, "Les grands itinéraires," 87–97; Sombart, *Jews*, 24–25; Anoyatis-Pelé, "Aperçu," 75–112; Phillips and Phillips Jr., *Spain's golden fleece*, 214–17.

53. Kellenbenz, *Rise*, 278–79.

54. Kamen, *Empire*, 434–36; Phillips and Phillips Jr., *Spain's golden fleece*, 261–62; McGowan, *Economic life*, 15–27.

55. Bautier, *Economic development*, 209–26.

56. Vassberg, *Village*, 50–53; Philips and Phillips Jr., *Spain's golden fleece*, ch. 10; Stoianovich, "Commerce et industrie," 329–52; Genç, "18. yüzyılda Osmanlı," 231–33.

57. Vries and Woude, *First modern economy*, 290–93.

58. Simpson, *Spanish agriculture*, pt. I.

59. Delano Smith, *Western Mediterranean Europe*, 157; Philips and Philips Jr., *Spain's golden fleece*, ch. 5; McNeill, "Tragedies of privatization," 222–34; Grossi, *Alternative to private property*, ch. 8; Latron, *La vie rurale*, 183–203; Firestone, "Land-equalizing," 91–129.

60. On the "involution" of Italian agriculture: Romano, "Italy in the crisis," 185–98; on the *siesta* of Spanish agriculture: Simpson, *Spanish agriculture*, 33–57; also Gerber, *Social origins*, ch. 4.

61. See the articles in Peristiany, ed., *Mediterranean family structures.*

62. For a critical assessment of this line of analysis: Aymard, "From feudalism," 131–208, and also his "Rendement et productivité," 99–121; Sella, *Crisis and continuity*, 105–34; Petrusewicz, *Latifundium*, 4.

63. On the slow renaissance of the mountains of the Mediterranean: McNeill, *Mountains*, 86–93.

64. On the expansion of the arable during the long sixteenth century: Abel, *Agricultural fluctuations*, 99–146; Le Roy Ladurie and Goy, *Tithe*, 114; Montanari, *Culture of food*, 51–55.

65. Sereni, *History*, 285–87; Philips and Phillips Jr., *Spain's golden fleece*, 75; Gerber, *Social origins*, 31–33, 61–62; on the significance of livestock holdings in rural households: Güran, *19. Yüzyılda Osmanlı tarımı*, 100–110.

66. Cf. Faroqhi, "Wealth and power," 77–95; Sereni, *History*, 213–15; Capatti and Montanari, *Italian cuisine*, 140–46; Valensi, *Tunisian peasants*, 116–20.

67. Evliyâ Çelebi, *Seyahatnâme*, 7: 102; 8: 119; Atasoy and Raby, *İznik*, 23; Delano Smith, *Western Mediterranean Europe*, 207.

68. Braudel, *Memory*, 9.

69. In Mediterranean France, for instance, vineyards came to establish themselves in the cereal fields on the lowlands. On the *descente de la vigne*: Delano Smith, *Western Mediterranean Europe*, 171–72. On changes in ecological conditions and settlement patterns in the Aegean: Foss, *Ephesus*, 185–86.

70. Galloway, *Sugar cane industry*, 31–47.

71. Appleby, "Grain prices," 865–87.

72. Le Roy Ladurie, *Peasants*, 289–311; Cameron, "Europe's second logistic."

73. Le Roy Ladurie, *Times of feast* and *Histoire humaine*; Lamb, *Climate history*; Luterbacher and Xoplaki, "500-Year winter temperature," 133–53.

74. Planhol, *L'eau de neige*, ch. iv; David, *Harvest of the cold months*, 41–65; Rabbath, *Documents inédits*, 2: 203; Pascual, "Une neige à Damas," 57–81.

75. Houston, *Western Mediterranean world*, 33.

76. On natural catastrophes: Bennassar, ed., *Les catastrophes naturelles*; Zachariadou, ed., *Osmanlı imparatorluğu'nda doğal afetler*. On climatic variability and shifts between "torrid summers and subzero winters" during the Little Ice Age: Fagan, *Floods, famines*, 183.

77. On the repairing of bridges destroyed by heavy floods as well the replacement of wood with stone in bridge construction: Nader, *Liberty*, 166; Gökyay, *Kâtip Çelebi*, 216–17; Antoine et al., "Les grands *aygats*," 246; Grove, "Little Ice Age," 124–26.

78. Nader, *Liberty*, 23.

79. On deforestation: Devèze, "L'équilibre agro-sylvo-pastoral," 341–42; Fernández-Armesto, *Civilizations*, 139–45; on rain and spring flooding: Sereni, *History*, 187–88; Vaughan, "Campaigns in the Dauphiny," 230–31; d'Arrigo and Cullen, "A 350-year reconstruction," 169–77; Moriceau, *Terres mouvantes*, 62–63.

80. Rackham and Grove, *Nature*, 334–37; Horden and Purcell, *Corrupting sea*, 187; McNeill, *Mountains*, 312.

81. Dienne, *Histoire du desséchement*, 4.

82. Faroqhi, "Tarımsal değişimin," 283–84.

83. Bogucka, "Amsterdam," 433–47.

84. Herlihy, "Russian wheat," 45–68.

85. Vries, *Economy of Europe*, ch. 2.

86. On the impact of the seventeenth-century crisis on agriculture: Wallerstein, *Modern world-system*, 2: 13–34; Slicher van Bath, *Agrarian history*, 206–20.

87. On the historical significance of lesser grains and legumes in the Mediterranean: Sarpaki, "Palaeoethnobotanical approach"; Horden and Purcell, *Corrupting sea*, 203.

88. Houston, *Western Mediterranean world*, 4.

89. Delano Smith, *Western Mediterranean Europe*, 13; Horden and Purcell, *Corrupting sea*, 61.

90. Rogers, "To and fro," 57–74; on Ottoman imports of Venetian glass: McCray, *Glassmaking*, 135; and of mirror: Bağcı, "Gerçeğin saklandığı yer," 26–29; and of course, Goldthwaite, *Wealth*.

91. This was the second phase of the price revolution of the sixteenth century: Walker, "Capitalism," 14–19; on the narrowing of the price differential between the Baltic and the Mediterranean during the long sixteenth century: Braudel and Spooner, "Prices."

92. Tielhof, '*Mother of all trades*,' 43–50.

93. Watson, *Agricultural innovation*.

94. Wallerstein, *Modern world-system*, 1: 48–49; Dunin-Wąsowicz, "Environnement et habitat," 1026–45.

95. See Sclafert, *Cultures en Haute-Provence*.

96. Serre-Bachet and Guiot, "Summer temperature," 94–96.

97. "The Middle Ages, in the Mediterranean as elsewhere, were more fortunate: the waterways were more abundant and became once more the source of good-river valley soil . . . It was

in about the sixteenth century that the balance swung back the other way. Erosion began once more, the rivers cut channels through the ancient flats (sometimes forty meters deep) and carried off into the sea all the sand and mud accumulated there. The deltas expanded, but their fertile land was not easy to bring under cultivation." Braudel, *Memory*, 11; also Vita-Finzi, *Mediterranean valleys*.

98. Lane, "Venetian shipping," 219–39; Appuhn, "Inventing nature," 861–89.

99. Busch-Zantner, *Agrarverfassung, gesellschaft*, 80–90; Doumani, *Rediscovering Palestine*, 97–105; Stoianovich, *Balkan worlds*, 99–100.

100. Stoianovich, "Commerce et industrie," 349; Owen, *Middle East*, 7; Hütteroth, "Role," 343–54.

101. Davis, *Late Victorian holocausts*, chs. 9–10; Friedmann, "World market," 545–86.

102. Kasaba, Keyder, and Tabak, "Eastern Mediterranean," 125–26.

103. Gerber, *Social origins*, ch. 5; McNeill, "Tragedies of privatization," 222–34. On the decline of pastoralism: McNeill, *Mountains*, 234–36.

104. Fagan, *Floods, famines*, ch. 12; on North Africa: Curtin, *Death by migration*, 65–67, 124.

105. Jarcho, *Quinine's predecessor*.

106. Headrick, *Tools of empire*, ch. 3.

107. Rocco, *Quinine*, 259–80; Snowden, "Mosquitos," 176–209; Lewis, "Malaria, irrigation," 286–88; Dozier, "Establishing a framework," 491–92.

108. Celli, *History of malaria*.

109. Dienne, *Histoire du desséchement*, 361–71; Saraçoğlu, *Akdeniz*, 444–45, 483–84.

110. See, for instance, Gentilcore, "Reclamation of Agro Pontino," 301–27; Dozier, "Establishing a framework," 491–93.

111. Cf. Marino, "Economic structures," 58: "In Italy in 1500, 100 agriculturalists could produce a grain surplus to feed themselves and thirty-three others, whereas in the Low Countries at that time the surplus reached some seventy-five other people. By 1800, with Italian yields at the same level they had been for centuries, the agricultural revolution in northern Europe allowed for the grain surplus to feed an additional 150 mouths above the number of agricultural workers."

112. McNeill, *Mountains*, 5–7; see also Morilla Critz et al., "Horn of plenty," 316–52.

113. On how these so-called marginal environments created by wetter conditions, such as the wetlands, were an integral part of the Mediterranean landscape and life: Horden and Purcell, *Corrupting sea*, ch. 6.

One • Empires and Empire-Building City-States

1. Fossier, "Europe's second wind," 399–453; Slicher van Bath, *Agrarian history*, 132–44.

2. Pryor, *Geography, technology*, 130–34; Ashtor, *Les métaux précieux*, 533–86; Spufford, *Money*, 353–55; Wittek, "Ankara," 557–89.

3. "The Italian notarial contracts show how massively trade expanded between 1150 and 1250": Balard, "Christian Mediterranean," 196; also Lombard, *L'Islam*.

4. Braudel, *Out of Italy*, 20–44.

5. The classic text is of course by Ranke, *Turkish and Spanish empires*.

6. That Venetian convoy navigation reached Bruges via the Straits of Gibraltar did not

change the nature of this partitioning since the trading world of the western Mediterranean was already solidly placed under the command of the Genoese merchants: Abu-Lughod, *Before European hegemony*, 123; Tenenti and Vivanti, "Les galères," 83–86.

7. Hess, "Battle of Lepanto," 57.

8. Braudel, *Civilization and capitalism*, 3: 175.

9. Pike, "Genoese in Seville," 348–78; Birmingham, *Trade and empire*, 5–24; Kamen, *Empire*, 40–41, 69–70.

10. Braudel, *Mediterranean*, 1: 564–65; Lybyer, *Government*, 181; Alexandrescu-Dersca-Bulgaru, "Une relation vénetienne," 137. Gross imperial product estimates range from 30 to 40 million ducats; the revenue of the central government constituted roughly one-fourth of it: Goldsmith, *Premodern financial systems*, 82–83.

11. Lopez, "Market expansion," 446–47. Genoese Chios, for instance, was not directly governed by the mother-city but was administered by a chartered company. This was not the case in the Venetian empire, where the state was all: Miller, "Genoese in Chios," 418–32. On the differences in commercial practices between the eastern and western halves of the Mediterranean and between Venice and Genoa: Dotson, "Problem of cotton," 58–59.

12. Epstein, "Cities, regions," 44–45; on the Florentine merchants in Ottoman silk trade: İnalcık, "Ottoman state," 230–43.

13. Even when it annexed Pisa in 1406 and Livorno in 1422, Florence found itself in a world where the Venetian and Genoese merchants had already carved up and secured their distinct spheres of operation. In any event, the state galleys were abandoned in the 1480s: Mallett, *Florentine galleys*. The commercial weakness did not rule out the financial sway Florence held over the warring empires of the Hundred Years' War: Arrighi, *Long twentieth century*, 96–109. Still, what financial might of the city of St. Giovanni offered the empires was different from what the merchant republics provided.

14. Valensi, *Birth*, 20; on how Venice served as the main window to the world for the Ottoman ruling elite: Arbel, "Maps," 19–29.

15. Gentil da Silva, *Banque et crédit*, 1: 485–95; on the abandonment of Paris: Spufford, *Power and profit*, 136–39.

16. It was only after the withdrawal of the Genoese and Florentine money-lenders and bankers from the Lyon fairs that Paris emerged as a rival economic center partly because the Florentines of Lyon turned to public finance and firmly established themselves in Paris: Braudel, *Civilization and capitalism*, 3: 326–31. Ironically, it was the defeat of Francis I by Charles V that ended France's imperial ambitions, a defeat that did a "greater service to France than to Castile." Wallerstein, "Charles V," 391.

17. Anderson, *Lineages*, 60–76, 85–93, 361–84.

18. Most historical accounts of Venice and Genoa invariably end in the sixteenth century. It was then the Venetian colonies in the Mediterranean as well as the Genoese holdings along the shores of North Africa finally fell prey to Ottoman expansionism. Recent examples are Epstein who, in *Genoa*, ends his account in 1528, the year of reformation of the Republic under Andrea Doria, the city's Augustus, and Crouzet-Pavan, in *Venice triumphant*, at the close of the renaissance.

19. Kamen, *Empires*, 434–37; İnalcık, "Ottoman state," 364–76.

20. For an approach that sees these two entities as two clearly demarcated and competing polities: Spruyt, *Sovereign state*, 149–50.

21. Crouzet-Pavan, *Venice triumphant*, 91; Day, " 'General crisis,' " 26.

22. Pike, *Enterprise;* Arrighi, *Long twentieth century*, 97.

23. The seminal work on the topic is of course by Davis, *Decline;* Benadusi, "Career opportunities," 89–91; Rapp, *Industry*, 144–48.

24. Heers, *Gênes*, 367; Crouzet-Pavan, *Venice triumphant*, 131–32.

25. Cahen, "Quelques problèmes," 381–432.

26. Braudel, *Mediterranean*, 2: 657–61; Barker, "Late Byzantine Thessalonike," 5–33; Magdalino, "Maritime neighborhoods."

27. Cahen, *Orient et Occident*, 109–13; Laiou, "Byzantine trade," 157–92; Abu-Lughod, *Before European hegemony*, 102–31.

28. On the role the cities played not only in the Mediterranean but also in the Baltic region at the turn of the millennium and the bipolar nature of economic growth after c. 1000: Braudel, *Civilization and capitalism*, 3: 96–115.

29. Wink, *Al-Hind*, 2: ch. 2; Tabakoğlu, *Türk iktisat tarihi*, 140–58; Cahen, *Pre-Ottoman Turkey*, chs. 7–11; Chaudhuri, *Trade and civilization*, 58–60.

30. Steven Epstein and Thomas Madden both see the twelfth century as the turning point in the history of Genoa and Venice: Epstein, *Genoa*, and Madden, *Enrico Dandolo;* Verlinden, "Italian influence," 199–211.

31. Lopez, *Commercial revolution*, 166; also Spufford, *Power and profit*, 64–67.

32. Lopez, "Silk industry," 40–41; Dowd, "Economic expansion," 155–56.

33. Martin, "Venetian-Seljuk treaty," 321–30; Turan, *Türkiye-İtalya*, 182.

34. Bensch, *Barcelona*, 325–35. Bensch asserts that the shift of resources away from agricultural investments largely undermined Barcelona's chances of success à la Venice or Genoa. Yet, the patricians' inability to carve a niche for themselves in the rich trades of the Levant, *Ultramar*, played a not so insignificant role in their emergence as the financiers to the throne, since this was the most fruitful strategy of putting their capital to work.

35. Abulafia, *Western Mediterranean kingdoms*, xv–xvi; Kafadar, *Between two worlds*, 138–50.

36. Day, *Genoa's response*, ch. 7.

37. Yver, *Le commerce*, 245–88.

38. Wallerstein, "West, capitalism," 600–616.

39. On the political ramifications of this secular downturn: Pirenne, *History of Europe*, ch. 8.

40. Arrighi, *Long twentieth century*, 91–92; Spruyt, *Sovereign state*, 175–76.

41. Cox, *Foundations*, 179–80.

42. The Genoese merchants' rule over the Aegean and the Black Sea was confirmed, as it were, by the naval victories the Republic won against the Venetians in the 1290s as did the finalization of Venetian supremacy over its omnipresent rival a century later, in the 1380s: Epstein, *Genoa*, 182–84, 236–42; Lane, *Venice*, 189–96; Matschke, "Commerce, trade," 791.

43. Adshead, *China*, 195–200.

44. Braudel, *Civilization and capitalism*, 3: 119–24. On the collapse of states during the 1350–1450 downturn: Wallerstein, "West, capitalism," 601–4.

45. Crouzet-Pavan, *Venice triumphant*, 128–31; Guarini, "Center and periphery," S86–87.

46. Spufford, *Power and profit*, 67.

47. MacKay, *Money, prices*, appendix C; Pamuk, *Monetary history*, 47–50; Babinger, *Mehmed the Conqueror*, 456–58.

48. Ashtor, *Social and economic history*, 324; Day, "Problem," 83–94.

49. Lewis, "Closing," 475–83.

50. Rose, "Islam versus Christianity," 561–78; Holt, "Treaties," 67–76; Turan, "Osmanlı imparatorluğu," 79–81. The Byzantine fleet survived in the shade of the Genoese and the Venetian naval forces after Andronicus II reduced the Byzantine navy to place emphasis on field campaigns: Ahrweiler, *Byzance et la mer*.

51. Laiou-Thomadakis, "Byzantine economy," 188.

52. Bresc, *Un monde méditerranéen*, 1: 540; Day, "Peuplement, cultures," 700–701; Pelizzon, "Grain flour," 98.

53. Abulafia, *Mediterranean emporium*, 228–31.

54. Nicol, *Byzantium and Venice*, 410; Braudel, *Civilization and capitalism*, 3: 119–21.

55. On the impact of the severing of commercial networks that penetrated deep into Asia on the Genoese and Venetian merchants following the fall of the Mongol empire: Kedar, *Merchants in crisis*, 130; Adshead, *Central Asia*, ch. 5.

56. Ascherson, *Black Sea*, 92–95; Epstein, *Genoa*, 262–70; Galloway, "Mediterranean sugar," 190.

57. İnalcık, "Ottoman state," 194; Balard, "Le système portuaire," 34–39.

58. On "circuits of trade" underpinning the thirteenth-century world-system: Abu-Lughod, *Before European hegemony*, 34; also Adshead, *China*, 197–200. On the centrality of Mamluk lands in the formation of Venetian "orientalism": Raby, *Venice, Dürer*; Behrens-Abouseif, "European arts," 52–53.

59. Mitler, "Genoese in Galata," 71–91.

60. Heers, *Gênes*, ch. 2; Lopez, "Market expansion," 445–64.

61. In 1293, customs farmers in the port of Genoa "anticipated the transit of merchandise to the value of about 4 million Genoese pounds; in 1334 the value had dropped to less than 2 million pounds, and totals seldom went above this in the second half of the century": Martines, *Power and imagination*, 170. The Catalan expansion in the Maghreb largely took place when the Genoese empire was centered in the Pontic Sea: Dufourcq, *L'Espagne catalan*, 572–79.

62. Turan, *Türkiye-İtalya*, 184.

63. Kafadar, *Between two worlds*, 138–39.

64. Ashtor, *Social and economic history*, 299–300; Wiet, "Les marchands d'épices," 81–147; Dols, *Black Death*, ch. V.

65. On the prosperity of Burgos from the late fifteenth century to 1569: Lovett, *Early Habsburg*, 254.

66. Wallerstein, *Modern world-system*, 1: 129.

67. On the symbiotic relationship between empires and city-states: Arrighi, *Long twentieth century*, 96–126.

68. Lewis, *Nomads and crusaders*, 113–37.

69. Mazzaoui, "Cotton industry," ch. 7.

70. Crouzet-Pavan, *Venice triumphant*, 119–26.

71. On how the continental warfare benefited the Florentine high finance in the 1350–1450 period: Arrighi, *Long twentieth century*, 106–9.

72. Balard et al., "Le transport," 94, 170, 173; Delano Smith, *Western Mediterranean Europe*, 15. Before the crash, Tuscany took in close to 70 percent of Sicily's grain exports: Hunt, *Medieval super-companies*, 130.

73. On Giovanni di Francesco Maringhi's dealings in Constantinople and Bursa: Richards, *Florentine merchants*.

74. Mallett, *Florentine galleys*.

75. Hunt and Murray, *History of business*, 102–5.

76. Hunt, *Medieval super-companies*, 48–49; Abulafia, "Southern Italy," 385.

77. Goldthwaite, "Medici Bank," 17–19.

78. Balard et al., "Le transport," 152.

79. Hunt and Murray, *History of business*, 135.

80. Yver, *Le commerce*, 335–94.

81. Abulafia, "Southern Italy," 386–87.

82. Lapidus, *Mamluk cities*, 31; Ashtor, "Wheat supply," 283–95.

83. Martines, *Power and imagination*, 200.

84. "It has been calculated that a third of the grain in Venice warehouses came from Crete." Crouzet-Pavan, *Venice triumphant*, 88; Alexandrescu-Dersca-Bulgaru, "Une relation véneti-enne," 135; on the Genoese merchants in Alhóndiga del Pan in Seville, the hub for grain traffic in Andalusia: Constable, *Housing the stranger*, 337–38.

85. Planhol, *Les fondements*, 197–209; Bishko, "Castilian as plainsman," 47–69. "In the first half of the fifteenth century, the amount of grain exported from Sicily remained in the neigh-borhood of 25,000 *salme*, half of the amount shipped in the second half of the century": Epstein, *Island for itself*, 274.

86. Zachariadou, *Trade and crusades*, 171–73; Fleet, "Ottoman grain exports," 283–92.

87. Heers, *Gênes*, 233–47; Hocquet, *Le sel*, ch. 10; Brummett, *Ottoman seapower*, 124–25.

88. Brummett, *Ottoman seapower*, 124–39, 151–55; also Balard, *La Romanie genoise*; Thi-riet, *La Romanie vénetienne*.

89. Zachariadou, *Trade and crusades*, 163–65, 171–73; Emecen, *İlk Osmanlılar*, 109–12.

90. Abulafia, *Mediterranean emporium*, 214–15.

91. Verlinden, "Italian influence," 203–8.

92. Serrão, "Le blés des îles," 338.

93. Mazzaoui, *Italian cotton*, 142.

94. On the significance of Levantine cotton: Mazzaoui, "Cotton industry," 266; Turan, *Türkiye-İtalya*, 334; Constable, *Housing the stranger*, 320, 341–44; Lane, *Venice*, 286.

95. Pedani, "Ottoman empire," 597–98; Krekić, "Le port de Dubrovnik," 653–73.

96. Bojović, "Dubrovnik et les Ottoman," 119–37, 153–64, 271–76; Braudel, *Mediterranean*, 1: 121, 446–47.

97. Constable, *Trade and traders*, ch. 8; Busch, *Medieval Mediterranean*, 20–25; Abulafia, "Levant trade," 192–93; Ashtor, "Catalan cloth."

98. Dollinger, *German Hansa*.

99. On Valencian merchants: Guiral-Hadziiossif, *Valence*; on Catalan merchants: Abulafia, *Western Mediterranean kingdoms*, 127–30; and on merchants of Montpellier: Reyerson, "Mont-pellier"; also Heers, *Jacques Coeur*.

100. Pirenne, *Histoire économique*, ch. vii.

101. On Veneto-Ottoman relations: Goffman, *Ottoman empire*, ch. 5; Murphey, "Ottoman resurgence," 186–200.

102. Braudel, *Civilization and capitalism*, 3: 159–62.

103. Fleet, *European and Islamic trade*, 80–94.

104. Casale, "Ottoman administration," 170–98.

105. Hunt and Murray, *History of business*, 180.

106. Braudel, *Mediterranean*, 1: 389; Ricci, *Türk saplantısı*, 32; Pistarino, "Genoese in Pera," 63–85; Hocquet, "Ships, sailors," 533–67.

107. Pitcher, *Osmanlı*, 169–71; Scammel, *World encompassed*, 366.

108. Özbaran, *Yemen'den Basra'ya*; Hess, "Evolution," 1918–19. On the progressive decline of the Syrian markets in the affairs of a Venetian businessman in the mid-sixteenth century before the revival of the spice trade: Tucci, *Lettres d'un marchands*.

109. Braudel, *Out of Italy*, 114.

110. Braudel, *Mediterranean*, 1: 564; İnalcık, "Ottoman state," 345; Hanna, *Making big money*, 60–61.

111. İnalcık, "Ottoman state," 319–44.

112. Pike, "Genoese in Seville," 349.

113. Epstein, *Island for itself*, 274.

114. Kirk, *Genoa*, 35–36.

115. Braudel, *Mediterranean*, 1: 121; Singer, *Constructing Ottoman beneficence*, 121–25.

116. Braudel, *Civilization and capitalism*, 2: 444; Lopez, "Market expansion." On the blurriness of the line that separates the public from the private: Lane, "Oceanic expansion."

117. Scammel, *World encompassed*, chs. 3–4.

118. On the Venetian state's acquisition of functions that were fulfilled by private enterprise elsewhere: Katele, "Piracy."

119. Bautier, *Economic development*, 189–209.

120. Schick, *Un grand homme d'affaire*; Ehrenberg, *Capital and finance*.

121. On the extent of sugar production and consumption: Ouerfelli, "Les migrations," 485–500; Yusuf, *Economic survey*, 34–35, 39; Lagardère, *Campagnes et paysans*; Oral, "Selçuk devri yemekleri," 23–24; Lewicki, *West African food*, 114–16.

122. Byrne, "Genoese trade," 216–18; Krueger, "Wares of exchange," 57–71; Epstein, *Genoa*, 97; Mintz, *Sweetness and power*, 82. On *galee de trafego* interconnecting the shores of North Africa: Pryor, *Geography, technology*, 141–42. By the time the Marinid and the Hafsid dynasties were in place, the Maghrib had become deeply embedded in the history of the western Mediterranean: Abu Nasr, *History of the Maghrib*, 103–34. The prohibition of Fatimid dinar under Nûr al-Din was an early sign of this: Cahen, "Monetary circulation," 326.

123. Mazzaoui, *Italian cotton*, 28–55; Galloway, *Sugar cane industry*, 45–46; Lewicki, *West African food*, 114–16.

124. Curtin, *Rise and fall*, ch. 6.

125. Wee, "Structural changes," 14–33; Rapp, *Industry*, 6–7; Molà, *Silk industry*, ch. 1; MacFarlane and Martin, *Glass*, 21–24; Melchior-Bonnet, *Mirror*, ch. 1; Bloom, *Paper before print*.

126. On waves of ruralization of manufacturing: Tabak, "Informalization," 1–19.

127. Raymond, *Artisans et commerçants*, 1: 121–28, 136–42; Hathaway, "Wealth and influence," 305–6.

128. Lopez, Miskimin, and Udovich, "England to Egypt," 126–27.

129. McGowan, *Economic life*, ch. 1; Morineau, "Naissance d'une domination," 297–98.

130. Braunstein, "A propos de l'Adriatique," 1275.

131. Barrett, "World bullion flows," 251–52.

132. Braudel, *Civilization and capitalism*, 3: 170–73; Kirk, *Genoa*, 167–80.

133. Kamen, "Decline of Spain," 43 n. 85.

134. Anderson, *English consul*, 53; Cipolla, *Neşeli öyküler*, 25–31; Pamuk, *Monetary history*, ch. 9.

135. McGowan, *Economic life*, 15–27; Pagano de Divitiis, *English merchants*, 182–85.

136. Arbel, *Trading nations*, ch. 2.

137. Matthee, *Politics of trade*, ch. 3; Aghassian and Kévonian, "Le commerce arménien," 155–81; Dale, *Indian merchants*, 97.

138. On the growth of *caravane* trade in the eastern and central Mediterranean in the seventeenth and eighteenth centuries: Panzac "Le contrat d'affrètement," 359–61; and *contra* Green who sees country trade as a sign of economic ebullience in "Beyond the northern invasion": Wallerstein, *Modern world-system*, 3: 137, who sees it as a major weakness. The growing importance of barter, usually textiles in return for cereals, and debased coinage trade in the region is highlighted by Morineau, "Naissance d'une domination," 301–2; Pamuk, *Monetary history*, 151–53; Cipolla, *Neşeli öyküler*, 25–31.

139. Jeannin, "Sea borne trade," 5–59.

140. Lane, *Venice*, 423–25.

141. "Industry survived, if only in the form of silks and Neapolitan goods, while Italian commerce survived as more than a 'passive' activity, as they said in the eighteenth century. In short Italy maintained an equilibrium that cannot have been as unsatisfactory as is sometimes alleged": Braudel, *Out of Italy*, 206; also Georgelin, *L'Italie*.

142. Braudel, *Out of Italy*, 117.

143. İnalcık, "Ottoman state," 364–79.

144. Valensi, *Birth*, 69–87.

145. Devèze, *L'Espagne*, 1: 189.

146. Braudel, *Mediterranean*, 1: ch. 3.

147. Scammel, *World encompassed*, ch. 6; Pitcher, *Osmanlı*, ch. 7; also Lane, "Public debt," 72–82.

148. Crouzet-Pavan, *Venice triumphant*, 89.

149. Mantran, *17. Yüzyılda*, 2: 113–47; Ringrose, "Impact," 761–91.

150. Houtte, *Economic history*, 140–55.

151. Wallerstein, *Modern world-system*, 1: 42–43; on cotton exports from the Levant: Arbel, "Venetian trade," 245–51.

152. Malowist, *Croissance et régression*. Cook, in *Population pressure*, argues that Ottoman exports dwindled, because with a growing number of mouths to feed and declining productivity, there was less of it to export. But the Baltic grain did not replace its Mediterranean brethren because there was "more" of it to "export." A single town in the receiving end of the grain trade consumed "more than the whole realm of Poland": Braudel, *Civilization and capitalism*, 1: 126.

153. Hunt and Murray, *History of business*, 135–36; Semple, *Geography*, 344; Güçer, *XVI.–XVII. Asırlarda Osmanlı*, 10.

154. Braudel and Spooner, "Prices in Europe," 393; Georgelin, *Venise*; Hütteroth, "Role," 343–54.

155. Le Roy Ladurie, *Peasants*, 91–97; McGowan, *Economic life*, 67–79; Çizakça, "Tax farming," 219–50; Vassberg, "Sale of *tierras baldías*," 629–54; Day, "Peuplement, cultures," 700–701.

156. "In the north, huge areas moved back into the age of serfdom: from the Baltic to the Black Sea, the Balkans, the kingdom of Naples, and Sicily, and from Muscovy (a very special case) by way of Poland and Central Europe as far as a line running approximately from Hamburg to Vienna and Venice": Braudel, *Mediterranean*, 2: 265–66.

157. Munro, "Monetary origins," 1–34.

158. Flynn and Giráldez, "Cycles of silver," 399–405.

159. Watson, *Agricultural innovation.*

160. Godinho, "La guerres du blé," 227–59.

161. Kamen, *Empire*, 66, 294–96.

162. Heers, *Gênes*, 258–69.

163. Curtin, *Rise and fall*, chs. 1–2.

164. Ashtor, "Levantine sugar," 91–132; Spufford, *Power and profit*, 316–32.

165. For historical precedents: Burns, "Immigrants from Islam," 21–42; also Crosby, *Ecological imperialism*, 80–103.

166. Mintz, *Sweetness and power*, 28–35.

167. On Venice's import-substitution strategy: Wee, "Structural changes," 23–27.

168. On Sicily's declining fortunes during the crisis of the seventeenth century: Epstein, *Island for itself*, 402–12.

169. On falling profits in merchandise trade in the mid-sixteenth century: Arrighi, *Long twentieth century*, 123–26.

170. Lane, *Venice*, 312–18; Braudel, *Civilization and capitalism*, 2: 284–87.

171. Sella, "Crisis and transformation," 111–17.

172. To be sure, this was not the first time that the Venetian merchants resorted to putting-out operations, as we discuss in ch. 3: Mazzaoui, "Cotton industry," 279–80.

173. On Venice's ability to orchestrate bullion flows: Day, "Outline," 1–6.

174. Rapp, "Unmaking," 499–525.

175. Molà, *Silk industry*, 261–98.

176. Marino, "Economic structures," 58–62; Domenico Sella, *Italy*, 41–46.

177. On cotton cultivation in Venetian dominions: Jacoby, "Changing economic patterns," 219.

178. Ogilvie, *State corporatism.*

179. Mazzaoui, "Cotton industry," 282.

180. Philips and Philips Jr., *Spain's golden fleece*, 182: tab. 9.3.

181. Mazzaoui, *Italian cotton*, 129–53.

182. Stromer, "Une clé du succès," 29–49; also Wee, ed., *Rise and fall.*

183. Mottu-Weber, *Économie et refuge.*

184. Kriedte, Medick, and Schlumbohm, *Industrialization before industrialization*, 107–17, 136–45; Sella, *Crisis and continuity*, 99, 113–16.

185. For an analysis that demonstrates the full scope and sway of ruralization of manufacturing: Kellenbenz, "Industries rurales," 833–82; also Deyon, "Variations de la production textile," 939–55.

186. Eldem, *French trade*, 92–97.

187. On the fate of Segovia's wool industry: Weisser, "Agrarian depression," 149–54; on the fate of Ottoman textiles: Çizakça, "Incorporation," 353–78.

188. "Industry," Braudel asserts, was then often "a way of making up the poverty and harsh

existence of these regions," rather than a sign of economic opulence: Braudel, *Civilization and capitalism*, 3: 338.

189. On the Levantine wool trade: Carter, "Commerce," 379–83; on the Spanish wool trade: Phillips, "Spanish wool," 775–95; and on Ottoman exports: Çizakça, "Price history," 533–50; Güçer, *XVI–XVII. Asırlarda*, 12–19.

190. Blanchard, "Long sixteenth century."

191. Atasoy et al., *İpek*, 177: fig. 32.

192. Paskaleva, "Contribution," 277; Kellenbenz, "From Melchior Manlich," 611–22.

193. Houtte, "Les grands itinéraires," 89–90.

194. Kellenbenz, "Le déclin," 109–83.

195. Braudel, *Civilization and capitalism*, 2: 157, 165.

196. Blanchard, "Continental European cattle trades," 427–60.

197. Pounds, *Historical geography*, 284–85.

198. See the articles in Zachariadou, ed., *Via Egnatia*.

199. Philips and Phillips Jr., *Spain's golden fleece*, 179–82.

200. Felloni, *Gli investimenti*, 249–59, 265–76.

201. Coleman, "Innovation," 417–29; Munro, "Medieval woolens: West European woolen industries," 228–324.

202. Braude, "International competition," 437–51; Le Flem, "Vraies et fausses splendeurs," 525–36.

203. Braudel, "Remarques," 175–97.

204. Wallerstein, *Modern world-system*, 3: ch. 1.

205. Braudel, *Civilization and capitalism*, 2: 191; Curtin, *Rise and fall*, ch. 6.

206. On sugar and grain trade: Braudel, *Civilization and capitalism*, 2: 224–25.

207. Braudel, *Civilization and capitalism*, 1: 126.

208. Delano-Smith, *Western Mediterranean Europe*, 195, 399; Volney, *Voyage*, 2: 409.

209. On the growing popularity of lesser cereals as well as non-cereal components of agriculture during periods of economic downturn: Thirsk, *Alternative agriculture*, pt. ii.

210. Wallerstein, *Modern world-system*, 3: ch. 3.

211. Lewis, "Proto-industrialization," 150–64; Thomson, *Distinctive industrialization*, 36–49; Wolff, "Esquisse d'une histoire"; Genç, "18. Yüzyıl," 64–65.

212. Cf. Kriedte, Medick, and Schlumbohm, *Industrialization before industrialization*, 107–11.

213. Carter, "Commerce," 379–83; McGowan, *Economic life*, 38–41; Eldem, *French trade*, 92–97.

214. Kellenbenz, *Rise*, 246, 260–67.

215. On the Ottoman empire: Faroqhi, "Anayol kavşağında Bursa," 119–22; on Catalonia: Vilar, *La Catalogne*, 357–58.

216. İnalcık, "Osmanlı pamuklu," 19–29.

217. Stoianovich, "Balkan peasants," 26–27.

218. Masson, *Histoire du commerce français*.

219. Kellenbenz, "Le déclin," 154–60; Tilly, *Capital, coercion*, 64–65.

220. Stoianovich, "A route type."

221. Hanna, *Making big money*, 84–95.

222. Gentil da Silva, *Banque et crédit*, 1: 710–12.

223. Ciriacono, "Silk manufacturing," 196–97.

224. After all, the ducat remained almost stable in weight and value from the mid-fourteenth century to the fall of the Republic: Crouzet-Pavan, *Venice triumphant*, 91; on the assertion of the Venetian ducat in the eighteenth century: Pamuk, *A monetary history*, 169.

225. For connections between German and east European Jews and the Jewish community in Amsterdam: Bloom, *Economic activities*.

226. Bloch, *French rural history*, 5–12.

227. Palairet, *Balkan economies*, 36.

228. On the grain trade: Herlihy, *Odessa*.

229. Ciriacono, "Silk manufacturing," 199.

230. Braudel, *Civilization and capitalism*, 3: 175.

231. Cox, *Foundations*, 35–38; Braudel, *Civilization and capitalism*, 3: 107–9.

232. Lopez, "Market expansion," 447.

233. Weber, *General economic history*, 215.

234. On the nature of this world, as seen from China, and the circuits that connected it to the Levant and beyond, in the 1000–1350 period: Adshead, *China*, 126–62; Chaudhuri, *Trade and civilization*, ch. 8; and as seen from the Indian Ocean in the 1400–1650 period: Reid, *Southeast Asia*, 2: ch. 1; Hourani, *Arab seafaring*, 61–84.

235. Allsen, *Commodity and exchange*, 46–71.

236. Arrighi, *Long twentieth century*, 118.

Two • City-States and the Inner Sea

1. Harreld, *High Germans*.

2. Houtte, *Economic history*, 175–90.

3. Florence experienced food shortages in 1590, 1591, 1594, 1595, 1596, 1601, 1602, and 1607: Berner, "Florentine society," 214; Delumeau, *Vie économique*, 2: 521–646.

4. Tracy, "Habsburg grain policy," 293–319; Braudel, *Mediterranean*, 1: 640–41; 2: 726.

5. Vassberg, *Village*, 124–28; Seikaly, "Land tenure," 406; Andreasyan, "Celâlilerden kaçan," 45–53; Akdağ, *Türkiye'nin iktisadî*, 2: 460–61; *Şer'iye sicilleri*, 2: 165–67; Le Roy Ladurie, *French peasantry*, 239–50; Özel, "Population changes," 189.

6. Braudel, *Civilization and capitalism*, 2: 100; Vries and Woude, *First modern economy*, 414–16.

7. Braudel and Spooner, "Prices in Europe," 138–39; Barkan, "Price revolution," 3–28; Goldstone, "Urbanization."

8. Abel, *Agricultural fluctuations*, 147–49; Le Roy Ladurie and Goy, *Tithe*, 111–16, 125–27, 147–51; Özel, "Population changes," 190.

9. Braudel et al., "Le déclin," 29–35; Faroqhi, *Towns and townsmen*, 209: fig. 2; Slicher van Bath, *Agrarian history*, 195–206.

10. Lovett, "Castilian bankruptcy," 899–911; Elliott, *Imperial Spain*, 286–87; Braudel, *Mediterranean*, 2: 701–3; Le Roy Ladurie and Goy, *Tithe*, 126–27. The Ottoman budget deficit reached 400 million *akçes* in 1597–98, from a surplus of 127 million in 1567–68: Tabakoğlu, *Gerileme dönemine*, 16–17. This has been referred to as "the century's economic illness": Barkan, "Price revolution," 3–28.

11. Israel, "Trade, politics," 12–15; Lane, "Public debt," 80–81.

12. Le Roy Ladurie, *Peasants*, 71–72.

13. Huertz de Lemps, *Vignobles et vins*, 1: 199–200; Marquez, *Daily life*, 27–28; Vilar, *History of gold*, 87; Edwards, *Christian Córdoba*, 98–100.

14. Day, "Introduction," 10–11; Pamuk, *Monetary history*, 74–76.

15. Goldstone, "Urbanization," 1122–60; Brandse, "Trade and state," 188–91; İslamoğlu and Faroqhi, "Crop patterns," 401–36; on the monetary instability and attempts by the Sublime Porte to exact compliance in matters regarding the strict imposition of administered prices in the 1585–1640 period: Pamuk, *İstanbul*, 34–36.

16. On declining yields: Weisser, *Peasants*, 64–65; Öz, "15–16. Yüzyıllarda," 1643–51; and on the opening of relatively infertile land to cultivation in the sixteenth century: Cook, *Population pressure*, 1–9.

17. Bennassar, *Valladolid*, 1: 49–51; Chaunu, *L'Espagne*, 1: 211–14; Abdel Nour, *Introduction à l'histoire*, 217–18, 247; Güçer, *XVI–XVII. Asırlarda*, 8–9; Laoust, *Les gouverneurs*, 209; Rosenberger, "Famines et épidémies."

18. Ciriacono, "Venetian economy," 120–35.

19. Salvemini, "Florence," 318–19. This was in the aftermath of the Black Death, when the city's population plummeted more catastrophically than in its rural hinterland that provisioned it. The food deficit all but declined despite the new rural-urban ratio, which was favorable to the latter.

20. Rapp, *Industry*, 159–64. On the price differentials in the Mediterranean at the close of the sixteenth century: Stoianovich, *Balkan worlds*, 192: fig. 5.1.

21. Aymard, *Venise, Raguse*, 125–40.

22. Davis, *Decline*, 39–45.

23. Davies, "Changes," 381–97.

24. Benadusi, *Provincial elite*, ch. 7.

25. Sereni, *History*, 148–49; Le Roy Ladurie and Goy, *Tithe*, 114.

26. Georgelin, *Venise*, 232–40, 242–44.

27. Braudel, *Civilization and capitalism*, 2: 284; Davies, "Changes," 382.

28. Hufton stresses that the shift to lucrative crops was due to the availability of Baltic and Levantine grain: "Social conflict," 308–9.

29. Simon, "Le blé," 282–86; Brummet, *Ottoman seapower*, 138–39; Epstein, *Island in itself*, 274–75.

30. Le Roy Ladurie and Goy, *Tithe*, 106–8, 110–11.

31. Salomon, *La campagna*, 257–66; Vassberg, *Village*, 67–85; Defourneaux, *Daily life*, 97; Barkan, "Osmanlı imparatorluğu'nda kuruluş," 281–90; İnalcık, "Ottoman state," 103–19, 145–53. In the peak period of 1500–1650, close to 427,000 Spaniards and around 100,000 Portuguese (in the 1500–1700 period) migrated overseas: Kamen, *Empire*, 130.

32. Aymard, *Venise, Raguse*, 7; in the 1570s, only one-third of the population in Portugal was engaged in agriculture: Anderson, *Lineages*, 73.

33. On the two phases of the price revolution of the long sixteenth century: Walker, "Capitalism and Reformation," 1–19. On how prices, which had remained relatively stable during the 1450–1560 period, increased thereafter: Pamuk, *Monetary history*, 123; Dennis and Giráldez, "Silver," 10–18; and on wheat exports: McGowan, *Economic life*, 34–37.

34. On news and speculations in Venice about the Mediterranean grain markets: Sardella, *Nouvelles et spéculations*, 19–20; on Castile and its "obsession with wheat": Bennassar, *Valladolid*, 1: 65–69; and on Lisbon: Godinho, "La guerres du blé."

35. Le Roy Ladurie, *French peasantry*, 316; Vilar, *History of gold*, 86–88.

36. Bishko, "Castilian," 47–69; Phillip and Phillip Jr., *Spain's golden fleece*, 43–60; Planhol, "Geography, politics," 525–31; İnalcık, "Ottoman state," 40–41; Lindner, *Nomads and Ottomans*, 51–66; Refik, *Anadolu'da Türk*, 67–71.

37. Le Roy Ladurie, *Peasants*, 231.

38. Delano Smith, *Western Mediterranean Europe*, 194.

39. In the kingdom of Naples, the 1447–94 and 1550–1611 periods and in Iberia the 1450–1580 period were boom periods for livestock husbandry: Marino, *Pastoral economics*, 68–71; Phillip and Phillip Jr., *Spain's golden fleece*, ch. 4; in Anatolia, too, good times for livestock and meat industry came to an end at the close of the sixteenth and the turn of the seventeenth centuries: Faroqhi, *Towns and townsmen*, ch. 9; also Greenwood, "Istanbul's meat," 9–15; Tabakoğlu, *Türk iktisat tarihi*, 362–63.

40. Braudel, *Out of Italy*, 22; Cahen, "Quelques problèmes," 381–432; Hocquet, "Ships, sailors," 533–67; Turan, *Türkiye-İtalya*, 325–26.

41. Pryor, *Geography, technology*, 123.

42. Laiou-Thomadakis, "Byzantine economy," 177–222.

43. Jacoby, "Foreigners," 95–96; Crouzet-Pavan, *Venice triumphant*, 47–62.

44. Sakaoğlu, *Çeşm-i cihan Amasra*, 47–62; Matschke, "Commerce, trade," 789–93; Weatherford, *Genghis Khan*, 220–26.

45. Lopez, *Commercial revolution*, 93–94; Giurescu, "Genoese," 587–600.

46. On bullion flows reaching the Mongolian world by way of Cyprus and Lesser Armenia: Spufford, *Money*, 153–57. Also Constable, *Housing the stranger*, 223; Pamuk, *Monetary history*, 23–24.

47. Registering the economic activity heightened under Mongolian epoch was the close to thirtyfold increase in taxes collected in central and eastern Anatolia, from 200,000 *dinar*s in 1256 to 5,645,000 *dinar*s in 1336: Köprülü, *Osmanlı devleti'nin*, 55–56.

48. It was only when, and surely not by mere happenstance, the spice trade returned to the Red Sea that the kingdom of Lesser Armenia fell to the Mamluks, in 1375. On the Mediterranean trade in this period seen from the Seljuk empire: Cahen, *Pre-Ottoman Turkey*, 321–23; Togan, *Umumî Türk tarihi*, 244–45; Lopez and Raymond, *Mediterranean trade*, 171–72, 224–25; Lapidus, *Muslim cities*, 11–16.

49. Adshead, *China*, 195–200.

50. Salt was one of the staples of the Mediterranean trade, as emphasized by Heers, *Gênes*, 233–58, and Hocquet, *Le sel*, 256–72. In the fourteenth and fifteenth centuries, 30 to 50 percent of the volume of goods carried by the Venetian merchants was salt: Spufford, *Power and profit*, 297. The Venetian salt market handled close to 33,000 tons of salt a year. Unlike sugar and cotton, however, it was relatively easily available throughout the Mediterranean. So when the Genoese moved west, they came to rely on salt from Ibiza, whereas Venice's reliance on Egyptian salt-pans increased: Adshead, *Salt and civilization*, 85–90. On the copper trade: Brummett, *Ottoman seapower*, 151–55.

51. Raymond and Lopez, *Mediterranean trade*, 129; Ashtor, *Social and economic*, 318–19; Spufford, *Money*, 149. See also Howard, "Death in Damascus," 143–57.

52. Jacoby, "Silk production," 55, 59–60; Jacoby, "Genoa, silk trade," 27–29; Day, "Levant trade," 811–12; Mazzaoui, *Italian cotton*, 44–45.

53. On the growing role of Venetian merchants in Thessalonike Salonika from the first half of the fourteenth century: Jacoby, "Foreigners," 106–11, 115–16; on cotton imported from the Pontic Sea: Mazzaoui, *Italian cotton*, 43–44, 175 n. 6.

54. Hocquet, "Ships, sailors," 533–67; Cahen, *Orient et occident*, 183.

55. Kirk, *Genoa*, 15; Epstein, *Genoa*, 232; Molà, *Silk industry*, 3–4.

56. Mintz, *Sweetness and power*, 82; Ashtor, "Catalan cloth," 227–57; Thomson, *Distinctive industrialization*, 29–31.

57. Yusuf, *Economic survey*, 39–40; Planhol, *Les fondements*, 76. During the crusades, the sweet-tasting sugarcane made it possible for the inhabitants of Tripoli to sustain themselves during the siege of the city: Mintz, *Sweetness and power*, 28.

58. Mazzaoui, *Italian cotton*, 37; Ziadeh, *Urban life*, ch. 3; Schefer, ed., *Le voyage d'outremer*, 42–44.

59. Labib, "Egyptian commercial policy," 74; Dols, *Black Death*, 197–98; for the recovery during the fifteenth century in the Mamluk lands: Lapidus, *Muslim cities*, 33–35; Chapoutot-Remadi, "L'agriculture," 30–31.

60. The number of villages in Egypt increased from 2,071 in 1210 to 2,454 in 1315. Canals were repaired, new villages founded when the spice trade was at a lull: Ashtor, *Social and economic history*, 303, 318.

61. Yusuf, *Economic survey*, 40–41; Merçil, *Türkiye Selçuklularında*, 23–24, 25, 27; Turan, *Türkiye-İtalya*, 107; Mazzaoui, *Italian cotton*, 38–43.

62. Rabie, "Some technical aspects," 59–81.

63. Lane, *Venice*, 400–402; Léon, "La réponse," 231; Suraiya, "Notes," 405–17.

64. McNeill, *Venice*, 76.

65. Galloway, *Sugar cane industry*, 43–47; Miquel, *Le géographie*, 3: 407–8, 432–43.

66. Curtin, *Rise and fall*, 5–10.

67. Verlinden, "La Crète," 593–669; Scammel, *World encompassed*, 106–8. The Aegean- and Mediterranean-bound slave trade was but only a small proportion of the slave trade. On the growing significance of easterly flow of slaves following the establishment of new imperial bureaucracies and armies in lands formerly ruled by the Mongol empire and the subsequent inflow of slaves from the Black Sea: Wink, *Al-Hind*, 2: 70–73; Turan, *Türkiye-İtalya*, 101–2. On the slave trade conducted in Anatolian fairs: Doğru, *XVIII. Yüzyıla kadar*, 89.

68. Labib, "Egyptian commercial," 75. It was on the back of this growing trade that the cotton beaters in Venice established their own guild in 1279, leaving the cotton guild that controlled all stages of the production process: Mazzaoui, "Cotton industry," 277.

69. Wallerstein, "West, capitalism," 600–612.

70. Galloway, "Mediterranean sugar," 189–90.

71. During the fourteenth and the early fifteenth centuries, slaves were omnipresent in rural Cyprus. The king claimed in 1373 that Cyprus had "grant besoing de laboreurs qui laborassent les terres pour faire sucre." Arbel, "Slave trade," 160–61.

72. Dembiński, *Food and drink*, 38–43.

73. Arbel, "Cypriot population," 208, 213.

74. On the relocation of the Genoese empire in the western Mediterranean: Kirk, *Genoa*, 15–20; also Carrère, "Marseille, Aigues-Mortes," 161–72.

75. Arbel, "Slave trade," 165; Lane, *Venice*, 349–50; Crosby, *Ecological imperialism*, 78–79.

76. In the twelfth century, no local sugar was utilized in Andalusian cooking; honey was used more often: Bolens, *La cuisine andalouse*, 221–24, 219–20; Mintz, *Sweetness and power*, 82.

77. Vanacker, "Géographie économique," 677: maps 13 and 15; Constable, *Trade and traders*, 233, 246; on silk and sugarcane: Lagardère, *Campagnes et paysans*, pt. iii. For Barbary sugar in Bruges: Toussaint-Samat, *History of food*, 556; Watson, *Agricultural innovation*, 24–30. For the impact of this floral migration on the diet of the region: Ashtor, "An essay on the diet," 125–62.

78. For Genoa's growing role in slave trade: Epstein, *Genoa*, 262–70; in 1470, when a Spanish fleet captured the Genoese colony of Santiago in Cape Verdes and took Antonio da Noli hostage, demanding ransom, it was the Portuguese who rescued him and returned him to his sugar plantation. Birmingham, *Trade and empire*, 19–20.

79. Bresc, *Un monde méditerranéen*, 1: ch. v.

80. On the search for dyes and foodstuffs: Godinho, "Portugal."

81. Heers, *Gênes*, 495–97; Rau, "Settlement of Madeira," 3–12.

82. Braudel, *Civilization and capitalism*, 1: 224–27; Crosby, *Ecological imperialism*, 77–79.

83. Cretan sugar industry "flourished until destroyed in the 1450s by the competition of that of the Portuguese Atlantic possessions": Scammel, *World encompassed*, 105, 118–19.

84. García-Arenal and Wiegers, *Man of three worlds*, 24–25. William Lithgow, who visited Cyprus in 1610, reported that the island yielded "infinite canes of sugar": Jennings, *Christians and Muslims*, 302; on the decline of sugar industry in Sicily during the post-1650 period: Epstein, *Island in itself*, 407.

85. Galloway, "Mediterranean sugar," 189.

86. Harreld, "Atlantic sugar," 162–63.

87. Evliyâ Çelebi, *Seyahatnâme*, 4: 295; 8: 244; 9: 148, 167, 196, 215; Toussaint-Samat, *History of food*, 554; Miquel, *Le géographie*, 3: 407–8.

88. Merçil, *Türkiye Selçukluları'nda*, 63–65; Cahen, *Pre-Ottoman Turkey*, 158; Bilgin, "Osmanlı sarayında," 290–96. Honey and molasses were used as sweeteners. Curiously enough, sugarcane was displayed as a ceremonial article, perhaps too due to its scarcity, at the Sublime Porte in the sixteenth century: Schweigger, *Sultanlar kentine yolculuk*, 26, 66.

89. Watson, *Agricultural innovation*, 31–41; Vanacker, "Géographie économique," map 13.

90. Birmingham, *Trade and empire*, 39–40.

91. Mazzaoui, *Italian cotton*, 53.

92. McIntosh, *Urban decline*, 4–6.

93. Lane, *Venice*, 298.

94. On *funduq*s devoted to cotton trade: Constable, *Housing the stranger*, 242, 244 n. 34.

95. Dembińsky, *Food and drink*, 38–43; Jennings, *Christians and Muslims*, 297–99.

96. Lane, *Venice*, 298.

97. Jennings, *Christians and Ottomans*, 298–307, 325–29; Altan, *Belgelerle Kıbrıs*, 1: 17.

98. Spufford, *Money*, 368: map 35.

99. Phillips, "Spanish wool," 779–84; Mazzaoui, "Cotton industry," 282–86; Doğru, *XVIII. Yüzyıla kadar*, 153–54; Fleet, *European and Islamic trade*, 99–101.

100. Constable, *Housing the stranger*, 334.

101. Epstein, *Genoa*, 230–34.

102. Abulafia, *Mediterranean emporium*, 229.

103. This was the case in Byzantium as well, hammered by the continuous inflow of Turkish

tribes; both Byzantine and Turkish populations swelled the ranks of transhumant populations: Hopwood, "Manavgat valley," 459. On the logic of investing in animals in politically insecure regions: Sivignon, *Les pasteurs.*

104. Mollat, *Jacques Coeur,* 168.

105. Ashtor, *Levant trade,* 478, 485.

106. Wake, "Changing pattern," 361–403; Ashtor, "Venetian supremacy," 5–53.

107. On how sugar consumption increased in the fourteenth and fifteenth centuries: Mintz, *Sweetness and power,* 80–87.

108. Braudel, *Civilization and capitalism,* 3: 135–36.

109. Ashtor, "Levantine alkali," 475–522; Jacoby, "Changing economic patterns," 232–33.

110. Braudel, *Out of Italy,* 39; Dowd, "Economic expansion," 143–60.

111. Martines, *Power and imagination,* 277–96.

112. The Venetians were famed as people who sowed not, plowed not, and harvested not: Lopez, *Commercial revolution,* 64.

113. Quaglia, "Controls over food," 449–57.

114. Heers, *Gênes,* 248; Lopez, "Market expansion," 448.

115. Bresc, *Un monde méditerranéen,* 1: 128, 526; see also Laiou-Thomadakis, *Peasant society,* 24; Braudel, *Mediterranean,* 1: 579; Magdalino, "Grain supply," 36.

116. Balard, *La romanie genoise;* Thiriet, *La romanie venetienne.*

117. Crosby, *Ecological imperialism,* 73; Fernández-Armesto, *Before Columbus,* 161–66.

118. Topping, "Le régime agraire," 267–68.

119. On the provisioning of the Venetian empire in this period: Arenson, "Food," 177–85.

120. Lewis, "Closing," 475–83.

121. Lefort, "Rural economy," 265; Hendy, *Studies,* 70.

122. Cahen, *Pre-Ottoman Turkey,* 88–89; Şeşen, *İslam coğrafyacılarına göre,* 215; Hendy, *Studies,* 70: map 13; Vryonis Jr., "Nomadization," 41–71; Brauer, "Boundaries and frontiers," 53–60; on "marchlands": Wittek, *Rise,* 23–32.

123. Constable, *Housing the stranger,* 247–48.

124. On how "Thrace and eastern Macedonia and western Asia Minor" came to replace the Byzantium's former bread lands (i.e., Sicily, Egypt, and North Africa after they were lost to the advancing Abbasid empire): Teall, "Grain supply," 87–139. On the Black and White Sheep Turks: Uzunçarşılı, *Anadolu beylikleri,* 180–234.

125. Vernet, "Les relations céréalières," 333–35. A century later, this spatial division of labor was still in place: Laiou-Thomadakis, "Byzantine economy," 183–85, 214–15, 218–20.

126. Of course, there was also the re-export of grain to places where it fetched higher prices. In 1511–12, the amount of wheat imported into Venice was around 60,000 tons, enough to feed more than 300,000 people, more than twice the city's own population. Lane, *Venice,* 306; Sardella, *Nouvelles et spéculations,* 19–20.

127. Jacoby, "Foreigners," 110; Harvey, "Economic conditions," 117–18; Gertwagen, "Port of Modon," 192–94; on Genoa's grain re-export trade: Lopez, "Market expansion," 456.

128. Birmingham, *Trade and empire,* 7–8.

129. On the availability of flour (and rice and butter), transported in leather bags, in the eastern Mediterranean in the fourteenth century: İbn Battûta Tancî, *İbn Battûta seyahatnâmesi,* 1: 205, 421, 523. Also Derschwam, *İstanbul ve Anadolu'ya,* 149, 222–23, 229.

130. Laiou-Thomadakis, "Byzantine economy," 183–85.

131. Balard et al., "Le transport," 94; Guiral-Hadziiossif, *Valence*, 282.

132. Romano, "A propos du commerce du blé," 159–61; on the nature of the catchment area in the Aegean that served the provisioning of Rhodes: Vatin, *Rodos şövalyeleri*, 36–39.

133. On grain: Wolff, "Un grand commerce médieval," 147–64; on wine: Pirenne, "Un grand commerce," 225–43; and on textiles: Chorley, "Cloth exports," 349–79.

134. Braudel, *Civilization and capitalism*, 1: 26; On the dispatch of "good money," i.e., silver, from Castile for purchase of wheat: Weisser, "Rural crisis," 311–13. In 1506, the guarantee for the Sicilian shipment of grain via Seville to Córdoba, extended to the Genoese merchants by the town council, was the enormous sum of 10,000 ducats. Edwards, *Christian Córdoba*, 111. On the wealth of those engaged in licit or illicit grain trade in Ottoman and Mamluk lands: Brummett, *Ottoman seapower*, 139; İlgürel, "Hububat kaçakçılığı," 361–69; Lapidus, "Grain economy," 1–15.

135. Sicilian wheat was still so highly valued at the turn of the sixteenth century that Leo Africanus related that "the Arabs handed over their children as pledges to obtain Sicilian wheat." Braudel, *Mediterranean*, 1: 580.

136. Ashtor, "Wheat supply," 283–95; Day, "Prix agricoles," 629–56; Mayerson, "Role of flax," 201–7.

137. Arbel, "Venetian trade," 629–56; Romano, "A propos du commerce du blé," 149–61.

138. Constable, *Trade and traders*, 232–33; Dufourcq, *L'Espagne catalan*, 536–41.

139. The Genoese merchants attempted to bypass the Alhóndiga in Seville, and were later exempted from the Alhóndiga rule: Constable, *Housing the stranger*, 338.

140. Bresc, *Un monde méditerranéen*, 1: 545, 549; Carrère, *Barcelone*, 2: 628.

141. Vernet, "Les relations céréalières," 334–45; Balard et al., "Le transport," 156–57; Edwards, *Christian Córdoba*, 110–11.

142. On Languedoc: Le Roy Ladurie, *Peasants*, 15; on Mediterranean Spain: Bull, "Olive industry," 136–37. During the Mamluk era, olive oil was imported into Syria, normally one of the major oil-producing regions of the Levant: Ashtor, *Social and economic history*, 319.

143. Constable, *Trade and traders*, 185; Goitein, *Mediterranean society*, 1: 122; Pounds, *Historical geography*, 184.

144. Duby, *Rural economy*, 138–41; Phillips, *Short history*, 84–92; Abel, *Agricultural fluctuations*, 72; Slicher van Bath, *Agrarian history*, 143–44; Craeybeckx, *Un grand commerce*.

145. For an excellent account of this northerly march of the vineyards: Lachiver, *Vins, vignes*, chs. 3–4.

146. On the expansion of the vineyards in the Iberian *huerta*: Glick, *Irrigation*, 11–30; Le Roy Ladurie, *Peasants*, 144.

147. Phillips, *Short history*, 99–100; Enjalbert, "Comment naissent."

148. Le Roy Ladurie, *French peasantry*, 130–31; Le Roy Ladurie, *Peasants*, 65–66.

149. Spufford, *Power and profit*, 293–95.

150. Kalligas, "Monemvasia," 888–90; Tucci, "Le commerce venetien," 199–211; on Cretan malmsey: Scammel, *World encompassed*, 105.

151. Braudel, *Identity*, 2: 321. The "regression" of viticulture is underlined by Le Roy Ladurie, *Peasants*, 60–66.

152. Randsborg, *First millennium*, 128–29; Laiou, "Agrarian economy," 353; Beldiceanu, "Les biens du monastère," 54. In the eleventh and twelfth centuries, when monasteries provided the needy with the necessities of life, they served them wheat, wine, and legumes: Laoui-

Thomadakis, *Peasant society*, 29 n. 18 (and not with olives as was the case previously or after the sixteenth century). It remained as a product only the rich could afford; on the significance of olive oil in the Levant at the turn of the millennium: Miquel, *Le géographie humaine*, 3: 450–52.

153. Erevnìdes, "Politika mutfağında," 67, 70; Ashtor, *Social and economic history*, 319; Ashtor, "Wheat supply," 283–95.

154. Lefort, "Rural economy," 248–49; Spufford, *Power and profit*, 302–4; Yerasimos, *Sultan sofraları*, 13–17.

155. Stouff, *La table provençale*, 193–201; Bolens, *La cuisine andalouse*, 174–75; Constable, *Trade and traders*, 213–15; Ünsal, *Ölmez ağacın*, 46; on the "carnivorous" diet of the era: Montanari, *Culture of food*, 71–78.

156. Lefort, "Rural economy," 264.

157. Boulanger, *Marseille*, 66–72.

158. Spufford, *Power and profit*, 272–73; Sakellariou, "City of Puglia," 105.

159. Balard et al., "Le transport," 171.

160. Laiou, "Agrarian economy," 322, 324, 359.

161. In 1517, the cardinal of Aragón attributed the abundance of lepers in Flanders and Germany to consumption therein of butter and dairy products.

162. Day, *Les douane*, 1: 215, 232, 240, 251, 386, 512, 515.

163. Constable, *Housing the stranger*, ch. 9.

164. Spufford, *Money and its use*, 354–55.

165. Day, "Levant trade," 808–14.

166. Bloch, "Problem of gold," 214–17.

167. Crosby, *Ecological imperialism*, ch. 4.

168. Day, "Levant trade," 813–14. This was one of the reasons the Genoese fleets, before the Republic established its foothold in Castilian finance, were comprised of very large ships: Kirk, *Genoa*, 33–34.

169. On Florentine merchants' search for Persian silk in Bursa: Hoshino and Mazzaoui, "Ottoman markets," 17–24; on silk destined to Charles V's *arte della seta*: Gentil da Silva, "L'histoire économique," 201–3; and on textile *fondacos* in Florence, Pisa, and Genoa: Constable, *Housing the stranger*, 341–44. Also Goodman, "Financing," 415–35.

170. Watson, "Arab and European agriculture," 65.

171. Bautier, *Economic development*, 213; Molà, *Silk industry*, 4–5; Kamen, *Empire*, 154–55; Planhol, *Les fondements*, 132.

172. Jacoby, "Foreigners," 114.

173. On the Black Sea, the chief granary was the Crimea: Karpov, "Grain trade," 57.

174. Mantran, *17. Yüzyılın*, 1: 45–48. The janissaries of Istanbul alone consumed in the neighborhood of 150,000 *stera*, or 9,300 tons, of grain a year: Brummett, *Ottoman seapower*, 136; the city consumed 128,000 tons in the mid-eighteenth century: Aynural, *İstanbul değirmenleri*, 4.

175. Murphey, "Provisioning Istanbul," 232–34.

176. For the treaties signed between the Venetians and the emirs of Menteşe and Aydın between 1331 and 1414: Zachariadou, *Trade and crusade*; also Matschke, "Commerce, trade," 775–76; Jacoby, "Foreigners," 115–16.

177. From the 1550s, Istanbul reestablished its hold over the Aegean and the Black Sea: Goffman, *İzmir*, 33–35.

178. On the tapering off of grain exports from the Pontic Sea: Kortepeter, "Black Sea," 88–96.

179. Epstein, *Island in itself,* 72–73.

180. Bresc, *Un monde méditerranéen,* 1: 130. Surely, the situation changed from the 1450s; some of the decline was conjunctural rather than structural, but the direction of change was unmistakably toward lesser diversification.

181. This was a development all too common in most parts of France: Le Roy Ladurie, *French peasantry,* 19–20.

182. Vernet, "Les relations céréalières," 334–45; Rosenberger, *Société, pouvoir,* 79.

183. Braudel, *Mediterranean,* 1: 339–40; Herlihy and Klapisch-Zuber, *Tuscans,* 106–8; Martines, *Power and imagination,* 277–87.

184. Molà, *Silk industry,* 217–36.

185. Camporesi, *Magic harvest,* 99–100.

186. Spufford, *Money,* 369.

187. Chaunu and Chaunu, "Atlantic economy."

188. Epstein, *Genoa,* 316.

189. Kirk, *Genoa,* 29–32.

190. Raymond, *Osmanlı döneminde,* 18–21; Barendse, *Arabian Seas,* 187; Stoianovich, "L'Espace maritime," 49; İnalcık, "Osmanlı pamuklu," 1–42.

191. As the share of spice imports to Lisbon from Asia fell from the mid-sixteenth century, that of textiles (and in private trade, cottons and silks) increased. Subrahmanyam and Thomaz, "Evolution of empire," 308–10. This did eventually reduce the demand for textiles originating in the western Mediterranean.

192. Braudel, *Out of Italy,* 114.

193. Kirk, *Genoa,* 33.

194. It was at this time Ottoman merchants made their appearance in Venice. See Kafadar, "Death in Venice," 191–217; Turan, "Venedik'te Türk," 247–83; Braudel, *Mediterranean,* 1: 544.

195. İnalcık, "Ottoman state," ch. 12.

196. Earle, "Commercial development," 38–39.

197. Braudel, *Mediterranean,* 1: 581.

198. On the competition between wheat and wool: Vassberg, *Land and society,* 153; Gerbet, *L'élevage;* Marino, *Pastoral economics,* ch. 7.

199. Cook, *Population pressure,* 68; Vassberg, *Land and society,* 202; Bennassar, *Valladolid,* 1: 240; Nolte, "Andalusia," 68; Le Roy Ladurie, *French peasantry,* 111; Edwards, *Christian Córdoba,* 94. There was little progress in yields in the following two to three centuries. See Slicher van Bath, *Agrarian history,* 130–33, 180–81. In the mid-nineteenth century, yields were up to 1:7 in parts of Ottoman empire, especially in the case of barley: Güran, *19. Yüzyıl Osmanlı,* 98–99. In fourteenth-century Macedonia, yields were a little over 1:5: Lefort, "Une exploitation," 369.

200. Braudel, *Mediterranean,* 1: 587–88; Vincent, "Consummation alimentaire"; Horden and Purcell, *Corrupting sea,* 116.

201. Romano, "A propos du commerce du blé." On the Baltic trade grain before 1350: Hybel, "Grain trade"; and on English wheat shipped to Castile, particularly in the 1470s and the 1480s: Childs, *Anglo-Castilian trade.*

202. Fossier, "Europe's second wind," 404–12; Goy and Le Roy Ladurie, *Tithe,* ch. 6; Barkan, *XV. ve XVI. Asırlarda;* Jennings, "Urban population," 21–57; on the phenomenal expansion of

viticulture and horticulture in the mid-fifteenth century in the Quadalquivir valley: Cabrera, "Landed estates," 480–81.

203. Lindner, *Nomads and Ottomans*, 51–66; Phillips and Phillips Jr., *Spain's golden fleece*, 43–60. On the travails of cohabitation between the settled and nomadic populations at both ends of the Mediterranean at the turn of the sixteenth century: Gerbet. *L'élevage;* Şahin and Emecen, *II. Bâyezid*, 13, 19, 36, 41, 88; Heyd, *Studies*, 125.

204. On the resettlement of the coastal plains in the Italian peninsula: Le Roy Ladurie and Goy, *Tithe*, 114.

205. Slicher van Bath, *Agrarian history*, 144; Abel, *Agricultural fluctuations*, 99–106; Wallerstein, *Modern world-system*, 1: ch. 2; Braudel, *Identity*, 2: 172; Faroqhi, "Sixteenth-century periodic markets."

206. On peasants moving to better lands without hindrance in the 1530s in Castile: Vassberg, *Village*, 68, 77; in the Ottoman empire: Şahin and Emecen, *II. Bâyezid*, 3: 10, 46: 163.

207. Le Roy Ladurie and Goy, *Tithe*, 93–119; Gökbilgin, *XV–XVI. Asırlarda Edirne;* Yediyıldız, *Ordu kazası*, 106–34; Doğru, *XV. ve XVI. Yüzyıllarda Sivrihisar*, 122–25; Göyünç, *XVI. Yüzyılda Mardin*, 136–37; Kurt, *Çukurova*, xliii–xlvii; Gümüşçü, *XVI. Yüzyıl Larende*, 42–47; Vassberg, *Land and society*, 101.

208. İnalcık, "Ottoman state," 40.

209. Balard et al., "Le transport," 171; García-Arenal and Wiegers, *Man of three worlds*, 32.

210. Birmingham, *Trade and empire*, 51–52.

211. Braudel, *Mediterranean*, 1: 586.

212. Vilar, *Gold and silver*, 82; Chaunu and Chaunu, "Atlantic economy," 120; see also Nader, *Liberty*, 57.

213. Vassberg, *Land and society*, 163–64.

214. Vilar, *Gold and silver*, 86–88.

215. Le Roy Ladurie and Goy, *Tithe*, 105–7.

216. Houte, *Economic history*, 180–81.

217. On how grain tithes in two lowland villages in the Gulf of Cádiz remained below the levels reached in 1550 until the 1730s: Ponsot, "En Andalusie occidantale," 97–107.

218. Nader, *Liberty*, 57, 63; Vassberg, *Village*, 54–55; González, "Andalusia," 125.

219. Hoszowski, "L'Éurope centrale," 441–56; Samsonowicz, "Changes in the Baltic," 655–72; Walker, "Capitalism and the Reformation," 1–19.

220. Samsonowicz, "Relations commerciales," 537–45. On the rise of wheat prices in Lwów and Cracow: Braudel and Spooner, "Prices in Europe," 397–98.

221. Simon, "Le blé," 281–86; Epstein, *Island in itself*, 409.

222. Öz, "Agriculture," 27–36; Gümüşçü, *XVI. Yüzyıl Larende*, 151–53.

223. Except for the relatively small debasement of 7 percent in 1566: Pamuk, *Monetary history*, 58, 122; also Akdağ, *Türkiye'nin iktisadî*, 437–43.

224. Öz, "XVI. Yüzyıl Anadolu'sunda," 86; in Spain, the percentage of the harvest that remained in the hands of the peasant was close to 50: Defourneaux, *Daily life*, 102.

225. Adanır, "Tradition," 134–35.

226. Dàvid, "16–17. Yüzyıllarda," 349–51.

227. Moutafchieva, *Agrarian relations*, 79–80; Gökbilgin, *XV–XVI. Asırlarda Edirne*.

228. Kiel, "Vakıfnâme of Rakkas," 16–19.

229. Lapidus, *Muslim cities*, 39.

230. Barkan, "Essai sur les données," 25.

231. Kennedy, *Crusader castles*, 15–17, 185. Delort ties the rise of *incastellamento* to the advance of malaria: *Les animaux*, 205.

232. Brumont, *La Bureba;* Huertz de Lemps, *Vignobles et vins.*

233. Lovett, "Golden age," 745.

234. Casey, *Kingdom*, 85–87; on silk in France: Ciriciano, "Silk manufacturing."

235. Bautier, *Economic development*, 192–94.

236. Le Roy Ladurie and Goy, *Tithe*, 115–16; İslamoğlu and Faroqhi, "Crop patterns," 427–36.

237. On the growing disparity between market and official sheep prices in Istanbul: Greenwood, "Istanbul's meat," 142: tab. 14.

238. Pounds, *Historical geography*, 228–34.

239. Greene, *Shared world*, 114–15; Arıkan, "Osmanlı imparatorluğu'nda," 279–306.

240. Braudel, *Mediterranean*, 2: 725–34.

241. Lovett, "Golden age," 747–48; Devèze, *L'Espagne*, 626; Nader, "Noble income."

242. Vassberg, *Land and society*, 163–64; Salomon, *La campagne*, 82; Braudel, *Mediterranean*, 1: 588; Le Roy Ladurie, *French peasantry*, 121; Le Roy Ladurie and Goy, *Tithe*, 126. On harvest failures and resultant epidemics from the late sixteenth century to the eighteenth: Appelby, "Epidemics and famine," 654–63.

243. Le Roy Ladurie, *Peasants*, 293.

244. Stoianovich, *Balkan worlds*, 192.

245. After all, meat prices jumped by 300 percent between 1460 and 1560, whereas wheat prices rose by 200 percent: Akdağ, *Türkiye'nin iktisadî*, 2: 437–43.

246. Goldstone, *Revolution and rebellion*, 363–68, 375–79.

247. Cook, *Population pressure*, 7; Güçer, *XVI–XVII. Asırlarda*, 7–12.

248. Faroqhi, "Sixteenth-century periodic markets."

249. İslamoğlu and Faroqhi, "Crop patterns," 410–11; Emecen, *XVI. Asırda Manisa*, 242–46.

250. On grain shortages in Istanbul in the 1580s: *Osmanlı'da bir köle.*

251. Osmanlı Arşivi Daire Başkanlığı, *438 numaralı.*

252. Yüksel, *Osmanlı sosyal*, pt. 3; Yediyıldız, *Institution du vaqf*, ch. 5.

253. VGMA (Archives of the Directorate of Pious Foundations, Ankara): *586* (1417), 215; *618* (1467–68), 1; *584* (1469), 3; *581* (1477), 306; *584* (1476–77), 60; *583* (1507–8), 293; *584* (1509), 2; *614* (1522), 81; *491* (1539), 174; *594* (1575), 229; also Ipşirli and al-Tamimi, *Muslim pious foundations.*

254. Issawi, *Economic history*, 100.

255. Powers, "Revenues of public," 163–202; Yinanç and Elibüyük, *Maraş tahrir.*

256. On the changing economic climate in the Pamphilian plains in the sixteenth century: Erdoğru, "Antalya ve havalisi," 93; Planhol, *De la plain pamphyliennes*, 11–23

257. Hütteroth, *Ländliche siedlungen:* maps 1 and 2; Hütteroth and Abdal Fattah, *Historical geography*, 55–59; on Amasya: Özel, "Population changes," 187–88.

258. Cook, *Population pressure*, 10–11.

259. Emecen, *XVI. Asırda Manisa*, 239–50.

260. Gökçe, *XVI ve XVII. Yüzyıllarda Lâzikiyye*, 356–57; also Özdeğer, *Onaltıncı asırda Ayıntâb*, 72–75; Öz, *XV–XVI. Yüzyıllarda Canik*, 87–88; Gümüşçü, *XVI. Yüzyıl Larende*, 216.

261. İslamoğlu, *State and peasant*, 150.

262. Jennings, "Population, society," 149–250; Venzke, "Sixteenth-century Ottoman."

263. Cook, *Population pressure*, 85, 91, 98; Gümüşçü, *XVI. Yüzyıl Larende*, 203.

264. Barkan and Meriçli, *Hüdavendigâr livası*, 1: 64–67.

265. Gentil Da Silva, *En Espagne*, 26–31. A Spanish captive who served in the Ottoman navy from 1552 to 1557 relates that the victuals on Ottoman ships were much more plentiful than on Spanish ships: *Pedro'nun zorunlu*, 31; for the abundance and inexpensiveness of bread and food in general, see the memoirs of Manuel Serrano y Sanz who sojourned in Istanbul in 1552–56: Ünsal, *Nimet geldi ekine*, 61.

266. Salomon, *La campagne*, 83; Brumont, *La Bureba*.

267. McGowan, "Food supply," 139–96; Goffman, *İzmir*, 20.

268. Akdağ, *Türkiye'nin iktisadî*; on shortages and famines in Anatolia at the turn of the seventeenth century: Öz, *Anadolu'da alevi*, 46–48; Özkaya, *XVIII. Yüzyılda Osmanlı*, 241; and in Istanbul in 1585–86: *Osmanlı'da bir köle*.

269. Aymard, *Venise, Raguse*, 185; İlgürel, "Hububat kaçakçılığı," 366–69.

270. This was the case in Bursa: Gerber, *Social origins*, 30; Cook, *Population pressure*, 11. In late nineteenth-century Anatolia, the majority of peasant plots varied from 20 to 40 *dönüms*: Güçer, *19. Yüzyıl Osmanlı*, 82.

271. Moutaftchieva, *Le vakıf*, 48; Erdoğru, "Konya Mevlevî," 49.

272. Aynur, *İstanbul değirmenleri*, 8–12.

273. Barkan, "Edirne askerî," 16.

274. Veinstein, "On the *çiftlik* debate," 43, 48; Schilcher, *Families in politics*, 136–44; Nagata, *Tarihte âyânlar*, 136–42; Gerber, *Social origins*, 62–63.

275. Keyder, "Small peasant ownership," 62–63.

Three • Eclipse of the City-States and the Resurfacing of the Mediterranean

1. Kamen, "Mediterranean," 30–55.

2. On the prominent position held by wool dealers and manufacturers of cloth in the imperial parades of the late sixteenth century at the opposite ends of the basin: Phillips and Phillips Jr., *Spain's golden fleece*, 202; Atasoy, *1582 Surname-i hümayûn*, 58–59. This prominence was still on display in the eighteenth century under Ahmed III: İrepoğlu, *Levni*, 145.

3. İnalcık, "Ottoman state," 267; Mazower, *Salonica*, 46–64.

4. Veinstein, "Sur la draperie," 55–62; Cohen and Lewis, *Population and revenue*, 62–63, 156.

5. Arbel, *Trading nations*; Sombart, *Jews*, 24–25; Roth, *House of Nasi*.

6. On the movement of cloth, wool, and silk prices: Faroqhi, "Crisis and change," 452–60.

7. Emecen, *Unutulmuş bir cemaat*, 35–36.

8. Çizakça, "Price history," 534.

9. Lane, *Venice*, 309.

10. Braudel, *Civilization and capitalism*, 3: 153–54; Gentil da Silva, *Banque et crédit*, 1: 713–15; Piuz, "Marchands genevois," 460–62; Ciriciano, "Silk manufacturing," 174–76.

11. Arıkan, "Osmanlı," 289–93.

12. Sahillioğlu, "Yeniçeri çuhası," 415–23.

13. Pamuk, *Monetary history*, 131–48.

14. Braude, "Rise and fall," 216–36.

15. Goffman, *İzmir*, 81–84; in a similar context, Wyczánski argues that the steep rise in

prices during the second sixteenth century played a part in forcing urban artisans to rural stations: "La consommation alimentaire," 47.

16. Emecen, "Manisa," 578–80.

17. See the excellent account by Emecen, *Unutulmuş bir cemaat*, 37; Goffman, "İzmir," 97–100.

18. Lovett, "Golden age," 741, 745.

19. Gerber, *Economy and society*, 64–65; on the financing of sericulture: Çiftçi, *Bursa'da vakıfların*, 199–201, 227–28.

20. Gentil da Silva, *Banque et crédit*, 1: 711–12.

21. Wolff, "Esquisse d'une histoire."

22. Fustians of Milan and Piacenza were easily available in the markets of Mediterranean France and Spain: Mazzaoui, *Italian cotton*, 269.

23. Belfanti, "Town and country," 300–306; Molà, *Silk industry*, 304–5: Mazzaoui, *Italian cotton*, 153; Rapp, *Industry*, 105, 142.

24. Phillips and Phillips Jr., *Spain's golden fleece*, 202–3. On the decline of Ávila's once prosperous woolen industry in the mid-seventeenth century and its social and religious ramifications: Bilinkoff, *Avila of Saint Teresa*. On the religious ramifications in the Jewish circles of the growing presence of northern merchants in the Levant and the popularity of Sabbatarianism: Goffman, *İzmir*, 90–92.

25. Petmezas, "Patterns of industrialization," 575–601; Genç, "18. Yüzyılda," 233; Mazower, *Salonica*, 122; Kostis, "Salonique," 139–40.

26. Palairet, *Balkan economies*, 56.

27. Thomson, *Distinctive industrialization*, 39–40; La Force Jr., "Royal textiles factories," 349–59.

28. Braudel, *Identity*, 2: 530.

29. Genç, "Ottoman industry," 65.

30. Lovett, *Early Habsburg Spain*, 250–51; Çizakça, "Incorporation of the Middle East," 353–78.

31. "For in wool the [western European producers] competed against each other and were fairly certain that innovations would be rapidly copied. In cotton, however, western Europe collectively competed against India, and was eventually able to ensure politically that innovations did not diffuse there": Wallerstein, *Modern world-system*, 3: 23–25.

32. Gutmann, *Toward the modern economy*, ch. 4.

33. Braudel, *Civilization and capitalism*, 2: 309–11.

34. Kriedte, *Peasants, landlords*, 70–78; Pounds, *Historical geography*, 291–92.

35. Todorov, *Balkan city*, 209–13.

36. Reher, *Town and country*, 26–33.

37. Gerber, *Economy and society*, ch. 4; Evliyâ Çelebi, *Seyahatnâme*, 2: 226; Casey, *Kingdom of Valencia*, 90–91; Özdemir, *XIX. Yüzyılın ilk yarısında Ankara*, 237.

38. Sánchez Léon, "Town and countryside," 283–86; Vries, *European urbanization*, 66–69; Gelabert, "Urbanization," 204–5.

39. Thomson, *Distinctive industrialization*, ch. 6; Abdel Nour, *Introduction à l'histoire*, ch. 9; Fukasawa, *Toilerie et commerce*, 116–37; Johnson, *Life and death*, 7–9; Hubert, "Les *qaysariyya* de textile," 127–35; Pascual and Establet, *Des tissues*, ch. 1.

40. Braudel, *Civilization and capitalism,* 2: 307.

41. Fortea Pérez, "Textile industry," 137; Dalsar, *Türk sanayi,* 295.

42. "Lace making, embroidery, and stocking knitting were cottage industries and in the seventeenth and eighteenth centuries, the list was long and growing": Pounds, *Historical geography,* 294–97.

43. Sereni, *History,* 210–12; on southern France, Tuscany, and Catalonia: Vries, *Economy of Europe,* ch. 2.

44. Braudel, *Identity,* 2: 249–50.

45. Braudel, *Identity,* 2: 276–77; İslamoğlu, *State and peasant,* 164–65. At times, the residents of countryside fared comparatively better than those of sizeable cities in terms of meat consumption: Vedel, "La consommation alimentaire," 482; also Maczak, "Un voyageur," 335.

46. Torras, "Old and the new," 96; Gerber, *Social origins,* 47–48; Öztürk, *Tânzîmat döneminde,* 62–64.

47. Reher, *Town and country,* ch. 7; Tamur, *Ankara keçisi,* ch. 6; Leiser, "Travelers' accounts," 14.

48. On peasant mobility: Vassberg, *Village,* ch. 4; Reher, *Town and country,* ch. 7.

49. Stoianovich, "Conquering Balkan orthodox merchant," 248–54; Şen, *Osmanlı panayırları,* 8–16; Panzac, "International and domestic maritime trade," 199–200; Reimer, "Ottoman Alexandria," 128–34; Ringrose, *Spain, Europe,* 89–90.

50. Carter, "Cracow's wine trade," 551–53; Braudel, *Mediterranean,* 1: 199–214.

51. On Genoa's silk trade: Kirk, *Genoa,* 115–16, 127–28.

52. Lane, *Venice,* 418–21; Kirk, *Genoa,* ch. 6.

53. Sella, "Rise and fall," 108–12; on Genoese woolen trade: Heers, "La mode," 1107–14.

54. Day, "Levant trade," 809–10; Spufford, *Money,* 367–69.

55. Reid, *Southeast Asia,* 2: 24–26; Barendse, *Arabian Seas,* 224–31.

56. Putting-out was only one of the strategies, however. Specializing in luxury goods and making up for the shortage of labor by using hydraulically powered machines were equally deployed by the Venetian businessmen to increase their competitive edge.

57. On Venetian and Genoese trade in northern fabrics in the thirteenth century: Chorley, "Cloth exports of Flanders," 350–53.

58. Spufford, *Power and profit,* 255–58, 269–74. On the difficulties experienced by these industries in the Levant as a result: Ashtor, "Venetian supremacy," 5–53.

59. The figure is from the 1420s: Mazzaoui, *Italian cotton,* 46.

60. On the growth of Florentine wool exports to the Ottoman lands in the latter half of the fifteenth century: Hoshino and Mazzaoui, "Ottoman markets," 19–21; Atasoy et al., *İpek,* 182–90; on the fairs of Champagne and Brie where the exchange took place: Braudel, *Civilization and capitalism,* 3: 111–15.

61. Lane, *Venice,* 309–12.

62. Molà, *Silk industry,* ch. 11; Crouzet-Pavan, *Venice triumphant,* 178.

63. Kriedte, *Peasants, landlords,* 72–77; Lewis, "Proto-industrialization," 155–58.

64. Spufford, *Power and profit,* 116–19, 248–55; Pounds, *Historical geography,* 136–37, 172–74, 234–37; Epstein, *Genoa,* 275–76.

65. As a luxury item, silk was more vulnerable to competition than cotton, which was mostly used as undercloth, house linen, and light clothing: Bautier, *Economic development,* 212–16.

66. Owen, *Middle East*, 6–7; McGowan, *Economic life*, 21–24.

67. Gerber, *Social origins*, 46–50; Phillips and Phillips Jr., *Spain's golden fleece*, ch. 10; Wolff, "Esquisse"; Fukasawa, *Toilerie et commerce*, 21–28.

68. These two waves are compared in Wee, ed., *Rise and decline of urban industries*.

69. Vries, *Economy of Europe*, 103–4. It needs to be mentioned that in the thirteenth and fourteenth centuries the production of silk and cotton in the western Mediterranean also rendered these much-needed raw materials available to artisans in the basin: Mazzaoui, *Italian cotton*, 138.

70. Munro, "Medieval woolens," 228–324; Pirenne, *Histoire économique*, 390–94; Allsen, *Culture and conquest*, 101–2.

71. Bautier estimates that the wool available to the weavers and cloth manufacturers in the fourteenth and fifteenth centuries was roughly three to five times greater than had been the case in the golden age of the Frenco-Flemish industry: *Economic development*, 210.

72. Abulafia, *Mediterranean emporium*, 216–19.

73. Mazzaoui, *Italian cotton*, 134–35; Thomson, *Distinctive industrialization*, 29–31; Alpine and Apennine wool was not sufficient to satisfy the needs of the industries in Tuscany and Lombardy.

74. Hunt, *Medieval super companies*, 57–58.

75. Bautier, *Economic development*, 213; on the jump in demand for luxury clothing from the thirteenth century and on the linen industry in Swabia: Spufford, *Power and profit*, 116–19, 254–55.

76. "In the 1400s, Venice enforced a monopoly over cotton supplies to many hinterland cities by obligating them to buy only from her": Mazzaoui, *Italian cotton*, 46.

77. Wright, "Nature," 196–98.

78. Mazzaoui, *Italian cotton*, 158.

79. On the Flanders–Milan/Florence axis of the woolen textile manufacturing in the thirteenth century: Lopez, *Commercial revolution*, 130–37; on the impact of demographic decline in Egypt on industrial production: Lopez et al., "England to Egypt," 116–17.

80. Chassagne, *Coton et ses patrons*.

81. Ashtor, *Social and economic history*, 319; McNeill, *Venice*, 76; Faroqhi, "Notes," 405–17; Evliyâ Çelebi, *Seyahatnâme*, 8: 54, 59, 83, 94, 124, 140, 146, 165–66, 247, 262–63, 337.

82. Mazzaoui, *Italian cotton*, 44–45.

83. Thomson, *Distinctive industrialization*, 26–31; Ashtor, "Catalan cloth," 227–57.

84. Ashtor, *Social and economic history*, 310.

85. Turan, *Türkiye-İtalya*, 100.

86. Mazzaoui, "Cotton industry," 267.

87. Laiou, "Agrarian economy," 324, 327.

88. Wee, "Structural changes," 24–25; Melchior-Bonnet, *Mirror*, ch. 1; Crouzet-Pavan, *Venice triumphant*, 171–82; McCray, *Glassmaking*; Bloom, *Paper*. On the development of the Genoese woolen industry in the fifteenth century: Heers, "La mode," 1114–17.

89. Mazzaoui, *Italian cotton*, ch. 7.

90. Bautier, *Economic development*, 213; in 1470, the yearly gross value of business conducted via the Fondaco dei Tedeschi was a million ducats: Crouzet-Pavan, *Venice triumphant*, 119–20; Stromer, "Un clé du succès des maisons de commerce," 29–49.

91. Mazzaoui, *Italian cotton*, 137.

92. Spufford, *Power and profit*, 332; Stromer, "Nuremberg," 216–17.

93. Thomson, *Distinctive industrialization*, 29–31.

94. Epstein, *Genoa*, 232.

95. Dotson, "Problem of cotton," 58–59.

96. Lane, *Venice*, 309–10.

97. Braude, "International competition," 437–51; Le Flem, "Vraies et fausses splendeurs," 525–36; Todorov, *Balkan city*, ch. 13.

98. Houtte, *Economic history*, 180–81; Phillips and Phillips Jr., *Spain's golden fleece*, 181–82.

99. Le Flem, "Vraies et fausses splendeurs," 525–36; Wilson, "Cloth production." From about 1610, the Spanish woolen manufacturers found it difficult to market their goods abroad as competition faced at home from imported cloths made matters worse: Phillips and Phillips Jr., *Spain's golden fleece*, 202.

100. Rapp, "Unmaking," 502–3; Munro, "Spanish *merino* wools," 475–76.

101. Braudel et al., "Le déclin," 58–62; Faroqhi, "Crisis and change," 455–56.

102. Kriedte, *Peasants, landlords*, 75; Çizakça, "Price history," 259–60.

103. Çizakça, "Incorporation of the Middle East," 35–78.

104. Kostis, "Structure sociales," 110.

105. For prices in Bursa: Çizakça, "Price history," 259: fig. 10.3; in Salonika: Braude, "International competition," 442, 446: graph II; and in Segovia: Phillips and Phillips Jr., *Spain's golden fleece*, 310–12: tab. A4.1.

106. Kriedte, *Peasants, landlords*, 72; Vries, *Economy of Europe*, 18–19: fig. 3.

107. Braudel, *Civilization and capitalism*, 2: 309–14; Thomson, *Distinctive industrialization*, 35.

108. Kriedte, *Peasants, landlords*, 74.

109. Thomson, *Distinctive industrialization*, 40; on the transformation of the Genevan silk industry into a rural cottage industry: Mottu-Weber, *Économie et refuge*.

110. Blanchard, "Long sixteenth century," 26.

111. Braudel, *Identity*, 2: 544; Le Roy Ladurie, *Peasants*, 70–71.

112. Phillip and Phillip Jr., *Spain's golden fleece*, 204.

113. La Force Jr., *Development*, 22–23; Thomson, *Distinctive industrialization*, 39.

114. La Force Jr., *Development*, 15–16.

115. Braudel, *Identity*, 2: 511.

116. Davis, "English imports," 202.

117. Braudel et al., "Le declin," 28–35.

118. Slicher van Bath, *Agrarian history*, 206–21; Pamuk, *Monetary history*, 123: graph 7.2; Berov, "Changes in price conditions," 168–78.

119. Genç, "Ottoman industry," 68–78.

120. Sereni, *History*, 206–9.

121. Wallerstein, *Modern world-system*, 2: 194–95; on the extent of cottage industries: Kellenbenz, "Industries rurales," 833–82.

122. For the sudden turnaround in the fortunes of the woolen industry during the seventeenth century: Bilinkoff, *Avila of Saint Teresa*; Kostos, "Structure sociales," 108–9.

123. Braudel, *Identity*, 2: 556.

124. Braudel indeed identifies the waning of the Mediterranean with widespread industrialization on its shores: Braudel, "Remarques," 427–30.

125. Panova, "Zur frage der handelsbeziehungen," 103–14.

126. Johnson, *Life and death*, 12; Thomson, *Clermont-de-Lodève*, 6–7.

127. Kostis, "Salonique," 131–42; Palairet, *Balkan economies*, 38–39; Todorov, *Balkan city*, ch. 13.

128. Gelabert, "Urbanization," 202 n. 70.

129. Ciriacono, "Silk manufacturing," 125.

130. It is, in fact, from the vantage-point of rural industry and trade, and not from that of the cities of Spanish Lombardy, that Sella depicts the region as still economically vibrant in the seventeenth century: Sella, *Crisis and continuity*. With the widespread dissemination of manufacturing, former boundaries between guilds suffered an erosion. Complaints of guilds against the infringement of their privileges by other guilds was an increasingly frequent feature of the seventeenth and eighteenth centuries: candle-makers complained of butchers selling tallow; the makers of vinegar complained of others who had recently embarked on production and sale of vinegar; the weavers of *boğasi* and *dimi*, two varieties of silk cloth, tried to prevent the weavers of *boğasi* from producing *dimi*, and vice versa: Baer, *Fellah and townsman*, 178–79.

131. Nomads, after all, lived in a universe "draped" in textiles, where walls, furniture, and doors of permanent habitats were replaced by carpets, curtains, cushions, and tents. They were a "moving feast" for the textile industry, for the textiles wrapped not only their bodies, but their homes: Golombek, "Draped universe"; also Żygulski Jr., *Ottoman art*, 160, 173.

132. Todorov, *Balkan city*, ch. 13; Casey, *Kingdom of Valencia*, 80.

133. Braudel, *Identity*, 2: 448–49.

134. Thomson, "Take-off," 703–5.

135. Quataert, *Ottoman manufacturing*, 27–29, 49–52.

136. Léon, "La réponse," 231; Owen, *The Middle East*, 7.

137. Fukasawa, *Toilerie et commerce*, 52–53; Ergenç, "XVIII. Yüzyılda Osmanlı sanayi," 519–22; Palairet, "Désindustrialisation," 254–65; Oğuz, *Türkiye halkının*, 4: 278, 284–87.

138. Genç, "17.–19. Yüzyıllarda sanayi," 276–87; Doğru, *XVIII. Yüzyıla kadar*, 153–55.

139. Nieuwenhuis, *Politics and society*, 103.

140. Issawi, *Economic history*, 152–53.

141. Stoianovich, "Balkan peasants," 26–27; Abdel Nour, "Notes," 321–25; Lawless, "Economy and landscapes," 521–22.

142. McNeill, *Venice*.

143. Archives Nationales, Paris [AN], Affaires Étrangères [Aff. Étr.], B¹ 1028, Seyde, le 26 avril 1751; B¹ 1029, Seyde, le 31 décembre 1753; B¹ 1039, Seyde, le 26 juillet 1782; and B¹, Tripoly, le 25 mars 1745.

144. AN, Aff. Étr., Bi 1041, Seyde, le 31 décembre 1789; Cohen, *Palestine*, 23. On the role played by the janissaries in securing the flow of cotton to urban industries in the eighteenth century: Bodman Jr., *Political factions*, 63.

145. Genç, "XVIII. Yüzyılda Osmanlı sanayi," 212–22; La Force, "Royal textile factories," 338–39; Wallerstein, *Modern world-system*, 2: 28–29.

146. Heyberger, *Les chrétiens*, 24; Dalsar, *Türk sanayi*, 162, 365; Yılmaz, "XVIII. Yüzyılda Osmanlı," 1662–64; Demir, *Antakya*, 130, 133–34, 139, 142–43; Oğuz, *Türkiye halkının*, 4: 140–43.

147. Affaires Étrangères (Paris), Correspondance commerciales, Tripoly 15, le 22 juin 1811; Tripoly 15, fin mai 1812; Tripoly 15, le 11 novembre 1812; Tripoly 18, le 25 janvier 1827.

148. Bodman Jr., *Political factions*, 28.

149. Ali Bey, *Travels*, 2: 307.

150. Casey, *Kingdom of Valencia*, 85; Ciriacono, "Silk manufacturing," 178–99.

151. Yüksel, *Osmanlı sosyal*, 103–12; Yediyıldız, *Institution du vaqf*, 95–96.

152. Braudel, *Identity*, 1: 44–49.

153. Landsteiner, "Crisis," 323–34.

154. Phillips, *Short history*, ch. 4.

155. Pounds, *Historical geography*, 170.

156. Houston, *Western Mediterranean world*, 92–93, 213–14, 221.

157. Brown, *In the shadow of Florence*.

158. Lamb, *Climatic history*, 436–37, 460–61.

159. Sereni, *History*, 214. In the Ottoman empire, the mixing of usufruct and proprietary rights over the arable and the tree crops planted on them created a series of legal problems: *Fetava-i Feyzullah*, 398, 416–18.

160. Houston, *Western Mediterranean world*, 532.

161. Fagan, *Little Ice Age*, ch. 1.

162. Craeybeckx, *Un grand commerce*; Pirenne, "Un grand commerce," 588–609.

163. Carter, "Cracow's wine trade," 553–70.

164. Phillips, *Short history*, ch. 4; Lachiver, *Vins, vignes*, 57–110; Hunt and Murray, *History of business*, 48.

165. Pounds, *Historical geography*, 279–80; Pfister et al., "Documentary evidence," 55–110; on the increasing share of Spanish, Portuguese, and French wine in Cracow, an important relay center: Carter, "Cracow's wine trade," 553.

166. On the rise of beer consumption: Montanari, *Culture of food*, 115; Unger, *Beer*, 127–33.

167. Montanari, *Culture of food*, 123–24.

168. Houtte, *Economic history*, 196; Fagan, *Fish on Friday*, 241–44.

169. Childs, *Anglo-Castilian trade*.

170. Rebora, *Çatal kültürü*, 96–97; Flandrin, "Le goût," 369–401.

171. Laurioux, "Medieval cooking," 301.

172. Olive oil was used not for cooking only, but also for illumination, soap-making, and cloth manufacture. Likewise, vineyards were not put to use solely for viniculture. Grape syrup, vinegar, dried raisins, and varieties of grape juice (such as the popular *müselles* and *şıra*) were equally important in Ottoman diets: Goffman, *İzmir*, 42–43; Üçel-Aybet, *Avrupalı seyyahların*, 151; Busbecq, *Turkish letters*, 53–54. In the sixteenth century, slaves on Ottoman ships were given vinegar and dried raisins, along with biscuit and olive oil: *Osmanlı'da bir köle*, 170; Ünsal, *Ölmez ağacın*, 48; Arıkan, "Ayvalık," 585–86.

173. Nader, *Liberty*, 63; Olagüe, *L'Espagne*, 277–78.

174. Phillips, *Cuidad Real*, 41–43; in the eighteenth century, Livorno's wine imports originated in Catalonia, Languedoc, Provence, Naples, and Corsica: Filippini, "La mer," 298–99; Asdrachas, "Le surplus rural," 39.

175. Houston, *Western Mediterranean world*, 143; the spread of olive trees suffered drawback but did not slow down despite the hard freeze that killed forty thousand trees in Tendilla in 1571: Nader, *Liberty*, 166.

176. Roth, *House of Nasi*, 81–82, 183; Tavernier, *Topkapı Sarayında*, 138; Thévenot, *1655–56'da Türkiye*, 218; on the increasing prominence of *şarapdâr*, the imperial *sommelier*, under

Suleiman the Lawmaker in the second half of the sixteenth century: Bilgin, *Osmanlı saray mutfağı*, 22–23. On Spain: Hamilton, *American treasure*, 242.

177. "The olive of Cádiz, the Levant, the Morea and Crete invaded the kitchens and soapworks of Provence." Le Roy Ladurie, *Peasants*, 229–30; Triantafyllidou, "L'industrie du savon," 85–87; Ze'evi, *Ottoman century*, 106–68. In Crete in the eighteenth century, olive oil came to replace wine as the island's principal export: Green, *Shared world*, 118. On the share of oil imported into Marseille from the Ottoman lands: Boulanger, *Marseille*, appendix, lix–lxii.

178. For the extent of oleiculture in the Aegean in the seventeenth and eighteenth centuries: Evliyâ Çelebi, *Seyahatnâme*, 8: 294; Tournefort, *Tournefort seyahatnamesi*, 1: 61–65, 74, 103, 126, 135–36, 167, 200, 219, 234, 244; 2: 9, 254, and in New Castile: Olagüe, *Histoire d'Espagne*, 277–78; Rebora, *Çatal kültürü*, 96. On olive cultivation and vine-growing in the Black Sea coast: Sakaoğlu, *Çeşm-i cihan*, 75; and on olive oil, currant, and silk imports by the English merchants: Pagano de Divitiis, *English merchants*, ch. 4.

179. Camperosi, *Magic harvest*, 88; Chorley, *Oil, silk*, 19–20, 33; Romano, *Commerce et prix de blé*, pt. I.

180. Grandin, "La savonnerie traditionelle," 145.

181. On olive oil and wine trade: Constable, *Trade and traders*, 182–83, 213–15; Constable, *Housing the stranger*, 141–44, 276–78.

182. Wallerstein, *Modern world-system*, 2: 13–14.

183. Valensi, *Tunisian peasants*, 114.

184. Le Roy Ladurie, *Peasants*, 223–27; on the role of mulberry and olive trees and vines in foothill villages in the Balkans: Lawless, "Economy and landscapes of Thessaly," 520; on growing currant exports from the Greek peninsula and the Aegean: Vergopoulos, *Le capitalisme difforme*, 117; Goffman, *İzmir*, 46–48.

185. Le Roy Ladurie and Goy, *Tithe*, 112.

186. Romano, "Italy in the crisis," 188; Georgelin, *L'Italie*, 81–85.

187. Houston, *Western Mediterranean world*, 593–94, 435, 214; Braudel, *Identity*, 1: 45; Sereni, *History*, 251, 261; on Apulia: Galt, "Social class," 424–25. On oil exports from Spain and the Italian peninsula to Marseille: Boulanger, *Marseille*, 49–59.

188. On the marriage of vines to trees: Sereni, *History*, 96, 98, 136; Planhol, "Grandeur et décadance," 314–29.

189. Braudel, *Mediterranean*, 2: 249.

190. Yüksel, *Osmanlı sosyal*, 113; by the eighteenth century, most of these holdings had changed hands: Yediyıldız, *Institution du vaqf*, 95.

191. Slicher van Bath, *Agrarian history*, 217, 236.

192. Yüksel, *Osmanlı sosyal*, 112.

193. Gerber, *Social origins*, 33–36; Özdeğer, *1463–1640 yılları Bursa*.

194. Orhonlu, *Osmanlı imparatorluğu'nda aşiretlerin*, 4–6.

195. Camperosi, *Magic harvest*, 88, 103.

196. Yerasimos, *Sultan sofraları*, 13–17; Ünsal, *Ölmez ağacın*, 49–55; Saraçoğlu, *Bitki örtüsü*, 67–74; Artan, "Aspects," 145–53.

197. Leeuwen, *Notables and clergy*, 11 n. 9; on the recurrence of soap buying in the settlement of debts: Rafeq, "Impact of Europe," 431; Rafeq, "Damascus," 674–75; Ashtor, "Profits from trade," 251–53. The city of Antioch had a Gate of Olives: Demir, *Antakya*, 146, 150, also 134, 142–43.

198. Ze'evi, *Ottoman century*, 132.

199. Notables based on or at the environs of the Anatolian plateau, Amasya and Yozgat, enjoyed modest fortunes in the amounts of 28,000 and 82,000 *guruş*, respectively, the latter more so because livestock occupied a significant part in their fortunes. On the western coast of Anatolia, politically less known Müridoğlu, whose wealth was based on olive oil trade, commanded over 690,000 *guruş*: Sakaoğlu, *Anadolu derebeyi*, 109–13; Cezar, "Bir âyanın muhallefatı"; Faroqhi, "Wealth and power," 92–93. He owned over 19,000 olive trees, young and old. Even lesser known notables had, in 1723–24, come in possession of significant numbers of olive trees: Abdülkadir Efendi of Hama owned a total of 11,598 olive trees, not to mention countless mulberry and pomegranate trees. The Zuhrâwî family of Aleppo thrived thanks to their part in olive oil production and trade: Archives of the Directorate of Pious Foundations: *736*: 151 (n° 75); Salati, "Urban notables," 193–94. Panayot Benakis, notable of Kalamatas in Greece, owed his wealth to his orchards (of olive, mulberry, and fig trees) and gardens. His arable land, which was 37 hectares, was smaller than his orchards, 54 hectares: Veinstein, "Le patrimoine foncier," 228–31. In silk-producing Mount Lebanon, Yusuf Karam left an estate worth 100,000 *guruş* behind in the 1860s; those who were engaged in silk trade in the eighteenth century were wealthier: AE, CC, Registre no 18, "Inventaire des biens que possède cheikh Youssef Karam," le 12 mars 1867; Chevallier, "Que possédait un cheikh" and also his *La société du Mont Liban*, 147. One thing common among them was they all engaged in money-lending, and in fact, in most instances, over half of their revenues originated from this activity.

200. Owen, *Middle East*, 65–66, 92; on the relative rise in agricultural prices versus industrial prices to the benefit of the latter at the turn, and the conjuncture of the early half, of the nineteenth century: Rostow, *World economy*, 113, 123–24; Schilcher, *Families in politics*, 75–78.

201. Herr, *Rural change*, 736–37.

202. Marino, *Pastoral economics*, 68, 71; Orhonlu, *Osmanlı imparatorluğu'nda aşiretlerin*. The decline of livestock industry was more pronounced in the Mediterranean than in the northern regions: Braudel, *Identity*, 2: 194–97.

203. Phillips and Phillips Jr., *Spain's golden fleece*, ch. 3.

204. Gonzáles, "Andalusia," 134–35.

205. Erder and Faroqhi, "Population rise," 322–45; Pascual, "Janissaries," 359–65.

206. Le Roy Ladurie and Goy, *Tithe*, ch. 8; Neveux, "Dîme et production," 512–18.

207. Berov, "Changes in price conditions," 175–76; Weisser, "Agrarian depression," 154; Wallerstein, *Modern world-system*, 2: ch. 1; Braudel et al., "Le decline," 29: tab. 1, 31.

208. McGowan, *Economic life*, 38–39; Carter, "Commerce," 376–83.

209. Gerber, *Social origins*, 64.

210. Hütteroth, "Role," 343–54; Weisser, "Agrarian depression," 149–54; on Nasuh Pasha's livestock holdings: Öz, *Osmanlı'da alevi*, 54–55.

211. Braudel, *Identity*, 2: 293–95.

212. Braudel, *Identity*, 2: 194–97.

213. Archives of the Directorate of Pious Foundations: *328* (1731), 124; *355* (1710), 19; *335* (1760), 208–9; *335* (1760), 267–68; *335* (1760), 213–14; *335* (1760), 266–67; *377* (1786), 121; *376* (1819), 269.

214. Phillips and Phillips Jr., *Spain's golden fleece*, ch. 5; Gerber, *Social origins*, 31–33; Raymond, *Osmanlı döneminde*, 132–33.

215. Archives of the Directorate of Pious Foundations: *355* (1714), 233; *352* (1774–75), 97–

98; *352* (1784), 76; *352* (1774–75), 97–98; *352:* (1784), 76; *369* (1786), 40–41; *369* (1786), 45; *377* (1786), 114–15; also Klein, *Mesta,* 341–42.

216. Phillips and Phillips Jr., *Spain's golden fleece,* 72; Darling, *Revenue-raising,* 110; Marino, *Pastoral economics,* 137–39.

217. Slicher van Bath, "Les problèmes fondamentaux," 33.

218. Orhonlu, *Osmanlı imparatorluğu'nda aşiretlerin,* 7–11; Klein, *Mesta.*

219. Wallerstein, *Modern world-system,* 2: 13–17.

220. On the salience of forage crops from the sixteenth century: Ambrosoli, *Wild and the sown,* 101–2, 202–3.

221. On terrestrial routes fanning north and northwest from Istanbul and Edirne: Atasoy et al., *İpek,* 177.

222. The provision and storage of animal feed is diligently spelled out in most foundation deeds: Archives of the Directorate of Pious Foundations: *736* (1723–24), 151; *580* (1753–54), 23; *512* (1764), 391; *513* (1769), 668; *368* (1775), 387–88; *587* (1813), 39.

223. Stoianovich and Haupt, "Le maïs"; Sereni, *History,* 219.

224. Sarpaki, "Palaeoethnobotanical approach"; Wright, *Mediterranean feast,* 11–12; Glick, *Irrigation,* 11–30.

225. On the organization of rice production: İnalcık, "Rice cultivation," 69–141.

226. Aymard, "Rendements et productivité," 475–97; Sereni, *History,* 235.

227. Braudel, *Civilization and capitalism,* 2: 265–72.

228. Wallerstein, *Modern world-system,* 2: 135.

229. Kirk, *Genoa,* 34–35.

230. Goffman, *Ottoman empire,* 199–206.

231. Rapp, *Industry,* 10; Braudel et al., "Le déclin," 60–61.

232. Phillip and Phillip Jr., *Spain's golden fleece,* 183–85; Mathers, "Family partnerships."

233. Braudel, *Identity,* 2: 558.

234. İnalcık, "Ottoman state," 196–99. On the resultant increase in piratic activity: Fontenay, "La place de la course," 1321–47.

235. Braudel, *Civilization and capitalism,* 3: 328, 331.

236. Goffman, *Ottoman empire,* 192.

237. Boyajian, *Portuguese trade,* 177–79; Kamen, *Empire,* 343.

238. Kirk, *Genoa,* 254 n. 95.

239. Kirk, *Genoa,* ch. 5.

240. Faroqhi, "Crisis and change," 584–85; McGowan, *Economic life,* 32–44; Raymond, *Artisans et commerçants,* 2: ch. 5.

241. Barbir, *Ottoman rule,* ch. 3; Faroqhi, *Pilgrims and sultans,* 79–82; Flynn and Giráldez, "Cycles of silver." The significance of pilgrimage trade on the region's urban economies is underlined by Raymond, *Osmanlı döneminde,* 21–22.

242. Pamuk, *Monetary history,* 99–101.

243. Eldem, *French trade;* Crecelius and 'Abd al-Aziz Badr, "French ships," 251–86; Dogo, "Merchants," 87–89.

244. Braudel, *Identity,* 2: 565; on the implications of the insufficient volume of exportable surplus: Keyder, *State and class,* 29–32.

245. Stoianovich, "Conquering Balkan orthodox merchant"; Ringrose, *Spain, Europe,* 380–81; Braudel, *Identity,* 2: 561–62.

246. Kasaba, *Ottoman empire*, ch. 6.

247. Eldem, *French trade*, chs. 5–6.

248. Kafadar, "Death in Venice," 191–218; Earl, "Commercial development," 28–44; Carter, "Commerce," 370–94; Braunstein, "Le commerce," 1270–78.

249. Godinho, "Le repli vénetien," 283–300.

250. Malowist, "Le commerce du Levant," 349–56; Montanari, *Culture of food*, 110–16.

251. For the silver flows of the 1550–1650 period: Flynn and Giráldez, "Cycles of silver," 399–405.

252. Pagano de Divitiis, *English merchants*, ch. 3.

253. Phillips and Phillips Jr., *Spain's golden fleece*, 188.

254. Ringrose, *Spain, Europe*, 89–90.

255. Braudel, *Identity*, 2: 558.

256. Carter, *Trade and urban development*, 302–15.

257. Steensgaard, *Asian trade*, ch. 9; Malowist, "Commerce du Levant," 349–56; Atasoy et al., *İpek*, 177; İnalcık, "Ottoman state," ch. 12.

258. Herzig, "Iranian raw silk," 73–89.

259. Wake, "Changing patterns," 361–403. The number of spices consumed in Istanbul rose from the sixteenth to the seventeenth centuries: Barkan, "İstanbul saraylarına"; Yücel, *1640 tarihli es'âr defteri*; Artan, "Aspects," 121.

260. Kołodziejczyk, "Export of silver," 105–12; Stoianovich, "Conquering Balkan orthodox merchant," 248–54.

261. Stoianovich, "A route type"; Kellenbenz, "From Melchior," 611–22.

262. Braudel, *Identity*, 2: 478, 557.

263. İnalcık, "Osmanlı pamuklu," 16–37.

264. Goffman, *İzmir*, 25–49; Faroqhi, "Crisis and change," 505–6.

265. Braudel, *Identity*, 2: 564.

266. Ringrose, "Impact," 761–91.

267. Braudel, *Civilization and capitalism*, 3: 331.

268. Panzac, *La peste*, 119–28. See Yerasimos's introduction to Tavernier, *Tavernier seyahatnamesi*, 40–42.

269. Tadić, "Le commerce en Dalmatie," 250–54; Berov, "Transport costs," 89–90.

270. Kafadar, "Death in Venice," 197–98.

271. Hagen, "Seventeenth-century crisis," 302–35.

272. "Part of the surplus of the Baltic trade must have reached Europe by a series of linked exchanges between eastern, central and western Europe, but in this case overland through Poland and Germany. While the western balance was in deficit with the northern ports, it may have been partially compensated by a favorable balance in overland traffic—the payments being effected by way of Leipzig fairs." Braudel, *Mediterranean*, 2: 214–15, 93; also Beachy, *Soul of commerce*, 32–37.

273. Braudel, *Out of Italy*, 16.

274. Şen, *Osmanlı panayırları*, 69–84; Palairet, "Désindustrialisation," 255–56; Paskaleva, "Contribution," 275–77; Ciercerska-Chapowa, "Échanges commerciaux," 261–87.

275. Braudel, "En Méditerranée," 452; Lawless, "Economy and landscapes," 520–21; Braudel, *Civilization and capitalism*, 2: 157.

276. Dogo, "Merchants," 85–96.

277. Barrett, "World bullion flows," 250–51; on these two waves of silver flow to China via the Mediterranean (1540–1640 and 1700–1750): Flynn and Giráldez, "Silver," 25–28, 36–37.

278. Şen, *Osmanlı panayırları*, ch. 4; Stoianovich, "A route type," 210–14.

279. Baroque spread north from the Italian peninsula, and "turquerie" spread west: Braudel, *Out of Italy*, 122–36; Desmet-Grégoire, *Büyülü divan;* also Arel, "Gothic towers."

280. Felloni, *Gli investimenti*, ch. 8.

281. Boulanger, *Marseille*, 197.

Four • Reversal in the Fortunes of the Plains

1. Mack Smith, *Medieval Sicily*, 1: 183–86; Bennassar, *Valladolid*, 1: 39–51; Berthier, *Sucreries de Maroc*, 277.

2. Cohn Jr., "Inventing Braudel's mountains," 387; Houston, *Western Mediterranean world*, 125; Samsonowicz, "Relations commerciales," 38–39; Obuchowska-Pysiowa, "Trade between Cracow and Italy," 633–53.

3. Nencini, *Florence*, 126–33; Braudel, *Mediterranean*, 1: 63–68.

4. Grove and Rackham, *Nature*, 337; also Walling and Webb, "Patterns of sediment yield," 74: fig. 4.2; McNeill, *Mountains*, 340–41.

5. Sereni, *History*, 139.

6. Rodrigo and Esteban-Parra, "Attempt," 397–418; Emecen, *XVI. Asırda Manisa*, 168, 171, 174, 204, 206, 209, 227–28.

7. Pfister and Glazer, "Climatic variability," 5–53.

8. On the increase in the number of floods during the Little Ice Age: Grove and Rackham, *Nature*, 131–33; Braudel, *Identity*, 2: 239.

9. Sereno, "Ecology," 310.

10. Appuhn, "Inventing nature," 865–66; on the extent of deforestation in France in the first half of the sixteenth century: Devèze, *La vie*, 2: 9–55.

11. Nader, *Liberty*, 166.

12. Delano Smith, *Western Mediterranean Europe*, 162; Yusuf, *Economic survey*, 28–29; İnalcık, "Ottoman state," 120–26; Saraçoğlu, *Akdeniz*, 247–48.

13. Thirgood, *Man and the Mediterranean*, 47–48; Musgrave, *Land and economy*, 102–3. It needs to be underlined here that the search for naval timber is admittedly not a good index of general deforestation, because navies were in search of special sorts of timber. Commercial shipbuilders were less particular.

14. Appuhn, "Inventing nature," 879–88; Braudel, *Mediterranean*, 1: 131; Lane, *Ships and shipbuilders*, 230–33.

15. Phillips, *Six galleons*, 21, 79, 81; Bennassar, *Vallodolid*, 1: 34–35.

16. Houston, *Western Mediterranean world*, 113–14; Goodman, *Spanish naval power*, ch. 2.

17. Imber, "Navy of Süleyman," 230–31, 269–75; Bostan, *Osmanlı bahriye*, 102–24; Planhol, *L'Islam et la mer*, 208–11 T. C. Çevre ve Orman Bakanlığı; *Osmanlı ormancılığı*, 1: 3–11. İbn Battûta Tanci, *İbn Battûta*, 1: 402, refers to the timber trade. In the eighteenth century, timber from Karaman was shipped to Crete and Egypt. Triantafyllidou, "L'industrie du savon," 85–87; Faroqhi, *Towns and townsmen*, 147. It was during the Seljuk era that the "Pontic and Tauric coastal ranges of Anatolia" were opened up to lumbering: Redford, *Landscape*, 107.

18. Galloway, "Mediterranean sugar," 187–88; Houston, *Western Mediterranean world*, 113.

19. Phillips and Phillips Jr., *Spain's golden fleece*, 49–51; Thirgood, *Man and the Mediterranean*, 32. Pastoralists leveled the landscape by burning, but also helped "retain that form, despite pressures by forests to generate." Pyne, *Vestal fire*, 94, 99, 102–3.

20. İnalcık, "Ottoman state," 34–41; Orhonlu, *Osmanlı imparatorluğu'nda aşiretler*, 17; Emecen, *İlk Osmanlılar*, 180–85.

21. Gökçe, *XVI ve XVII. Yüzyıllarda Lâzıkiyye*, 286–304; Özel, "Population changes," 187–88.

22. Refik, *Onuncu asr-ı hicri'de*, 113; Bostan, *Osmanlı bahriye*, 102–18.

23. Bostan, *Beylikten imparatorluğa*, 198.

24. Emecen, *İlk Osmanlılar*, 110–12; Refik, *Anadolu'da Türk aşiretleri*, 14 n. 26.

25. Öz, "Agriculture in the Ottoman empire," 36; Emecen, *Manisa kazası*, 154–57, 226–27; see also Gökçe, *XVI ve XVII. Yüzyıllarda Lâzıkiyye*, 308–14.

26. Gökçe, *XVI ve XVII. Yüzyıllarda Lâzıkiyye*, 343.

27. Barkey, *Bandits and bureaucrats*, 245–46.

28. Orhonlu, *Osmanlı imparatorluğu'nda aşiretler*, 37 n. 33.

29. On Spain: Nader, *Liberty*, 23; Braudel, *Identity*, 2: 239–41.

30. Arıkan, "XV.–XVI. Yüzyıllarda Anadolu'da çeltük," 479–80; also Kurt and Erdoğru, *Çukurova*, xxxvii; Ener, *Tarih boyunca*, 212.

31. Emecen, *XVI. Asırda Manisa*, 258–60; Gökçe, *XVI ve XVII. Yüzyıllarda Lâzıkiyye*, 383–85; Evliyâ Çelebi, *Seyahatnâme*, 8: 60.

32. Kütükoğlu, "XVI. Yüzyıl İzmir," 497–504; Emecen, *XVI. Asırda Manisa*, 125, 226, 237; Cook, *Population pressure*, 87–95; İnalcık, "Ottoman state," 34–35.

33. Foss, *Ephesus*, 171–72; for similar developments that took place in the southern Alps after the 1550s: Sclafert, *Cultures en Haute Province*, 178–79.

34. "Malaria is largely responsible for the decline of the city [Aya Solug/Ephesus] under the Ottomans": Foss, *Ephesus*, 186.

35. Arıkan, "XIV.–XVI. Yüzyıllarda Ayasulug," 121–77; Goffman, *İzmir*, 50–76.

36. On the details of the emergence of Kuşadası: Kiel, "Kuşadası," 403–16.

37. Arıkan, "XIV.–XVI. Yüzyıllarda Ayasulug," 121–77; Emecen, "Ayasuluk," 226–27.

38. Foss, *Ephesus*, 178–79.

39. On the historical frequentation by malaria of the Roman Campaign: Celli, *History of malaria*.

40. Horden and Purcell, *Corrupting sea*, 178–82.

41. Crouzet-Pavan details the process of the city's disappearance in her *La mort lente de Torcello*.

42. On the impact of soil erosion in north-central Italy: Musgrave, *Land and economy*, 80.

43. Grove and Rackham, *Nature*, 335; Russell, "Geomorphology," 198–201.

44. See the excellent discussion in Grove and Rackham, *Nature*, 135–36, 336; Horden and Purcell, *Corrupting sea*, 192; on the impact of this fluvial environment on the Vistula region: Dunin-Wąsowicz, "Environnement et habitat," 1026–42.

45. See the articles in *Villages désertés*; Vryonis Jr., "Nomadization," 41–71.

46. McNeill, "Great powers," 192; Duby, *Rural economy*, 297–300.

47. On the increase in the number of rural settlements in the former Roman world: Randsborg, *First millennium*, 68–69. In Macedonia, farming the slopes pushed back the forest between the eleventh and fourteenth centuries: Lefort, "Rural economy," 270.

48. Wickham, *Mountains*, 6; Lefort, "Rural economy," 265; Davies, "Patterns of transhumance," 155–68.

49. Appuhn, "Inventing nature," 865.

50. Lamb, *Climate history*, ch. 13; Grove, *Little Ice Age*, 415–19; Xoplaki et al., "Variability of climate," 604–69.

51. Delano Smith, *Western Mediterranean Europe*, 190.

52. Jennings, "Population, society," 149–250; Ogilvie, "Physiography and settlements," 191–93; on the alluviation of the lower Danube and Carpathian valleys, and on Konya plains, from the seventeenth to the nineteenth centuries: Grove, *Little Ice Age*, 388; Ali Bey, *Travels*, 1: 312.

53. Arıkan, "Osmanlı imparatorluğu'nda ihracı." The decrease in tithes collected in religious foundations during this period is underlined by Güran, *Ekonomik ve malî*, 65.

54. Le Conte de Dienne, *Histoire du desséchement*, 364–411; Braudel, *Mediterranean*, 1: 61–62.

55. Gerber, *Social origins*, 86; Ener, *Tarih boyunca Adana*, 228; Lawless, "Economy and landscapes," 520.

56. Sereni, *History*, 192; Delano Smith, *Western Mediterranean Europe*, 13.

57. Simpson, *Spanish agriculture*, 63.

58. Grove and Rackham, *Nature*, 133.

59. Braudel, *Mediterranean*, 1: 399.

60. Ener, *Tarih boyunca Adana*, 219–28; Volney, *Travels*, 2: 90–91; Beaufort, *Karamania*.

61. Evliyâ Çelebi, *Seyahatnâme*, 9: 162; Tournefort, *Tournefort*, 2: 246; Arıkan, "1821 Ayvalık isyanı," 571–600; Ancel, *La macédoine*, 38–55; Atasoy and Raby, *İznik*, 23; Ali Bey, *Travels*, 1: 324.

62. Venzke, "Rice cultivation," 175–271; Murphey, "Tobacco cultivation," 205–26; Rabbath, *Documents inédits*, 1: 42.

63. Delille, *Croissance*, 179–87; Camperoni, *Magic harvest*, 6–9.

64. The best depiction of this change in the landscape is by Sereni, *History*.

65. Bloch, *French rural history*, 18–19; on the pace of settlement in Ottoman lands: Barkan, "Osmanlı," 305–86; Doğru, *Eskişehir*, 41–43; İnalcık, "Ottoman state," ch. 7.

66. Jordan, *Great famine*, 159–60; Brodman, *Charity*, 33.

67. Le Roy Ladurie, *Histoire humaine*, 1: 91–181; Nader, *Liberty*, 165.

68. In the following two-and-a-half centuries, the Pontic Sea was not spared of repeated episodes of deep freeze; the final episode took place in 1862: Tchihatchef, *İstanbul ve Boğaziçi*, 146–66.

69. Sereni, *History*, 157.

70. Nader, *Liberty*, 165–66; Catalonia and north and central Italy, on the skirts of the Alps and the Pyrenees, suffered from the first half of the sixteenth century, and Andalusia and Garonne from the latter half: Brazdil et al., "Flood events," 276–78.

71. Davis, *Pursuit of power*, 1–29; Braudel, *Civilization and capitalism*, 3: 120–22.

72. Hütteroth and Abdulfattah, *Historical geography*, 1–16; Öz, "Agriculture," 32–40; Salomon, *La campagne*, 9–15; Gentil da Silva, *En Espagne*, ch. 1.

73. Gentil da Silva, "L'histoire économique," 196–203.

74. Wallerstein and Tabak, "Ottoman empire," 120–27.

75. Grossi, *Alternative to private property*, ch. 8.

76. McGowan, *Economic life*, 58–73.

77. McNeill, "Tragedies of privatization," 222–34.

78. Le Roy Ladurie, *Histoire humaine*, 1: 17–20.

79. Lamb, *Climatic history*, 435–49; Le Roy Ladurie equates the 1500–1560 period with "the return of four seasons": *Histoire humaine*, 1: 158–62.

80. On the misfortunes of the eastern Mediterranean in the 1700–1710 period: Desmet-Grégoire, *Büyülü divan*, 156; Le Roy Ladurie, *Histoire humaine*, 2: ch. 18.

81. Fagan, *Floods, famines*, 186.

82. Variability was high in the 1600–1740 period: Serre-Bachet, "Middle Ages temperature," 221; also Serre-Bachet, "Dendroclimatic evidence," 361–62.

83. Delano Smith, *Western Mediterranean Europe*, 207.

84. Ener, *Tarih boyunca Adana*, 212, 228; Hinderink and Kıray, *Social stratification*, 11–13; Akgündüz et al., *Arşiv belgeleri*, 284–89; Eberhard, "Nomads and farmers," 37.

85. Georgelin, *Venise*.

86. On the contrasting features of Apulia, of large estates and grain, and Campania, of olives, vines, and fruits: Delille, *Famille et propriété*.

87. On cycles of hegemony: Braudel, *Civilization and capitalism*, 3; Wallerstein, *Modern world-system*, 2: ch. 1, 3: ch. 1.

88. İnalcık, "Ottoman state," ch. 14.

89. Farnie, *East and west of Suez*.

90. Ciriacono, "Land reclamation"; Delano Smith, *Western Mediterranean Europe*, 185–91.

91. Houston, *Western Mediterranean world*, 144–62; Sapancalı, *Karaman*, 75–76.

92. For a thorough discussion: Vries, *Economy of Europe*, ch. 1; on Spain: Elliott, "Decline of Spain," 52–75; Yun Casalilla, "Spain," 301–21.

93. Goldstone, *Revolution*, 25–26.

94. On the basic trends of this era: Abel, *Agricultural fluctuations*, 158–93; Slicher van Bath, *Agrarian history*, 206–20; Wallerstein, *Modern world-system*, 3: 13–34.

95. Fagan, *Little ice age*, 23–44. The Little Ice Age came on the throes of increasing seigneurial oppression; the villages were already being deserted. On the emptying of villages in the inland plains in Old Castile: Ruiz, *Crisis and continuity*, 319–22.

96. Dunin-Wąsowicz, "Environnement et habitat," 1026–45.

97. Bautier, *Economic development*, 171, 188–89.

98. Le Roy Ladurie, *Histoire humaine*, 1: 41–62; Lamb, *Climate history*, 195–99.

99. For a comprehensive history of this episode: Jordan, *Great famine*, 7–23; also Lucas, "Great European famine," 343–77; Alexandre, *Le climat*, 805–8.

100. Le Roy Ladurie, *Peasants*, 13; also Larenaudie, "Les famines."

101. Lamb, *Climate history*, 207.

102. Cohn Jr., "Inventing Braudel's mountains," 386–88. The Camargue became increasingly unhealthy: Georges, *La région du Bas-Rhône*, 606; in the French Alps, the mountains opened up economically in the fourteenth and fifteenth centuries, and reverted back to closure at the end of the fifteenth: Bergier, "Le cycle médiéval," 163–264.

103. Herlihy, *Medieval and renaissance Pistoia*, 43, 46–47; Yver, *Le commerce*, 140–41; on measures enacted against erosion in France: Sclafert, *Cultures en Haute Province*, ch. 4.

104. Lefort, "Rural economy," 270.

105. Glick links the crisis of *huerta* agriculture to climatic variability: *Irrigation and society*, 135, 137–45.

106. Grove, "'Little Ice Age,'" 128–29.

107. Tabakoğlu, *Türk iktisat tarihi*, 177, 184; Lamb, *Climatic history*, 145; Lefort, "Rural economy," 269.

108. Geyer, "Physical factors," 43–44.

109. Doğru, *XVI. Yüzyılda Eskişehir*, 90–91; Tunçdilek, "Orta Sakarya," 183–84; Yediyıldız, *Ordu kazası*, 36–37; Bryer, "Late Byzantine rural," 53–59.

110. Montanari, *Culture of food*, 71–82; Braudel, *Civilization and capitalism*, 1: 190–94.

111. Ünver, *Fatih külliyesi*, 80–87; Barkan, "Edirne ve civarındaki," 246–47.

112. Bergier, "Le cycle médiévale."

113. Cohn Jr., "Inventing Braudel's mountains," 368; McNeill, *Mountains*, 86–93.

114. Le Roy Ladurie, *Peasants*, 15; Evliyâ Çelebi, *Seyahatnâme*, 7: 102, 119.

115. Le Roy Ladurie, *Peasants*, 250.

116. Cohn Jr., *Creating the Florentine state*, ch. 3.

117. Barriendos Vallvé and Martín-Vide, "Secular climatic oscillations," 473–87; on the distribution of cold years: Le Roy Ladurie, *Times of feast*, 238; Luterbacher and Xoplaki, "500-Year winter temperature," 138–39: fig. 2; Touchan et al., "Reconstructions of spring/summer precipitation," 95–96; Grove and Conterio, "Climate," 276–84.

118. Petrusewicz, *Latifundium*; Stoianovich, "Land tenure," 398–411; Lawless, "Economy and landscapes," 518–19.

119. Barker, *Biferno*, 292; Sella, "Au dossier des migrations," 547–54; Braudel, *Mediterranean*, 1: 69.

120. Unclaimed land "was to be found mainly in open, swampy areas." Hütteroth, "Role," 348; *çiftliks* and sodden and untended plains went hand in hand: Ancel, *La macédoine*, 12.

121. Gerber, *Social origins*, 67–90; Phillips and Phillips Jr., *Spain's golden fleece*, 89.

122. Pounds, *Historical geography*, 331.

123. Rabbath, *Documents inédits*, 1: 42.

124. Braudel, *Mediterranean*, 2: 100.

125. Greene, "Northern invasion," 51–52; Panzac, "Le contrat d'affrètement maritime," 359–62; Morineau, "Naissance d'une domination," 144–84.

126. Rueda Hernanz, "Disentailment in Spain," 1–31; İslamoğlu and Keyder, "Agenda," 31–55; Owen, *Middle East*, 1–23.

127. Issawi, *Economic history*, 44.

128. Scammel, *World encompassed*, 147.

129. Braudel, *Identity*, 2: 375–78; Frank, "Multilateral merchandise trade," 407–38; Friedmann, "World market," 545–86

130. Day, "Monetary crisis," 43–58.

131. Chaunu, *European expansion*, 290–92; on the depression of cereal prices by the Baltic grain: Tits-Dieuaide, *La formation des prix*.

132. Israel, *Dutch Republic*, 315–18.

133. Öztürk, *Askeri kassama*, 249; for the general direction of prices: Pamuk, *Ottoman monetary*, 123; on Spain: García Sanz, "Castile 1580–1650," 20–21.

134. In the Aegean region, wheat prices started to gradually climb after the mid-eighteenth century: Balta, "Bread in Greek territories," 185: tab. 1, 186: diagram 1.

135. On the vast holdings of the Sursuq family: Owen, *Middle East*, 267–68; on Andalusian large estates: Simpson, *Spanish*, 41–44.

136. Woude, "Future," 243–56; Kasaba and Tabak, "Fatal conjuncture," 88–89.

137. Slicher van Bath, *Agrarian history*, 206–20.

138. On the change in economic conjuncture in the eighteenth century: Revel, "Au XVIIe siècle," 348–61.

139. Delano Smith, *Western Mediterranean Europe*, 295. Provence was known as the *région du feu:* Pyne, *Vestal fire*, 110.

140. See, for instance, Rosenthal, *Fruits*, 52; Braudel, *Mediterranean*, 1: 68–69.

141. Braudel, *Mediterranean*, 1: 64.

142. Delano Smith, *Western Mediterranean Europe*, 194–96. Years of subsistence crisis in Pampólona were 1605, 1606, 1614, 1630, 1631, 1643, 1684, 1692, 1710, 1712, 1793, 1794, and 1796.

143. On how tithes collected from mountain villages in Atlantic Andalusia were consistently higher, from 1550 to the late eighteenth century, than those in the plains: Ponsot, "En Andalousie occidentale," 105–7.

144. In Mediterranean Iberia, for instance, its high points were reached in 1570–1638, 1760–1800, and 1830–70: Barriendos Vallvé and Martín-Vide, "Secular climatic oscillations," 473–87; Le Roy Ladurie argues that a period of cooling that started in 1551 lasted until 1809, with a warm episode at the beginning of the eighteenth century: Le Roy Ladurie, *Times of feast*, ch. 5.

145. Braudel, *Mediterranean*, 1: 63; on droughts and deluges in Crete: Grove and Conterio, "Climate of Crete," 241–42.

146. Çulpan, *Türk taş köprüleri*, pt. ii: ch. 4.

147. Matar, ed., *In the lands of Christians*, 130–31, 141.

148. Sereni, *History*, ch. 56; Enggass, "Land reclamation," 125–43.

149. On the expansion of the 950–1350 period: Bartlett, *Making of Europe*, 292–306; also Slicher van Bath, *Agrarian history*, 132–45; Devroey, "Le céréaliculture," 221–53.

150. The classical account of this environmental change is, of course, by Braudel, *Mediterranean*, 1: 60–85, 399.

151. This is masterfully documented by Hütteroth, who calls inner Anatolia the pioneer fringe: "Influence," 19–47; Delano Smith, *Western Mediterranean Europe*, 13.

152. Briffa, "European tree rings," 151–68; Jones and Bradley, "Climatic variations," 652–53; Lamb, *Climatic history*, 145–46; Grove and Conterio, "Climate of Crete," 241–42.

153. Barriendos Vallvé and Martín-Vide, "Secular climatic oscillations," 473–87; Braudel, *Mediterranean*, 1: 272–73. The plains of the Caudine valley were all ridden with paludial problems in the seventeenth and eighteenth centuries, and this was confirmed by place names such as Paluda, Acquara, and Acqua Vivala: Delille, *Croissance*, 49. On Crete and its malaria lowlands: Brumfield, "Osmanlı Girit'inde tarım," 63. On the relationship between flooding and agricultural production: McCloy, "Flood relief and control," 1–18.

154. Planhol, *De la plaine pamphylienne*, 111; Soysal, "Onaltıncı yüzyılda Adana," 177–78; Amiran, "Pattern of settlement," 198.

155. Houston, *Western Mediterranean world*, 158–62; Curtin, *Death by migration*, 132–40; Zulueta, "Malaria," 7; Bruce-Chwatt, "History of malaria," 20–21.

156. Houston, *Western Mediterranean world*, 152–54.

157. Nader, *Liberty*, 166; Bennassar, *Valladolid*, 1: 39–51.

158. Vilar, *La Catalogne*, 1: 573.

159. Hanna, *Making big money*, 84–97; Raymond, *Artisans et commerçants*, 1:129–56; see also Faroqhi, *Towns and townsmen*, ch. 10.

160. Pascual, "Janissaries," 357–69; Masters, *Origins*, 157–58. In Saruhan, between the end of the sixteenth century and the second half of the seventeenth, even though the number of *timar*s remained unchanged, at 674, the number of *ze'amet*s almost trebled, from 15 to 41: Arıkan, "Un sancak anatolien," 331.

161. Vassberg, *Village*, 164; Nader, "Noble income," 425–26; Casey, *Kingdom*.

162. Le Roy Ladurie, *Peasants*, ch. 3.

163. Telling is that even Ximenez, the Portuguese merchant who was one of the main players in the arrival of northern grain to the Mediterranean in 1590 as we mentioned in ch. 2, repatriated his fortunes to Florence to invest in agriculture. They in fact became Florentines: Braudel, *Mediterranean*, 2: 725–34; Sereni, *History*, ch. 58; Le Roy Ladurie, *Tithe*, 114.

164. Braudel et al., "Le déclin," 28–30.

165. Wallerstein, *Modern world-system*, 2: 14–15.

166. Özel, "Population changes," 190–91; on the numbers of nomadic populations in the Aleppo province in the eighteenth century: Murphey, "Some features of nomadism," 194–96.

167. Darling, *Revenue-raising*, chs. 5–6; Nader, *Liberty*, ch. 4; Murphey, *Ottoman warfare*, 188–89.

168. Dienne, *Histoire du desséchement*, 198–99.

169. On the draining of the marshes of Fucecchio in the lower Arno by the ducal governments, under Ferdinando I, in the late sixteenth century: McArdle, *Altopascio*.

170. Ciriacono, "Land reclamation," 282, 291; Le Roy Ladurie, *Paysans*, 270–71; Holland's farmland grew by one-third, roughly 100,000 hectares, between the late sixteenth and nineteenth centuries. Most of this land was reclaimed between 1600 and 1650: Fagan, *Little Ice Age*, 107.

171. Sereni, *History*, 187–88; Evliyâ Çelebi, *Seyahatnâme*, 9: 78, 109, 162, 207. Evliyâ Çelebi diligently took note of where the air was good and there was no malaria.

172. Kıray, "Social change," 179.

173. Lovett, "Golden age," 745.

174. Houston, *Western Mediterranean world*, 597.

175. Bruce-Chwatt, "History of malaria," 11–19; Malaria hit clerics as well as laypersons. In 1623 as in 1387, seven cardinals died of malaria: Delort, *Les animaux*, 195–96.

176. Braudel, *Civilization and capitalism*, 2: 287. One-tenth of the Roman countryside was under tillage: Romano, "Italy," 189.

177. Bloch, *French rural history*, 19–20.

178. Stoianovich, "Balkan peasants," 26–27.

179. This was not the case in the mid-thirteenth century according to the description of the region by 'Izz al-Dîn ibn Šaddâd in his *Description de la Syrie*, 91–101.

180. Goyau, *Un précurseur*, 56; Rabbath, *Documents inédits*, 1: 221.

181. For the malarial conditions of inland plains: Jennings, "Population, society," 156; see also his *Christians and Muslims*, 188–91; Pyne, *Vestal fire*, 132; Pascual, "Une neige," 59.

182. Evliyâ Çelebi, *Seyahatnâme*, 9: 78, 109, 162, 207; Volney, *Voyage*, 2: chs. 8–13; Swinburne, *Travels*, 85–86.

183. İbn Battûta Tanci, *İbn Battûta*, 1: chs. 10–12, 32; Constable, *Housing the stranger*, 69; İbni Cübeyr, *Endülüsten kutsal topraklara*, 224–32; Le Strange, *Palestine*, 11–24.

184. Crosby, *Ecological imperialism*, 65–67.

185. Volney, *Travels*, 1: 374.

186. See, for instance, Ali Bey's impressions of Palestine: *Travels*, 2: 209.

187. Swinburne, *Travels*, 65.

188. Simpson, *Spanish agriculture*, 65–66.

189. Ringrose, *Spain, Europe*, 190–93.

190. Delano Smith, *Western Mediterranean Europe*, 195.

191. On rural indebtedness along the Syrian littoral and inland: Archives of the Directorate of Pious Foundations: *355* (1710), 19; *355* (1710), 20, *355* (1710), 35; *355* (1710), 43; *355* (1710), 83; *355* (1711), 96; *355* (1711), 128: *355* (1712), 142: *355* (1712), 285; *364* (1717), 88; *364* (1717), 88; *364* (1718), 101; *364* (1718), 175; *364* (1719–20), 258; *360* (1762), 131; *347* (1769–70), 47; *360* (1769–70), 73. See also Le Roy Ladurie, *Peasants*, 252–56; Elliott, "Decline of Spain," 378.

192. Le Roy Ladurie, *Peasants*, 257, 237–38.

193. Gerber, *Social origins*, 24–25; Kula, *Measurements and men*, 30–31; Latron, *La vie rurale*, ch. 1.

194. Sereni, *History*, 234–35.

195. Sereni, *History*, 191–92.

196. Sella, *Crisis*, 106–7; Camporesi, *Magic harvest*, 7–8.

197. In 1710, grain exported from Sicily amounted to one-tenth of the level reached in 1580: Revel, "Au XVIIe siècle," 358; Morineau, "Les mancenilliers de l'Europe," 143–48; Romano, *Commerce et prix du blé*, 65, 142–43; Carrière and Buti, "Un aspect du commerce international," 116; Bonaccorso et al., "Spatial variability of drought," 273–96.

198. Pelizzon, "Grain flour," 166.

199. Murphey, "Provisioning Istanbul," 249 n. 9.

200. Sauvaget, *Alep*, 1: 193 n. 707.

201. Eldem, *French trade*, 109.

202. Stoianovich, "Mode de production maghrébin," 92.

203. Valensi, *Tunisian peasants*, 220–23.

204. In Andalusia, there were plans to establish a public granary, *pósito*, in the 1680s: Pérez, "Cordoban textile," 149–50. Again at the end of the seventeenth century, bread was sold at fixed prices in Palermo: Mack Smith, *History of Sicily*, 2: 274. On the changing nature of grain markets in the seventeenth and eighteenth centuries: Pelizzon, "Grain flour," 154–80; Filippini, "La mer," 297–300; Quaglia, "Controls over food," 449–57.

205. Semple, *Geography*, 366.

206. Braudel, *Civilization and capitalism*, 1: 125–27; Braudel, *Identity*, 2: 375–77; Pelizzon, "Grain flour," 160–80; Tielhof, '*Mother of all trades.*'

207. Faroqhi, *Pilgrims and sultans*, 106–7; Raymond, *Artisans et commerçants*, 1: 307–11.

208. Braudel, *Mediterranean*, 1: 127; Güran, *19. Yüzyıl*, 26–27; Hezârfen Hüseyin Efendi, *Telhîsü'l-beyân*, 248.

209. McGowan, *Economic life*, 27–32.

210. On rye: Weisser, *Peasants*, 67–68; Musgrave, *Land and economy*, 72–73; Sella, *Crisis*, 107; on rice: Sereni, *History*, 210–12; Arıkan, "XV.–XVI. yüzyıllarda," 479–80; on millet: Emecen, *XVI. Asırda Manisa*, 242.

211. Beldiceanu and Beldiceanu-Steinherr, "Riziculture," 9–28.

212. Braudel, *Mediterranean*, 1: 62; Houston, *Western Mediterranean world*, 151.

213. Delano Smith, *Western Mediterranean Europe*, 294.

214. For instance, peasants in the province of Aleppo who were forced to leave their tempo-

rary settlements had 150 *camus,* water buffaloes, in 1725: Archives of the Directorate of Pious Foundations: *328* (1725), 54–55. Buffalo is one of the most often mentioned animals in *waqf* records: on Adana and Misis: *335* (1760), 257–58; *359* (1762), 87; *360* (1762), 124; *367* (1766), 233; *377* (1785), 214–15; on Antioch: *578* (1773–74), 81; *363* (1775–76), 59–60; *363* (1176–77), 114; on Aleppo and Tripoli: *323* (1735), 172; *352* (1774–75), 97–98; and on Damascus: *352* (1785), 282–83; in the late sixteenth century, the wealthy of Konya were investing in better draught animals, and buffaloes were replacing oxen in their farmyards. Faroqhi, "*Vakıf* administration," 167; Evliyâ Çelebi, *Seyahatnâme,* 8: 60; Ali Bey, *Travels,* 275, 305, 310, 319. On the use of buffaloes in the eleventh century near Miletus: Lefort, "Rural economy," 235.

215. Delano Smith, *Western Mediterranean Europe,* 222.

216. Nagata, *Tarihte âyânlar,* 136–42.

217. Sereni, *History,* 196–98.

218. Glamann, "European trade," 467.

219. Veinstein, "On the *çiftlik* debate," 48.

220. On the *çiftlik* debate: Keyder and Tabak, eds., *Landholding.*

221. Yediyıldız, *Institution du vaqf,* ch. 5.

222. Petrusewicz, *Latifundium,* 12.

223. Le Roy Ladurie and Goy, *Tithe,* 93–119.

224. Le Roy Ladurie, *Peasants,* 12–13. In the eastern Mediterranean as well, deadlands and swamps were granted to tribes or individuals for reclamation. Yusuf, *Economic survey,* 28–29.

225. Hunt, *Medieval super-companies,* 132–37; Morgan, *Merchants of grain.*

226. For a long-term evaluation, from c. 1000 to 1800, of the growing attraction and later the overshoot of the mountainous zones: McNeill, *Mountains,* ch. 7.

227. Wagstaff, *Evolution,* 242–46.

228. Stoianovich, "Land tenure"; Sereni, *History,* 235, 264; Lawless, "Economy and landscapes," 519–20.

229. On the dissolution of common lands in the Iberian and Italian peninsulas: McNeill, "Tragedies of privatization," 223–24.

230. On the turning of abandoned or waste lands into agricultural estates in the Ottoman empire: İnalcık, "Emergence of big farms," 19–22.

231. On "refeudalization": Marino, *Pastoral economics,* 137–39; on localized herding: Phillips and Phillips Jr., *Spain's golden fleece,* 73–90; on the extent of livestock husbandry in the large agricultural estates in the Ottoman lands: Veinstein, "On the *çiftlik* debate," 49.

232. Le Roy Ladurie and Goy, *Tithe,* 71–92, 120–53.

233. Hütteroth, "Pattern"; also his *Ländliche siedlungen,* ch. 8. The same can be said of the Cilician plain on the Mediterranean coast: the population of the Tarsus region was higher in the 1550s, totaling 50,000, than in 1890, when it stood at 38,000. Adana, to its west, too, was settled, and the coastal plain brought partly under cultivation, despite the presence of vast marshlands on the coast: Akgündüz, *Arşiv belgeleri,* 247–50; Soysal, "Onaltıncı yüzyılda," 179; Sümer, "XIX. Yüzyılda Çukurova'da," 231–35.

234. Baehrel, *Une croissance,* 109–10; Sereni, *History,* 248–49, 336–37; Rosenthal, *Fruits,* 42–43; in Nîmes, 2,300 hectares of sodden land was drained from in the first part of the eighteenth century: Blanchemanche, *Batisseurs de paysages,* 189.

235. Braudel, *Identity,* 2: 263–64.

236. Rosenthal, *Fruits,* 52.

237. Phillips and Phillips Jr., *Spain's golden fleece*, 57; Sereni, *History*, 248–49; Georgelin, *L'Italie*, 76–78.

238. On deforestation in Ottoman lands in the seventeenth and eighteenth centuries, and on assart lands (*baltalık orman*): T. C. Çevre ve Orman Bakanlığıu, *Osmanlı ormancılığı*, 1: 105–15, 2: 11–19. On broadcast burning of trees in Cyprus: Pyne, *Vestal fire*, 132–33; on declining timber supply in France from its forests to its navy: Bamford, *Forests*.

239. Richards, "Land transformation," 164: tab. 10–1.

240. Karmon, "Settlement," 9. Most of the marshes were still not drained in the 1860s: *Rambles in the deserts*, 148.

241. Rosenthal, *Fruits*, 52.

242. Vries, *Economy of Europe*, 33–36; Raymond, *Artisans et commerçants*, 2: 307–16.

243. Sereni, *History*, 221; Le Roy Ladurie, *French peasantry*, 412.

244. Le Roy Ladurie, *French peasantry*, 329–30; Sereni, *History*, 221; Yediyıldız, *Institution du vaqf*, 26, 124. By the early nineteenth century, religious foundations owned two-thirds of the land in the empire: Barnes, *Introduction*, 42–43.

245. İnalcık, "Emergence," 25; McGowan, *Economic life*, 145–46.

246. Braudel, *Identity*, 1: 117.

247. Vergopoulos, *Le capitalisme difforme*.

248. On the periodic redistribution of land in Portugal, the Ottoman empire, and Spain: Silbert, *Portugal méditerranéen*; Owen, *Middle East*, 256–59; McNeill, "Tragedies of privatization," 222–34.

249. Herr, *Eighteenth century*, 110.

250. Gilmore, "Class consciousness," 151.

251. Moriceau, *Terres mouvantes*, 72–74.

252. See Ali Bey's account in 1816: *Travels*, 2: 306; Kinnear, *Journey*, 169; Saraçoğlu, *Akdeniz*, 475–76, 484–85.

253. Sereni, *History*, 248–49; Snowden, "Mosquitos."

254. Ciriacono, "Venise et la Hollande," 295–320; Ciriacono, "Venetian economy," 120–35.

255. Rosenthal, *Fruits*, 41.

256. On attempts to drain the marshes of Alexandretta: *Accounts & Papers* (1859), sessions, 2, vol. 30: 820.

257. Braudel, *Mediterranean*, 1: 66, 68–69.

258. Rosenthal, *Fruits*, 39.

259. Braudel, *Mediterranean*, 1: 41 n. 9.

260. Rocco, *Quinine*, 78–79.

261. Hobhouse, *Seeds of change*, 3–40.

262. Emecen, *İlk Osmanlılar*, 121–32; Le Roy Ladurie and Goy, *Tithe*, 71–92; Bartlett, *Making of Europe*, ch. 6.

263. Delano Smith, *Western Mediterranean Europe*, 185; Herlihy, *Pistoia*, 50.

264. Dienne, *Histoire du desséchement*, 77–78.

265. Glick, *Irrigation*, pt. I.

266. Dienne, *Histoire du desséchement*, 262.

267. Sereni, *History*, 92–93.

268. Braudel, *Mediterranean*, 1: 52; McNeill, *Mountains*, 85–86.

269. Wickham, *Mountains*, 24–25.

270. Duby, *Rural economy*, 81–84; Bois, *La grande depression*, 15–21; Lefort, "Rural economy," 270.

271. Wickham, *Mountains*, 6.

272. Moreno, "La colonizzazione," 986–1006.

273. Grove and Rackham, *Nature*, 334–36.

274. Herlihy, *Medieval and Renaissance*, 40–43.

275. Cohn Jr., "Inventing Braudel's mountains," 409–10; Barker, *Mediterranean valley*, 286–87.

276. Forests in some regions of the Byzantine dominion were exploited heavily: Crete, Cyprus, Levantine Syria, the Taurus, Macedonia, and possibly the northeastern part of Asia Minor and the Albanian coastline. Dunn, "Exploitation," 258–61.

277. Rosenthal, *Fruits*, 40–41.

278. Le Roy Ladurie, *Peasants*, 246–50; Sereni, *History*, 267–72.

279. Arel, "Gothic towers," 212–18; Arel, "Ege bölgesi," 787–98; Kasaba, "Migrant labor," 116.

280. Hanson, *Trade in transition*, ch. 6.

281. Rostow, *World economy*, 147; Palairet, *Balkan*, 36.

282. Gerber, *Social origins*, 86–90; Ancel, *La macédoine*, ch. 5.

283. Richard, "Land transformation," 163–78: tabs. 10–1 and 10–2.

284. Lewis, "Frontier," 267.

285. Houston, *Western Mediterranean world*, 151, 152–54.

286. Phillips and Phillips Jr., *Spain's golden fleece*, 57; Braudel, *Identity*, 2: 263–64. In Rumelia, the pace of *çiftlik* formation increased, attracting new hands. The number of *çiftlik*s in the Gazi Evrenos *waqf* increased from 10 percent of their agricultural holdings in 1702–4 to over 50 percent at the end of the nineteenth century, absorbing some of the migration from the mountains. There is evidence that laborers used to come to the plain of Salonica from as far as the Pindus mountains, despite the pending malaria threat around the lake of Yenice-i Vardar and in the lower course of the river Vardar: Demetriades, "Problems of land-owning," 44–45, 55.

287. Mulhall, *Dictionary of statistics*, 8.

288. Stoianovich, "Balkan peasants," 26–27.

289. *Accounts & papers*, (1906), 93, "Diplomatic and consular reports" from southeastern Anatolia, 9.

290. *Accounts & papers* (1858), 30, "Report on the trade of Aleppo during the year 1857," 429.

291. Owen, *Middle East*, 267–68; Gerber, *Social origins*, 79. The hills were exempt from closure.

292. On how the winter quarters of the nomads in Erdemli, on the Mediterranean shores of Anatolia, were transformed into arable: Szyliowicz, *Political change*, 23–24.

293. Simpson, *Spanish agriculture*, 67.

294. Rosenthal, *Fruits*, 52–53; on the launching of projects dealing with irrigation and drainage along Ribera del Ebro in the eighteenth century: Perez Sarrión, "Hydraulic policy," 138–42; Simpson, *Spanish agriculture*, ch. 6; and in the twentieth century: Naylon, "Irrigation and internal colonization," 178–91.

295. Braudel, *Identity*, 1: 45.

296. Grigg, *Agricultural systems*, 141.

297. Lewis, "Malaria," 286–88.

298. Simpson, *Spanish agriculture*, 135–36.

299. Delano Smith, *Western Mediterranean Europe*, 185.

300. Houston, *Western Mediterranean world*, 152–54.

301. Sereni, *History*, 248–49.

302. Braudel, *Mediterranean*, 1: 61–62.

Five • New World of the Hills

1. On the waxing and waning of political movements and their demographic basis: Goldstone, *Revolution*, 159–69, 356–59.

2. Griswold, *Great Anatolian rebellion*, ch. 2; Faroqhi, "Crisis and change," 416–19; Elliot, *Imperial Spain*, 333–49; Braudel, *Identity*, 2: 387–89. Over time, extraordinary taxes turned ordinary: on the *millones* tax: Elliott, *Imperial Spain*, 285–86; on the Ottoman *tekâlif-i şakka:* McGowan, *Economic life*, 155–57. On millenarian movements that colored the period of turbulence in the Ottoman empire in the late sixteenth and early seventeenth centuries: Stoianovich, "Prospective," 94–104.

3. Braudel, *Mediterranean*, 1: 323; Appleby, "Epidemics and famine," 643–63; also Pirenne, "Stages." The direction of change did not remain confined to the late sixteenth and early seventeenth centuries. Again at the end of the seventeenth century, bread was sold at fixed prices in Palermo: Mack Smith, *History of Sicily*, 2: 274. In Andalusia, there were plans to establish a public granary, *pósito*, in the 1680s: Pérez, "Cordoban textile," 149–50. In 1571, there were 3,371 public and 2,865 charitable *pósitos*, most of which were located "in the two Castiles, Léon, and Andalusia." Herr, *Rural change*, 31–33; also Herr, "Villages désertés," 212–13.

4. Braudel, *Identity*, 2: 398–99; Hufton, "Social conflict," 308–9.

5. Pelizzon, "Grain flour," 144–46; Geremek, *Poverty*, 151; Abdel Nour, *Introduction à l'histoire*, 232–35.

6. Faroqhi, "Crisis and change," 438–39; Orhonlu, *Osmanlı imparatorluğu'nda aşiretlerin*, 39–52. Deterring diversion of food supplies away from designated nodes of delivery was another popular method, however unsuccessful it often proved to be: Refik, *Onuncu asr-i hicrî'de*, 81–82, 83–84, 91–92.

7. Vries, *Economy of Europe*, 148–64. The misfortunes that struck the countryside went on to inflate the populations of the main and secondary cities. In the eastern Mediterranean, the ratio of urban population to the total population ranged from 10 to 20 percent at the end of the eighteenth century. Aleppo, for instance, was dealt a series of deadly blows during the seventeenth and eighteenth centuries by crop shortages and plagues that at times wiped out close to one-third of its population. But the city never failed to bounce back, and recovered each time almost instantaneously, its population reaching 80,000 to 100,000 souls, and at times even 120,000: Issawi, *An economic history*, 100; Abdel Nour, *Introduction à l'histoire*, 66–74. In Spain, despite emigration to the Indies, the share of urban population fell only slightly, from 11.4 percent in 1600 to 11.1 percent in 1800: Gelabert, "Urbanization and deurbanization," 182, 190.

8. Tilly, *Coercion, capital*, 88–91.

9. Salibi, *House of many mansions*, 104–7; Heyberger, *Les chrétiens*, 30.

10. Chehab, "Reconstructing," 117–24.

11. Fukasawa, *Toileries et commerce*, 52–53; Abdel Nour, *Introduction à l'histoire*, 276–79; Chevallier, *La société*, 210–14; Steensgaard, *Asian trade*, 175–93.

12. Dalsar, *Türk sanayi*, 162, 295; artisan and craft work became an "auxiliary activity" for mountain populations: McNeill, *Mountains*, 119–23.

13. Frêche shows that in the seventeenth and eighteenth centuries wheat was grown in upper Languedoc in a commercial fashion: Frêche, *Toulouse et la région*.

14. Le Roy Ladurie and Goy, *Tithe*, 183–84; Braudel, *Civilization and capitalism*, 3: 330, 338.

15. "In the late fifteenth and sixteenth centuries Castile had made Spain. Now in the late seventeenth century there was for the first time a possibility that Spain might remake Castile." Elliott, *Imperial Spain*, 371.

16. Ringrose, "Impact," 761–91.

17. Elliott, *Imperial Spain*, 370–72.

18. Lovett, *Early Habsburg*, 275.

19. Herr, *Rural change*, 18.

20. Sereni, *History*, 178; Chorley, *Oil, silk*, 19–24.

21. Casey, "Moriscos," 36, 37–39; on the alternation of phases of dryness with phases of wetness: Lamb, *Climatic history*, 468–69.

22. Phillip and Phillip Jr., *Spain's Golden fleece*, ch. 10.

23. Casey, *Kingdom of Valencia*, 60.

24. Ringrose, *Spain, Europe*, 194–96, 210, 215; Delano Smith, *Western Mediterranean Europe*, 83–84.

25. Casey, *Kingdom of Valencia*, 56. "There were attempts by the authorities to control or restrict the expansion of rice growing in Spain. Much of this stemmed from the fear of aggravating the malarial infestations associated with the marshes and increasingly attributed to the new paddies." On rice cultivation in Valencia and Spain: Delano Smith, *Western Mediterranean Europe*, 207.

26. When the threat looming over the countryside was lifted or brought under control, however, the exodus from the countryside was all but reversed. In the first half of the seventeenth century, substantial numbers of prebendal holdings stood vacant and unassigned, without even a petitioner: Howard, "Ottoman timar system," 221–22.

27. Le Roy Ladurie, *French peasantry*, 239–50; Le Roy Ladurie, *Peasants*, 68, 91–94; Braudel, *Mediterranean*, 1: 67; Wright, *Mediterranean feast*, 479, 615.

28. Besides chestnuts, turnips and potatoes proved popular in making up for the deficit in cereals: Molinier, *Stagnations et croissance*; also McNeill, *Mountains*, 130.

29. Moreno, "La colonizzazione," 1002–3. On the significance of chestnuts in the diets of the poor in the Apennines: Camperosi, *Magic harvest*, 7; Montanari, *Italian cuisine*, 85; Huffton, "Social conflict," 307–8. In Anatolia, too, chestnut flour was one among many used as a substitute for wheat: Ünsal, *Nimet*, 123.

30. Epstein, *Genoa*, 219; Planhol, *Historical geography*, 219–22.

31. Population densities as high as 46 persons per square kilometer were not unlikely in the chestnut zone: Brueton-Governatori, "Alimentation et idéologie," 1181–89; on Corsican Castagniccia: Planhol, *Historical geography*, 219.

32. On the planting of mulberry trees and the expansion of vineyards, and the effect of this dual development on the economic life of Vivarais: Molinier, *Stagnations et croissance*.

33. Tenenti, *Piracy*, 3–31; Fontenay, "La place de la course," 1321–47; Grove and Rackham, *Nature*, 77.

34. On plague: Delille, *Croissance*, 101–47; Biraben, *Les hommes et la peste*, 1: pt. ii.

35. Dols, "Second plague pandemic," 162–89. On plague in Edirne in the mid-seventeenth century: Bobovius, *Topkapı Sarayı'nda yaşam*, 49; on Cornelius de Bruyn's account of loss of population to plague in the Manisa region in the summer of 1677: Üçel-Aybet, *Osmanlı dünyası*, 373. People moved from İzmir to higher ground in its proximity due to threat of pestilence, to Belen—which was blessed by clean and good water as well as fruit unlike any other, and from Tripoli to Mount Lebanon: Tavernier, *Tavernier seyahatnamesi*, 118, 165–66; Rahme, "Some socio-economic observations," 427–28.

36. Delille, *Famille et propriété*. Piri Reis relates that mountains were commonly named after their most valuable *yayla*s, summer pastures: Piri Reis, *Kitab-ı bahriye*, 1: 267.

37. Randsborg, *First millennium*, 71; Potter, "Valley and settlement," 287; Moreno, "La colonizzazione," 1000–1002; Grove and Rackham, *Nature*, 133. On the role of wheat in Haut Languedoc during the seventeenth and eighteenth centuries: Vedel, "La consummation alimentaire," 479–82.

38. See Butzer's masterful investigation of Aín: "Realm of cultural-human ecology," 685–701.

39. These demographic trends were fully in line with the overall trends in Valencia: Casey, *Kingdom of Valencia*, 4–34; Cabrillana, "Villages désértes," 510.

40. Butzer, "Realm of cultural-human ecology," 687.

41. See also Whited, *Forests and peasant*, 38–39.

42. As a result of the recovery and the modest growth in the eighteenth century in the mountainous regions of the peninsula and "the progressive breaking up of the forests," the inundations reached their peak in the nineteenth century, seventeen in separate years: Thirgood, *Man*, 50.

43. Ringrose, *Spain, Europe*, 286.

44. Butzer, "Realm of cultural-human ecology," 692: tab. 42–1.

45. McNeill, *Mountains*, 1–11; Terzi, "Güzelhisar-ı Aydın," 158–60. On the descent from the Pontic mountains in the same period: Planhol, "Les migrations."

46. Moreno, "La colonizzazione," 1000–1002; Rhodes, "Tobacco cultivation," 208; Weulersse, *Le pays des Alaouites*, 232, 324–26; Hourani, "Ideologies," 170–79.

47. Owen, *Middle East*, 7, 110–12, 171. A substantial extension of cultivated area took place in Ottoman Anatolia and Syria from the 1840s to the 1880s: Issawi, *An economic history*, 133. In Spain, "approximately 10 million hectares of church and municipal land" were sold between 1836 and 1900: Simpson, *Spanish agriculture*, 67.

48. On the impact of railroads on commercial agriculture: Ringrose, *Spain, Europe*, 303–4; Sereni, *History*, 291–94; Issawi, *Economic history*, 53–60.

49. For the similar trajectory of a hilltop village at the other end of the Mediterranean: Touma, *Un village de montagne*. Also on Bilecik: Öztürk, *Tanzîmât döneminde*.

50. Crosby, *Columbian exchange*, ch. 5.

51. The best depiction of this wave is by Watson, *Agricultural innovation*.

52. Braudel, *Mediterranean*, 1: 103–4.

53. McNeill, *Mountains*, 93–101; Tabak, *"Ars longa,"* 23–48.

54. Bechmann, *Trees and men*, ch. 4.

55. Hobhouse, *Seeds of change*, 43–91, 141–87.

56. In China, too, the hilly and poor soils of the realm were devoted to the advance of sweet potatoes, maize, and peanuts, which were referred to as "famine relief crops." Anderson, *Food of China*, 96–98.

57. On the "elusive trail" of these crops: Warman, *Corn & capitalism*, 112–31.

58. The seminal work that depicts this process in all its richness is Sereni, *History*, pt. 6.

59. Planhol, *Historical geography*, 348–49.

60. Braudel calls this "shadowy zone" of monopoly the "favored domain of capitalism": *Civilization and capitalism*, 1: 24; see also Wallerstein, "Braudel and capitalism," 354–61.

61. Le Roy Ladurie, *Peasants*, 250; McGowan, "Age of the *ayans*," 658–79.

62. Goubert, *Conquest of water*, 21–26.

63. Planhol, *Historical geography*, 148; Fuller, *Buarij*, 3–10.

64. Brandes, "Maize," 331–36.

65. Montanari, *Culture of food*, 135.

66. Le Roy Ladurie and Zysberg, "Géographie des hagiotoponymes."

67. Montanari, *Culture of food*, 26–30, 35–35.

68. Wright, *Mediterranean feast*, 60–61.

69. Khanzadian, *Atlas*, 56; Tabak, "Agrarian fluctuations," 143–44. On North Africa: Rosenberger, *Société, pouvoir*, 117–18.

70. Fagan, *Floods, famines*, 200.

71. Lunde, "New world foods," 53–54; Touissant-Samat, *History of food*, 172–73; Rebora, *Çatal kültürü*, 124–25.

72. Pelizzon, "Grain flour," 154–60; on the popularity of mixing up the grain for sowing, such as *meteil, basjalade, bréchet,* and *mélarde:* Braudel, *Identity*, 2: 268. On historical precedents: Bolens, "Pain quotidien," 462–74. Also Delano Smith, *Western Mediterranean Europe*, 206–7; Braudel, *Civilization and capitalism*, 1: 167–68.

73. Capatti and Montanari, *Italian cuisine*, 49–51; Camporesi, *Magic harvest*, 100; McCann, *Maize and grace*, 23–27; Wright, *Mediterranean feast*, 615–16.

74. Sereni, *History*, 181.

75. McNeill, "American food crops," 48–52; Valensi, *Tunisian peasants*, 127. On the role of the introduction of new crops in altering the overall face of agriculture in the receiving regions: Thirsk, *Alternative agriculture*, pt. ii; on the role of horticulture and polyculture: Sereni, *History*, chs. 62–64; Braudel, *Identity*, 2: 267–68.

76. Crosby, *Columbian exchange*, 172.

77. Burckhardt, *Travels*, 15–16; Sereni, *History*, 192.

78. Montanari, *Culture of food*, 130–31; Braudel, *Identity*, 2: 266–74.

79. Semple, *Geography*, 342.

80. This was not necessarily the case north of the Alps and the Pyrenees, where ergot fungus afflicted rye due to cooler and wetter conditions: Matossian, *Poisons*, chs. 4–5.

81. Rice was also used as an instrument of land reclamation: Houston, "Social geography," 36–40; Rosenberger, "Arab cuisine," 218.

82. Montanari, *Culture of food*, 131.

83. Capatti and Montanari, *Italian cuisine*, 50–51, 43–44; Morineau, "La pomme de terre," 1781; Geremek, *Poverty*, 220–29. Before 1650, the potato was a garden crop, used mostly as cattle feed. Between its introduction in the 1530s and adoption on a wider scale in the 1650s, it

made a gradual yet inexorable transition toward being a field crop. Its graduation from the gardens to the fields was full-heartedly supported by the political seats of power for the reprieve it provided for the populace.

84. Blum, *Lord and peasant,* 334–35; Stoianovich and Haupt, "Le maïs," 84–93.

85. Beldicaunu-Steinherr, "Fiscalité et formes de possession," 279–80; Slicher van Bath, *Agrarian history,* 330–31.

86. Maize was almost always cheaper than barley, which in turn was cheaper than wheat. On the Black Sea coast, in the province of Ordu in 1547, it commanded 60 *akçe*s per *kile,* as opposed to 80 for barley and 100 for wheat. On the eastern Mediterranean, in Saida in 1812, its price was half that of wheat and less than that of barley (40 *guruş* per *garara* to be precise, when barley fetched 50 and wheat 80 *guruş*). That both figures date back to times of expansion when prices of lesser grains increased faster than that of wheat indicates that it was much cheaper during the seventeenth and eighteenth centuries. On Ordu: Yediyıldız, *Ordu kazası,* 199; and on Saida: Adel, *Documents diplomatiques,* pt. 1, 3: 107.

87. Brandes, "Maize," 331–36.

88. McNeill, "American food crops," 52.

89. Crosby, *Columbian exchange,* 176.

90. Pollen analysis demonstrates that maize was widely grown in Anti-Lebanon in the eighteenth century: Bottema, "A pollen diagram," 259–68.

91. The Ottoman government actively supported experimentation with new crops with promises of light taxation: Tabakoğlu, *Türkiye iktisat tarihi,* 360; Montanari, *Culture of food,* 137–40.

92. Stoianovich, "Le maïs," 273.

93. Montanari, *Culture of food,* 107, 134–35.

94. Long, "Mexican contribution," 37–49.

95. Crosby, *Columbian exchange,* 176–78, 189; Lunde, "New world foods," 48–55; on legumes, broad beans, and spring crops: Glick, *Irrigation,* 11–30.

96. Braudel, *Identity,* 2: 269.

97. Zhukovsky, *Türkiye'nin ziraî bünyesi,* ch. xi; Hoogasian and Matossian, *Armenian village life,* 45; Appleby, "Agrarian capitalism," 574–94.

98. The Mediterranean already had a comfortable familiarity with cultivating the family of legumes, for historically it had been a home to it, and oftentimes, the diets of its inhabitants centered around it: May, *Ecology of malnutrition,* 476–77; Stoianovich, "Le maïs," 1029–30; Turkowski, "Peasant agriculture," 110–11.

99. Long, "Mexican contribution," 37–49; Crosby, *Columbian exchange,* 168.

100. Sereni, *History,* 181.

101. Appleby, "Grain prices," 882–84.

102. Khanzadian, *Atlas de géographie,* 56: the introduction of maize allowed biennial cropping without fallow.

103. Warman, *Corn & capitalism,* 105; McNeill, *Mountains,* 89–90; Hobhouse, *Seeds of change,* 192.

104. Braudel, *Identity,* 2: 269; Morineau, "Malthus," 154–55; peasants who worked on large cotton plantations in Egypt depended on maize for subsistence: Richards, "Political economy of Gutswirtschaft," 501–2; Stoianovich, "Land tenure," 403–4.

105. Hohenberg, "Maize in French agriculture," 79–80.

106. McNeill, "American food crops," 51.

107. Lunde, "New world foods," 47–55.

108. Watson, *Agricultural innovation.*

109. Horden and Purcell, *Corrupting sea,* 203, 222.

110. Hemardinquer, "*Turcicum frumentum,*" 221–28; Lunde, "New world foods," 51.

111. The earliest sighting of maize in Ottoman lands was by a botanist in the 1570s in the environs of Tripoli of Syria: Dannenfeldt, *Leonhard Rauwolf.* There are also accounts that maintain that maize was grown, experimentally, on the Cilician lowlands as early as the late fifteenth or the turn of the sixteenth century when the region was under Mamluk rule; it was then administratively attached to Tripoli. Also Touissant-Samat, *History of food,* 172–73.

112. Murphey, "Tobacco cultivation." Also McGowan, "Age of the *ayan*s," 689; Sanders, *Balkan village,* 107; Benedict, *Ula,* 67–71.

113. Bilbao, "L'expansion," 594–608; McNeill, "American food crops," 51.

114. Stoianovich and Haupt, "Le maïs," 84–93; Rosenberger, "Cultures complémentaires," 487–89.

115. Sereni, *History,* 181.

116. AE, CC, Beyrouth, Reg no 1, le 31 décembre 1826; Cunningham, ed., *Early correspondence,* 52; Vital Cuinet also relates that the cultivation of maize was much more common than that of millet in his *La Turquie d'Asie.*

117. Braudel, *Identity,* 2: 267–68.

118. Ayvansarâyi, *Mecmuâ-i tevârih,* 18.

119. Griswold, "A sixteenth-century Ottoman foundation," 190.

120. Murphey, "Tobacco cultivation," 205–6.

121. Tzounis, *Les tabacs turcs,* 3–4; Philip, *Der Türkische tabak.*

122. Tzounis, *Les tabacs turcs,* 13–19.

123. Özkaya, *XVIII. Yüzyılda Osmanlı,* 243.

124. Bodman Jr., *Political factions,* 40–41.

125. Ironically, it was this new cash crop, the cultivation and consumption of which grew expeditiously to constitute one of the major sources of revenue for the Ottoman state, which was later placed under the aegis of the Public Debt Administration at the end of the nineteenth century: duties on it were singled out and earmarked to be collected by the agency with the express purpose of paying back the debts the Ottoman state owed to world capital markets.

126. Whereas bread consumption per capita in most parts of the northern Mediterranean revolved around 500 grams in the seventeenth and eighteenth centuries as the share of complementary ingredients increased, it jumped to 700 to 800 grams in the nineteenth century: Bennassar and Goy, "Consummation alimentaire," 419–25.

127. In Bilecik, for example, where 65 percent of the land was given over to mulberry trees, only 17.2 percent of the 553 hectares (5,530 *dönüm*s) of land was left fallow: Öztürk, *Tanzîmât döneminde,* 73–75; see also Ron, "Agricultural terraces," 1: 33–49; 2: 111–22; Issawi, *Economic history,* 215–16.

128. Braudel, *Civilization and capitalism,* 1: 118.

129. Galassi, "Reassessing Mediterranean."

130. The cultivation of specialized vineyards, in particular, of vines festooned on trees, is also mentioned by Sereni, *History,* 213.

131. Dandini, *Voyage,* 34.

132. Planhol, "Grandeur et décadance," 314–29.

133. For the opening of the seventeenth century: Akgündüz, *Osmanlı kanunnâmeleri,* 8: 395–96, 398, 405, 411; for the eighteenth century: Feyzullah Efendi, *Fetava-i feyziyye.*

134. Ze'evi, *Ottoman century,* 130–31; Singer, *Palestinian peasants,* 84–85.

135. The wetter clime of the era served millet well, for it needs moisture. According to Eremya Çelebi, the three principal grains that filled the warehouses and were used in bread-making in Istanbul in 1657 were wheat, barley, and millet: Ünsal, *Nimet,* 67.

136. Musgrave, *Land and economy,* 80.

137. Musgrave, *Land and economy,* 129, 101; Vassberg, *Village,* 55–57.

138. Moreno, "La colonizzazione," 986–1006.

139. Semple, *Geography,* 343.

140. Rosenberger, "Cultures complémentaires," 485–87.

141. Faroqhi, "Rural society, 2," 170; on the rise and decline of millet cultivation in Moldavia: Neamtu, *Le technique,* 216–18. Millet showed a noticeable increase in the province of Manisa, where its production increased from 336 *kile* to 1,016 between 1531 and 1575. The increases were more spectacular in Erzincan and Kemah: in the latter, from 6,894 *kile* in 1530 to 12,887 in 1568, and 17,032 in 1591; and in the former, from 436 *kile* to 13,201 between 1530 and 1591: Emecen, *XVI. Asırda Manisa,* 242; Miroğlu, *Kemah sancağı,* 191–92.

142. Algar, "Food," 300–301.

143. Semple, *Geography,* 344.

144. Volney, *Travels,* 1: 296.

145. See the articles in Cohn Jr. and Epstein, *Portraits.*

146. Braudel, *Mediterranean,* 1: 55–60; Pitt-Rivers, *People of the Sierra,* 42.

147. Delano Smith, *Western Mediterranean Europe,* 160; Güçer, *XVI.–XVII. Asırlarda,* 3–4.

148. Semple, *Geography,* 344.

149. Tanoğlu, "Türkiye'nin irtifa," 47–48; Phillips and Phillips Jr., *Spain's golden fleece,* 9–11.

150. Güran, *Osmanlı dönemi,* 6–16.

151. Orman ve Maden ve Ziraat Nezareti, *1325 senesi,* 7; Akalın, "Türkiye'nin arazi varlığı," 3–14.

152. On the intensification of exploitation of valleys since the turn of the seventeenth century: Potter, "Valleys and settlement," 213–14; Vita-Finzi, *Mediterranean valleys,* 101–2; on "nucleation": Grove and Rackham, *Nature,* 73–74.

153. Neither did the temporary settlements established by nomads: they too blanketed wide stretches of land and were dispersed along altitudinal lines. Summer and winter pastures as well as permanently and temporarily settled fields were geographically dispersed, but structurally speaking, closely interconnected: Kolars, "Tradition, season." Inhabitants of Mount Lebanon descended to work in soap factories on a seasonal basis: Rahme, "Some socio-economic observations," 421–26.

154. Houston, *Western Mediterranean world,* 592; Amiran, "Pattern of settlement," 73–74.

155. Planhol, *Historical geography,* 242.

156. Delort and Walter, *Histoire de l'environnement,* 228–32. When the level of economic activity along the shores of the Inner Sea declined in the sixth and seventh centuries, there was a "forest famine": Montanari, *Culture of food,* 27.

157. Sereni, *History,* 139.

158. Phillips and Phillips Jr., *Spain's golden fleece*, ch. 3; Marino, *Pastoral economics*, 69; there were 2 million sheep wintering in the lowlands of northern Syria alone in the sixteenth century and half a million in Mount Lebanon: Orhonlu, *Osmanlı*, 21 n. 52; Touma, *Paysans*, 610.

159. Lefort, "Rural economy," 265.

160. Delano Smith, *Western Mediterranean Europe*, 35, 60–61; Devèze, "L'équilibre agro-sylvo-pastoral," 333–43.

161. On the lowering of agricultural limits on the mountainous regions due to climatic changes: Sereno, "Ecology," 307–11.

162. Warman, *Corn & capitalism*, ch. 8.

163. Grove and Rackham, *Nature*, 66–67.

164. McNeill, *Mountains*, ch. 3; Planhol, *Historical geography*, 148.

165. Bechmann, *Trees and man*, 295–96; Delano Smith, *Western Mediterranean Europe*, 60.

166. Lane, *Ships and shipbuilders*, ch. xii.

167. Herlihy, *Pistoia*, 39, 50–51. The process that encouraged the sliding down of mountain villages to relocate in the plains since the turn of the millennium slowed down in the fourteenth and fifteenth centuries: Delille, *Croissance*, 49. A similar movement whereby settlements moved from river valleys and marshy areas into higher land occurred in the central European lowland: Dunin-Wąsowicz, "Natural environment," 92–103.

168. For similar arrangements in France at the time: Sclafert, *Cultures en Haute Province*, 37–40.

169. Montanari, *Culture of food*, 34–35, 77.

170. On *Alpi Fiorentine:* Cohn Jr., *Creating*, chs. 1 and 3.

171. Ourfelli, "Le Maghreb."

172. McNeill, *Mountains*, 86–87.

173. Geyer, "Physical factors," 42–43; Dunn, "Exploitation," 244; Wickham, "European forests," 540–41.

174. On the degradation of woodlands: Bechmann, *Trees and men*, 295–96.

175. Wickham, *Mountains*, 147.

176. Planhol, *Les fondements*, 225–29.

177. Le Roy Ladurie, *Montaillou*, 133–49; Braudel, *Mediterranean*, 1: 40.

178. Cohn Jr., *Creating*, 26.

179. On the territorial compass of village territory in Mount Lebanon: Touma, *Un village*, 28–30; on terracing: Touma, *Paysans*, 589, 636.

180. Riley-Smith, *Feudal nobility*, 44; Prawer, "Étude," 34–35; Kaplan, *Les hommes*, 95–101; Lefort, "Les villages," 294; Cahen, "La communauté rurale," 25.

181. Cahen, *Pre-Ottoman Anatolia*, 159.

182. Braudel, *Identity*, 1: 133; Derlange, "Hameaux," 167–73.

183. Adanır, "Mezra'a."

184. Pesez and Le Roy Ladurie, "Le cas français," 164–66.

185. Osmanlı Arşivi Daire Başkanlığı, *387 Numaralı Muhâsebe*, 2: 19.

186. Grossman and Safrai, "Satellite settlements," 446–61; Lindner, *Nomads and Ottomans*, 87; Koç, *XVI. Yüzyılda*, 37; Özel, "Population changes," 187.

187. Koç, *XVI. Yüzyılda*, 37. In Tarsus, too, the number of tribal organizations (*cema'at*) declined from 142 to 50, and *mazra'a*s, agriculturally used sites, from 219 to about 90: Faroqhi,

"Tarsus and the *tahrir*," 81. The *mazra'a*s, which were used as pastures by the Horse Drovers of Karaman at the turn of the sixteenth century, had turned into cultivated fields by 1523: Lindner, *Nomads and Ottomans*, 87.

188. İnalcık, *Hicrî 835 tarihli*, xxix; Adanır, "Mezra'a."

189. The figure, given by Sir Thomas Roe, English ambassador to the Porte in the 1620s, is for the latter half of the sixteenth century: Creasy, *History of the Ottoman Turks*, 1: 322.

190. For the 1620s, see the information given by the author of *Kitâb-i Müstetâb* in Yücel, *Osmanlı devlet teşkilâtina*, 19–29; the latter figure is cited by Creasy. Volney asserts that the number of villages in the Aleppo province had declined by the end of the eighteenth century from its peak of 3,200 to 400: *Travels*, 2: 90–91; d'Arvieux relates that of the 1,200 villages, 300 were ruined and forfeited: *Mémoires*, 6: 144. In Cyprus, of the 850 villages that had existed in the sixteenth century, only 550 remained in the eighteenth: Luke, *Cyprus*, 26.

191. Houston, *Western Mediterranean world*, 242–43.

192. Burckhardt, *Travels*. On the Mediterranean coast, agricultural tithes of abandoned lands remained attached to the lands of Megri, presumably to be paid by whomever cultivated them: Faroqhi, "Rural society, 1," 105–6.

193. Grossman and Safrai, "Satellite settlements," 446–61.

194. Tanoğlu, "İskân coğrafyası," 1–32; Tunçdilek, "Types," 52–70; İnalcık, "Ottoman state," 162–67; Burckhardt, *Travels*, 129.

195. Venzke, "Question of declining," 251–64.

196. İnalcık, "Ottoman state," 143–53.

197. İnalcık, "Ottoman state," 162–67.

198. Braudel, *Identity*, 1: 130–31.

199. On "mixed production strategies" that put to use "vertical biotic zonation" in the Alps: Viazzo, *Upland communities*, 17–24.

200. Houston, *Western Mediterranean world*, 240–42; Braudel, *Identity*, 1: 133–35; Delano Smith, *Western Mediterranean Europe*, 60–62; Sereno, "Ecology," 312.

201. Benedict, *Ula*, 58–60; Kasaba, "Migrant labor," 119–21.

202. Delano Smith, *Western Mediterranean Europe*, 171.

203. Amiran, "Pattern of settlement," 73–74.

204. İnalcık, "Ottoman state," 164; Benedict, *Ula: An Anatolian town*, 58–59; Khalidi, ed., *All that remains*, 124–26, 159–60, 508–9. The double village analyzed by Gulick was established at the beginning of the seventeenth century: *Social structure*, 31–32, 146.

205. Planhol, *Historical geography*, 148.

206. Pitt-Rivers, *People of the Sierra*, 42.

207. Aladdin, "Deux *fatwâ*-s du shayh," 18–20; Burckhardt, *Travels*, 299.

208. Vassberg, *Village*, 171–75; Salomon, *Campagne*.

209. Delano Smith, *Western Mediterranean Europe*, 176.

210. Rafeq, "Economic relations."

211. Baer, *Fellah and townsman*, 271–72.

212. It is this mobility, Suraiya Faroqhi argues, which was the underlying reason large-scale rural rebellions were absent in the Ottoman empire: Faroqhi, "Crisis and change"; see also Heyberger, *Les chrétiens*, 26–37; Touma, *Paysans*, 588–90; on Spain: Vassberg, *Village*; and on the "culture of mobility" that contributed to the stability of the countryside: Reher, *Town and country*.

213. Petrusewicz, *Latifundium*, 14; Aymard, "L'Europe moderne," 426–35.

214. Semple, *Geography*, 342.

215. Even in times of plenty, close to 50 percent of the economic activities of these villages were conducted on differing altitudes of the mountainous range: Touma, *Un village de montagne*, 28–30; Kolars, "Tradition, season."

216. Grossman and Safrai, "Satellite settlements," 451.

217. Simpson, *Spanish agriculture*, 65–66.

218. Pesez and Le Roy Ladurie, "Le cas français."

219. Delille, *Famille et propriété*, 377–78.

220. Sereni, *History;* Delano Smith, *Western Mediterranean Europe*, 180–81; Evliyâ Çelebi, *Seyahatnâme*, 8: 60.

221. The best register of this phenomenon can be seen in the multiplication in the number of water mills on the Taurus Mountains as well as Mount Lebanon: Archives of the Directorate of Pious Foundations: *2113* (1609–10), 2; *589* (1616–17), 211; *779* (1688–89), 140; *355* (1710), 24; *578* (1715), 10; *639* (1718–19), 1164; *609* (1720–21), 38; *735* (1720–21), 46; *735* (1737–38), 211; *737* (1741), 207; *803* (1745), 29–30; *608* (1745), 23; *580* (1753–54), 23; *512* (1764), 391; *513* (1770), 668; *608* (1791–92), 264. In late nineteenth-century Palestine, there were 137 places that had names starting with the *tahunah* or *tawahin*—mill(s): Conder and Kitchener, *Survey*, 75–76.

222. Hanson, *Trade in transition;* Ancel, *La macédoine*, ch. 6.

223. Horden and Purcell, *Corrupting sea*, 222.

224. On commercial cotton cultivation: Stoianovich, "Conquering Balkan," 260.

225. Akgündüz et al., *Arşiv belgeleri*, 72, 75; Refik, *On ikinci asr-ı hicride*, 102–3. Starting from the 1620s, cotton production on the Aegean coast, in particular in the Manisa region, started to expand, taking up the land previously sown to sesame: Uluçay, *XVII. Yüzyılda Manisa'da*, 41. In 1740, restrictions placed on the export of wax, cotton, and cotton yarn were lifted: Arıkan, "Osmanlı imparatorluğu'nda ihracı," 279–306.

226. Demetriades, "Problems of land-owning," 43–57.

227. Braudel, *Mediterranean*, 2: 725.

228. Stoianovich, "Land tenure," 398–411; also McGowan, *Economic life*, 75–79.

229. Owen, *Middle East*, 6–7.

230. Karmon, "Settlement," 9. Most of the marshes were still undrained in the 1860s: *Rambles in the deserts*, 148.

231. Keyder, "Small peasant," 53–107; for earlier attempts: Ener, *Tarih boyunca*, 219; Sümer, "XIX. Yüzyılda Çukurova," 232.

232. Faroqhi, "Notes," 405–17.

233. Venzke, "Rice cultivation," 179–89.

234. On rice cultivation in Valencia and Spain: Delano Smith, *Western Mediterranean Europe*, 207.

235. Montanari, *Culture of food*, 130–31; Toussaint-Samat, *History of food*, 162.

236. Delano Smith, *Western Mediterranean Europe*, 206–8.

237. On the northerly march of rice (and silk): Sereni, *History,* 134, 188, 210, 235; Romano, "Italy in the crisis," 192, identifies the rice field as the single, solid change in seventeenth-century Lombardy.

238. Faroqhi, "Tarsus," 79.

239. Murphey, *Regional structure*, 228; *Accounts & Papers* (1869–70), vol. 67, pt. II.

240. Delano Smith, *Western Mediterranean Europe*, 207–8.

241. Braudel, *Civilization and capitalism*, 1: 117.

242. Vergopoulos, *Le capitalisme difforme*, 69–70.

243. Vergopoulos, *Le capitalisme difforme*, 62–64; Planhol, "Les migrations," 583–600; on migration from the Pontic range: Planhol, "Aspects," 302; Fuller, *Buarij*, 9.

244. Petmezas, "Patterns," 581.

245. Palairet, *Balkan economies*, 56.

246. Perevolotsky and Seligman, "Role of grazing"; Kolars, "Locational aspects."

247. Barker, *Mediterranean valley*, 286.

248. Sereni, *History*, 251–58; Sereno, "Ecology," 310.

249. Houston, *Western Mediterranean world*, 116.

250. Slicher van Bath, *Agrarian history*, 207.

251. Delano Smith, *Western Mediterranean Europe*, 352; Vogt, "Aspects," 83–91.

252. Houston, *Western Mediterranean world*, 220, 462.

253. Thirgood, *Man and forest*, 52–53, 114; Richards, "Land transformation," 164.

254. Appleby, "Grain prices," 884–87.

255. This was related by Grigor, an Armenian priest who lived during the period of Mehmet III (1595–1640): Öz, *Osmanlı'da alevi*, 47–48; Andreasyan, "Celâlilerden kaçan," 45–49.

256. Stoianovich, "Prospective: Third and fourth," 99.

257. Braudel, *Civilization and capitalism*, 1: 66; also Marks, *Tigers, rice, silk*.

258. Laoust, *Les gouverneurs de Damas*, 219; on fauna: Rabbath, *Documents inédits*, 2: 196; Adel, *Documents diplomatiques*, 1: 320.

259. In addition to the wild animals, the forests of the Mediterranean mountains contained tigers. In Kütahya, summer pastures were named after them: "Kaplan Alanı Yaylağı," where tigers roamed: Refik, *Anadolu'da Türk aşiretleri*, 3; both Tournefort and Evliyâ Çelebi mention the presence of tigers in the Mediterranean and on Mount Ararat; also Leake, *Journal*, 112.

260. Redford, *Landscape*, 38–39.

261. Braudel, *Identity*, 2: 265–66.

262. McNeill, *Mountains*, 311–25; Eyice, "J. von Hammer-Purgstall," 544; Mikesell, "Deforestation," 22–23.

263. Biger and Liphschitz, "Historical geography," 4–18.

264. Planhol, "Expansion," 94; Beals, "Remnant cedar forests," 679–80.

265. Margalit, "Some aspects," 18. Palestine, which lacked forest resources, was supplied with timber from within the eastern Mediterranean until the 1870s, from without thereafter: Biger and Liphschitz, "Historical geography."

266. Lamb, *Climatic history*, 463.

267. Grove and Rackham, *Nature*, 84; Issawi, *An economic history*, 119; Alouche, *Évolution d'un centre*, 129–30.

268. Mulhall, *Dictionary of statistics*, 8.

269. Stoianovich, "Balkan peasants," 26–27.

270. Demetriades, "Problems of land-owning," 44–45, 55.

271. Barnes, *Introduction*, 91.

272. Baer, *Fellah and townsman*, 259–60.

273. Houston, *Western Mediterranean world*, 592.

274. Thirgood, *Man and forest*, 100; Issai, *An economic history*, 54–57.

275. A good account of this mass migratory movement is by Clay, "Labor migration."

276. Planhol, *Les fondements*, 100–103.

277. Güran, *19. Yüzyıl Osmanlı*, 68; Quataert, "Agricultural trends," 76–77; Lewis, *Nomads and settlers*, 105; also Latron, *La vie rurale*.

278. Issawi, *Economic history of Turkey*; Hütteroth, *Ländliche siedlungen*, 72–80.

279. Salibi, *House of many mansions*, 104–7; Alouf, *Histoire de Baalbek*, 96.

280. Cahen, "La communauté rurale," 17–18; on the dynamics of *mushâ'* during the late nineteenth and early twentieth centuries: Firestone, "Land-equalizing *mushâ'*," 91–129; Gulick, *Social structure*, 64.

281. Weulersse, *Les paysan*, 109–19.

282. Tökin, *Türkiye köy iktisadiyatı*.

283. Baer, "Dismemberment of *awqaf*," 306; also Keyder, "Small peasant," 94–98.

284. Baer, *Fellah and townsman*, 272; also Bouron, *Les Druses*.

285. Saâdé, *L'Agriculture à Lattaquié*, 29–31; and on the political consequences of this migration: Weullersse, *Le pays des Alaouites*, 1: 113–15.

286. Hütteroth, "Influence," 22–25; Thoumin, "Le Ghab," 517–18.

287. Braudel, *Identity*, 1: 69.

288. Viazzo, *Upland communities*, 289–90; Montemayor, "Les migrations"; Sella, "Au dossier des migrations."

289. Baer, *Fellah and townsman*, 182–83; Ochsenwald, *Hijaz railroad*, 34.

290. Orhonlu, *Osmanlı imparatorluğu'nda aşiretlerin*, 113–20; Gould, "Burning of the tents"; Dumont, "La pacification," 108–30.

291. Braudel, *Identity*, 1: 45.

292. Grigg, *Agricultural systems*, 141.

293. Houston, *Western Mediterranean world*, 593.

294. Delano Smith, *Western Mediterranean Europe*, 171–72.

295. *Boule de neige* and *coulée* are terms employed by Delille in his analysis of the Caudine valley in the seventeenth and eighteenth centuries: *Croissance*, 49.

296. Le Roy Ladurie, "Le cas français," 165; Braudel, *Identity*, 1: 45.

Conclusion

1. Febvre, "Patate et pomme de terre," 643–45.

2. Curtin, *Rise and fall*, ch. 1.

3. On the development of hilltop settlements late in the first millennium: Randsborg, *First millennium*, 180; Planhol, *Historical geography*, 147–48.

4. This was not surprising, because the Americas produced some of the most important food crops; the Mediterranean in return has given few cultivated plants to the world: Crosby, *Columbian exchange*, 170; Grove and Rackham, *Nature*, 66–67.

5. On the Mongolian explosion and afterward: Adshead, *Central Asia*, 53–126.

6. Braudel, *Mediterranean*, 1: 103–10.

7. Watson, *Agricultural innovation*, 103–4.

8. On the epochal confrontation between the heathen and the Christian, between beer- and wine-drinking, after the Barbarian invasions: Montanari, *Culture of food*, 18–21.

9. Pounds, *Historical geography*, 279–80.

10. Thirgood, *Man and the Mediterranean*, 72–80.

11. See the articles in *Villages desertés*.

12. Thirgood, *Man and the Mediterranean*, 33–34.

13. Crosby, *Ecological imperialism*, 70–103.

14. Lunde, "Pillars Of Hercules," 7.

15. Musgrave and Musgrave, *Empire of plants*, 15–33; McGowan, "Age of the *ayans*," 736, 738.

16. Crosby, *Columbian exchange*, 180, 189–90.

17. Sereni, *History*, 300.

18. Crosby, *Ecological imperialism*, 146–96; Crosby, *Columbian exchange*, ch. 5.

19. Stoianovich, "Land tenure," 404–6.

20. Lunde, "Leek-Green Sea," 12–19.

21. Montanari, *Culture of food*, 133–40.

22. Morineau, "La pomme de terre," 1767–85.

23. Headrick, *Tools of empire*, ch. 3.

24. Curtin, *Death by migration*, 62–68.

25. Abernethy, *Dynamics*, 53.

26. On the 1870–1914 period: Curtin, *Death by migration*, 132–40.

27. Vergopoulos, *Capitalisme difforme*; Keyder, "Cycle of sharecropping," 130–45.

Abdel Nour, Antoine. "Habitat et structures sociales à Alep aux XVIIe et XVIIIe siècles," in Abdelwahab Bouhdiba and Dominique Chevallier, eds. *La ville arabe dans l'Islam*. Tunis, 1982.

———. *Introduction à l'histoire urbaine de la Syrie ottomane (XVIe–XVIIIe siècle)*. Beirut, 1981.

———. "Notes sur quelques questions de métrologie concernant la Syrie à l'époque ottomane," *Arabica* 24 (3), 1977.

Abel, William. *Agricultural fluctuations in Europe: From the thirteenth to the twentieth centuries*. New York, 1980.

Abernethy, David. *The dynamics of global dominance: European overseas empires, 1415–1980*. New Haven, 2002.

Abou el-Rousse Slim, Souad. *Le métayage et l'impot au Mont-Liban (XVIIIe et XIXe siècles)*. Beirut, 1987.

Abulafia, David. *The Western Mediterranean kingdoms, 1200–1500: The struggle for dominion*. New York, 1997.

———. *A Mediterranean emporium: The Catalan kingdom of Majorca*. New York, 1994.

———. "The Levant trade of the minor cities in the thirteenth and fourteenth centuries: Strengths and weaknesses," *Asian and African Studies* 22 (1–3), 1988.

———. "Southern Italy and the Florentine economy, 1265–1370," *Economic History Review* 34 (3), 1981.

Abu-Lughod, Janet. *Before European hegemony: The world system A.D. 1250–1350*. New York, 1989.

Abun-Nasr, Jamil M. *A history of the Maghrib in the Islamic period*. New York, 1987.

Adanır, Fikret. "Tradition and rural change in southeastern Europe during Ottoman rule," in Daniel Chirot, ed. *The origins of backwardness in eastern Europe*. Berkeley, 1989.

———. "Mezra'a: Zu einem problem der siedlungs- und agrargeschichte südosteuropas im ausgehenden mittelalter und in der frühen neuzeit," in Ralph Melville et al., eds. *Deutschland und Europa in der Neuzeit*. Stuttgart, 1988.

Adel, Ismail. *Documents diplomatiques et consulaires relatifs à l'histoire du Liban et des pays du Proche-Orient du XVIIe siècle à nos jours*. Beirut, 1975–.

Adshead, S. A. M. *Central Asia in world history*. New York, 1993.

. *Salt and civilization*. New York, 1992.

———. *China in world history*. New York, 1988.

Aghassian, Michel, and Kéram Kévonian. "Le commerce arménien dans l'Océan Indien aux 17e et 18e siècles," in Denys Lombard and Jean Aubin, eds. *Marchands et hommes d'affaires asiatiques dans l'Océan Indien et la mer de Chine, 13e–20e siècles.* Paris, 1987.

Ahrweiler, Hélène. *Byzance et la mer. La marine de guerre, la politique et les institutions de Byzance au VIIe–XVe siècles.* Paris, 1966.

Akalın, İlhan. "Türkiye'nin arazi varlığı ve toprak potansiyeli," in *Tarım topraklarının amaç dışı kullanılmasının önlenmesi semineri.* Ankara, 1984.

Akdağ, Mustafa. *Türkiye'nin iktisadî ve içtimaî tarihi.* 2 vols. Ankara, 1971.

Akgündüz, Ahmed. *Osmanlı kanunnâmeleri ve hukukî tahlilleri.* Vol. 8. İstanbul, 1996.

Akgündüz, Ahmet, et al. *Arşiv belgeleri ışığında Tarsus ve eshâb-ı kehf.* İstanbul, 1993.

Aladdin, Bakri. "Deux *fatwâ*-s du shayh abd al-Ganî al-Nâbulusî (1143/1731)," *Bulletin d'études orientales* 39–40, 1987–88.

Alexandre, Pierre. *Le climat en Europe au moyen âge.* Paris, 1987.

Alexandrescu-Dersca-Bulgaru, Marie-Mathilde. "Une relation vénetienne sur l'Empire ottoman à l'époque de Soliman le Magnifique," in J.-L. Bacqué-Grammont and Emeri van Donzel, eds. *Comité international d'études pré-Ottomanes et ottomanes.* Istanbul, 1987.

Algar, Ayla. "Food in the life of the tekke," in Raymond Lifchez, ed. *The Dervish lodge: Architecture, art, and Sufism in Ottoman Turkey.* Berkeley, 1992.

Ali Bey. *Travels of Ali Bey in Morocco, Tripoli, Cyprus, Egypt, Arabia, Syria, and Turkey between the Years 1803 and 1807.* 2 vols. London, 1816.

Allsen, Thomas. *Commodity and exchange in the Mongol Eurasia: A cultural history of Islamic textiles.* New York, 1997.

Alouche, Richard. *Évolution d'un centre de villégiature au Liban (Broummana).* Beirut, 1970.

Alouf, Michel M. *Histoire de Baalbek.* Beirut, 1970.

Altan, Mustafa Haşim. *Belgelerle Kıbrıs Türk vakıflar tarihi (1571–1974).* 2 vols. Kıbrıs, 1986.

Ambrosoli, Mauro. *The wild and the sown: Botany and agriculture in Western Europe, 1350–1850.* New York, 1997.

Amiran, D. H. K. "The pattern of settlement in Palestine," *Israel Exploration Journal* 3 (2–3), 1953.

Ancel, Jacques. *La macédoine: son évolution contemporaine.* Paris, 1930.

Anderlind, Leo. "Ackerbau und thierzucht in Syrien, insbesondere in Palästina," *Zeitschrift des Deutchen Palästina-Vereins* 9, 1866.

Anderson, E. N. *Food of China.* New Haven, 1988.

Anderson, Perry. *Lineages of the absolutist state.* London, 1974.

Anderson, Sonia P. *An English consul in Turkey: Paul Rycault at Smyrna, 1667–1678.* Oxford, 1989.

Andreasyan, Hrand D. "Celâlilerden kaçan Anadolu halkının geri gönderilmesi," in *İsmail Hakkı Uzunçarşılı'ya armağan.* Ankara, 1976.

Andrews, Jean. "Diffusion of Mesoamerican food complex to southeastern Europe," *Geographical Review* 83 (2), 1993.

Anes, Gonzalo. "The agrarian depression in Castile in the seventeenth century," in I. A. A. Thompson and Bartolomé Yun Casilla, eds. *The Castilian crisis of the seventeenth century.* New York, 1994.

Anon. *Rambles in the deserts of Syria and among the Turkomans and Bedaweens.* London, 1864.

Anoyatis-Pelé, D. "Aperçu sur le coût du transport terrestre dans les Balkans au XVIIIe siècle

(1715–1820)," in *Économies méditerranéennes: équilibres et intercommunications XIIIe–XIXe siècles.* Vol. 2. Athens, 1985.

Antoine, Jean-Marc, Bertrand Desailly, and Jean-Paul Métaillié. "Les grands *aygats* du XVIIIe siècle dans les Pyrénés," in B. Bennassar, ed. *Les catastrophes naturelles dans l'Europe médiévale et moderne.* Toulouse, 1996.

Appleby, Andrew B. "Epidemics and famine in the Little Ice Age," *Journal of Interdisciplinary History* 10 (4), 1980.

———. "Grain prices and subsistence crises in England and France, 1590–1740," *Journal of Economic History* 39 (4), 1979.

———. "Agrarian capitalism or seigneurial reaction? The Northwest of England, 1500–1700," *American Historical Review* 80 (3), 1975.

Appuhn, Karl. "Inventing nature: Forests, forestry, and state power in Renaissance Venice," *Journal of Modern History* 72, 2002.

Arbel, Benjamin. "Maps of the world for Ottoman princes? Further evidence and questions concerning 'The "Mappomondo" of Hajji Ahmed,'" *Imago Mundi* 54, 2004.

———. "Cypriot population under Venetian rule (1473–1571)," reprinted in Benjamin Arbel. *Cyprus, the Franks and Venice (13th–16th centuries).* London, 2000.

———. *Trading nations: Jews and Venetians in the early modern eastern Mediterranean.* Leiden, 1995.

———. "Slave trade and slave labor in Frankish Cyprus," *Studies in Medieval and Renaissance History* 14, 1993.

———. "Venetian trade in fifteenth-century Acre: The letters of Francesco Bevilaqua (1471–1472)," *Asian and African Studies* 22 (1–3), 1988.

Arel, Ayda. "Gothic towers and baroque mihrabs: The post-classical architecture of Aegean Anatolia in the eighteenth and nineteenth centuries," *Muqarnas* 10, 1993.

———. "Ege bölgesi âyânlık dönemi mimarisi hakkında bir ön araştırma," in *IV. Araştırma sonuçları toplantısı.* Ankara, 1986.

Arenson, Sarah. "Food for a maritime empire: Venice and its basis in the middle ages, 1250–1350," in Klaus Friedland, ed. *Maritime food transport.* Cologne, 1994.

Arıkan, Zeki. "Un *sancak* anatolien au temps de Soliman le Magnifique: la situation économique et sociale du Saruhan," in Gilles Veinstein, ed. *Soliman le Magnifique et son temps.* Paris, 1992.

———. "Osmanlı imparatorluğu'nda ihracı yasak mallar (memnu meta)," in *Prof. Dr. Bekir Kütükoğlu'na armağan.* Istanbul, 1991.

———. "XIV.–XVI. Yüzyıllarda Ayasuluğ," *Belleten* 54 (209), 1990.

———. "XV.–XVI. Yüzyıllarda Anadolu'da çeltük üretimi," in *V. Milletlerarası Türkiye sosyal ve iktisat tarihi kongresi.* Ankara, 1989.

———. "1821 Ayvalık isyani," *Belleten* 52 (203), 1988.

Arrighi, Giovanni. *The long twentieth century: Money, power, and the origins of our times.* New York, 1994.

Artan, Tülay. "Aspects of Ottoman elite's food consumption," in Donald Quataert, ed. *Consumption studies in the history of the Ottoman empire.* Albany, 2000.

Arvieux, Laurent d'. *Mémoires du chevalier d'Arvieux.* 6 vols. Paris, 1735.

Ascherson, Neal. *Black Sea.* New York, 1995.

Asdrachas, Spyros. "Le surplus rural dans les régions de la Méditerranée orientale," in *Actes du*

IIe Colloque international d'histoire. Économies méditerranéennes: équilibres et intercom-munications. Vol. 2. Athens, 1985.

Ashtor, Eliyahu. "Catalan cloth in the late medieval Mediterranean markets," *Journal of Euro-pean Economic History* 17 (2), 1988.

———. "The wheat supply of the Mamluk kingdom," *Asian and African Studies* 18 (3), 1984.

———. "Levantine alkali ashes and European industries," *Journal of European Economic His-tory* 12 (3), 1983.

———. "Levantine sugar industry in the later Middle Ages: A case of technological decline," in Abraham L. Udovitch, ed. *The Islamic Middle East, 700–1900: Studies in economic and social history.* Princeton, 1981.

———. *A social and economic history of the Near East in the middle ages.* Berkeley, 1976.

———. "An essay on the diet of various classes in the medieval Levant," in R. Forster and O. Ranum, eds. *Biology of man in history.* Baltimore, 1975.

———. "Profits from trade with the Levant in the fifteenth century," *Bulletin of the School of Oriental and African Studies* 38 (2), 1975.

———. "The Venetian supremacy in Levantine trade: Monopoly or pre-colonialism?" *Journal of European Economic History* 3 (1), 1974.

———. *Les métaux précieux et la balance des payements de Proche-Orient à la basse époque.* Paris, 1971.

Atasoy, Nurhan. *1582 Surname-i hümayûn: Düğün kitabı.* İstanbul, 1997.

Atasoy, Nurhan, W. B. Denny, L.W. Mackie, and H. Tezcan. *Ipek: Imperial Ottoman silks and velvets.* London, 2001.

Atasoy, Nurhan, and Julian Raby. *İznik seramikleri.* London, 1989.

Aymard, Maurice. "L'Europe moderne: feodalité ou feodalités," *Annales E.S.C.* 36 (3), 1981.

———. "From feudalism to capitalism in Italy: The case that doesn't fit," *Review* 6 (2), 1977.

———. "Rendements et productivité agricole dans l'Italie moderne," *Annales E.S.C.* 28 (2), 1973.

———. *Venise, Raguse et le commerce du blé pendant la seconde moitié du XVIe siècle.* Paris, 1966.

Aynural, Salih. *İstanbul değirmenleri ve fırınları: Zahire ticareti 1740–1840.* İstanbul, 2002.

Ayvansarâyi, Hâfiz Hüseyin. *Mecmuâ-i tevârih.* Ed. Fahri Ç. Derin and Vâhid Çabuk. Istanbul, 1985.

Babinger, Franz. *Mehmed the Conqueror and his time.* Princeton, 1978.

Baehrel, René. *Une croissance. La Basse-Provence rurale de la fin du seizième siècle à 1789.* Paris, 1988.

Baer, Gabriel. "The dismemberment of *awqaf* in early nineteenth-century Jerusalem," in Gad G. Gilbar, ed. *Ottoman Palestine 1800–1914.* Leiden, 1990.

———. *Fellah and townsman in the Middle East.* London, 1982.

Bağcı, Serpil. "Gerçeğin saklandığı yer: Ayna," in *Sultanın aynaları.* İstanbul, 1998.

Balard, Michel. "A Christian Mediterranean: 1000–1500," in David Abulafia, ed. *The Mediter-ranean in history.* Los Angeles, 2003.

———. "Le système portuaire génois d'outre-mer (XIIe–XVe siècle)," in Simonetta Cavaciocchi, ed. *I porti come impresa economica.* Florence, 1988.

———. *La Romanie génoise (XIIe–début du XVe siècle).* 2 vols. Rome, 1978.

Balard, Michel, et al. "Le transport des denrées alimetaires en Mediterranée à la fin du moyen âge," in Klaus Friedland, ed. *Maritime food transport.* Cologne, 1994.

Balta, Evangelia. "Bread in Greek territories during Ottoman rule," in E. Balta. *Problèmes et approches de l'histoire ottoman.* Istanbul, 1997.

———. "Du document fiscal à l'économie agricole: les cultures à Santorin au XVIIIe s.," *Historia* 6, 1986.

Bamford, Paul W. *Forests and French sea power, 1660–1789.* Toronto, 1956.

Barbir, Karl K. *Ottoman rule in Damascus, 1708–1758.* Princeton, 1980.

Barendse, R. J. *The Arabian seas: The Indian Ocean world of the seventeenth century.* Armonk, NY, 2002.

———. "Trade and state in the Arabian Seas: A survey from the fifteenth to the eighteenth century," *Journal of World History* 11 (2), 2000.

Barkan, Ömer L. "İstanbul saraylarına ait muhasebe defterleri," *Belgeler* 9 (13), 1979.

———. "The price revolution of the sixteenth century: A turning point in the economic history of the Near East," *International Journal of Middle Eastern Studies* 6 (1), 1975.

———. "Edirne askerî kassamı'na âit tereke defterleri (1545–1659)," *Belgeler* 3 (5–6), 1966.

———. "Edirne ve civarındaki bazı imâret tesislerinin yıllık muhasebe bilânçoları," *Belgeler* 1 (2), 1964.

———. "Essai sur les données statistiques des registres de recensement dans l'Empire ottomane au XVe et XVIe siècles," *Journal of the Economic and Social History of the Orient* 1 (1), 1957.

———. *XV. ve XVI. Asırlarda Osmanlı imparatorluğu'nda ziraî ekonominin hukuki ve mali esasları.* İstanbul, 1943.

Barkan, Ömer L., and Enver Meriçli. *Hüdavendigâr livasi tahrir defterleri.* Vol. 1. Ankara, 1988.

Barker, Graeme. *A Mediterranean valley: Landscape archeology and* Annales *history in the Biferno valley.* New York, 1995.

Barker, John W. "Late Byzantine Thessalonike: A second city's challenges and responses," *Dumbarton Oaks Papers* 57, 2003.

Barkey, Karen. *Bandits and bureaucrats: The Ottoman route to state centralization.* Ithaca, 1994.

Barnes, John Robert. *An introduction to religious foundations in the Ottoman empire.* New York, 1986.

Barrett, W. "World bullion flows, 1450–1800," in James Tracy, ed. *The rise of merchant empires: Long-distance trade in the early modern world, 1350–1750.* New York, 1990.

Barriendos Vallvé, Mariano, and Javier Martín-Vide. "Secular climatic oscillations as indicated by catastrophic floods in the Spanish Mediterranean coastal area (14th–19th centuries)," *Climatic Change* 38 (4), 1998.

Bartlett, Robert. *The making of Europe: Conquest, colonization and cultural change, 950–1350.* Princeton, 1993.

Bates, Daniel G. *Nomads and farmers: A study of the yörük of southeastern Turkey.* Ann Arbor, MI, 1973.

Bautier, Robert-Henri. *The economic development of medieval Europe.* New York, 1971.

Baykara, Tuncer. *Hinis ve Malazgird sancakları yer adları (XVI. yüzyıl).* Ankara, 1991.

Beachy, Robert. *The soul of commerce: Credit, property, and politics in Leipzig, 1750–1840.* Boston, 2005.

Beals, E. W. "The remnant cedar forests of Lebanon," *Journal of Ecology* 53 (3), 1965.

Beaufort, Francis. *Karamania, or, a brief description of the south coast of Asia-Minor and of the remains of antiquity.* London, 1818.

Bechmann, Roland. *Trees and man: The forest in the Middle Ages.* New York, 1990.

Behrens-Abouseif, Doris. "European arts and crafts at the Mamluk court," *Muqarnas* 21, 2004.

Beldiceanu, N., and Irène Beldiceanu-Steinherr. "Riziculture dans l'Empire ottoman (XIVe–XVe siècles)," *Turcica* 9–10 (2), 1978.

Beldiceanu, N., and P. Ş. Năsturel. "Les biens du monastère Sainte-Sophie de Trébizonde dans plusieurs bandon du pays à la charnière de la conquête (1461)," *Byzantion* 60, 1990.

Beldiceanu-Steinherr, Irène. "Les laboureurs associés en Anatolie (XVe–XVIe siècles)," in J.-L. Bacqué-Grammont and Paul Dumont, eds. *Contributions à l'histoire économique et sociale de l'Empire ottoman.* Paris, 1983.

———. "Fiscalité et formes de possession de la terre arable," *Journal of the Social and Economic History of the Orient* 19 (3), 1976.

Belfanti, C. M. "Town and country in central and northern Italy, 1400–1800," in S. R. Epstein, ed. *Town and country in Europe, 1300–1800.* New York, 2001.

Benadusi, Giovanna. *A provincial elite in early modern Tuscany: Family and power in the creation of the state.* Baltimore, 1996.

———. "Career strategies in early modern Tuscany: The emergence of a regional elite," *Sixteenth Century Journal* 25 (1), 1994.

Benedict, Peter. *Ula: An Anatolian town.* Leiden, 1974.

Beniston, M., et al. "Climatic change at high elevation sites: An overview," *Climatic Change* 36 (3–4), 1997.

Bennassar, B., and J. Goy. "Consommation alimentaire (XVIe–XIXe siècle)," *Annales E.S.C.* 30 (2–3), 1975.

Bennassar, Bartolomé. *Valladolid au siècle d'or, une ville de Castille et sa campagne au XVIe siècle.* 2 vols. Paris, 1967.

Bennassar, Bartolomé, ed. *Les catastrophes naturelles dans l'Europe médiévale et moderne.* Toulouse, 1996.

Bensch, Stephen P. *Barcelona & its rulers, 1096–1291.* New York, 1995.

Bergier, Jean-François. "Le cycle médiévale: des sociétés féodales aux Etats territoriaux," in Paul Guichonnet, ed. *Histoire et civilisations des Alpes,* vol. 1: *Destin historique.* Toulouse, 1980.

Berindei, Mihnea, and Gilles Veinstein. *L'Empire ottoman et les pays roumains, 1544–1545: étude et documents.* Paris, 1987.

Berner, Samuel. "Florentine society in the late sixteenth and early seventeenth centuries," *Studies in the Renaissance* 18, 1971.

Berov, Ljuben. "Transport costs and their role in trade in the Balkan lands in the 16th–19th centuries," *Bulgarian Historical Review* 3 (4), 1975.

———. "Changes in price conditions in trade between Turkey and Europe in the 16th–19th century," *Etudes balkaniques* 10 (2–3), 1974.

Berthier, P. *Les anciennes sucreries du Maroc.* Rabat, 1966.

Biger, Gideon, and Nili Liphschitz. "Foreign tree species as construction timber in nineteenth-century Palestine," *Journal of Historical Geography* 21 (3), 1995.

———. "Historical geography and botany: Timber in the nineteenth-century old city of Jerusalem," *Palestine Exploration Quarterly* 123, 1991.

Bilbao, Luis María. "L'expansion de la culture de maïs et le déplacement des centres de gravité économique dans le Pays Basque-Espagnol," in Annalisa Guarducci, ed. *Agricoltura e trans-formazione dell'ambiente, secoli XIII–XVIII.* Florence, 1984.

Bilgin, Arif. *Osmanlı saray mutfağı (1453–1650).* İstanbul, 2004.

———. "Osmanlı sarayında tüketilen iki zıt tadın baş temsilcileri: Tuz, şeker ve bal," in E. G. Naskalı and Mesut Şen, eds. *Tuz kitabı.* İstanbul, 2004.

Bilinkoff, Jodi. *The Avila of Saint Teresa: Religious reform in a sixteenth-century city.* Ithaca, 1989.

Biraben, Jean Noël. *Les hommes et la peste en France et dans les pays européens et méditerranéens.* 2 vols. Paris, 1975–76.

Birmingham, David. *Trade and empire in the Atlantic, 1400–1600.* New York, 2000.

Bishko, J. "Castilian as plainsman: The medieval ranching frontier in La Mancha and Extremadura," in Archibald R. Lewis, ed. *The new world looks at its history.* Austin, 1963.

Blanchard, Ian. " 'The long sixteenth century,' circa. 1450–1650," paper presented at the XIIth International Congress held in Madrid, August 24–28, 1998.

———. "The continental European cattle trades, 1400–1600," *Economic History Review* 39 (3), 1986.

Blanchemanche, Philippe. *Batisseurs de paysages: terrassement, épierrement et petite hydraulique agricoles en Europe, XVIIe–XIXe siècles.* Paris, 1990.

Bloch, Marc. "The problem of gold in the Middle Ages," in M. Bloch, *Land and work in Medieval Europe.* Berkeley, 1967.

———. *French rural history: An essay on its basic characteristics.* Berkeley, 1966.

Bloom, Herbert I. *The economic activities of the Jews in Amsterdam in the seventeenth and eighteenth centuries.* Williamsport, PA, 1937.

Bloom, Jonathan M. *Paper before print: The history and impact of paper in the Islamic world.* New Haven, 2001.

Blum, Jerome. *Lord and peasant in Russia, from the ninth to the nineteenth century.* Princeton, 1961.

Bobovius, Albertus. *Topkapı Sarayı'nda yaşam.* İstanbul, 2002.

Bodman, Herbert L., Jr. *Political factions in Aleppo, 1760–1826.* Chapel Hill, 1963.

Bogucka, Maria. "Amsterdam and the Baltic in the first half of the seventeenth century," *Economic History Review* 26 (3), 1973.

Bois, Guy. *La grande dépression médiéval XIVe et XVe siècles. Le précédent d'une crise systémique.* Paris, 2000.

Bojović, Boško. "Dubrovnik et les Ottoman (1430–1472)," *Turcica* 19, 1987.

Bolens, Lucie. *La cuisine andalouse, un art de vivre XIe–XIIIe siècle.* Paris, 1990.

———. "Pain quotidien et pains de disette dans l'Espagne musulmane," *Annales E.S.C.* 35 (3–4), 1980.

Bonaccorso, B., et al. "Spatial variability of drought: An analysis of the SPI in Sicily," *Water Resources Management* 17 (4), 2003.

Bostan, İdris. *Beylikten imparatorluğa Osmanlı denizciliği.* İstanbul, 2006.

———. *Osmanlı bahriye teşkilatı: XVII. yüzyılda Tersâne-i Âmire.* Ankara, 1992.

Bottema, S. "A pollen diagram from the Syrian Anti Lebanon," *Paleorient* 3, 1975–77.

Boubaker, Sadok. "L'économie de traite dans la régence de Tunis au début du XVIIIe siècle: le comptoir du Cap-Nègre avant 1741," *Revue d'histoire maghrébine* 16 (53–54), 1989.

Boulanger, Patrick. *Marseille, marché international de l'huile d'olive: un produit et des hommes de 1725 à 1825.* Marseille, 1996.

———. "Marines marchands en Méditerranée occidentale: le commerce de l'huile d'olive," in *La Méditerranée au XVIIIe siècle.* Aix-en-Provence, 1987.

Boyajian, James C. *Portuguese trade in Asia under the Habsburgs, 1580–1640.* Baltimore, 1993.

———. *Portuguese bankers at the court of Spain, 1626–1650.* New Brunswick, 1983.

Brandes, Stanley. "Maize as a culinary mystery," *Ethnology* 31 (4), 1992.

Braude, Benjamin. "Rise and fall of Salonica woollens, 1500–1650: Technology transfer and western competition," *Mediterranean Historical Review* 6 (2), 1991.

———. "International competition and domestic cloth in the Ottoman empire, 1500–1650: A study in underdevelopment," *Review* 2 (3), 1979.

Braudel, Fernand. *Memory and the Mediterranean.* New York, 2001.

———. "Le siècle des Génois s'achève-t-il en 1627?" in Fernand Braudel, *Autour de la Méditerranée.* Paris, 1996.

———. *Out of Italy, 1450–1650.* Paris, 1991.

———. *The identity of France.* 2 vols. New York, 1988–90.

———. *Civilization and capitalism, 15th–18th century.* 3 vols. New York, 1981–84.

———. *The Mediterranean and the Mediterranean world in the age of Philip II.* 2 vols. New York, 1972.

———. "Remarques sur la Méditerranée au XVIIIe siècle," *Cahiers de Tunisie* 4, 1956.

Braudel, Fernand, Pierre Jeannin, Jean Meuvret, and Ruggiero Romano. "Le déclin de Venise au XVIIIe siècle," in *Aspetti e cause della decadenza economica veneziana nel secolo XVII.* Venice, 1961.

Braudel, Fernand, and Frank Spooner. "Prices in Europe from 1450 to 1750," in M. M. Postan, ed. *The Cambridge economic history of Europe.* Vol. 4. Cambridge, 1967.

Brauer, Ralph W. "Boundaries and frontiers in medieval Muslim geography," *Transactions of the American Philosophical Society* 85 (6), 1995.

Braunstein, Philippe. "A propos de l'Adriatique entre le XVIe et le XVIIIe siècle," *Annales E.S.C.* 16 (5), 1970.

Brázdil, Rudolph, et al. "Flood events of selected European rivers in the sixteenth century," *Climatic Change* 43 (1), 1999.

Bresc, Henri. *Un monde méditerranéen. Économie et société en Sicile 1300–1450.* 2 vols. Rome, 1986.

Briffa, R. "European tree rings and climatic change in the 16th century," *Climatic Change* 43 (1), 1999.

Brimblecombe, Peter. "Climate conditions and population development in the Middle Ages," *Saeculum* 39 (2), 1988.

Brodman, James William. *Charity and welfare: Hospitals and the poor in medieval Catalonia.* Philadelphia, 1998.

Brown, Judith C. *In the shadow of Florence: Provincial society in Renaissance Pescia.* New York, 1982.

Bruce-Chwatt, L. J. "History of malaria from prehistory to eradication," in Walther H. Wernsdorfer and Ian McGregor, eds. *Malaria: Principles and practice of malariology.* New York, 1988.

Brueton-Governatori, Ariane. "Alimentation et idéologie: le cas de châtaigne," *Annales H.S.S.* 39 (6), 1984.

Brumfield, Allaire. "Osmanlı Girit'inde tarım ve yerleşme, 1669–1898," in U. Baram and Lynda Carroll, eds. *Osmanlı arkeolojisi*. Istanbul, 2004.

Brummett, Palmira. *Ottoman seapower and Levantine diplomacy in the age of discovery*. Albany, 1994.

Brumont, Francis. *La Bureba à l'époque de Philippe II: essai d'histoire rurale quantitative*. New York, 1977.

Brunelle, Gayle K. "Immigration, assimilation, and success: Three families of Spanish origin in sixteenth century Rouen," *Sixteenth Century Journal* 2 (2), 1989.

Bryer, Anthony. "Byzantine porridge," in H. Mayr-Harting and R. I. Moore, eds. *Studies in medieval history presented to R.H.C. Davis*. London, 1985.

Bull, W. E. "The olive industry of Spain," *Economic Geography* 12 (2), 1936.

Burckhardt, John Lewis. *Travels in Syria and the Holy Land*. London, 1812.

Burns, Robert I. "Immigrants from Islam: The Crusaders' use of Muslims as settlers in thirteenth-century Spain," *American Historical Review* 80 (1), 1975.

Busbecq, Ogier Ghislain de. *The Turkish letters of Ogier Ghiselin de Busbecq, imperial ambassador at Constantinople, 1554–1562*. Oxford, 1968.

Busch, Silvia Orvietani. *Medieval Mediterranean ports: The Catalan and Tuscan coasts, 1100 to 1235*. Boston, 2001.

Busch-Zantner, Richard. *Agrarverfassung, gesellschaft und siedlung in Südosteuropa*. Leipzig, 1938.

Butzer, Karl. "The realm of cultural-human ecology: Adaptation and change in historical perspective," in B. L. Turner et al. *Earth as transformed by human action: Global and regional change in the biosphere over the past 300 years*. New York, 1990.

Byrne, Eugene H. "Genoese trade with Syria in the twelfth century," *American Historical Review* 25 (2), 1920.

Cabrera, B. E. "The medieval origins of the great landed estates of the Guadalquivir valley," *Economic History Review* 42 (4), 1989.

Cabrillana, N. "Villages désértes en Espagne," in *Villages désertés et histoire économique, XIe–XVIIe siècles*. Paris, 1967.

Cahen, Claude. *Orient et Occident au temps des croisades*. Paris, 1983.

——. "La communauté rurale dans le monde musulman médiéval," in *Recueils de la Société Jean Bodin*, vol. 42: *Les communautés rurales*. Paris, 1982.

——. "Monetary circulation in Egypt at the time of the Crusades and the reform of Al-Kâmil," in A. Udovitch, ed. *The Islamic Middle East, 700–1900: Studies in economic and social history*. Princeton, 1981.

——. *Pre-Ottoman Turkey: A general survey of the material and spiritual culture and history c. 1071–1330*. New York, 1968.

——. "Quelques problèmes concernant l'expansion économique musulmane au haut moyen âge," *Settimani de Spolete*, 1964.

Cameron, Rondo. "Europe's second logistic," *Comparative Studies in Society and History* 12 (4), 1970.

Camporesi, Piero. *The magic harvest: Food, folklore, and society*. Cambridge, MA, 1993.

Capatti, Alberto, and Massimo Montanari. *Italian cuisine: A cultural history.* New York, 2003.

Carrère, Claude. "Marseille, Aigues-Mortes, Barcelone, et la competition en Méditerranée occidentale au XIIIe siècle," *Anuario de estudios medievales* 10, 1980.

———. *Barcelone. Centre économique à l'époque des difficultés,1380–1462.* 2 vols. Paris, 1967.

Carrière, Charles, and G. Buti. "Un aspect du commerce international du blé: les échanges entre Marseille et l'Andalousie pendant la crise frumentaire de 1753," in Jean-Louis Miège, ed. *Les céréales en Méditerranée.* Marseille, 1993.

Carter, F. W. *Trade and urban development in Poland: An economic geography of Cracow, from its origins to 1795.* New York, 1994.

———. "Cracow's wine trade (fourteenth to eighteenth centuries)," *Slavonic and East European Review* 65 (4), 1987.

———. "The commerce of the Dubrovnik Republic, 1500–1700," *Economic History Review* 24 (3), 1971.

Casale, Giancarlo. "The Ottoman administration of the spice trade in the sixteenth century Red Sea and Persian Gulf," *Journal of the Social and Economic History of the Orient* 49 (2), 2006.

Casey, James. *The Kingdom of Valencia in the seventeenth century.* New York, 1979.

———. "Moriscos and the depopulation of Valencia," *Past and Present* 50, 1971.

Celli, Angelo. *The history of malaria in the Roman campagna from ancient times.* London, 1933.

Cezar, Yavuz. "Bir âyanın muhallefatı: Havza ve Köprü kazaları ayânı Kör İsmail-Oğlu Hüseyin (müsadere olayı ve terekenin incelenmesi)," *Belleten* 41 (161), 1977.

Chapoutot-Remadi, Mounira. "L'agriculture dans l'empire mamluk au moyen âge d'apres al-Nuwayri," *Les cahiers de Tunisie* 12 (85–86), 1974.

Chassagne, Serge. *Coton et ses patrons: France, 1760–1840.* Paris, 1991.

Chaudhuri, K. N. *Trade and civilization in the Indian Ocean: An economic history from the rise of Islam to 1750.* New York, 1985.

Chaunu, Pierre. *European expansion in the later Middle Ages.* New York, 1979.

———. *L'Espagne de Charles Quint.* 2 vols. Paris, 1973.

Chaunu, Pierre, and Huguette Chaunu. "The Atlantic economy and the world economy," in Peter Burke, ed. *Essays in European economic history 1500–1800.* Oxford, 1974.

Chehab, Hafez. "Reconstructing the Medici portrait of Fakhr al-Din al-Ma'ani," *Muqarnas* 11, 1994.

Chevallier, Dominique. *La société du Mont Liban à l'époque de la révolution industrielle en Europe.* Paris, 1971.

———. "Que possédait un cheikh maronite en 1859? Un document de la famille el-Khazen," *Arabica* 7, 1960.

Childs, Wendy R. *Anglo-Castilian trade in the later middle ages.* Manchester, 1978.

Chorley, G. P. H. "The agricultural revolution in northern Europe, 1750–1880: Nitrogen, legumes, and crop productivity," *Economic History Review* 34 (1), 1981.

Chorley, Patrick. "*Rascie* and the Florentine cloth industry during the sixteenth century," *Journal of European Economic History* 32 (3), 2003.

———. "The cloth exports of Flanders and Northern France during the thirteenth century: A luxury trade?" *Economic History Review* 40 (3), 1987.

———. *Oil, silk, and Enlightenment: Economic problems in XVIIIth century Naples.* Naples, 1965.

Ciercerska-Chapowa, Teresa. "Échanges commerciaux entre la Pologne et la Turquie au XVIIIe siècle," *Folia orientalia* 14, 1972–73.

Çiftçi, Cafer. *Bursa'da vakıfların sosyo-ekonomik işlevi.* İstanbul, 2004.

Cipolla, Carlo. *Neşeli öyküler.* İstanbul, 2000.

———. "The decline of Italy: The case of a fully matured economy," *Economic History Review* 5 (2), 1952.

Ciriacono, Salvatore. "Land reclamation: Dutch windmills, private enterprises, and state intervention," *Review* 18 (2), 1995.

———. "The Venetian economy and its place in the world-economy of the 17th and 18th centuries: A comparison with the Low Countries," in Hans-Jürgen Nitz, ed. *The early-modern world-system in geographical perspective.* Stuttgart, 1993.

———. "Venise et la Hollande, pays de l'eau (XVe–XVIIIe siècle)," *Revue historique* 285 (2), 1991.

———. "Silk manufacturing in France and Italy in the XVIIth century: Two models compared," *Journal of European Economic History* 10 (1), 1981.

Çizakça, Murat. "Cash *waqf*s of Bursa, 1555–1823," *Journal of the Economic and Social History of the Orient* 38 (3), 1995.

———. "Tax farming and financial decentralization in the Ottoman economy, 1520–1697," *Journal of European Economic History* 22 (2), 1993.

———. "Incorporation of the Middle East into the European world-economy," *Review* 8 (3), 1985.

———. "Price history and Bursa silk industry: A study in Ottoman industrial decline," *Journal of Economic History* 40 (3), 1980.

Clay, Christopher. "Labor migration and economic conditions in nineteenth-century Anatolia," *Middle Eastern Studies* 34 (4), 1998.

Clayburn, La Force, Jr. *The development of the Spanish textile industry, 1750–1800.* Berkeley, 1965.

———. "Royal textile factories in Spain, 1700–1800," *Journal of Economic History* 24 (3), 1964.

Cohen, Amnon. *Economic life in Ottoman Jerusalem.* New York, 1989.

———. *Palestine in the 18th century.* Jerusalem, 1973.

Cohen, Amnon, and Bernard Lewis. *Population and revenue in the towns of Palestine in the sixteenth century.* Princeton, 1978.

Cohn, Samuel K., Jr. *Creating the Florentine state: Peasants and rural rebellion 1348–1434.* New York, 1999.

———. "Inventing Braudel's mountains: The Florentine Alps after the Black Death," in Samuel K. Cohn Jr. and Steven A. Epstein, eds. *Portraits of medieval and renaissance living: Essays in memory of David Herlihy.* Ann Arbor, MI, 1996.

Coleman, D. C. "An innovation and its diffusion: The 'new draperies,'" *Economic History Review* 12 (3), 1969.

Conder, C. R., and R. E. Kitchener. *The survey of western Palestine.* London, 1881.

Constable, Olivia Remie. *Housing the stranger in the Mediterranean world: Lodging, trade and travel in late Antiquity and the Middle Ages.* New York, 2003.

———. *Trade and traders in Muslim Spain: The commercial realignment of the Iberian peninsula, 900–1500.* New York, 1996.

Cook, Michael. *Population pressure in rural Anatolia, 1450–1600.* London, 1972.

Cox, Oliver. *The foundations of capitalism.* New York, 1959.

Craeybeckx, Jan. *Un grand commerce d'importation: les vins de France aux anciens Pays-Bas, XIIIe–XVIe siècles.* Paris, 1958.

Creasy, E. S. *History of the Ottoman Turks.* London, 1854.

Crecelius, Daniel, and Hamza 'Abd al-Aziz Badr. "French ships and their cargoes: Sailing between Damiette and Ottoman ports 1777–1781," *Journal of the Economic and Social History of the Orient* 37 (1), 1994.

Crosby, Alfred. *Ecological imperialism: The biological expansion of Europe, 900–1900.* New York, 1986.

———. *Columbian exchange: Biological and cultural consequences of 1492.* Westport, CT, 1972.

Crouzet-Pavan, Elisabeth. *Venice triumphant: The horizons of a myth.* Baltimore, 2002.

———. *La mort lente de Torcello.* Paris, 1995.

Cuinet, Vital. *La Turquie d'Asie.* Paris, 1890.

Çulpan, Cevdet. *Türk taş köprüleri: Ortaçağdan Osmanlı devri sonuna kadar.* Ankara, 1975.

Cunningham, A. B., ed. *The early correspondence of Richard Wood, 1831–1841.* London, 1966.

Curtin, Philip. *The rise and fall of the plantation complex.* New York, 1991.

———. *Death by migration: Europe's encounter with the tropical world in the nineteenth century.* New York, 1989.

Cvetkova, Bistra A. *Les institutions ottomanes en Europe.* Wiesbaden, 1978.

Dale, S. *Indian merchants and Eurasian trade, 1600–1750.* New York, 2002.

Dalsar, Fahri. *Türk sanayi ve ticaretinde Bursa'da ipekçilik.* İstanbul, 1960.

Dandini, Jerome. *A voyage to Mount Libanus.* London, 1698.

Dannenfeldt, Karl H. *Leonhard Rauwolf: Sixteenth-century physician, botanist, and traveler.* Cambridge, MA, 1968.

Darling, Linda. *Revenue-raising and legitimacy: Tax collection and finance administration in the Ottoman empire, 1550–1650.* Leiden, 1996.

d'Arrigo, R., and H. Cullen. "A 350-year reconstruction of Turkish precipitation," *Dendrochronologia* 19 (2), 2001.

David, Elizath. *Harvest of the cold months: The social history of ice and ices.* New York, 1995.

Dàvid, Géza. "16–17. Yüzyıllarda Macaristan'ın demografik durumu," *Belleten* 59 (225), 1995.

Davies, Elwyne. "The patterns of transhumance in Europe," *Geography* 26, 1941.

Davis, James C. "Venetian shipbuilders and the fountain of wine," *Past and Present* 156, 1997.

———. *Pursuit of power: Venetian ambassadors' reports on Spain, Turkey, and France in the age of Philip II, 1560–1600.* New York, 1970.

———. *The decline of the Venetian nobility as a ruling class.* Baltimore, 1962.

Davis, Mike. *Late Victorian holocausts: El Niño famines and the making of the third world.* New York, 2001.

Davis, Ralph. "English imports from the Middle East, 1580–1780," in M. A. Cook, ed. *Studies in the economic history of the Middle East.* New York, 1970.

Davis, Timothy. "Changes in the structure of the wheat trade in the seventeenth century Sicily and the building of new villages," *Journal of European Economic History* 12 (2), 1983.

Day, Gerald W. *Genoa's response to Byzantium, 1155–1204: Commercial expansion and factionalism in a medieval city.* Urbana, 1988.

Day, John. "The Levant trade in the middle ages," in Angeliki E. Laiou, ed. *The economic history of Byzantium, from the seventh through the fifteenth century.* Vol. 2. Washington, DC, 2002.

———. "The 'general crisis' and the great depression," in John Day. *Money and finance in the age of merchant capitalism.* New York, 1999.

———. "The monetary crisis of the seventeenth century crisis on colonial Sardinia," in John Day. *Money and finance in the age of merchant capitalism.* New York, 1999.

———. "An outline of money from the Middle Ages to the industrial revolution," in John Day. *Money and finance in the age of merchant capitalism.* New York, 1999.

———. "Problem of the standard in preindustrial Europe," in John Day. *Money and finance in the age of merchant capitalism.* New York, 1999.

———. "Peuplement, cultures et régimes fonciers en Trexenta (Sardaigne), XIII–XVIIIe siècles," in Annalisa Guarducci, ed. *Agricoltura e transformazione dell'ambiente secoli XIII–XVIII.* Florence, 1988.

———. "Introduction," in John Day, ed. *Etudes d'histoire monetaire, XIIe–XIXe siècles.* Lille, 1984.

———. *Les douanes de Gênes, 1376–1377.* 2 vols. Paris, 1963.

———. "Prix agricoles en Méditerranée à la fin du XIVe siècle," *Annales E.S.C.* 16 (4), 1961.

Defourneaux, Marcelin. *Daily life in Spain in the golden age.* Stanford, 1979.

Delano Smith, Catherine. "Climate or man? The evidence of sediments and early maps for the agents of environmental change in post-medieval period," in Catherine Delano Smith and Martin Perry, eds. *Consequences of climatic change.* Nottingham, 1981.

———. *Western Mediterranean Europe: A historical geography of Italy, Spain and southern France since the Neolithic.* London, 1979.

Delbet, M. E. "Paysans en communauté et en polygamie de Bousrah (Esky Cham) dans le pays de Haouran," in F. LePlay, ed. *Les ouvriers de l'Orient.* Tours, 1877.

Delille, Gérard. *Famille et propriété dans le Royaume de Naples, XVe–XIXe siècle.* Rome, 1985.

———. *Croissance d'une société rurale: Montesarchio et la vallée Caudine aux XVIIe et XVIIIe siècles.* Naples, 1973.

Delort, Robert. *Les animaux ont une histoire.* Paris, 1993.

Delort, Robert, and François Walter. *Histoire de l'environnement européen.* Paris, 2001.

Delumeau, J. *Vie économique et sociale de Rome dans la seconde moitié de XVIe siècle.* 2 vols. Paris, 1962.

———. *L'alun de Rome, XVe–XIXe siècles.* Paris, 1962.

Dembińska, Maria. *Food and drink in medieval Poland: Rediscovering a cuisine of the past.* Philadelphia, 1999.

Demetriades, Vassilis. "Problems of land-owning and population in the area of Gazi Evrenos Bey's *wakf,*" *Balkan Studies* 22 (1), 1981.

Demir, Ataman. *Antakya through the ages.* Istanbul, 1996.

Demirel, Ömer. *II. Mahmud döneminde Sivas'ta esnaf teşkilâtı ve üretim-tüketim ilişkileri.* Ankara, 1989.

Derlange, Michel. "Hameaux et communautés d'habitants en Provence orientale sous l'Ancien-Régime," in *Bastides, bories, hameaux: l'habitat dispersé en Provence.* Mouans-Sartoux, 1986.

Dernschwam, Hans. *İstanbul ve Anadolu'ya seyahat günlüğü.* Ankara, 1992.

Desmet-Grégoire, Hélène. *Büyülü divan. 18 yüzyıl Fransasında Türkler ve Türk dünyası.* İstanbul, 1991.

Devèze, Michel. "L'équilibre agro-sylvo-pastoral du XIIIe au XVIIIe siècle en Europe moyenne et Europe méridionale," in A. Guardicci, ed. *Agricoltura e transformazione dell'ambiente, secoli XIII–XVIII.* Firenze, 1984.

———. *L'Espagne de Philippe IV (1621–1665).* Paris, 1970.

———. *La vie de la forêt.* 2 vols. Paris, 1961.

Devroey, Jean Pierre. "Le céréaliculture dans le monde franc," in *L'ambiente vegetale nell'alto medioevo.* Spoleto, 1990.

Deyon, Pierre. "Variations de la production textile aux XVIe et XVIIe siècles: sources et premiers résultats," *Annales E.S.C.* 18 (5), 1963.

Dienne, Louis Édouard Marie Hippolyte, comte de. *Histoire du desséchement des lacs et marais avant 1789.* Paris, 1891.

Dogo, Marco. "Merchants between two empires: The Ottoman colonies of Trieste in the XVIIIth century," *Études balkaniques* 33 (3–4), 1997.

Doğru, Halime. *XV. ve XVI. Yüzyıllarda Sivrihisar nahiyesi.* Ankara, 1997.

———. *XVIII. Yüzyıla kadar Osmanlı kentlerinin sosyal ve ekonomik görüntüsü.* Eskişehir, 1995.

———. *XVI. Yüzyılda Eskişehir ve Sultanönü sancağı.* İstanbul, 1992.

Dollinger, Philippe. *The German Hansa.* Stanford, 1970.

Dols, Michael. "The second plague pandemic and its recurrence in the Middle East: 1347–1894," *Journal of the Economic and Social History of the Orient* 22 (2), 1979.

———. *The Black Death in the Middle East.* Princeton, 1977.

Dotson, John E. "A problem of cotton and lead in medieval Italian shipping," *Speculum* 57 (1), 1982.

Doumani, Beshara. *Rediscovering Palestine: Merchants and peasants in Jabal Nablus, 1700–1900.* Berkeley, 1995.

Dowd, Douglas. "The economic expansion of Lombardy, 1300–1500: A study in political stimuli to economic change," *Journal of Economic History* 21 (2), 1962.

Dozier, Craig L. "Establishing a framework for development in Sardinia: The Campadino," *Geographical Review* 47 (4), 1957.

Duby, Georges. *Rural economy and country life in the medieval West.* Columbia, 1968.

Dufourcq, Charles-Emmanuel. *L'Espagne catalan et le Maghrib aux XIIIe et XIVe siècles.* Paris, 1966.

Dumont, Paul. "La pacification du sud-est anatolien en 1865," *Turcica* 5, 1975.

Dunin-Wąsowicz, Teresa. "Natural environment and human settlement over the central European lowlands in the 13th century," in P. Brimblecombe and C. Pfister, eds. *The silent countdown: Essays in European environmental history.* Berlin, 1990.

———. "Environnement et habitat: la rupture d'équilibre du XIIIe siècle dans la grande plaine européenne," *Annales E.S.C.* 35 (5), 1980.

Dunn, Archie. "Exploitation of woodland and scrubland in the Byzantine World," *Byzantine and Modern Greek Studies* 17 (1993).

Earle, Peter. "The commercial development of Ancona, 1479–1551," *Economic History Review* 22 (1), 1969.

Eberhard, Wolfram. "Nomads and farmers in southeastern Turkey," *Oriens* 6, 1953.

Edwards, Camilla, Karen Livingstone, and Andrew Petersen. "Dayr Hanna: An eighteenth century fortified village in Galilee," *Levant* 25, 1993.

Edwards, John. *Christian Córdoba: The city and its region in the late Middle Ages.* Cambridge, 1982.

Ehrenberg, Richard. *Capital and finance in the age of the Renaissance: A study of the Fuggers and their connections.* London, 1928.

Eldem, Edhem. *French trade in Istanbul in the eighteenth century.* Boston, 1999.

Elliott, J. H. *Imperial Spain, 1469–1716.* London, 1970.

———. "The decline of Spain," *Past and Present* 20, 1961.

Emecen, Feridun. *İlk Osmanlılar ve Batı Anadolu beylikler dünyası.* İstanbul, 2003.

———. "Manisa," in *İslam Ansiklopedisi.* Vol. 7. İstanbul, 2003.

———. *Unutulmuş bir cemaat: Manisa yahudileri.* İstanbul, 1997.

———. "Ayasuluk," in *İslam Ansiklopedisi.* Vol. 4. İstanbul, 1991.

———. *XVI. Asırda Manisa kazası.* Ankara, 1989.

Ener, Kasım. *Tarih boyunca Adana ovasına bir bakış.* İstanbul, 1964.

Enggass, Peter M. "Land reclamation and resettlement in the Guadalquivir Delta-Las Marismas," *Economic Geography* 44 (2), 1968.

Enjalbert, Henri. "Comment naissent les grands crus, Bordeaux, Porto, Cognac," *Annales E.S.C.* 8 (3–4), 1953.

Epstein, S. R. "Cities, regions, and the late medieval crisis: Sicily and Tuscany compared," *Past and Present* 130, 1991.

———. *An island for itself: Economic development and social change in late medieval Sicily.* New York, 1991.

Epstein, Steven A. *Genoa and the Genoese, 958–1528.* Chapel Hill, 1996.

Erder, Leila, and Suraiya Faroqhi. "Population rise and fall in Anatolia, 1550–1620," *Middle Eastern Studies* 15 (3), 1979.

Erdoğru, M. Akif. "Konya Mevlevî Dergâhi'nın malî kaynakları ve idaresi üzerine düşünceler ve belgeler," *Tarih Dergisi,* 1996.

———. "Antalya ve havalisi tarihi için bir kaynak: Defter-i Evkâf-ı Livâ-i Teke," *Tarih İncelemeleri Dergisi* 10, 1995.

———. "Akşehir sancağındaki dirliklerin III. Murad devrindeki durumu ve 1583/991 tarihli Akşehir Sancağı İcmal Defteri," *OTAM* 1 (1), 1990.

Erevnìdis, Pàvlos. "Politika mutfağında tavuk göğsü," *Yemek ve kültür* 1, 2005.

Ergenç, Özer. "XVIII. Yüzyılda Osmanlı sanayi ve ticaret hayatına ilişkin bazı bilgiler," *Belleten* 52 (203), 1988.

Establet, Colette, Jean-Paul Pascual, and André Raymond. "La mesure de l'inégalité dans la société ottomane: utilisation de l'indice de gini pour Le Caire et Damas vers 1700," *Journal of the Economic and Social History of the Orient* 37 (2), 1994.

Evliyâ Çelebi bin Derviş Mehemmed Zillî. *Evliyâ Çelebi seyâhatnâmesi.* 9 vols. Istanbul, 1995–2005.

Eyice, Semavi. "J. von Hammer-Purgstall ve seyahatnâmesi," *Belleten* 46 (183), 1983.

Fagan, Brian. *Fish on Friday: Feasting, fasting, and the discovery of the New World.* New York, 2006.

———. *The Little Ice Age: How climate made history, 1300–1850.* New York, 2000.

———. *Floods, famines and emperors: El Niño and the fate of civilizations.* New York, 1999.

Farnie, D. A. *East and west of Suez: The Suez Canal in history, 1854–1956.* Oxford, 1969.

Faroqhi, Suraiya. "Anayol kavşağında Bursa: İran ipeği, Avrupa rekabeti ve yerel ekonomi," in Suraiya Faroqhi, *Osmanlı dünyasında üretmek, pazarlamak, yaşamak*. İstanbul, 2003.

———. "Tarımsal değişimin bir göstergesi olarak doğal afet: Edirne bölgesinde sel, 1100/1688–89," in Elizabeth Zachariadou, ed. *Osmanlı imparatorluğu'nda doğal afetler*. İstanbul, 2001.

———. "Crisis and change, 1590–1699," in Halil İnalcık and Donald Quataert, eds. *An economic and social history of the Ottoman empire*. New York, 1994.

———. *Pilgrims and sultans*. London, 1994.

———. "Wealth and power in the land of olives: Economic and political activities of Müridzade Hacı Mehmed Ağa, notable of Edremit," in Çağlar Keyder and Faruk Tabak, eds. *Landholding and commercial agriculture in the Middle East*. Albany, 1991.

———. "Notes on the production of cotton and cotton cloth in sixteenth- and seventeenth-century Anatolia," in Huri İslamoğlu-Inan, ed. *The Ottoman empire and the world-economy*. New York, 1987.

———. *Towns and townsmen of Ottoman Anatolia*. New York, 1984.

———. "Tarsus and the tahrir," *Journal of Ottoman Studies* 13, 1983.

———. "Rural society in Anatolia and the Balkans during the sixteenth century, II," *Turcica* 11, 1979.

———. "Sixteenth-century periodic markets in various Anatolian sancaks: Içel, Hamid, Kara-hisar-ı Sahib, Kütahya, Aydın and Menteşe," *Journal of the Economic and Social History of the Orient* 22 (1), 1979.

———. "Rural society in Anatolia and the Balkans during the sixteenth century, I," *Turcica* 9 (1), 1977.

———. "*Vakıf* administration in sixteenth century Konya: The *zaviye* of Sadreddin-i Konyevî," *Journal of the Economic and Social History of the Orient* 17 (2), 1973.

Febvre, Lucien. "Patate et pomme de terre," *Annales d'histoire social* 29 (2), 1940.

Felloni, Guiseppe. "The population dynamics and economic development of Genoa, 1750–1939," in Richard Lawton and Robert Lee, eds. *Population and society in western European port-cities*. Liverpool, 2002.

———. *Gli investimenti finanziari genovesi in Europa tra il Seicento e la Restaurazione*. Milan, 1971.

Fernández-Armesto, Felipe. *Civilizations: Culture, ambition, and the transformation of nature*. New York, 2001.

———. *Before Columbus: Exploration and colonization from the Mediterranean to the Atlantic 1229–1492*. Philadelphia, 1987.

Feyzullah Efendi. *Fetava-i feyziyye*. Istanbul, 1266/1848.

Filippini, Jean Pierre. "Les nations à Livourne XVIIe–XVIIIe siècles," in Simonetta Cava-ciocchi, ed. *I porti come impresa economica*. Florence, 1988.

———. "La mer et le ravitaillement de Livourne au XVIIIe siècle," in Micheline Galley and Leïla Ladjimi Sebai, eds. *L'homme mediterranéen et la mer*. Tunis, 1985.

Firestone, Ya'akov. "The land-equalizing *mushâ*'village: A reassessment," in Gad G. Gilbar, ed. *Ottoman Palestine, 1800–1914*. Leiden, 1990.

Flandrin, Jean-Louis. "Le goût et la nécessité: sur l'usage des graisses dans les cuisines d'Europe occidentale (XIVe–XVIIIe siècle)," *Annales E.S.C.* 38 (2), 1983.

Fleet, Kate. *European and Islamic trade in the early Ottoman state: The merchants of Genoa and Turkey*. New York, 1999.

———. "Ottoman grain exports from western Anatolia at the end of the fourteenth century," *Journal of the Economic and Social History of the Orient* 40 (3), 1997.

Flynn, Dennis, and Arturo Giráldez. "Silver and Ottoman monetary history in global perspective," *Journal of European Economic History* 31 (1), 2002.

———. "Cycles of silver: Global economic unity through the mid-eighteenth century," *Journal of World History* 13 (2), 2002.

———. "Silk for silver: Manila-Macao trade in the 17th century," *Philippine Studies* 44 (1), 1996.

Fontenay, Michel. "La place de la course dans l'économie portuaire: l'exemple de Malte et des ports barbaresques," *Annales E.S.C.* 43 (6), 1988.

Fortea Pérez, José Ignatio. "The textile industry in the economy of Cordoba at the end of the seventeenth and the start of the eighteenth centuries: A frustrated recovery," in I. A. A. Thompson and Bartolomé Yun Casilla, eds. *The Castilian crisis of the seventeenth century.* New York, 1994.

Foss, Clive. *Ephesus after antiquity: A late antique, Byzantine and Turkish city.* New York, 1979.

Fossier, Robert. "Europe's second wind," in Robert Fossier, ed. *The middle ages, 1250–1520.* New York, 1986.

Franc, J. *La colonisation de la Mitidja.* Paris, 1928.

Frêche, Georges. *Toulouse et la région et la région Midi-Pyrénées au siècle des lumières (vers 1670–1789).* Paris, 1974.

Friedmann, Harriet. "World market, state, and family farm: Social bases of household production in the era of wage labor," *Comparative Studies in Society and History* 20 (4), 1978.

Fukasawa, Katsumi. *Toilerie et commerce du Levant d'Alep à Marseille.* Paris, 1987.

Fuller, Anne H. *Buarij, portrait of a Lebanese Muslim village.* Cambridge, MA, 1961.

Galassi, F. "Reassessing Mediterranean agriculture: Retardation and growth in Tuscany, 1870–1914," *Revista di storia economica* 3, 1986.

Galloway, John H. *The sugar cane industry: An historical geography from its origins to 1914.* New York, 1989.

———. "The Mediterranean sugar industry," *Geographical Review* 67 (2), 1977.

Galloway, Patrick R. "Long-term fluctuations in climate and population in the preindustrial era," *Population and Development Review* 12 (1), 1986.

Galt, Anthony H. "Social class in a mid-eighteenth-century Apulian town: Indications from the Catasto Onciario," *Ethnohistory* 33 (4), 1986.

García-Arenal, Mercedes, and Gerard Wiegers. *A man of three worlds: Samuel Pallache, a Moroccan Jew in Catholic and Protestant Europe.* Baltimore, 2003.

García Sanz, Angel. "Castile 1580–1650," in I. A. A. Thompson and Bartolomé Yun Casilla, eds. *The Castilian crisis of the seventeenth century.* New York, 1994.

Gascon, Richard. *Grand commerce et vie urbaine au XVIe siècle: Lyon et ses marchands (environs de 1520–environs de 1580).* 2 vols. Paris, 1971.

———. "Quelques aspects du rôle des Italiens dans la crise du foires de Lyon du dernier tiers du XVIe siècle," *Cahiers d'histoire* 5 (1), 1960.

Gelabert, Juan E. "Urbanization and deurbanization," in I. A. A. Thompson and Bartolomé Yun Casallila, eds. *The Castilian crisis of the seventeenth century.* New York, 1994.

Genç, Mehmet. "18. Yüzyılda Türkiye'den Fransa'ya yapılan pamuklu ihracatı ile ilgili bir araştırma hakkında düşünceler," in Mehmet Genç. *Osmanlı imparatorluğu'nda devlet ve ekonomi.* İstanbul, 2000.

———. "Ottoman industry," in Donald Quataert, ed. *Manufacturing in the Ottoman empire and Turkey, 1500–1914.* Albany, 1994.

———. "17.–19. Yüzyıllarda sanayi ve ticaret merkezi olarak Tokat," in S. Hayri Bolay et al., eds. *Türk tarihinde ve Türk kültüründe Tokat.* Ankara, 1987.

———. "18. Yüzyılda Osmanlı ekonomisi ve savaş," *Yapıt* 4–5, 1984.

Gentilcore, R. Louis. "The reclamation of Agro Pontino, Italy," *Geographical Review* 60 (3), 1970.

Gentil Da Silva, José. *Banque et crédit en Italie au XVIIe siècle.* 2 vols. Paris, 1969.

———. *En Espagne. Développement économique, subsistance, déclin.* Paris, 1965.

———. "L'histoire économique et sociale de l'empire Ottoman et de la Méditerranée," *Association Internationale d'Études du Sud-Est Européen Bulletin* 12 (2), 1963.

Georgelin, Jean. *L'Italie à la fin du XVIIIe siècle.* Paris, 1989.

———. *Venise au siècle des lumières.* Paris, 1978.

Georges, Pierre. *La région du Bas-Rhône, étude de géographie régionale.* Paris, 1935.

Gerber, Haim. *Economy and society in an Ottoman city: Bursa, 1600–1700.* Jerusalem, 1988.

———. *Social origins of the modern Middle East.* Boulder, CO, 1987.

Gerbet, Marie Claude. *L'élevage dans la royaume de Castille sous les rois Catholiques (1474–1516).* Madrid, 1999.

Geremek, Bronislaw. *Poverty: A history.* Cambridge, MA, 1994.

Gertwagen, Ruth. "Port of Modon," in Alexander Cowan, ed. *Urban Europe, 1500–1700.* New York, 1998.

Geyer, Bernard. "Physical factors in the evolution of the landscape and land use," in Angeliki E. Laiou, ed. *The economic history of Byzantium, from the seventh through the fifteenth century.* Vol. 1. Washington, DC, 2002.

Gibb, H. A. R., and H. Bowen. *Islamic society and the West.* Vol. 1. London, 1962.

Gilmore, David. "Class consciousness of the Andalusian rural proletarians in historical perspective," *Ethnohistory* 24 (2), 1977.

Gioffrè, Domenico. *Gênes et les foires de changes, de Lyon a Besançon.* Paris, 1960.

Glahn, Richard von. *Fountain of fortune: Money and monetary policy in China, 1000–1700.* Berkeley, 1996.

Glamann, Kristof. "European trade 1500–1750," in Carlo M. Cipolla, ed. *The Fontana economic history of Europe,* vol. 2: *The sixteenth and seventeenth centuries.* London, 1974.

Glazer, Rüdiger. "On the course of temperature in central Europe since the year 1000 A.D.," *Historical Social Research* 22 (1), 1997.

Glick, Thomas F. *Irrigation and society in medieval Valencia.* Cambridge, MA, 1970.

Godinho, Vitorino Magalhães. "Portugal and the making of the Atlantic world: Sugar fleets and gold fleets, the seventeenth to the eighteenth centuries," *Review* 28 (4), 2005.

———. "La guerres du blé au Maroc," *Historia Economica y Social* 1, 1968.

———. "Le repli vénetien et égyptien et la route du Cap, 1496–1533," in *Éventail de l'histoire vivante: Hommage à Lucien Febvre.* Vol. 2. Paris, 1953.

Goffman, Daniel. *The Ottoman empire and early modern Europe, 1300–1700.* New York, 2002.

———. "İzmir," in Edhem Eldem, Daniel Goffman, and Bruce Masters. *The Ottoman city between East and West: Aleppo, İzmir, and Istanbul.* New York, 1999.

———. *İzmir and the Levantine world, 1550–1650.* Seattle, 1990.

Goitein, S. D. *Mediterranean society: The Jewish communities of the Arab world as portrayed in the documents of the Cairo Geniza.* Berkeley, 1967–.

Gökbilgin, Tayyib. "Venedik devlet arşivindeki vesikalar külliyatında Kanunî Sultan Süleyman devri belgeleri," *Belgeler* 1 (2), 1964.

———. *XV.–XVI. Asırlarda Edirne ve Paşa livası: Vakıflar, mülkler, mukataalar.* İstanbul, 1952.

Gökçe, Turan. *XVI. ve XVII. Yüzyıllarda Lâzıkıyye (Denizli) kâzası.* Ankara, 2000.

Gökyay, Orhan Şaik. *Kâtip Çelebi'den seçmeler.* İstanbul, 1968.

Goldsmith, Raymond. *Premodern financial systems: A historical comparative study.* New York, 1987.

Goldstone, Jack A. *Revolution and rebellion in the early modern world.* Berkeley, 1991.

———. "Urbanization and inflation: Lessons from the English price revolution of the 16th and 17th centuries," *American Journal of Sociology* 89 (5), 1984.

Goldthwaite, Richard A. "The Florentine wool industry in the late sixteenth century: A case study," *Journal of European and Economic History* 32 (3), 2003.

———. *Wealth and the demand for art in Italy, 1300–1600.* Baltimore, 1993.

———. "The Medici Bank and the world of Florentine capitalism," *Past and Present* 114, 1987.

Golombek, Lisa. "The draped universe of Islam," in Priscilla Soucek, ed. *Content and context of visual arts in the Islamic world.* University Park, PA, 1988.

Gonzáles, A. García-Baquero. "Andalusia and the crisis of the Indies trade, 1610–1720," in I. A. A. Thompson and Bartolomé Yun Casilla, eds. *The Castilian crisis of the seventeenth century.* New York, 1994.

Goodman, David. *Spanish naval power, 1589–1665: Reconstruction and defeat.* New York, 1997.

Goodman, Jordan. "Financing pre-modern European industry: An example from Florence, 1580–1660," *Journal of European Economic History* 10 (2), 1981.

Goubert, Jean-Pierre. *The conquest of water: The advent of health in the industrial age.* Princeton, 1989.

Goubert, Pierre. "Sociétés rurales françaises du XVIIIe siècle. Vingt paysanneries contrastées: quelques problèmes," in *Conjoncture économique structures sociales. Hommage à Ernest Labrousse.* Paris, 1974.

———. "Les campagnes françaises," in Fernand Braudel and Ernest Labrousse, eds. *Histoire économique et sociale de la France,* vol. 2: *1660–1789.* Paris, 1970.

Gould, Andrew G. "The burning of the tents: The forcible settlement of nomads in southern Anatolia," in Heath W. Lowry and Donald Quataert, eds. *Humanist and scholar: Essays in honor of Andreas Tietze.* Istanbul, 1993.

Goy, Joseph, and Anne-Lise Head-Köenig. "Une expérience. Les revenus décimaux en France méditerranéenne XVIe–XVIIIe siècle," *Études rurales* 36, 1969.

Goyau, Georges. *Un précurseur: François Picquet. Consul de Louis XIV en Alep.* Paris, 1942.

Göyünç, Nejat. *XVI. Yüzyılda Mardin sancağı.* Ankara, 1991.

Grandin, Thierry. "La savonnerie traditionnelle à Alep," *Bulletin d'études orientales* 36, 1984.

Grant, Christina Phelps. *The Syrian desert: Caravans, travel and exploration.* London, 1937.

Greene, Molly. "Resurgent Islam: 1500–1700," in David Abulafia, ed. *The Mediterranean in history.* Los Angeles, 2003.

———. "Beyond the northern invasion: The Mediterranean in the seventeenth century," *Past and Present* 174, 2002.

———. *A shared world: Christians and Muslims in the early modern Mediterranean.* Princeton, 2000.

Greenwood, Antony. "Istanbul's meat provisioning: A study of the celebkeşan system," Ph.D. diss., University of Chicago, 1988.

Grigg, D. B. *The agricultural systems of the world: An evolutionary approach.* Cambridge, 1974.

Griswold, William J. "Climatic change: A possible factor in the social unrest of seventeenth century Anatolia," in Heath W. Lowry and Donald Quataert, eds. *Humanist and scholar: Essays in honor of Andreas Tietze.* Istanbul, 1993.

———. *The great Anatolian rebellion, 1591–1611.* Berlin, 1983.

———. "A sixteenth-century Ottoman foundation," *Journal of the Economic and Social History of the Orient* 27 (2), 1983.

Grossi, Paolo. *An alternative to private property: Collective property in the juridical consciousness of the nineteenth century.* Chicago, 1981.

Grossman, David, and Zeev Safrai. "Satellite settlements in western Samaria," *Geographical Review* 70 (4), 1980.

Grove, A. T. "The 'Little Ice Age' and its geomorphological consequences in Mediterranean Europe," *Climatic Change* 48 (1), 2001.

Grove, Jean M. *The Little Ice Age.* New York, 1988.

Grove, Jean M., and Annalisa Conterio. "The climate of Crete in the sixteenth and seventeenth centuries," *Climatic Change* 30 (2), 1995.

———. "Climate in the eastern and central Mediterranean, 1675 to 1715," in Burkhard Frenzel et al., eds. *Climatic trends and anomalies in Europe, 1675–1715.* Stuttgart, 1994.

Guarini, Elena Fasano. "Center and periphery," *Journal of Modern History* 67, 1995.

Güçer, Lütfi. *XVI.–XVII. Asırlarda Osmanlı imparatorluğu'nda hububat meselesi ve hububattan alınan vergiler.* İstanbul, 1964.

———. "Onaltıncı yüzyılın sonlarında Osmanlı imparatorluğu dahilinde hububat ticaretinin tabi olduğu kayıtlar," *İktisat Fakültesi Mecmuası* 13 (1–4), 1951–52.

Guiral-Hadziiossif, Jacqueline. *Valence, port méditerranéen au XVe siècle (1410–1525).* Paris, 1986.

Guirescu, Constantin C. "The Genoese and the lower Danube in the XIIIth and XIVth centuries," *Journal of European Economic History* 5 (3), 1976.

Gulick, John. *Social structure and cultural change in a Lebanese village.* New York, 1955.

Gümüşçü, Osman. *XVI. Yüzyıl Larende (Karaman) kazasında yerleşme ve nüfus.* Ankara, 2001.

Güran, Tevfik. *Ekonomik ve mali yönleriyle vakıflar.* İstanbul, 2006.

———. *19. Yüzyıl Osmanlı tarımı.* İstanbul, 1998.

———. *Osmanlı dönemi tarım istatistikleri 1909, 1913 ve 1914.* İstanbul, 1997.

Gutmann, Myron P. *Toward the modern economy: Early industry in Europe, 1500–1800.* Philadelphia, 1988.

Hagen, William W. "Seventeenth-century crisis in Brandenburg: The Thirty Years' War, the destabilization of serfdom, and the rise of absolutism," *American Historical Review* 94 (2), 1989.

Hamilton, Alastair, Alexander H. de Groot, and Maurits H. van den Boogert, eds. *Friends and rivals in the East: Studies in Anglo-Dutch relations in the Levant from the seventeenth to the early nineteenth century.* Boston, 2000.

Hamilton, E. J. *American treasure and the price revolution in Spain, 1501–1650.* Cambridge, MA, 1934.

Hanna, Nelly. *Making big money in 1600: The life and times of Isma'il abu Taqiyya, Egyptian merchant.* Syracuse, 1998.

Hanson, J. R. *Trade in transition: Exports from the third world.* New York, 1980.

Harik, Ilya. *Politics and change in a traditional society: Lebanon, 1711–1845.* Princeton, 1968.

Harreld, Donald. *High Germans in the Low Countries: German merchants and commerce in golden age Antwerp.* Boston, 2004.

———. "Atlantic sugar and Antwerp's trade with Germany in the sixteenth century," *Journal of Early Modern History* 7 (1–2), 2003.

Harvey, A. "Economic conditions in Thessaloniki between the two Ottoman occupations," in Alexander Cowan, ed. *Urban Europe, 1500–1700.* New York, 1998.

Hathaway, Jane. "The wealth and influence of an exiled Ottoman eunuch in Egypt: The waqf inventory of 'Abbâs Agha," *Journal of the Economic and Social History of the Orient* 37 (4), 1997.

Hattox, Ralph S. "Some Ottoman *tapu defters* for Tripoli in the sixteenth century," *al-Abhath* 19, 1981.

Headrick, Daniel. *The tools of empire: Technology and European imperialism in the nineteenth century.* New York, 1981.

Heers, Jacques. *Jacques Coeur, 1400–1456.* Paris, 1997.

———. *Gênes au XVe siècle. Civilisation méditerranéenne, grand capitalisme, et capitalisme populaire.* Paris, 1971.

———. "La mode et les marchés des draps de laine: Gênes et la montagne à la fin du moyen âge," *Annales E.S.C.* 26 (5), 1971.

Hemardinquer, J. J. "*Turcicum frumentum:* une survivance du principe d'autorité?" *Candollea* 19, 1964.

Hendy, Michael F. *Studies in the Byzantine monetary economy, c. 300–1450.* New York, 1985.

Herlihy, David. *Medieval and Renaissance Pistoia: The social history of an Italian town, 1200–1430.* New Haven, 1967.

Herlihy, David, and Christiane Klapisch-Zuber. *Tuscans and their families: A study of the Florentine catasto of 1427.* New Haven, 1985.

Herlihy, Patricia. *Odessa: A history, 1794–1914.* Cambridge, MA, 1986.

———. "Russian wheat and the port of Livorno, 1794–1861," *Journal of European Economic History* 5 (1), 1976.

Herr, Richard. *Rural change and royal finances in Spain at the end of the old regime.* Berkeley, 1989.

———. "Villages désertés en Espagne au XVIIIe siècle," *Annales E.S.C.* 41 (1), 1986.

Herzig, Edmund. "The Iranian raw silk trade and European manufacture in the seventeenth and eighteenth centuries," *Journal of European Economic History* 19 (1), 1990.

Hess, Andrew. "The battle of Lepanto and its place in Mediterranean history," *Past and Present* 57, 1972.

———. "The evolution of the Ottoman seaborne empire in the age of oceanic discoveries, 1453–1525," *American Historical Review* 75 (7), 1970.

Heyberger, Bernard. *Les chrétiens de proche-orient au temps de la réforme catholique (Syrie, Liban, Palestine, XVIIe–XVIIIe siècles).* Rome, 1994.

Heyd, Uriel. *Studies in old Ottoman criminal law.* Oxford, 1973.

Heyd, William. *Histoire du commerce du Levant au moyen-âge.* 2 vols. Leipzig, 1885–86.

Hezârfen Hüseyin Efendi. *Telhîsü'l-beyân fî kavânîn-i Al-i Osmân*. Ed. Sevim İlgürel. Ankara, 1998.

Hinderink, Jan, and Mübeccel Kiray. *Social stratification as an obstacle to development: A study of four Turkish villages*. New York, 1970.

Histoire économique du monde méditerranéen, 1450–1650. Mélanges en l'honneur de Fernand Braudel. Toulouse, 1973.

Hobhouse, Henry. *Seeds of change: Five plants that transformed mankind*. New York, 1987.

Hocquet, Jean-Claude. "Ships, sailors, and maritime activity in Constantinople (1436–1440)," *Journal of European Economic History* 30 (3), 2001.

———. *Le sel et la fortune de Venise*. 2 vols. Villenuve d'Ascq, 1978–79.

Hohenberg, Paul H. "Maize in French agriculture," *Journal of European Economic History* 6 (1), 1977.

Holt, P. M. "The treaties of the early Mamluk Sultans with the Frankish states," *Bulletin of the School of Oriental and African Studies* 43 (1), 1980.

Hoogasian, Susie, and Mary Kilbourne Matossian. *Armenian village life before 1914*. Detroit, 1982.

Hopwood, Keith. "Manavgat valley: Perspectives on social history," in J.-L. Bacque-Grammont, İlber Ortaylı, and Emeri van Donzel, eds. *CIÉPO Osmanlı öncesi ve Osmanlı araştırmaları Komitesi VII. Sempozyumu bildirileri*. Ankara, 1994.

Horden, Peregrine, and Nicholas Purcell. *The corrupting sea: A study of Mediterranean history*. Malden, MA, 2000.

Hoshino, Hidetoshi, and Maureen Fennell Mazzaoui. "Ottoman markets for Florentine woolen cloth in the late fifteenth century," *International Journal of Turkish Studies* 3 (2), 1985–86.

Hoszowski, Stanislas. "L'Europe centrale devant la révolution des prix XVIe et XVIIe siècles," *Annales E.S.C.* 16 (3), 1961.

Hourani, Albert. "Ideologies of the mountain and the city," in Roger Owen, ed. *Essays on the crisis in Lebanon*. Ithaca, 1976.

Hourani, George. *Arab seafaring in the Indian Ocean in ancient and early medieval times*. Princeton, 1995.

Houston, J. M. *The western Mediterranean world: An introduction to its regional landscapes*. New York, 1967.

———. "Social geography of rice cultivation in Spain," *Indian Geographical Journal* 25, 1951.

Houtte, J. A. van. "Les grands itinéraires du commerce (XIIIe–XVIIIe siècles)," in Anna Vannini Marx, ed. *Transporti e sviluppo economico (secoli XIII–XVIII)*. Florence, 1986.

———. *An economic history of the Low Countries, 800–1800*. New York, 1977.

Howard, Debbie. "Death in Damascus: Venetians in Syria in the mid-fifteenth century," *Muqarnas* 20, 2003.

Howard, Douglas A. "The Ottoman *tımar* system and its transformation, 1563–1656," Ph.D. diss., Indiana University, 1987.

Hubert, Dominique. "Les *qaysâriyya* de textile: une équipment de la ville," *Bulletin d'études orientales* 36, 1984.

Huetz de Lemps, Alain. *Vignobles et vins du Nord-Ouest de l'Espagne*. 2 vols. Bordeaux, 1967.

Hufton, Olwan. "Social conflict and the grain supply in eighteenth century France," *Journal of Interdisciplinary History* 14 (3), 1983.

Hunt, Edwin S. *The medieval super-companies: A study of the Peruzzi company of Florence.* New York, 1994.

Hunt, Edwin S., and James M. Murray. *A history of business in Medieval Europe, 1200–1550.* New York, 1999.

Hütteroth, Wolf-Dieter. "The role of the Ottoman empire in the early modern world-system," in Hans-Jürgen Nitz, ed. *The early-modern world-system in geographical perspective.* Stuttgart, 1993.

———. "The pattern of settlement in Palestine in the sixteenth century," in Moshe Maoz, ed. *Studies on Palestine during the Ottoman period.* Jerusalem, 1975.

———. "The influence of social structure on land division and settlement in inner Anatolia," in Peter Benedict et al., eds. *Turkey: Geographic and social perspectives.* Leiden, 1974.

———. *Ländliche siedlungen im südlichen Inneranatolien in den letzten vierhundert jahren.* Göttingen, 1968.

Hütteroth, Wolf-Dieter, and Kamal Abdulfattah. *Historical geography of Palestine, Transjordan and Southern Syria in the late 16th century.* Erlangen, 1977.

Hybel, Nils. "The grain trade in northern Europe before 1350," *Economic History Review* 55 (2), 2002.

İbn Battûta Tancî. *İbn Battûta seyahatnâmesi.* 2 vols. İstanbul, 2000.

İbni Cübeyr. *Endülüsten kutsal topraklara.* İstanbul, 2003.

İlgürel, Mücteba. "Hububat kaçakçılığı," in Tuncer Baykara, ed. *XIV. CIÉPO Sempozyumu bildirileri.* Ankara, 2004.

Imber, Colin. "The navy of Süleyman the Magnificent," *Archivum Ottomanicum* 6, 1980.

İnalcık, Halil. "The Ottoman state: Economy and society, 1300–1600," in Halil İnalcık and Donald Quataert, eds. *An economic and social history of the Ottoman empire.* New York, 1994.

———. "The emergence of big farms, *çiftlik*s: State, landlords, and tenants," in J.-L. Bacqué-Grammont and Paul Dumont, eds. *Contributions à l'histoire économique et sociale de l'Empire ottoman.* Louvain, 1983.

———. "Rice cultivation and the *çeltükci-re'âyâ* system in the Ottoman empire," *Turcica* 14, 1982.

———. "Osmanlı pamuklu pazarı, Hindistan ve İngiltere: Pazar rekabetinde emek maliyetinin rolü," *ODTÜ Gelişme Dergisi,* 1979–80.

———. "The impact of the *Annales* school on Ottoman studies and new findings," *Review* 1 (3–4), 1978.

———. *Hicrî 835 tarihli sûret-i defter-i sancak-ı Arvanid.* Ankara, 1954.

İpşirli, Mehmed, and Mohammed Da'oud al-Tamimi. *The Muslim pious foundations and real estates in Palestine.* Istanbul, 1982.

İrepoğlu, Gül. *Levni.* Ankara, 1998.

İslamoğlu, Huri, and Suraiya Faroqhi. "Crop patterns and agricultural production trends in sixteenth-century Anatolia," *Review* 2 (3), 1979.

İslamoğlu, Huri, and Çağlar Keyder. "Agenda for Ottoman history," *Review* 1 (1), 1977.

İslamoğlu-İnan, Huri. *State and peasant in the Ottoman empire.* Leiden, 1994.

Israel, Jonathan. "Trade, politics, and strategy: The Anglo-Dutch wars in the Levant (1647–1675)," in Alastair Hamilton et al., eds. *Friends and rivals in the East: Studies in Anglo-Dutch relations in the Levant from the seventeenth to the early nineteenth century.* Boston, 2000.

———. *The Dutch Republic: Its rise, greatness, and fall, 1477–1806.* New York, 1995.

———. *Dutch primacy in world trade, 1585–1740.* New York, 1989.

Issawi, Charles. *The Fertile Crescent, 1800–1914: A documentary study.* New York, 1988.

———. *An economic history of the Middle East and North Africa.* New York, 1982.

———. *The economic history of Turkey, 1800–1914.* Chicago, 1980.

Izz al-Dîn ibn Šaddâd. *Description de la Syrie du Nord.* Trans. Anne-Marie Eddé-Terasse. Damascus, 1984.

Jacoby, David. "Foreigners and the urban economy in Thessalonike, ca. 1150–ca. 1450," *Dumbarton Oaks Papers* 57, 2003.

———. "Changing economic patterns in Latin Romania: The impact of the west," in Angeliki E. Laiou and Roy Parvez Mottahedeh, eds. *The crusades from the perspective of Byzantium and the Muslim world.* Washington, DC, 2001.

———. "Genoa, silk trade and silk manufacturing in the Mediterranean region (ca. 1100–1300)," in A. R. Calderoni Masetti, C. Di Fabio, and M. Mercenaro, eds. *Tessuti, oreficerie, miniature in Liguria, XIII–XV xecolo.* Bordighera, 1999.

———. "Silk production in the Frankish Peloponnese: The evidence of the fourteenth century surveys and reports," in H. A. Kalligas, ed. *Travellers, officials in the Peloponnese.* Monemvasia, 1994.

Jacquart, Jean. "Le rôle de la grande exploitation dans la formation des paysages des plaines limoneuses de la France du Nord (c. 1450–c. 1800)," in Annalisa Guarducci, ed. *Agricoltura e transformazione dell'ambiente secoli XIII–XVIII.* Florence, 1988.

Jarcho, Saul. *Quinine's predecessor: Francesco Tortini and the early history of cinchona.* Baltimore, 1993.

Jeannin, Pierre. "The sea borne and the overland routes to northern Europe in the XVIth and XVIIth centuries," *Journal of European Economic History* 11 (1), 1982.

Jennings, Ronald C. *Christians and Muslims in Ottoman Cyprus and the Mediterranean world, 1571–1640.* New York, 1993.

———. "Pious foundations in the society and economy of Ottoman Trabzon, 1565–1640," *Journal of the Economic and Social History of the Orient* 33 (3), 1990.

———. "The population, society, and economy of the region of Erciyes Dağı in the sixteenth century," in J.-L. Bacqué-Grammont and Paul Dumont, eds. *Contributions à l'histoire économique et sociale de l'Empire ottoman.* Paris, 1983.

———. "Urban population in Anatolia in the sixteenth century: A study of Kayseri, Karaman, Amasya, Trabzon and Erzurum," *International Journal of Middle Eastern Studies* 7 (1), 1976.

Johnson, Christopher. *The life and death of industrial Languedoc, 1700–1920.* New York, 1995.

Jones, P. D., and R. S. Bradley. "Climatic variations over the last 500 years," in Raymond S. Bradley and Philip D. Jones, eds. *Climate since A.D. 1500.* New York, 1992.

Jordan, William Chester. *The great famine: Northern Europe in the early fourteenth century.* Princeton, 1996.

Kafadar, Cemal. *Between two worlds: The construction of the Ottoman state.* Berkeley, 1995.

———. "A death in Venice (1575): Anatolian muslim merchants trading in the Serenissima," *Journal of Turkish Studies* 10, 1986.

Kal'a, Ahmet, ed. *İstanbul ahkâm defterleri: İstanbul tarım tarihi,* vol. 1: *1743–1757.* İstanbul, 1997.

Kalligas, Haris. "Monemvasia, seventh–fifteenth centuries," in Angeliki E. Laiou, ed. *The economic history of Byzantium, from the seventh through the fifteenth century.* Vol. 2. Washington, DC, 2002.

Kamen, Henry. *Empire: How Spain became a world power, 1492–1763.* New York, 2003.

———. "Mediterranean and the expulsion of Spanish Jews," *Past and Present* 119, 1988.

———. "The decline of Spain: A historical myth?" *Past and Present* 81, 1973.

Kaplan, Michel. *Les hommes et la terre à Byzance du VIe au XIe siècle.* Paris, 1992.

Karmon, Y. "The settlement of the northern Huleh valley since 1838," *Israel Exploration Journal* 3 (1), 1953.

Karpov, S. P. "The grain trade in the southern Black Sea region: The thirteenth to the fifteenth century," *Mediterranean Historical Review* 8 (1), 1993.

Kasaba, Reşat. "Migrant labor in Anatolia, 1750–1850," in Çağlar Keyder and Faruk Tabak, eds. *Landholding and commercial agriculture in the Middle East.* Albany, 1991.

———. *The Ottoman empire and the world-economy.* Albany, 1988.

Kasaba, Reşat, Çağlar Keyder, and Faruk Tabak. "Eastern Mediterranean port cities and their bourgeoisies: Merchants, political projects, and nation-states," *Review* 10 (1), 1986.

Katele, Irene. "Piracy and the Venetian state: The dilemma of maritime defense in the fourteenth century," *Speculum* 63 (4), 1988.

Kedar, Benjamin. *Merchants in crisis: Genoese and Venetian men of affairs and the fourteenth century depression.* New Haven, 1976.

Kellenbenz, Hermann. "From Melchior Manlich to Ferdinand Cron: German Levantine and Oriental trade relations (second half of XVIth and beginning of XVIIth centuries)," *Journal of European Economic History* 19 (3), 1990.

———. *The rise of the European economy: An economic history of continental Europe from the fifteenth to the eighteenth century.* New York, 1976.

———. "Industries rurales en Occident de la fin du moyen âge au XVIIIe siècle," *Annales E.S.C.* 18 (5), 1963.

———. "Le déclin de Venise et les relations économiques de Venise avec les marchés au nord des Alpes (fin du XVIème–commencement du XVIIIème siècle)," in *Aspetti e cause della decadenza economica veneziana nel secolo XVII.* Venice, 1961.

Kennedy, Hugh. *Crusader castles.* Cambridge, 1994.

Kévonian, Kéram. "Marchands arméniens au XVIIe siècle, à propos d'un livre arménien publié à Amsterdam en 1699," *Cahiers du monde russe et soviétique* 16 (2), 1975.

Keyder, Çağlar. *State and class in Turkey: A study in capitalist development.* New York, 1987.

———. "Small peasant ownership in Turkey: Historical formation and present structure," *Review* 7 (1), 1983.

———. "The cycle of sharecropping and the consolidation of small peasant ownership in Turkey," *Journal of Peasant Studies* 10 (2), 1983.

Khalidi, Walid, ed. *All that remains: The Palestinian villages occupied and depopulated by Israel in 1948.* Washington, DC, 1992.

Khanzadian, Z. *Atlas de géographie économique de Syrie et du Liban.* Paris, 1926.

Kiel, Machiel. "Kuşadası: Genoese colonial town of the 1300s or Ottoman creation of the 17th century?" in Tuncer Baykara, ed. *CIÉPO XIV. Sempozyumu bildirileri.* Ankara, 2004.

———. "H'rzgrad-Hezârgrad-Razgrad: The vicissitudes of a Turkish town in Bulgaria," *Turcica* 21–22, 1991.

————. "The *vakıfnâme* of Rakkas Sinân Beg in Karnobat (Karîn-âbâd) and the Ottoman colonization of Bulgarian Thrace (14th–15th century)," *Journal of Ottoman Studies* 1, 1980.

Kinnear, John MacDonald. *Journey through Asia Minor, Armenia, and Koordistan, in the years 1813 and 1814.* London, 1818.

Kıray, Mübeccel. "Social change in Çukurova: A comparison of four villages," in P. Benedict, E. Tümertekin, and F. Mansur, eds. *Turkey: Geographic and social perspectives.* Leiden, 1974.

Kirk, Thomas. *Genoa and the sea: Policy and power in an early modern maritime republic, 1559–1684.* Baltimore, 2005.

————. "A little country in a world of empires: Genoese attempts to penetrate the maritime trading empires of the seventeenth century," *Journal of European Economic History* 25 (2), 1996.

Klein, Julius. *The Mesta: A study in Spanish economic history.* Cambridge, MA, 1920.

Koç, Yunus. *XVI. Yüzyılda bir Osmanlı sancağının iskân ve nüfus yapısı.* Ankara, 1989.

Kolars, John. "Locational aspects of cultural ecology: The case of the goat in non-western agriculture," *Geographical Review* 56 (4), 1966.

————. "Tradition, season, and change in a Turkish village," Ph.D. diss., University of Chicago, 1963.

Kołodziejczyk, Dariusz. "The export of silver coin through the Polish-Ottoman border and the problem of the balance of trade," *Turcica* 28, 1996.

Komlos, John. "The New World's contribution to food consumption during the industrial revolution," *Journal of European Economic History* 27 (1), 1998.

Köprülü, Fuad. *Osmanlı devleti'nin kuruluşu.* Ankara, 1988.

Kortepeter, Carl M. "Ottoman imperial policy and the economy of the Black Sea region in the sixteenth century," *Journal of the American Oriental Society* 86 (2), 1996.

Kostis, Kostas. "Salonique et le commerce de la laine au XVIIIe siècle," in Louis Bergeron, ed. *La croissance régionale dans l'Europe méditerranéenne XVIIIe–XXe siècles.* Paris, 1992.

————. "Structure sociales et retard économique: Salonique et l'économie de la laine XVIe–XVIIIe siècles," *Etudes balkaniques* 26 (1), 1990.

Krekić, Barisa. "Le port de Dubrovnik (Raguse), entreprise d'Etat, plaque tournante du commerce de la ville (XIV–XVIe siècles)," in Simonetta Cavaciocchi, ed. *I porti come impresa economica.* Florence, 1988.

Kriedte, Peter. *Peasants, landlords and merchant capitalists: Europe and the world economy, 1500–1800.* New York, 1983.

Kriedte, Peter, Hans Medick, and Jürgen Schlumbohm. *Industrialization before industrialization: Rural industry in the genesis of capitalism.* New York, 1981.

Krueger, Hilmar G. "The wares of exchange in the Genoese-African traffic of the twelfth century," *Speculum* 12 (1), 1937.

Kula, Witold. *Measures and men.* Princeton, 1986.

Kurt, Yılmaz. *Çukurova tarihinin kaynakları,* vol. 1: *1525 tarihli Adana sancağı mufassal tahrir defteri.* Ankara, 2004.

————. "Ramazanoğulları'nın vakıfları," in *X. Türk Tarih Kongresi.* Ankara, 1991.

Kurt, Yılmaz, and Akif Erdoğru. *Çukurova tarihinin kaynakları,* vol. 4: *Adana evkaf defteri.* Ankara, 2000.

Kütükoğlu, Mübahat. "XVI. Yüzyıl İzmir kazasında cemaat adı taşıyan köyler," in *V. Milletlerarası Türkiye sosyal ve iktisat tarihi kongresi.* Ankara, 1989.

Labaki, Boutros. *Introduction à l'histoire économique du Liban. Soie et commerce extérieur en fin de période ottomane (1840–1914)*. Beirut, 1984.

Labib, Subhi. "Egyptian commercial policy in the Middle Ages," in M. A. Cook, ed. *Studies in the economic history of the Middle East*. New York, 1970.

Lachiver, Marcel. *Vins, vignes et vignerons*. Paris, 1988.

Ladero Quesada, Miguel Angel. "Aristocratie et régime seigneurieal dans l'andalousie du XVe siècle," *Annales E.S.C.* 38 (6), 1983.

Lagardère, Vincent. *Campagnes et paysans d'al-Andalus (VIIIe–XVe siècles)*. Paris, 1993.

Laiou, Angeliki E. "The agrarian economy, thirteenth–fifteenth centuries," in Angeliki E. Laiou, ed. *The economic history of Byzantium, from the seventh through the fifteenth century*. Vol. 1. Washington, DC, 2002.

———. "Byzantine trade with Christians and Muslims and the crusades," in Angeliki E. Laiou and Roy Mottahedeh. *The crusades from the perspective of Byzantium and the Muslim world*. Washington, DC, 2001.

Laiou-Thomadakis, Angeliki E. "The Byzantine economy in the Mediterranean trade system: Thirteenth–fifteenth centuries," *Dumbarton Oaks Papers* 34, 1980–81.

———. *Peasant society in the late Byzantine empire: A social and economic history*. Princeton, 1977.

Lamb, H. H. *Climate history and the modern world*. London, 1997.

———. *Climatic history and the future*. Princeton, 1977.

Landsteiner, Erich. "The crisis of wine production in late sixteenth-century Central Europe: Climatic causes and economic consequences," *Climatic Change* 63 (1), 1999.

Lane, Frederic. "Oceanic expansion: Force and enterprise in the creation of oceanic commerce," in F. Lane, *Profits from power: Readings in protection rent and violence-controlling enterprises*. Albany, 1979.

———. "Public debt and private wealth: Particularly in sixteenth-century Venice," in F. Lane, *Profits from power: Readings in protection rent and violence-controlling enterprises*. Albany, 1979.

———. *Venice: A maritime republic*. Baltimore, 1973.

———. "Venetian shipping during the commercial revolution," *American Historical Review* 38, 1937.

———. *Ships and shipbuilders of the Renaissance*. Baltimore, 1934.

Laoust, Henri. *Les gouverneurs de Damas sous les mamlouks et les premiers ottomans (658–1156/1260–1744). (Traduction de d'ibn Tulun et d'ibn Gum'a)*. Damascus, 1952.

Lapidus, Ira. "Grain economy of Mamluk Egypt," *Journal of the Social and Economic History of the Orient* 12 (1), 1969.

———. *Muslim cities in the later Middle Ages*. Cambridge, MA, 1967.

Larenaudie, M.-J. "Les famine en Languedoc aux XIVe et XVe siècles," *Annales du Midi* 64, 1952.

Latron, Andre. *La vie rurale en Syrie et au Liban*. Beirut, 1936.

Laurioux, Bruno. "Medieval cooking," in Jean-Louis Flandrin and Massimo Montanari, eds. *Food: A culinary history*. New York, 1999.

Lawless, Richard I. "The economy and landscapes of Thessaly during Ottoman rule," in Francis W. Carter, ed. *An historical geography of the Balkans*. New York, 1977.

Le Flem, Jean-Paul, "Vraies et fausses splendeurs de l'industrie ségovienne (vers 1460–vers

1650)," in Marco Spallanzi, ed. *Produzione, commercio, e consumo dei panni di lana (nei sècoli XII–XVIII)*. Florence, 1976.

Le Roy Ladurie, Emmanuel. *Histoire humaine et comparée du climat.* 2 vols. Paris, 2004–6.

———. *The French peasantry 1450–1660.* Berkeley, 1987.

———. *Montaillou: Cathars, Catholics in a French village.* London, 1979.

———. "Un concept: l'unification microbienne du monde (XIVe–XVIIe siècles)," *Revue suisse d'histoire 23*, 1973.

———. *The peasants of Languedoc.* Urbana, 1972.

———. *Times of feast, times of famine: A history of climate since the year 1000.* New York, 1971.

Le Roy Ladurie, Emmanuel, and Joseph Goy. *Tithes and agrarian history from the fourteenth to the nineteenth centuries.* New York, 1982.

Le Roy Ladurie, Emmanuel, and André Zysberg, "Géographie des hagiotoponymes en France," *Annales E.S.C. 38* (6), 1983.

Le Strange, G. *Palestine under the Moslems: A description of Syria and the Holy Land from A.D. 650 to 1500.* Beirut, 1965.

Leake, William Martin. *A journal of a tour in Asia Minor.* London, 1824.

Lees, Lynne H., and Paul M. Hohenberg. "Urban decline and regional economies: Brabant, Castile, and Lombardy, 1550–1750," *Comparative Studies in Society and History 31* (3), 1989.

Lefort, Jacques. "Les villages de Macédoine orientale au moyen âge," in Jacques Lefort, Cécile Morrison, and Jean-Pierre Sodini, eds. *Les villages dans l'empire byzantine (IVe–XVe siècle).* Paris, 2005.

———. "The rural economy, seventh–twelfth centuries," in Angeliki E. Laiou, ed. *The economic history of Byzantium, from the seventh through the fifteenth century.* Vol. 1. Washington, DC, 2002.

———. "Une exploitation de taille moyenne au XIIIe siècle en Chalcidique," in *Aphiérôma ston Niko Svôrôno.* Vol. 1. Réthymnon, 1986.

Leiser, Gary. "Travellers' accounts of mohair production in Ankara from the fifteenth through the nineteenth century," *The Textile Museum Journal*, 1993–94.

Léon, Pierre. "La réponse de l'industrie," in Fernand Braudel and Ernst Labrousse, eds. *Histoire économique et sociale de la France*, vol. 2: *Des derniers temps de l'âge seigneurial aux préludes de l'âge industriel (1660–1789).* Paris, 1970.

Leuween, Richard van. *Notables and clergy in Mount Lebanon: The Khazin sheikhs and the Maronite church (1736–1840).* New York: E. J. Brill, 1994.

Lewicki, Tadeusz. *West African food in the Middle Ages.* New York, 1974.

Lewis, Archibald. *Nomads and crusaders, A.D. 1000–1368.* Bloomington, 1988.

———. "The closing of the mediaeval frontier 1250–1359," *Speculum 33* (4), 1958.

Lewis, Gwynne. "Proto-industrialization in France," *Economic History Review 47* (1), 1994.

Lewis, Norman N. *Nomads and settlers in Syria and Jordan, 1800–1980.* New York, 1987.

———. "The frontier of settlement in Syria, 1800–1950," in Charles Issawi, ed. *The economic history of the Middle East 1800–1914.* Chicago, 1966.

———. "Malaria, irrigation and soil erosion in Central Syria," *Geographical Review 39* (2), 1949.

Lewison, Anthony, and Catherine Ungar. "Constructions en pièrre sèche dans un système agro-pastoral, Cipière, Alpes-maritimes," in *Bastides, bories, hameaux. L'habitat dispersé en Provence.* Mouans-Sartoux, 1986.

Lindner, Rudi Paul. *Nomads and Ottomans in medieval Anatolia.* Bloomington, 1983.

Lombard, Maurice. *L'Islam dans sa première grandeur VIII–XI siècle*. Paris, 1971.

Long, Janet. "The Mexican contribution to the Mediterranean world," *Diogenes* 159, 1992.

Lopez, Robert. *The commercial revolution of the Middle Ages, 950–1350*. New York, 1976.

———. "Market expansion: The case of Genoa," *Journal of Economic History* 24 (4), 1964.

———. "Silk industry in the Byzantine empire," *Speculum* 20 (1), 1945.

Lopez, Robert S., Harry Miskimin, and Abraham Udovich. "England to Egypt, 1350–1500: Long-term trends and long-distance trade," in M. A. Cook, ed. *Studies in the economic history of the Middle East*. New York, 1970.

Lopez, Robert S., and Irwin W. Raymond. *Medieval trade in the Mediterranean world*. New York, 2001.

Lovett, A. W. *Early Habsburg Spain 1517–1598*. New York, 1986.

———. "The golden age of Spain: New work on an old theme," *Historical Journal* 24 (3), 1984.

———. "The Castilian bankruptcy of 1575," *Historical Journal* 23 (4), 1980.

Lucas, H. S. "The great European famine of 1315–17," *Speculum* 5, 1930.

Luke, Harry. *Cyprus under the Turks, 1571–1878*. London, 1921.

Lunde, Paul. "The leek-green sea," *Aramco World* 56 (4), 2005.

———. "New world foods, old world diet," *Aramco World* 43 (3), 1992.

———. "Pillars Of Hercules," *Aramco World* 43 (3), 1992.

Luterbacher, J., and E. Xoplaki, "500-Year winter temperature and precipitation variability over the Mediterranean area and its connection to the large-scale atmospheric circulation," in Hans-Jurgen Bolle, ed. *Mediterranean climate: Variability and trends*. New York, 2003.

Lybyer, A. H. *The government of the Ottoman empire in the time of Suleiman the Magnificent*. Cambridge, MA, 1919.

MacFarlane, Alan, and Gerry Martin. *Glass: A world history*. Chicago, 2002.

Mack, Rosamond E. *Bazaar to piazza: Islamic trade and Italian art, 1300–1600*. Berkeley, 2001.

Mack Smith, Denis. *A history of Sicily*. 2 vols. London, 1968.

MacKay, Angus. *Money, prices and politics in fifteenth-century Castile*. London, 1981.

Maczak, Antoine. "Un voyageur temoin des prix," in *Histoire économique du monde méditerranean 1450–1650. Mélanges en l'honneur de Fernand Braudel*. Toulouse, 1973.

Madden, Thomas. *Enrico Dandolo and the rise of Venice*. Baltimore, 2003.

Magdalino, Paul. "The maritime neighborhoods of Constantinople: Commercial and residential functions, sixth to twelfth centuries," *Dumbarton Oaks Papers* 54, 2000.

———. "The grain supply of Constantinople, ninth–twelfth centuries," in Cyril Mango and Gilbert Dagron, eds. *Constantinople and its hinterland*. Brookfield, VT, 1995.

Makkai, László. "Grand domaine et petites exploitations, seigneur et paysan en Europe au Moyen Age et aux temps modernes," in Péter Gunst and Tamás Hoffmann, eds. *Eight International Economic History Congress: "A" Themes*. Budapest, 1982.

Mallett, Michael E., *Florentine galleys in the fifteenth century*. Oxford, 1967.

Małowist, Marian. "Le commerce du Levant avec l'Europe de l'Est au XVIe siècle: Quelques problèmes," in *Histoire économique du monde méditerranean 1450–1650. Mélanges en l'honneur de Fernand Braudel*. Toulouse, 1973.

———. *Croissance et régression en Europe: XIVe–XVIIe siècles*. Paris, 1972.

Mantran, Robert. *17. Yüzyılın ikinci yarısında İstanbul. Kurumsal, iktisadi, toplumsal tarih denemesi*. 2 vols. Ankara, 1990.

Margalit, Hanna. "Some aspects of the cultural landscape of Palestine during the first half of the nineteenth century," *Israel Exploration Journal* 13 (1), 1963.

Marino, John A. "Economic structures and transformations," in John A. Marino, ed. *Early modern Italy, 1550–1796*. New York, 2002.

———. *Pastoral economics in the Kingdom of Naples*. Baltimore, 1988.

Marks, Robert B. *Tigers, rice, silk, and silt: Environment and economy in late imperial South China*. New York, 1998.

Marques, A. H. de Oliveira. *Daily life in Portugal in the late Middle Ages*. Madison, 1971.

Martín, F. Ruiz. *Lettres marchandes échangées entre Florence et Medina del Campo*. Paris, 1965.

Martin, Jean-Marie. *La pouille du VIe au XIIe siècle*. Rome, 1993.

Martin, M. E. "The Venetian-Seljuk Treaty of 1220," *English Historical Review* 95 (375), 1980.

Martines, Lauro. *Power and imagination: City-states in Renaissance Italy*. New York, 1970.

Masson, Paul. *Histoire du commerce français dans le Levant au XVIIIe siècle*. New York, 1967.

Masters, Bruce. *The origins of western economic dominance in the Middle East: Mercantilism and the Islamic economy in Aleppo 1600–1750*. New York, 1988.

Matar, Nabil, ed. *In the lands of the Christians*. New York, 2003.

Mathers, Constance Jones. "Family partnerships and international trade in early modern Europe: Merchants from Burgos in England and France, 1470–1570," *Business History Review* 62 (3), 1988.

Matossian, Mary Kilbourne. *Poisons of the past: Molds, epidemics, and history*. New Haven, 1989.

Matschke, Klaus-Peter. "Commerce, trade, markets, and money: Thirteenth–fifteenth centuries," in Angeliki E. Laiou, ed. *The economic history of Byzantium, from the seventh through the fifteenth century*. Vol. 2. Washington, DC, 2002.

Matthee, Rudolph. *The politics of trade in Safavid Iran: Silk for silver, 1600–1730*. New York, 1999.

May, Jacques M. *The ecology of malnutrition in the Far and Near East*. New York, 1961.

Mayerson, Philip. "The role of flax in Roman and Fatimid Egypt," *Journal of Near Eastern Studies* 56 (3), 1997.

Mazower, Mark. *Salonica, city of ghosts: Christians, Muslims and Jews, 1430–1950*. New York, 2005.

Mazzaoui, Maureen Fennell. *The Italian cotton industry in the later middle ages, 1100–1600*. New York, 1981.

———. "Cotton industry of northern Italy in the late middle ages," *Journal of Economic History* 32 (1), 1972.

McArdle, Frank. *Altopascio: A study in Tuscan rural society, 1587–1784*. New York, 1978.

McCann, James. *Maize and grace: Africa's encounter with a new world crop, 1500–1200*. Cambridge, MA, 2005.

McCloy, Shelby T. "Flood relief and control in eighteenth-century France," *Journal of Modern History* 13 (1) 1941.

McCray, W. Patrick. *Glassmaking in Renaissance Venice: The fragile craft*. Brookfield, 1999.

McGowan, Bruce. "Age of the *ayans*," in Halil İnalcık and Donald Quataert, eds. *An economic and social history of the Ottoman empire*. New York, 1994.

———. *Economic life in Ottoman Europe: Taxation, trade and the struggle for land, 1600–1800*. New York, 1981.

————. "Food supply and taxation on the Middle Danube (1568–1579)," *Archivum Ottomanicum* 1 (1), 1976.

McIntosh, Terence. *Urban decline in early modern Germany: Schwäbisch Hall and its region, 1650–1750.* Chapel Hill, 1997.

McNeill, John. "The great powers and the global environment in the twentieth century," in F. Tabak, ed. *Allies as rivals: The U.S., Europe, and Japan in a changing world-system.* Boulder, CO, 2005.

————. "Tragedies of privatization: Land, liberty, and environmental change in Spain and Italy, 1800–1900," in John F. Richards, ed. *Land, property, and the environment.* Oakland, CA, 2002.

————. *The mountains of the Mediterranean world: An environmental history.* New York, 1992.

McNeill, William. "How the potato changed history," *Social Research* 66 (1), 1999.

————. "American food crops in the Old World," in Herman J. Viola and Carolyn Margolis, eds. *Seeds of change: A quincentennial commemoration.* Washington, DC, 1991.

————. *Venice: The hinge of Europe, 1081–1797.* Chicago, 1974.

Melchior-Bonnet, Sabine. *The mirror: A history.* New York, 2002.

Merçil, Erdoğan. *Türkiye Selçukluları'nda meslekler.* Ankara, 2000.

Meriwether, Margaret L. *The kin who count: Family and society in Ottoman Aleppo, 1770–1840.* Austin, 1999.

Mikesell, Marvin M. "The deforestation of Mount Lebanon," *Geographical Review* 59 (1), 1969.

Miller, William. "The Genoese in Chios, 1346–1566," *English Historical Review* 30 (119), 1915.

Mintz, Sidney. *Sweetness and power: The place of sugar in modern history.* New York, 1991.

Miquel, André. *Le géographie humaine du monde musulman jusqu'au au milieu de 11e siècle.* 3 vols. Paris, 1980.

Miroğlu, İsmet. *Kemah sancağı ve Erzincan kazası (1520–1566).* Ankara, 1990.

Mitler, Louis. "The Genoese in Galata: 1453–1682," *International Journal of Middle East Studies* 10 (1), 1979.

Molà, Luca. *The silk industry of Renaissance Venice.* Baltimore, 2000.

Molinier, Alain. *Stagnations et croissance: le Vivarais aux XVIIe–XVIIIe siècles.* Paris, 1985.

Mollat, Michel. *Jacques Coeur, ou, l'esprit d'entreprise au XVe siècle.* Paris, 1988.

Montanari, Massimo. *The culture of food.* Cambridge, MA, 1994.

Montemayor, J. "Les migrations françaises en Espagne au XVIIIe siècle: démographie, économie et société," in Irad Malkin, ed. *La France et la Méditerranée. Vingt-sept siècles d'interdépendance.* New York, 1990.

Morana, Antonio Maria. *Relazione del commercio d'Aleppo ed altre scale della Siria e Palestina.* Venice, 1797.

Moreno, Diego. "La colonizzazione dei 'Boschi Ovada,' XVI–XVII sec.," *Quaderni storici* 24 (3), 1973.

Morgan, Dan. *Merchants of grain.* New York, 1977.

Moriceau, Jean-Marc. *Terres mouvantes. Les campagnes françaises du féodalisme à la mondialization, 1150–1850.* Paris, 2002.

Morilla Critz, Jose, Alan L. Olmstead, and Paul W. Rhode. " 'Horn of plenty': The globalization of Mediterranean horticulture and the economic development of southern Europe, 1880–1930," *Journal of Economic History* 59 (2), 1999.

Morineau, Michel. "Malthus: There and back from the period preceding the Black Death to the 'industrial revolution,'" *Journal of European Economic History* 27 (1), 1998.

———. *Incroyable gazettes et fabuleux métaux: les retours des trésors americains d'après les gazettes hollandaises*. New York, 1985.

———. "Naissance d'une domination marchands européens, marchands et marché du Levant aux XVIIIe et XIXe siècles," in M. Morineau, *Pour une histoire économique vraie*. Lille, 1985.

———. "Les mancenilliers de l'Europe," in Pierre Deyon and Jacques Jacquard, eds. *Les hésitations de la croissance, 1580–1740*, Vol. 2 of Pierre Léon, ed. *Histoire économique et sociale du monde*. Paris, 1978.

———. "La pomme de terre au XVIIIe siècle," *Annales E.S.C.* 25 (6), 1970.

Mottu-Weber, Liliane. *Économie et refuge au Genève au siècle de la Réforme. La draperie et la soierie (1540–1630)*. Geneva, 1987.

Moutaftchieva, Vera P. *Agrarian relations in the Ottoman empire in the 15th and 16th centuries*. Boulder, CO, 1988.

———. *Le vakif-un aspect de la structure socio-économique de l'Empire ottoman (XVe–XVIIe s.)*. Sofia, 1981.

Mulhall, Michael G. *The dictionary of statistics*. London, 1892.

Munro, John. "Spanish *merino* wools and the nouvelles draperies: An industrial transformation in the late medieval Low Countries," *Economic History Review* 58 (3), 2005.

———. "The monetary origins of the 'price revolution': South German silver mining, merchant-banking and Venetian commerce, 1470–1540," in Dennis Flynn, Arturo Giráldez, and Richard von Glahn, eds. *Global connections and monetary history, 1470–1800*. Brookfield, VT, 2003.

———. "The monetary origins of the 'Price Revolution' before the influx of Spanish-American treasure: The south German silver-copper trades, merchant-banking, and Venetian commerce, 1470–1540," in Dennis O. Flynne, Arturo Giráldez, and Richard von Glahn, eds. *Global connections and monetary history, 1470–1800*. Aldershot, 2003.

———. "Medieval woolens: Textiles, textile technology, and industrial organization, c. 1000–1500," in David Jenkins, ed. *The Cambridge history of western textiles*. New York, 2002.

———. "Medieval woolens: West European woollen industries and their struggles for international markets, c. 1000–1500," in David Jenkins, ed. *The Cambridge history of western textiles*. New York, 2002.

Murphey, Rhoads. *Ottoman warfare, 1500–1700*. New Brunswick, 1999.

———. "The Ottoman resurgence in the seventeenth-century Mediterranean: The gamble and its results," *Mediterranean Historical Review* 8 (2), 1993.

———. "Provisioning Istanbul: The state and subsistence in the early modern Middle East," *Food and Foodways* 2 (1), 1988.

———. "Tobacco cultivation in northern Syria and conditions of its marketing and distribution in the late eighteenth century," *Turcica* 17, 1985.

———. "Some features of nomadism in the Ottoman empire: A survey based on tribal census and judicial appeal documentation from archives in Istanbul and Damascus," *Journal of Turkish Studies* 8, 1984.

———. *Regional structure in the Ottoman economy*. Wiesbaden, 1980.

Musgrave, Peter. *Land and economy in Baroque Italy: Valpolicella, 1630–1797*. Leicester, 1992.

Musgrave, Toby, and Will Musgrave. *An empire of plants*. New York, 2000.

Nader, Helen. *Liberty in absolutist Spain: The Habsburg sale of towns, 1516–1700.* Baltimore, 1990.

———. "Noble income in sixteenth-century Castile: The case of the Marquises of Mondejar, 1480–1580," *Economic History Review* 30 (3), 1977.

Naff, Alixa. "A social history of Zahle, the principal market town in nineteenth-century Lebanon," Ph.D. diss., University of California, Los Angeles, 1973.

Nagata, Yuzo. *Tarihte âyânlar: Karaosmanoğulları üzerine bir inceleme.* Ankara, 1997.

———. "16. Yüzyılda Manisa köyleri: 1531 tarihli Saruhan Sancağına ait bir tahrir defteri incelemesi," *İstanbul Üniversitesi Edebiyat Fakültesi Tarih Dergisi* 32, 1979.

Naylon, John. "Irrigation and internal colonization," *The Geographical Journal* 133 (2), 1967.

Neamtu, Vasile. *Le technique de la production céréalière en Valachie et en Moldavie jusqu'au XVIIIe siècle.* Bucharest, 1975.

Nencini, Franco. *Florence: The last days of the flood.* Florence, 1966.

Neveux, Hugues. "Dîme et production céréalière: l'exemple du Cambrésis (fin XIVe–début XVIIe siècle)," *Annales E.S.C.* 28 (2), 1973.

Nicol, Donald M. *Byzantium and Venice: A study in diplomatic and cultural relations.* New York, 1988.

Nieuwenhuis, Tom. *Politics and society in early modern Iraq.* The Hague, 1982.

Nolte, Christiane. "Andalusia, country of missed chances or paradise lost," in Hans-Heinrich Nolte, ed. *Internal peripheries in European history.* Göttingen, 1991.

Obuchowska-Pysiowa, H. "Trade between Cracow and Italy from the customs-house registers of 1604," *Journal of European Economic History* 9 (3), 1980.

Ochsenwald, William. *The Hijaz railroad.* Charlottesville, 1980.

Ogilvie, Alan A. "Physiography and settlements in southern Macedonia," *Geographical Review* 11 (2), 1921.

Ogilvie, Sheila. *State corporatism and proto-industry: The Württemberg Black Forest, 1580–1792.* New York, 1997.

Oğuz, Burhan. *Türkiye halkının kültür kökenleri,* vol. 4: *Dokuma ve giyim teknikleri.* İstanbul, 2004.

Olagüe, Ignacio. *Histoire d'Espagne.* Paris, 1958.

Oral, M. Zeki. "Selçuk devri yemekleri ve ekmekleri," in M. Sabri Koz, ed. *Yemek kitabı: Tarih-halk bilimi-edebiyat.* İstanbul, 2003.

Orhonlu, Cengiz. *Osmanlı imparatorluğu'nda aşiretlerin iskanı.* İstanbul, 1987.

Orman ve Maden ve Ziraat Nezareti Kalemi Mahsus Müdüriyyeti İstatistik Şubesi. *1325 senesi Asya ve Afrika-i Osmani zira'at istatistiği.* İstanbul, 1327/1909.

Osmanlı Arşivi Daire Başkanlığı. *387 numaralı Muhâsebe-i vilâyet-i Karaman ve Rûm defteri (937/1530).* Vol. 2. Ankara, 1997.

———. *438 numaralı Muhâsebe-i vilâyet-i Anadolu defteri (937/1530).* Vol. 2: *Bolu, Kastamonu, Kengiri ve Koca-ili livâları.* Ankara, 1994.

Osmanlı'da bir köle, Brettenli Michael Heberer'in anıları, 1585–1588. İstanbul, 2003.

Ouerfelli, Mohamed. "Les migrations liées aux plantations et à la production du sucre dans la mediterranée à la fin du Moyen Âge," in Michel Balard and Alain Ducellier, eds. *Migrations et diasporas méditerranéennes.* Paris, 2002.

———. "Le Maghreb: zone de passage et de diffusion des plantes cultivées d'origine orientale en Méditerranée occidentale au Moyen Âge," *Mésogeios* 8, 2000.

Owen, Roger. *The Middle East in the world economy, 1800–1914.* London, 1981.

Öz, Baki. *Osmanlı'da alevi ayaklanmaları.* İstanbul, 1992.

Öz, Mehmet. "16. Yüzyıl Osmanlı Anadolusu'nda tarımsal verimlilik meselesi," in *XIII. Türk Tarih Kongresi.* Vol. 3. Ankara, 2002.

———. "15–16. Yüzyıllarda Anadolu'nun sosyal tarihine dair araştırmalar: Genel bir değerlendirme," in *Uluslararası kuruluşunun 700. yıldönümünde bütün yönleriyle Osmanlı Devleti Kongresi.* Konya, 2000.

———. "Agriculture in the Ottoman classical period," in K. Çiçek, ed. *The Great Ottoman Turkish Civilisation.* Vol. 2. Ankara, 2000.

———. *XV–XVI. Yüzyıllarda Canik sancağı.* Ankara, 1999.

Özbaran, Salih. *Yemen'den Basra'ya sınırdaki Osmanlı.* İstanbul, 2004.

Özdağ, Mustafa. *Osmanlı imparatorluğu'nda dağlı isyanları.* Ankara, 1983.

Özdeğer, Hüseyin. *1463–1640 Yılları Bursa şehri tereke defterleri.* İstanbul, 1988.

———. *Onaltıncı asırda Ayıntâb livâsı.* İstanbul, 1988.

Özdemir, Rıfat. *XIX. Yüzyılın ilk yarısında Ankara.* Ankara, 1986.

Özel, Oktay. "Population changes in Ottoman Anatolia during the 16th and 17th centuries: The 'demographic crisis' reconsidered," *International Journal of Middle Eastern Studies* 36 (2), 2004.

Özkaya, Yücel. *XVIII. Yüzyılda Osmanlı kurumları ve Osmanlı toplum yaşantısı.* Ankara, 1985.

Öztürk, Nazif. *Türk yenileşme tarihi çerçevesinde vakıf müessesesi.* Ankara, 1995.

Öztürk, Rifat. "Osmanlı döneminde Antakya'nın fizikî ve demografik yapısı, 1709–1860," *Belleten* 63 (221), 1994.

Öztürk, Said. *Tanzîmât döneminde bir Anadolu şehri: Bilecik.* İstanbul, 1996.

———. *Askeri kassama ait onyedinci asır İstanbul tereke defterleri.* İstanbul, 1995.

Pagano de Divitiis, Gigliola. *English merchants in seventeenth-century Italy.* New York, 1997.

Palairet, Michael. *The Balkan economies c. 1800–1914: Evolution without development.* New York, 1997.

———. "Désindustrialisation à la périphérie: études sur la région des Balkans au XIXe siècle," *Histoire, économie et société* 4 (2), 1985.

Pamuk, Şevket. *İstanbul ve diğer kentlerde 500 yıllık fiyatlar ve ücretler, 1469–1998.* Ankara, 2000.

———. *A monetary history of the Ottoman empire.* New York, 2000.

Panova, Snezka. "Zur frage der handelsbeziehungen zwischen den Bulgarischen und den Rumänischen landen im XVII. jahrhundert," *Études balkaniques* 11 (1), 1975.

Panzac, Daniel. "Le contrat d'affrètement maritime en Méditerranée: Droit maritime et pratique entre Islam et Chrétienté (XVIIe–XVIIIe siècles)," *Journal of the Economic and Social History of the Orient* 45 (3), 2002.

———. "International and domestic maritime trade in the Ottoman empire during the 18th century," *International Journal of Middle Eastern Studies* 24 (2), 1992.

———. *La peste dans l'Empire ottoman, 1700–1850.* Leiden, 1985.

Pascual, Jean-Paul. "The janissaries and the Damascus countryside at the beginning of the seventeenth century according to the archives of the city's military tribunal," in Tarif Khalidi, ed. *Land tenure and social transformation in the Middle East.* Beirut, 1984.

———. "Une neige à Damas au XIXe siècle," *Bulletin d'Études orientales* 28, 1975.

Pascual, Jean Paul, and Colette Establet. *Des tissus et des hommes. Damas vers 1700*. Damascus, 2005.

Paskaleva, Virginia. "Contribution aux relations commerciales des provinces balkaniques de l'Empire ottoman avec les états européens au cours du XVIIIe et la première moitié du XIXe s.," *Études historiques* 4, 1968.

Pedani, Maria Pia. "The Ottoman empire and the Gulf of Venice," in Tuncer Baykara, ed. *CIÉPO XIV. Sempozyumu bildirileri*. Ankara, 2004.

Pedro'nun zorunlu İstanbul seyahati. İstanbul, 1996.

Pelizzon, Sheila. "Grain flour, 1590–1790," *Review* 23 (1), 2000.

Perevolotsky, Avi, and No'am G. Seligman. "Role of grazing in Mediterranean rangeland ecosystems," *BioScience* 48 (12), 1998.

Pérez Sarrión, G. "Hydraulic policy and irrigation works in Spain in the second half of the eighteenth century," *Journal of European Economic History* 24 (1), 1995.

Peristiany, J. G., ed. *Mediterranean family structures*. New York, 1976.

Pesez, Jean-Marie, and E. Le Roy Ladurie, "Le cas français: vue d'ensemble," in *Villages désertés et histoire économique, XIe–XVIIe siècles*. Paris, 1967.

Petersen, A. D. "Khirbat Ja'thun: An Ottoman farmhouse in western Galilee," *Palestine Exploration Quarterly* 127, 1995.

Petmezas, Socrates D. "Patterns of protoindustrialization in the Ottoman empire: The case of eastern Thessaly, ca. 1750–1860," *Journal of European Economic History* 19 (3), 1990.

Petrusewicz, Marta. *Latifundium: Moral economy and material life in a European periphery*. Ann Arbor, MI, 1996.

Pfister, Christian. "Fluctuations climatiques et prix céréalièrs en Europe du XVIe au XXe siècle," *Annales E.S.C.* 63 (1), 1988.

Pfister, Christian, and Rüdiger Glazer. "Climatic variability in sixteenth century Europe and its social dimension," *Climatic Change* 43 (1), 1999.

Philip, E. B. *Der Türkische tabak*. Munich, 1926.

Phillips, Carla Rahn. *Six galleons for the King of Spain: Imperial defense in the early seventeenth century*. Baltimore, 1986.

———. "Spanish wool trade, 1500–1780," *Journal of Economic History* 42 (4), 1982.

———. *Ciudad Real, 1500–1750: Growth, crisis, and readjustment in the Spanish economy*. Cambridge, MA, 1979.

———. "The Spanish wool trade, 1500–1700," *Journal of Economic History* 42 (4), 1977.

Phillips, Carla Rahn, and William D. Phillips Jr. *Spain's golden fleece: Wool production and the wool trade from the Middle Ages to the nineteenth century*. Baltimore, 1997.

Phillips, Rod. *A short history of wine*. New York, 2002.

Pike, Ruth. *Enterprise and adventure: The Genoese in Seville and the opening of the New World*. Ithaca, NY, 1966.

———. "The Genoese in Seville and the opening of the New World," *Journal of Economic History* 22 (4), 1962.

Pirenne, Henri. *Histoire économique et sociale du moyen âge*. Paris, 1969.

———. *A history of Europe*. New York, 1955.

———. "Un grand commerce d'exportation au Moyen Age: les vins de France," *Annales d'histoire économiques et sociale* 5, 1933.

———. "Stages in the social history of capitalism," *American Historical Review* 19 (3), 1914.

Pîrî Reis. *Kitab-ı bahriye*. 3 vols. Istanbul, 1988.

Pistarino, Geo. "The Genoese in Pera Turkish Galata," *Mediterranean Historical Review* 1, 1986.

Pitcher, Donald. *Osmanlı imparatorluğu'nun tarihsel coğrafyası*. İstanbul, 1999.

Pitt-Rivers, Julian. *The people of the Sierra*. Chicago, 1961.

Piuz, Anne-Marie. "Marchands genevois du monde méditerranéen," in *Histoire économique du monde méditerranean, 1450–1650. Mélanges en l'honneur de Fernand Braudel*. Toulouse, 1973.

Planhol, Xavier de. *Le paysage animal: l'homme et la grande faune: une zoo-géographie historique*. Paris, 2004.

———. *L'Islam et la mer. La mosquée et le matelot VIIe–XXe siècle*. Paris, 2000.

———. *L'eau de neige. Le tiède et le frais*. Paris, 1995.

———. *An historical geography of France*. New York, 1994.

———. "Grandeur et décadance du vignoble de Trébizonde," *Journal of the Economic and Social History of the Orient* 22 (3), 1979.

———. *Les fondements géographiques de l'histoire de l'Islam*. Paris, 1968.

———. "Aspects of mountain life in Anatolia and Iran," in S. R. Eyre and G. R. Jones, eds. *Geography as human methodology by example*. New York, 1966.

———. "Expansion et problèmes de l'agriculture turque," *Revue de géographie de Lyon* 35 (1), 1960.

———. "Geography, politics, and nomadism in Anatolia," *International Social Science Journal* 11, 1959.

———. *De la plain pamphyliennes aux lacs pisidiens: nomadisme et vie paysanne*. Paris, 1958.

———. "Les migrations de travail en Turquie," *Revue de géographie alpine* 40 (4), 1952.

Ponsot, Pierre. "En Andalousie occidentale. Les fluctuations de la production du blé sous l'Ancien Régime," *Études rurales* 34, 1969.

Potter, T. W. "Valleys and settlement: Some new evidence," *World Archaeology* 8 (2), 1977.

Pounds, N. J. G. *An historical geography of Europe*. New York, 1990.

Powers, D. S. "Revenues of public *waqfs* in sixteenth-century Jerusalem," *Archivum Ottomanicum* 9, 1984.

Prawer, Joshua. "Étude de quelques problèmes agraires et sociaux d'une seigneurie croisée au XIIIe siècle," *Byzantion* 22–23, 1952.

Pryor, John. *Geography, technology, and war: Studies in the maritime history of the Mediterranean, 649–1571*. New York, 1988.

Pyne, Stephen J. *Vestal fire: An environmental history told through fire, of Europe and Europe's encounter with the world*. Seattle, 1997.

Quaglia, A. M. Pult. "Controls over food supplies in Florence in the late XVIth and early XVIIth centuries," *Journal of European Economic History* 9 (2), 1980.

Quataert, Donald. *Ottoman manufacturing in the age of industrial revolution*. New York, 1993.

———. "Agricultural trends and government policy in Ottoman Anatolia, 1800–1914," *Asian and African Studies* 15 (1), 1981.

Rabbath, Antoine. *Documents inédits pour servir à l'histoire du Christianisme en Orient (XVI–XIX siècle)*. 2 vols. Paris, 1905–10.

Rabie, Hassanein. "Some technical aspects of agriculture in medieval Egypt," in A. Udovitch, ed. *The Islamic Middle East, 700–1900: Studies in economic and social history.* Princeton, 1981.

Raby, Julian. *Venice, Dürer, and the oriental mode.* Totowa, NJ, 1982.

Rackham, A. T., and Oliver Grove. *The nature of Mediterranean Europe: An ecological history.* New Haven, 2001.

Rafeq, Abdal Karim. "Impact of Europe on a traditional economy: The case of Damascus 1840–1870," in J.-L. Bacqué-Grammont and Paul Dumont, eds. *Économie et sociétés dans l'empire ottoman (fin du XVIIIe–debut du XXe siècle).* Paris, 1983.

———. "Economic relations between Damascus and the dependent countryside, 1743–71," in A. L. Udovitch, ed. *The Islamic Middle East, 700–1900: Studies in economic and social history.* Princeton, 1981.

Rahme, Joseph G. "Some socioeconomic observations on the relationship between the mountain and the coast in early 17th century Ottoman Syria," *ARAM* 10 (1–2), 1998.

Randsborg, Klavs. *The first millennium A.D. in Europe and the Mediterranean: An archeological essay.* New York, 1991.

Ranke, Leopold von. *The Turkish and Spanish empires in the 16th and 17th centuries.* London, 1845.

Rapp, Richard T. *Industry and economic decline in seventeenth-century Venice.* Cambridge, MA, 1977.

———. "The unmaking of the Mediterranean trade hegemony: International trade rivalry and the commercial revolution," *Journal of Economic History* 35 (3), 1975.

Rau, Virgínia, "The settlement of Madeira and the sugar cane plantation," *A.A.G. Bijdragen* 11 1964.

Raymond, André. *Osmanlı döneminde Arap kentleri.* İstanbul, 1995.

———. *Artisans et commerçants au Caire au XVIIIe siècle.* 2 vols. Damascus, 1973–74.

Rebora, Giovanni. *Çatal kültürü: Avrupa mutfağının kısa tarihi.* İstanbul, 2003.

Redford, Scott. *Landscape and the state in medieval Anatolia: Seljuk gardens and pavilions of Alanya, Turkey.* Oxford, 2000.

Refik, Ahmed. *Anadolu'da Türk aşiretleri (966–1200).* Istanbul, 1989.

———. *On ikinci asr-i hicri'de İstanbul hayatı.* İstanbul, 1988.

———. *Onuncu asr-i hicrî'de İstanbul hayatı.* İstanbul, 1988.

Reher, David Sven. *Town and country in pre-industrial Spain: Cuenca, 1550–1870.* New York, 1990.

Reid, Anthony. *Southeast Asia in the age of commerce, 1450–1680.* 2 vols. New Haven, 1988–93.

Reimer, M. J. "Ottoman Alexandria: The paradox of decline and the reconfiguration of power in eighteenth-century Arab provinces," *Journal of the Economic and Social History of the Orient* 37 (2), 1994.

Revel, Jacques. "Au XVIIe siècle: le déclin de la Méditerranée?" in Irad Malkin, ed. *La France et la Méditerranée. Vingt-sept siècles d'interdépendance.* Brill, 1990.

Reyerson, K. "Montpellier and Genoa: The dilemma of dominance," *Journal of Medieval History* 20, 1994.

Reynard, Pierre Claude. "Charting environmental concerns: Reactions to hydraulic works in eighteenth-century France," *Environment and history* 9, 2003.

Ricci, Giovanni. *Türk saplantısı: Yeniçağ Avrupa'sında korku, nefret ve sevgi.* İstanbul, 2005.

Richard, John F. "Land transformation," in B. L. Turner et al., eds. *The earth as transformed by human action: Global and regional changes in the biosphere over the past 300 years.* New York, 1990.

Richards, Alan. "The political economy of *Gutswirtschaft:* A comparative analysis of East Elbian Germany, Egypt, and Chile," *Comparative Studies in Society and History* 21 (4), 1979.

Richards, Gertrude, ed. *Florentine merchants in the age of Medici.* Cambridge, MA, 1936.

Riley-Smith, Jonathan. *The feudal nobility and the kingdom of Jerusalem, 1174–1277.* Hamdon, CT, 1973.

Ringrose, David. *Spain, Europe, and the "Spanish miracle," 1700–1900.* New York, 1996.

———. "The impact of a new capital city: Madrid, Toledo, and New Castile, 1560–1660," *Journal of Economic History* 33 (4), 1973.

Rocco, Fiametta. *Quinine: Malaria and the quest for a cure that changed the world.* New York, 2004.

Rodrigo, F. S., and M. J. Esteban-Parra. "An attempt to reconstruct the rainfall regime of Andalusia (Southern Spain) from 1601 A.D. to 1650 A.D. using historical documents," *Climatic Change* 27 (4), 1994.

Rogers, J. Michael. "To and fro: Aspects of Mediterranean trade and consumption in the 15th and 16th centuries," *Revue du monde musulman et de la méditerranée* 55–56 (1–2), 1990.

Romano, Ruggiero. "Italy in the crisis of the seventeenth century," in Peter Earl, ed. *Essays in European economic history, 1500–1800.* Oxford, 1974.

———. *Commerce et prix du blé à Marseille au XVIIIe siècle.* Paris, 1956.

———. "A propos du commerce du blé dans la Méditerranée des XIVe et XVe siècle," in *Éventail de l'histoire vivante. Hommage à Lucien Febvre.* Vol. 1. Paris, 1953.

Ron, Z. "Agricultural terraces in the Judean heights," *Israel Exploration Journal* 16 (1); 16 (2), 1966.

Rose, Susan. "Islam versus Christianity: The naval dimension, 1100–1600," *Journal of Military History* 63 (3), 1999.

Rosenberger, Bernard. *Société, pouvoir et alimentation. Nourriture et précarité au Maroc précolonial.* Rabat, 2001.

———. "Arab cuisine and its contribution to European culture," in Jean-Louis Flandrin and Massimo Montanari, eds. *Food: A culinary history.* New York, 1999.

———. "Cultures complémentaires et nourritures de substitution au Maroc (XV–XVIIIe siècle)," *Annales E.S.C.* 35 (3–4), 1980.

Rosenberger, Bernard, and Hamid Triki. "Famines et épidémies au Maroc aux XVIe et XVIIe siècles," *Hespéris tamuda* 14, 1973.

Rosenthal, Jean-Laurent. *The fruits of revolution: Property rights, litigation, and French agriculture, 1700–1860.* New York, 1992.

Rostow, W. W. *The world economy: History and prospect.* Austin, 1978.

Roth, Cecil. *The house of Nasi: The duke of Naxos.* Philadelphia, 1948.

Rueda Hernanz, Germán. "Disentailment in Spain, 1766–1854," *Iberian Studies* 19 (1–2), 1990.

Ruiz, Teofilo F. *Crisis and continuity: Land and town in medieval Castile.* Philadelphia, 1994.

Russell, Richard Joel. "Geomorphology of the Rhône delta," *Annals of the Association of American Geographers* (32) 2, 1942.

Saâdé, Edouard. *L'agriculture à Lattaquié.* Beauvais, 1905.

Sahillioğlu, Halil. "Yeniçeri çuhası ve II.Bayezıd'in son yıllarında yeniçeri çuha muhasebesi," *Güney-Doğu Avrupa Araştırmaları Dergisi* 2–3, 1973–74.

Şahin, İlhan, and Feridun Emecen. *II. Bâyezid dönemine ait 906/1501 tarihli Ahkâm Defteri.* İstanbul, 1994.

Sakaoğlu, Necdet. *Çeşm-i cihan Amasra.* İstanbul, 1999.

———. *Anadolu derebeyi ocaklarından Köse Paşa hanedanı.* İstanbul, 1984.

Sakellariou, Eleni. "The cities of Puglia in the fifteenth and sixteenth centuries: Their economy and society," in Alexander Cowan, ed. *Mediterranean urban culture, 1400–1700.* Exeter, 2000.

Salati, Marco. "Urban notables, private *waqf* and capital investment: The case of the 17th century Zahrâwî family of Aleppo," in *Le waqf dans l'espace islamique.* Damascus, 1995.

Salibi, Kemal. *A house of many mansions: The history of Lebanon reconsidered.* Berkeley, 1988.

Salomon, Noël. *La campagne de nouvelle Castille à la fin du XVIe siècle.* Paris, 1964.

Salvemini, Gaetano. "Florence in the age of Dante," *Speculum* 11 (3), 1936.

Samsonowicz, Henryk. "Changes in the Baltic zone in the XIII–XVI centuries," *Journal of European Economic History* 4 (3), 1975.

———. "Relations commerciales entre la Baltique et la Méditerranée," in *Histoire économique du monde méditerranéen, 1450–1650. Mélanges en l'honneur de Fernand Braudel.* Toulouse, 1973.

Sánchez Léon, Pablo. "Town and countryside in Castile, 1400–1650," in S. R. Epstein, ed. *Town and country in Europe, 1300–1800.* New York, 2001.

Sanders, Irwin T. *Balkan village.* Lexington, 1949.

San José, Carmen Trillo. "El paisaje vegetal en la Granada islámica y sus transformaciones tras la conquista castellana," *Historia agraria* 17, 1999.

Sapancalı, Hüseyin. *Karaman: Ahval-i içtimaiyye, coğrafiyye ve tarihiyyesi.* Ankara, 1993.

Saraçoğlu, Hüseyin. *Bitki örtüsü.* İstanbul, 1972.

———. *Akdeniz bölgesi.* İstanbul, 1968.

Sardella, Pierre. *Nouvelles et spéculations à Venise au début du XVIe siècle.* Paris, 1948.

Sarpaki, Anaya. "The palaeoethnobotanical approach: The Mediterranean triad or is it a quartet?" in Berit Wells, ed. *Agriculture in ancient Greece.* Stockholm, 1992.

Sauvaget, Jean. *Alep. Essai sur le développement d'une grande ville syrienne.* Vol. 1. Paris, 1941.

Savaş, Saim. *Bir tekkenin dinî ve sosyal tarihi: Sivas Ali Baba Zaviyesi.* İstanbul, 1992.

Scammel, G. L. *The world encompassed.* Berkeley, 1981.

Schefer, C., ed. *Le voyage d'outremer de Bertrandon de la Broquière.* Paris, 1892.

Schick, Léon. *Un grand homme d'affaire au début du XVIe siècle: Jacob Fugger.* Paris, 1957.

Schilcher, Linda Schatkowski. *Families in politics: Damascene factions and estates of the 18th and 19th centuries.* Wiesbaden, 1985.

Schweigger, Salomon. *Sultanlar kentine yolculuk 1578–1581.* İstanbul, 2004.

Sclafert, Thérése. *Cultures en Haute-Provence. Déboisements et patûrages au Moyen Age.* Paris, 1959.

Seikaly, S. "Land tenure in 17th century Palestine: The evidence from the al-Fatawa al-Khairiyya," in Tarif Khalidi, ed. *Land tenure and social transformation in the Middle East.* Beirut, 1984.

Sella, Domenico. *Italy in the seventeenth century.* New York, 1997.

———. *Crisis and continuity: The economy of Spanish Lombardy in the seventeenth century.* Cambridge, MA, 1979.

———. "Au dossier des migrations montagnardes: l'exemple de la Lombardie au XVIIIe siècle," in *Histoire économique du monde méditerranean, 1450–1650. Mélanges en l'honneur de Fernand Braudel.* Toulouse, 1973.

———. "Crisis and transformation in Venetian trade," in Brian Pullan, ed. *Crisis and change in the Venetian economy in the sixteenth and seventeenth centuries.* London, 1968.

———. "The rise and fall of the Venetian woollen industry," in Brian Pullan, ed. *Crisis and change in the Venetian economy in the sixteenth and seventeenth centuries.* London, 1968.

Semple, Ellen Churchill. *The geography of the Mediterranean region: Its relation to ancient history.* New York, 1931.

Şen, Ömer. *Osmanlı panayırları (18.–19. yüzyıl).* İstanbul, 1996.

Sereni, Emilio. *History of the Italian agricultural landscape.* Princeton, 1997.

Sereno, P. "Ecology and marginal lands: A former phase of retreat of settlement and agricultural limits in the western Italian Alps," in B. K. Roberts and R. E. Glasscock, eds. *Villages, fields, and frontiers.* Oxford, 1983.

Şer'iye sicilleri. Vol. 2. İstanbul, 1989.

Serrão, Joel. "Le blés des îles atlantiques: Madère et Açores aux Xve et XVIe siècles," *Annales E.S.C.* 9 (3), 1954.

Serre-Bachet, Françoise. "Middle ages temperature reconstructions in Europe: A focus on northeastern Italy," *Climatic Change* 26 (2–3), 1994.

Serre-Bachet, Françoise, and J. Guiot. "Summer temperature changes from tree rings in the Mediterranean area during the last 800 years," in W. H. Berger and L. D. Labeyrie, eds. *Abrupt climatic change: Evidence and implications.* Boston, 1985.

Serre-Bachet, Françoise, J. Guiot, and L. Tessier. "Dendroclimatic evidence from southwestern Europe and northwestern Africa," in Raymond S. Bradley, ed. *Climate since A.D. 1500.* New York, 1992.

Şeşen, Ramazan. *Islam coğrafyacılarına göre Türkler ve Türk ülkeleri.* Ankara, 2001.

Sezer, Hamiyet. "Tepedelenli Ali Paşa'nın çiftlikleri üzerine bir araştırma," *Belleten* 62 (233), 1998.

Silbert, Albert. *Le Portugal méditerranéen à la fin de l'Ancien regime du XVIIIe au début du XIXe siècles.* Paris, 1966.

Simon, Bruno. "Le blé et les rapports véneto-ottomans au XVIe siècle," in J.-L. Bacqué-Grammont and Paul Dumont, eds. *Contributions à l'histoire économique et sociale de l'Empire ottoman.* Louvain, 1983.

Simpson, James. *Spanish agriculture: The long siesta, 1765–1965.* New York, 1995.

Singer, Amy. *Constructing Ottoman beneficence: An imperial soup kitchen in Jerusalem.* Albany, NY, 2002.

———. *Palestinian peasants and Ottoman officials: Rural administration around sixteenth-century Jerusalem.* New York, 1994.

Sivignon, Michel. *Les pasteurs du Pinde septentrional.* Lyon, 1968.

Slicher Van Bath, B. H. "Les problèmes fondamentaux de la société pré-industrielle en Europe occidentale," *A.A.G. Bijdragen* 12, 1965.

———. *The agrarian history of western Europe.* London, 1963.

Snowden, Frank M. "Mosquitos, quinine, and the socialism of Italian women, 1900–1914," *Past and Present* 178, 2003.

Sombart, Werner. *Jews and modern capitalism.* New Brunswick, 1982.

Soysal, Mustafa. "Onaltıncı yüzyılda Adana ilinin "Mufassal Defteri'ne göre sosyal ve ekonomik yapısı üzerine bir araştırma," *Belleten* 52 (202), 1988.

Spruyt, Hendrik. *The sovereign state and its competitors.* Princeton, 1994.

Spufford, Peter. *Power and profit: The merchant in medieval Europe.* New York, 2003.

———. *Money and its use in medieval Europe.* New York, 1988.

Steensgaard, Niels. "Asian trade and world economy from the 15th to 18th centuries," in Teotonio R. de Souza, ed. *Indo-Portuguese history: Old issues, new questions.* New Delhi, 1985.

Stein, Stanley J., and Barbara H. Stein. *Silver, trade, and war: Spain and America in the making of early modern Europe.* Baltimore, 2000.

Stoianovich, Traian. "A route type: The Via Egnatia under Ottoman rule," in Elizabeth Zachariadou, ed. *The Via Egnatia under Ottoman rule (1380–1699).* Rethymnon, 1996.

———. "Prospective: Third and fourth levels of history" in *Between East and West: The Balkan and Mediterranean worlds,* vol. 4: *Material culture and mentalités: Land, sea and destiny.* New Rochelle, 1995.

———. *Balkan worlds: The first and last Europe.* Armonk, 1994.

———. "Balkan peasants and landlords and the Ottoman state: Familial economy, market economy, and modernization." in *Between East and West: The Balkan and Mediterranean worlds,* vol. 2: *Economies and societies: Land, lords, states and middlemen.* New Rochelle, 1992.

———. "Mode de production maghrébin de commandement ponctuel," in *Between East and West: The Balkan and Mediterranean worlds,* vol. 2: *Economies and societies: Land, lords, states and middlemen.* New Rochelle, 1992.

———. "L'Espace maritime segmentaire de l'Empire ottoman," in *Actes du IIe Colloque international d'histoire: Économies méditerranéennes équilibres et intercommunications.* Vol. 1. Athens, 1985.

———. "Commerce et industrie ottomans et maghrébins: pôles de diffusion et aires d'expansion," in J.-L. Bacqué-Grammont and Paul Dumont, eds. *Contributions à l'histoire économique et sociale de l'Empire ottoman.* Paris, 1983.

———. "Pour un modèle du commerce du Levant: économie concurrentielle et économie de bazar 1500–1800," *Bulletin de l'Association Internationale d'Etudes du Sud-Est Européen* 12 (2), 1974.

———. "Le maïs dans les Balkans," *Annales E.S.C.* 21 (2), 1966.

———. "The conquering Balkan orthodox merchant," *Journal of Economic History* 20 (2), 1960.

———. "Land tenure and related sectors of the Balkan economy, 1600–1800," *Journal of Economic History* 13 (4), 1953.

Stoianovich, Traian, and G. C. Haupt. "Le maïs arrive dans les Balkans," *Annales E.S.C.* 17 (1), 1962.

Stouff, L. *La table provençale: boire et manger en Provence à la fin du Moyen Age.* Avignon, 1996.

Stradling, R. A. *Spain's struggle for Europe, 1598–1668.* Rio Grande, 1994.

Stromer, Wolfgang von. "Une clé du succès des maisons de commerce d'Allemagne du Sud: le grand commerce associé au Verlagssystem," *Revue historique* 285 (1), 1991.

———. "Nuremberg in the international economics of the Middle Ages," *Business History Review* 44 (2), 1970.

Subrahmanyam, Sanjay, and Luís Filipe F. R. Thomaz, "Evolution of empire: The Portuguese in the Indian Ocean during the sixteenth century," in James Tracy, ed. *The political economy of merchants empires: State power and world trade, 1350–1750.* New York, 1991.

Sümer, Faruk. "XIX. yüzyılda Çukurova'da içtimaî hayat," in J.-L. Bacqué-Grammont et al., eds. *CIÉPO Osmanlı Öncesi ve Osmanlı Araştırmaları Uluslararası Komitesi, VII. Sempozyumu bildirileri.* Ankara, 1994.

Swinburne, Henry. *Travels through Spain in the years 1775 and 1776.* London, 1779.

Szyliowicz, Joseph. *Political change in rural Turkey: Erdemli.* Paris, 1966.

Tabak, Faruk. "Informalization and the long term," in Faruk Tabak and Michaelina Crichlow, eds. *Informalization: Process and structure.* Baltimore, 2000.

———. *"Ars longa, vita brevis?* A geohistorical perspective on *Pax Mongolica," Review* 19 (1), 1996.

Tabakoğlu, Ahmet. *Türk iktisat tarihi.* Ankara, 1986.

———. *Gerileme dönemine girerken Osmanlı maliyesi.* İstanbul, 1985.

Tadić, Jorjo. "Le commerce en Dalmatie et à Raguse et la décadence économique de Venise au XVIIème siècle," in *Aspetti e cause della decadenza economica veneziana nel secolo XVII.* Venice, 1961.

Tamur, Erman. *Ankara keçisi ve Ankara tiftik dokumacılığı.* Ankara, 2003.

Tanoğlu, Ali. "İskân cografyası: Esas fikirler, problemler ve metod," *Türkiyat mecmuası* 11, 1954.

———. "Türkiye'nin irtifa kuşakları," *Türk Coğrafya Dergisi* 9–10, 1947.

Tate, Jihane. *Une waqfiyya du XVIIIe siécle à Alep. La waqfiyya d'al Hâgg Mûsâ al-Amîrî.* Damascus, 1990.

Tavernier, Jean-Baptiste. *Tavernier seyahatnamesi.* Ed. Stefanos Yerasimos. İstanbul, 2006.

———. *Topkapı sarayında yaşam.* İstanbul, 1984.

T. C. Çevre ve Orman Bakanlığı. *Osmanlı ormancılığı ile ilgili belgeler.* 3 vols. Ankara, 1999–2003.

Tchihatchef, Pierre A. de. *İstanbul ve Boğaziçi.* İstanbul, 2000.

Teall, John L. "The grain supply of the Byzantine empire, 330–1025," *Dumbarton Oaks Papers* 13, 1959.

Tenenti, Alberto. *Piracy and the decline of Venice, 1580–1615.* Berkeley, 1967.

Tenenti, Alberto, and Corrado Vivanti. "Les galères marchands vénitiennes XIVe–XVIe siècles," *Annales E.S.C.* 16 (1), 1961.

Terzi, Arzu. "Güzelhisar-ı Aydın: Portrait of a western Anatolian town," in Hayashi Kayoko and Mahir Aydın, eds. *The Ottoman state and societies in change.* New York, 2004.

Thévenot, Jean. *1655–56'da Türkiye.* İstanbul, 1978.

Thijs, A. K. L. "Technologie et productivité de l'industrie textile à Anvers (XVIe–XVIIIe siècles)," in Sara Mariotti, ed. *Produttività e technologie nei secoli XII–XVII.* Florence, 1981.

Thirgood, J. V. *Man and the Mediterranean forest: A history of resource depletion.* New York, 1981.

Thiriet, Freddy. *La Romanie vénetienne au moyen age. Le développement et l'exploitation du domaine coloniale vénetien (XIIIe–XVe siècles).* Paris, 1975.

Thirsk, Joan. *Alternative agriculture: A history from the Black Death to the present day.* New York, 1997.

Thomson, J. K. J. "Explaining the 'take-off' of the Catalan cotton industry," *Economic History Review* 58 (4), 2005.

———. *A distinctive industrialization: Cotton in Barcelona, 1728–1832.* New York, 1992.

———. *Clermont-de-Lodève, 1633–1789.* New York, 1982.

Thoumin, Richard. "Le Ghab," *Revue de géographie alpine* 24, 1936.

Tielhof, Milja van. *The 'mother of all trades': The Baltic grain trade in Amsterdam from the late sixteenth to the early nineteenth century.* Boston, 2002.

Tilly, Charles. *Coercion, capital, and European states, AD 990–1990.* Cambridge, MA, 1990.

Tits-Dieuaide, M.-J. *La formation des prix céréalieres en Brabant et en Flandre au XVIe siècle.* Brussels, 1975.

Todorov, Nikolai. *The Balkan city, 1400–1900.* Seattle, 1983.

Todorova, Maria N. "Was there a demographic crisis in the Ottoman empire in the seventeenth century?" *Études balkaniques* 24 (2), 1988.

Togan, Zeki Velidî. *Umumî Türk tarihi'ne giriş.* İstanbul, 1981.

Tökin, Ismail Hüsrev. *Türkiye köy iktisadiyatı.* İstanbul, 1990.

Topping, Peter. "Le régime agraire dans le Péloponnèse latin au XIVe siècle," *L'Hellénisme contemporain* (2ème série) X, 1956.

Torras, J. "The old and the new: Marketing networks and textile growth in eighteenth century Spain," in Maxine Berg, ed. *Markets and manufacturers in early industrial Europe.* London, 1991.

Touchan, Ramzi. "Reconstructions of spring/summer precipitation for the eastern Mediterranean from tree-ring widths and its connection to large-scale atmospheric circulation," *Climate Dynamics* 25, 2005.

Touma, Toufic. *Paysans et institutions féodales chez les druses et les maronites du Liban du XVIIe siècle à 1914.* Vol. 2. Beirut, 1972.

———. *Un village de Montagne au Liban: Hadeth al-Jobbé.* Paris, 1958.

Tournefort, Joseph de. *Tournefort seyahatnamesi.* Ed. Stefanos Yerasimos. İstanbul, 2005.

Toussaint-Samat, Maguelonne. *History of food.* Cambridge, MA, 1992.

Tracy, James. "Habsburg grain policy and Amsterdam politics: The career of Sheriff Willem Dirkszoon Baerdes, 1542–1566," *Sixteenth Century Journal* 14 (3), 1983.

Triantafyllidou, Yolande. "L'industrie du savon en Crète au XVIIIe siècle: aspects économiques et sociaux," *Études balkaniques* 11 (4), 1975.

Tucci, Ugo. "Le commerce venetien du vin en Crète," in Klaus Friedland, ed. *Maritime food transport.* Cologne, 1994.

———. *Lettres d'un marchands vénitien: Andrea Berengo (1553–1556).* Paris, 1957.

Tunçdilek, Necdet. "Orta Sakarya vadisinin iktisadî tarihi hakkında," *İstanbul Üniversitesi İktisat Fakültesi Mecmuası* 17, 1960.

Tuncer, Hadiye. *Yabani bitkiler sözlüğü.* Vol. 1. Ankara, 1978.

Turan, Şerafettin. *Türkiye-İtalya ilişkileri: Selçuklular'dan Bizans'ın sona erişine.* İstanbul, 1990.

———. "Venedik'te Türk ticaret merkezi," *Belleten* 32 (126), 1968.

———. "Osmanlı imparatorluğu ile iki Sicilya Krallığı arasındaki ticaretle ilgili gümrük tarife defterleri," *Belgeler* 4 (7–8), 1967.

Turkowski, Lucian. "Peasant agriculture in the Judaean Hills," *Palestine Exploration Fund Quarterly* 101 (2), 1969.

Tzounis, Alexandre. *Les tabacs turcs.* Naples, 1905.

Üçel-Aybet, Gülgün. *Avrupalı seyyahların gözünden Osmanlı dünyası ve insanları, 1530–1699.* İstanbul, 2003.

Ülker, Necmi. "XVII. Yüzyılın ikinci yarısında İzmir'deki İngiliz tüccarlarına dair ticarî problemlerle ilgili belgeler," *Belgeler* 14 (18), 1989–92.

———. "XVII. ve XVIII. Yüzyıllarda ipek ticaretinde İzmir'in rolü ve önemi," in *Prof. Dr. Bekir Kütükoğlu'na armağan*. İstanbul, 1991.

Uluçay, Çağatay. *XVII. Yüzyılda Manisa'da ziraat, ticaret, ve esnaf teşkilatı*. İstanbul, 1942.

Unger, Richard. *Beer in the Middle Ages and the Renaissance*. Philadelphia, 2004.

Ünsal, Artun. *Nimet geldi ekine: Türkiye'nin ekmeklerinin öyküsü*. İstanbul, 2003.

———. *Ölmez ağacın peşinde: Türkiye'de zeytin ve zeytinyağı*. İstanbul, 1999.

Ünver, Süheyl. *Fatih külliyesi ve zamanı ilim hayatı*. İstanbul, 1949.

Uzunçarşılı, İsmail Hakkı. *Anadolu beylikleri ve Akkoyunlu, Karakoyunlu devletleri*. Ankara, 1988.

Valensi, Lucette. *The birth of the despot: Venice and the Sublime Porte*. Ithaca, 1993.

———. *Tunisian peasants in the eighteenth and nineteenth centuries*. New York, 1985.

Vanacker, Claudette. "Géographie économique de l'Afrique du Nord selon les auteurs arabes, du IXe siècle au milieu du XIIe siècle" *Annales E.S.C.* 28 (3), 1973.

Vassberg, David E. *Land and society in Golden Age Castile*. New York, 1983.

———. "The sale of *tierras baldías* in sixteenth-century Castile," *Journal of Modern History* 47 (4), 1975.

———. *The village and the outside world in Golden Age Castile: Mobility and migration in everyday rural life*. New York, 1966.

Vatin, Nicolas. *Rodos şövalyeleri ve Osmanlılar: Doğu Akdeniz'de savaş, diplomasi ve korsanlık*. İstanbul, 2004.

Vaughan, D. M. "Campaigns in the Dauphiny Alps, 1588–1747," *English Historical Review* 28 (110), 1913.

Vedel, Jacques. "La consommation alimentaire dans le Haut Languedoc aux XVIIe et XVIIIe siècles," *Annales E.S.C.* 30 (2–3), 1975.

Veinstein, Gilles. "Sur la draperie juive de Salonique (XVIème–XVIIème siècles)," *Revue du monde musulman et de la Méditerranée* 66, 1992.

———. "On the *çiftlik* debate," in Çağlar Keyder and Faruk Tabak, eds. *Landholding and commercial agriculture in the Middle East*. Albany, 1991.

———. "Le patrimoine foncier de Panayote Benakis, Kocabasi de Kalamata," *Journal of Turkish Studies* 8, 1984.

Venzke, Margaret L. "Rice cultivation in the plain of Antioch in the 16th century," *Archivum Ottomanicum* 12, 1987–92.

———. "The question of declining cereals' production in the sixteenth century: A sounding on the problem-solving capacity of the Ottoman cadasters," *Journal of Turkish Studies* 8, 1984.

———. "The sixteenth-century Ottoman *sanjaq* of Aleppo: A study of provincial taxation," Ph.D. diss., Columbia University, 1981.

Vergopoulos, Kostas. *Le capitalisme difforme et la nouvelle question agraire. L'exemple de la Grèce moderne*. Paris, 1979.

Verlinden, Charles. "La Crète, débouché et plaque tournante de la traite des esclaves aux XVIe et XVe siècle," *Studi in onore de Amintore Fanfani*, III. Medioevo, 1957.

———. "Italian influence in Iberian colonization," *Hispanic American Historical Review* 33 (2), 1953.

Vernet, Robert. "Les relations céréalières entre le Maghreb et la péninsule ibérique du XIIe au XVe siècle," *Anuario de estudios medievales* 10, 1980.

Viazzo, Pier Paolo. *Upland communities: Environment, population and social structure in the Alps since the sixteenth century.* New York, 1989.

Vilar, Pierre. *A history of gold and money, 1450–1920.* London, 1976.

———. *La Catalogne dans l'Espagne moderne.* 3 vols. Paris, 1962.

Vincent, Bernard. "Consummation alimentaire en Andalousie orientale (les achats de l'hôpital royal de Guadix, 1581–1582)," *Annales E.S.C.* 30 (2–3), 1975.

Vita-Finzi, Claudio. *The Mediterranean valleys: Geological changes in historical times.* London, 1969.

Vogt, J. "Aspects of historical soil erosion in western Europe," in P. Brimblecombe and C. Pfister, eds. *The silent countdown: Essays in European environmental history.* Berlin, 1990.

Volney, C.-F. *Voyage en Syrie et en Egypte pendant les années 1783, 1784 et 1785.* 2 vols. Paris, 1787.

Vries, Jan de. *European urbanization, 1500–1800.* Cambridge, MA, 1984.

———. *The economy of Europe in the age of crisis, 1600–1750.* New York, 1976.

Vries, Jan de, and Ad van der Woude. *The first modern economy: Success, failure, and perseverance of the Dutch economy, 1500–1815.* New York, 1997.

Vryonis, S., Jr. "Nomadization and Islamization in Asia Minor," *Dumbarton Oaks Papers* 29, 1975.

Wagstaff, J. M. *The evolution of Middle Eastern landscapes: An outline to A.D. 1840.* London, 1985.

Wake, C. H. "The volume of European spice imports at the beginning and the end of the fifteenth century," *Journal of European Economic History* 15 (3), 1986.

———. "The changing patterns of Europe's pepper and spice imports, c. 1400–1700," *Journal of European Economic History* 8 (2), 1979.

Walker, P. C. Gordon. "Capitalism and the reformation," *Economic History Review* 8 (1), 1937.

Wallerstein, Immanuel. "Charles V and the nascent world-economy," in Hugh Soly, ed. *Charles V, 1500–1558.* Antwerp, 1999.

———. "The west, capitalism, and the modern world-system," *Review* 15 (4), 1992.

———. "Braudel on capitalism, or everything upside down," *Journal of Modern History* 63 (2), 1991.

———. *The modern world-system.* 3 vols. New York, 1974–88.

Wallerstein, Immanuel, and Faruk Tabak, "The Ottoman empire, the Mediterranean, and the European world-economy," in K. Çiçek, ed. *The Great Ottoman Turkish Civilisation.* Vol. 2. Ankara, 2000.

Walling, D. E., and B. W. Webb. "Patterns of sediment yield," in K. J. Gregory, ed. *Background to palaeohydrology.* New York, 1983.

Warman, Arturo. *Corn & capitalism: How a botanical bastard grew to global dominance.* Chapel Hill, 2002.

Watson, Andrew M. "Arab and European agriculture in the Middle Ages: A case of restriction diffusion," in Del Sweeney, ed. *Agriculture in the Middle Ages: Technology, practice, and representation.* Philadelphia, 1995.

———. *Agricultural innovation in the early Islamic world: The diffusion of crops and farming techniques, 700–1100.* New York, 1983.

Weatherford, Jack. *Genghis Khan and the making of the modern world.* New York, 2004.

Weber, Max. *General economic history.* New Brunswick, 1981.

Wee, Herman van der. "Structural changes in European long-distance trade, and particularly in the re-export trade from south to north, 1350–1750," in James D. Tracy, ed. *The rise of merchant empires: Long-distance trade in the early modern world, 1350–1750*. New York, 1993.

———. "Anvers et les innovations de la technique financière aux XVIe et XVIIe siècles," *Annales E.S.C.* 22 (5), 1967.

Wee, Herman van der, ed. *The rise and fall of urban industries in Italy and in the Low Countries (late Middle Ages-early modern times)*. Leuven, 1998.

Weisser, Michael R. "Rural crisis and rural credit in XVIIth-century Castile," *Journal of European Economic History* 16 (2), 1987.

———. "The agrarian depression in seventeenth-century Spain," *Journal of Economic History* 41 (1), 1982.

———. *The peasants of the Montes: The roots of rural rebellion in Spain*. Chicago, 1972.

Weulersse, Jacques. *Les paysans de Syrie et du Proche Orient*. Paris, 1946.

———. *Le pays des Alaouites*. Paris, 1940.

Whited, Tamara. *Forests and peasant politics in modern France*. New Haven, 2000.

Wickham, Chris J. "European forests in the early middle ages: Landscape and land clearance," in *L'ambiente vegetale nell'alto medioevo*. Vol. 2. Spoleto, 1990.

———. *The mountains and the city: The Tuscan Appennines in the early Middle Ages*. New York, 1988.

Wiet, Gaston. "Les marchands d'épices sous les sultans mamlouks," *Cahiers d'histoire Égyptienne*, 8, 1955.

Wigley, T. M. L., P. D. Jones, and P. M. Kelly. "Scenario for a warm, high-CO_2 world," *Nature* 283 (3), 1980.

Wilson, C. H. "Cloth production and international competition in the seventeenth century," *Economic History Review* 13, 1960.

Wink, André. *Al-Hind: The making of the Indo-Islamic world*. 2 vols. Boston, 2002.

Wittek, Paul. "Ankara bozgunundan Istanbul'un zaptina (1402–1455)," *Belleten* 7 (27), 1943.

———. *The rise of the Ottoman Empire*. London, 1938.

Wolff, Phillipe. "Esquisse d'une histoire de la draperie en Languedoc du XIIe au début du XVIIe siècle," in Marco Spallanzani, ed. *Produzione, commercio e consumo dei panni di lana (nei secoli XII–XVIII)*. Florence, 1976.

———. "Un grand commerce médiéval: les céréales dans le bassin de la Méditerranée Occidentale: remarques et suggestions," in *VI Congreso de Historia de la Corona de Aragón*. Madrid, 1959.

Woude, A. van der. "The future of west European agriculture: An exercise in applied history," *Review* 15 (2), 1992.

Wright, Clifford A. *A Mediterranean feast: The story of the birth of the celebrated cuisines of the Mediterranean from the merchants of Venice to the Barbary Corsairs*. New York, 1999.

Wright, William J. "The nature of early capitalism," in Bob Scribner, ed. *Germany: A new social and economic history, 1450–1630*. New York, 1996.

Wyczánski, Andrzej. "La consommation alimentaire en Pologne au XVIe siècle," in Jean-Jacques Hémardinquer, ed. *Pour une histoire de l'alimentation*. Paris, 1970.

Xoplaki, Eleni, P. Maheras, and J. Lutherbacher. "Variability of climate in Meridional Balkans during the periods 1675–1715 and 1780–1830 and its impact on human life," *Climatic Change* 48, 2001.

Yediyıldız, Bahaeddin. "Hızır Paşa oğlu Mehmed Paşa vakfının mâhiyeti," in *X. Türk Tarih Kongresi.* Vol. 4. Ankara, 1993.

———. *Institution du vaqf au XVIIIe siècle en Turquie.* Ankara, 1985.

———. *Ordu kazası sosyal tarihi.* Ankara, 1985.

Yerasimos, Stefanos. *Sultan sofraları: 15. ve 16. yüzyılda Osmanlı saray mutfağı.* İstanbul, 2002.

Yılmaz, Serap. "XVIII. Yüzyılda Osmanlı imparatorluğu'nda ipekli dokuma endüstrisine genel bir bakış ve Fransa ile ipekli kumaş ticareti," in *X. Türk Tarih Kongresi.* Vol. 4. Ankara, 1993.

Yinanç, Refet, and Mesut Elibüyük. *Maraş tahrir defteri (1563).* 2 vols. Ankara, 1988.

Yolalıcı, M. Emin. *Samsun eşrafından Hazinedâr-zâde es-seyyid Abdullah Paşa'nın terekesi.* Samsun, 1987.

Yücel, Yaşar. *Osmanlı devlet teşkilâtina dair kaynaklar.* Ankara, 1988.

Yücel, Yaşar, ed. *1640 tarihli es'âr defteri.* Ankara, 1982.

Yüksel, Hasan. *Osmanlı sosyal ve ekonomik hayatında vakıfların rolü üzerine bir araştırma (1585–1683).* Sivas, 1998.

———. *Osmanlı taşra hayatına ilişkin olaylar: Berber Bedirî'nin günlüğü (1741–1762).* Sivas, n.d.

Yun Casalilla, Bartolomé. "Spain and the seventeenth-century crisis in Europe," in I. A. A. Thompson and Bartolomé Yun Casalilla, eds. *The Castilian crisis of the seventeenth century.* New York, 1994.

Yusuf, Muhsin D. *Economic survey of Syria during the tenth & eleventh centuries.* Berlin, 1985.

Yver, George. *Le commerce et les marchands dans l'Italie méridionale au XIIIe & au XIVe siècle.* New York, 1903.

Zachariadou, Elizabeth A. *Trade and crusades: Venetian empire and the emirates of Menteshe and Aydin (1300–1415).* Venice, 1983.

Zachariadou, Elizabeth, ed. *The Via Egnetia under Ottoman rule 1380–1699.* Rhtymnon, 1996.

Ze'evi, Dror. *An Ottoman century: The district of Jerusalem in the 1600s.* Albany, 1996.

Zhukovsky, P. *Türkiye'nin ziraî bünyesi.* Ankara, 1951.

Ziadeh, Nicola A. *Urban life in Syria under the early Mamlûks.* Beirut, 1953.

Zulueta, J. de. "Malaria and Mediterranean history," *Parassitologia* 15 (1–2), 1973.

Żygulski, Zdzisław, Jr. *Ottoman art in the service of the empire.* New York, 1992.

Abbas, Safavid shah (1588–1629), 2

Abbasid empire, 38, 162

Accicciuoli, firm, 47

Acre, 66, 68, 92, 111, 151, 198, 218, 222; cotton and sugarcane cultivation in, 96

Adrianople (Edirne), 17, 209, 221, 264, 265

Adriatic Sea, 12, 14, 72–73, 118–19, 142, 156, 177, 285; and Dubrovnik, 180–81; Gulf of, 51; timber, 191; and Venice, 34, 105–6, 181, 191

Aegean Sea and region, 19, 46, 54, 136, 198, 238, 265, 299; as economic center, 62, 67, 180, 184, 198, 225; erosion, 192–95; and Genoa, 44, 50, 100, 106, 152; grain trade, 64, 106, 107; large agricultural estates, 14, 173, 227, 285; northern merchants, 194; oleiculture, 111, 165–67; silk and cotton, 95, 148, 150–51, 152; and Venice, 34, 45, 49–50, 95–96, 106–7, 114–15, 152; wool trade, 58, 136

Africa, 22, 66, 92, 99, 101, 113, 177, 306; Genoese merchants in, 53, 56–57, 101, 118; gold trade, 1, 55, 64, 108, 109, 113, 116; Portuguese merchants in, 66, 101; slave trade, 99, 101, 155. *See also* North Africa

agriculture, arable/cereal: commercial, 11, 18, 25, 63–65, 75–76, 80, 95–96, 132, 173, 184, 190, 203, 216–41, 250, 269, 283; great agrarian cycle, 207, 254, 284, 297, 305; intensification, 205, 210, 246, 248–49, 266, 289; investment in, 85, 218, 219, 232, 246, 272, 350n163; the Little Ice Age and, 169, 173, 190, 196–97, 217, 225, 267; Mediterranean, 13, 24, 169, 199, 243, 265; mixed, 169, 205, 254–55, 265–68, 282; mountain, 161; new land codes, 204, 240, 278; prices, 48, 85–90, 108, 123, 154–57, 166–73, 205; retreat of, 216–30; revitalization of,

282–97; shifting, 233, 281–82; small holders, 26, 63, 90, 171, 219, 247; yields, 14, 108, 119, 121, 248, 258, 281, 304

Aigues-Mortes, 35; *akçe* (asper), 123, 130, 136

Akkerman, 118, 178

Alâ al-Dîn Kaykubâdh, Seljuk sultan, 39

Alaiyye, 222

'Alawis, 262, 293, 295

Albacete, 231

Albania, 233, 264; migration from, 296

Albufera (Valencia), 221, 240, 241

Aleppo, 85, 96, 130, 218, 224, 232, 259, 265; immigration into, 293–94; and its port city (Alexandretta), 221; textile industry, 140, 147, 160, 165, 170, 244; trade, 148, 160, 174, 179, 180, 211; Venetian merchants in, 82, 97

Alèria, plain of (Corsica), 63

Alexandretta, 221; Gulf of, 198

Alexandria, 47, 54, 111, 142, 152, 178

Alexius I, Byzantine emperor, 39

Algarve, 99

Algiers, 241

Alicante, 58, 103, 112

Aljarefe, 164

alkali, 113

Alpajurras, revolt of, 248

Alps: migration from, 292, 296; snowfall, 17; summer flooding, 190

alum, 53, 59

Amalfi, 92; merchants of, 39

America, 162, 183–84, 191, 222, 225, 250; cotton exports, 76, 90, 96, 152, 159, 250, 251; food crops, 15–16, 25, 28, 252, 255–69, 272, 282–83; mines, 3, 65; tobacco, 258, 263, 303

American silver, 8–9, 69–70, 82, 90, 116, 136, 177–78, 242; Genoese control over, 58–59, 85–87, 118, 120, 136, 174–75

Amiens, 76

Ammiana, 195

'Amq/Amik plain (Antioch), 232

Amsterdam, 62, 73, 81, 175, 181, 182, 183; and Baltic trade, 116, 122, 177, 179, 199; center of world-economy, 5, 12, 13, 26, 78, 163, 176–80; and grain trade, 26, 85, 116, 122, 199, 206, 211–13; rise of, 68, 81, 162; silver trade, 3; spice trade, 8

Ancona, 119

Angevins, 47, 48

Anjou, house of, 40, 50, 52, 200

Ankara, 140, 142, 159, 232

l'Annonerie (Marseille), 48

Antalya, 229, 294; Gulf of, 191

Antep, 130, 159, 239

Antioch, 173, 235, 239, 294; plain of, 229; riziculture in, 198

Antwerp, 85, 89, 100, 125, 164, 177; and Augsburg axis, 56–128; as financial center, 3–5, 68, 86, 116, 123, 135, 153; grain trade, 84, 122, 211; as industrial center, 7, 13; spice trade, 8

Apennines, 106, 250

Apulia, 18, 51, 63, 111, 112, 148, 210

Aragón, 51, 52, 166; kingdom of, 33, 40–42, 44, 46–48, 50–53, 109, 200

Ardennes, 13

Arles, 220, 231, 235

Armenia, merchants, 6, 59, 72, 78, 181, 182; kingdom of Cilicia, 54, 94, 198

Arno River and valley, 17, 19, 189

Arras, 76

arta della lana (Venice), 70, 77

arta della seta (Venice), 70, 77

artisans, 39, 77, 134–37, 154, 156

asientos, 4, 5–6, 175, 310n28

assarting, 20, 193, 194, 215, 228, 231, 288

Astrakhan, 78

Atlantic Ocean, 5–8, 24, 49, 74–75, 162, 180, 215, 256; and the Dutch, 174, 178, 213; and the Genoese, 61–68; Mediterranean Atlantic, 5, 50, 66, 86–88, 116–17, 123, 125, 251; oriental crops in, 68, 74–76, 90–96, 99–102, 143, 197, 250, 305; plantations, 258, 300; Seville and Cádiz, 178, 183, 211; trade, 55, 63, 67, 116–17, 159, 201, 245–46. *See also* America

Atlas Mountains, 109, 269

Auge, pays d', 171, 236

Augsburg, 70, 149, 151–52, 155, 156; and Antwerp axis, 56, 123, 128

Austria, 7, 73, 164

Ávila, 157, 334n24

Avlona, 51

Ayas, 94, 289

Aya Solug/Ayasoluk (Ephesus), 192–94, 265

Aydın, 192, 193

Ayyubid empire, 33

'Azm family, 132

Azores, 106, 121

Azov, Sea of, 92

Bafra, 264

Baghdad, 93, 117, 301

Bahlûliyya, al-, 160

Baisan, 232, 285

Baldwin II, king of Jerusalem, 96

Balearic Islands, 44, 65

Balkans, 114, 123, 131, 216, 225, 280, 289, 307; American crops in, 229, 254, 258, 263, 264; cotton cultivation in, 69, 96, 150, 261, 284; economic centrality of, 67, 172, 183; large agricultural estates, 21, 173, 228, 229, 254, 285, 291, 304; manufacturing in, 142, 147, 159; migration from, 293; overland trade, 60, 72, 73, 80, 118, 172; trade and merchants, 154–55, 176, 181, 184, 192; transhumance in, 107, 111, 148, 170, 272

Baltic grain, 90, 116, 126, 197, 206, 229, 234, 290; Dutch merchants, 11, 18, 122, 177, 179, 199, 213, 225; impact on Mediterranean agriculture, 84–85, 120–21, 126, 189, 202

Baltic Sea and region, 4, 6, 8, 9, 10, 64, 161

Bandar Abbas, 179

Barcelona, 38, 52, 92, 122, 146, 176, 222, 250; and Atlantic trade, 223, 246; conquest by Juan II, 53; and kingdom of Sicily, 109, 115; notables of, 40, 315n34; textile industry in, 76, 138, 140, 155, 156, 159; wool trade, 58

Bardi company, 41, 47, 148, 228

Barkan, Ömer Lütfi, 310

barley, 130, 171, 172, 257, 258, 261, 263

Basque region, 52, 191, 263

Bavaria, 12, 51, 73

beans, spread of cultivation of, 252, 256, 257, 259–61, 263, 304 beer, 164

Beirut, 96, 152, 179, 243, 265; and rich trades, 47, 92, 94

Bekaa, 294

Benedictine abbeys, 236

Bergamo, 116, 138

Bergen, 213

Biferno region (Italy), 210

Bilbao, 58, 122

Biscay, Bay of, 52

Bizerte, 166

Black Death, 45, 49, 51, 98, 133, 236–37, 247, 251; and human toll, 20, 43–44, 47–48, 96, 114–15, 222, 274; and putting-out system, 149, 151; reversion of land to nature, 121, 210, 237; and scarcity of labor, 99, 102, 105, 110; and slave labor, 99; and upward migration, 208, 209

Black Sea, 46, 64, 106, 180, 201, 264, 306, 307; Genoese merchants, 41, 43–44, 49, 50, 53, 93, 149, 151; grain trade, 22, 224; large estates, 14, 285, 291, 304; maize cultivation, 258, 263; *pax Mongolica*, 43–44, 50, 53, 58, 92, 93, 94, 95–97; rich trades, 41, 43–44, 149, 151, 178; silk trade, 95, 148; slave trade, 97, 99; timber from, 191, 192; under Ottoman rule, 114, 132, 200; Venetian merchants, 11, 49, 81

Black Sheep Turks, 107

Bohemia, 51, 68, 132, 142, 151, 179

Bologna, 165, 167

Bordeaux, 110, 170

Bourbon monarchy, 9, 183, 184

Bourges, 35

Brabant, 76, 103, 110, 150

Bradley, Humprey, 23, 235

brandy, 164

Brašov, 118, 178

Braudel, Fernand, 8, 9, 16, 61, 185, 310n41

Brazil, 67, 74, 100, 265

bread crops, 11, 14, 52, 80, 106–7, 257, 302, 333n265; lands, 106, 107, 120, 199–200, 207, 216, 247, 248; scarcity of, 75, 86, 218, 222

Brenner Pass, 11, 12, 27, 78, 142, 149, 151, 180

Brenta River, 195

Brescia, 116, 190

brewing, 62

Brie fairs, 43

Britain, 75, 92, 95, 121, 149, 163, 164, 225;

demand for wheat and cotton, 206, 238, 239, 290, 297; *pax Britannica*, 23, 75, 79–80, 199, 211–13, 238, 282, 306; repeal of Corn Laws, 21, 80, 168, 204, 213, 214; wool, 148, 155; and world-economy, 18, 21, 206

Brittany, 52

buckwheat, 172

Buda, 181

budgets, state, 41–42, 43, 86

bullion, 66, 86–90, 113, 116, 123, 126, 144, 182, 212; famine, 33, 104, 113, 124; flows and Genoese bankers, 3, 5, 10, 175

Burgos, 45, 103, 122, 125, 174, 219

Burgundy, 11, 163

Bursa, 166, 178; silk industry, 137, 139, 140, 147, 153, 154; silk trade, 47, 77

butter, 164, 167

Byzantine empire (Byzantium), 38–39, 43, 107, 111, 274, 275; currency, 64; and Genoa, 42, 49, 92–93; transhumance, 208, 237, 272; and Venice, 34–35, 39, 45, 49–50, 82, 95

Cádiz, 17, 58, 74, 178, 211; silver trade, 7, 37, 175, 183

Caffa (Crimea), 41, 52

Cairo, 35, 39, 74, 93, 111, 160, 170, 180; rich trades, 36, 41, 44, 54, 93–94, 177–79, 301; Venetian merchants in, 97, 117–18, 178–79

Calabria, 95, 182, 223

Cambrais, 76

camel caravans, 172, 179, 221

camelot, 59, 135; industry in Ankara, 140, 142

Camera frumenti (Venice), 48, 256

Campidano (Sardinia), 63

Canary Islands, 39, 50, 121

Candia, 50, 111, 166

Cantabria (Spain), 84, 125, 191

Cape Bon, 166

Cape of Violent Storms (Cape of Good Hope), 58

Cape Verde, 50, 121

caravans, 176, 181

Cargill, 228

Caribbean sugar, 57, 74, 87, 121, 252

carpets, 151

Cartagena, 103, 191

Casalmaggiore, 166

Caspian Sea, 95; silk, 95, 114, 148, 159, 244

Castelltersol, 156

Castile, 122, 131, 158, 171, 183, 246, 280; bankruptcy of the throne of, 7, 86, 175; crop shortages, 87, 128; and Genoa, 116; grain trade, 84, 115, 120; kingdom of, 2, 9, 33, 42, 46, 50, 53, 134; and Madrid, 62; New Castile, 126, 165, 166, 217, 245; Old Castile, 277

Catalonia, 108, 125, 127, 166, 184, 218, 242, 245; grain scarcity, 75, 222; grain trade, 115; merchants, 33, 39, 44, 45, 52, 112, 148, 176; textiles, 138, 148, 150, 152

cattle, 72, 226

Caucasia, 58, 118, 263, 293

caviar, 58

Caymans, 23

celâlî rebellions, 131, 277

cereals: area devoted to, 18, 162, 163, 214–15, 227; lesser, 163, 168, 171–72, 255, 258, 260, 268, 272; substitute, 16, 255, 257, 268, 303, 304. *See also* bread crops

Cévennes, 190, 220, 247

Chambéry fair, 4

Champagne fairs, 43, 45, 150

Charles II, king of Naples (1285–1309), 40

Charles II, king of Spain (1661–1700), 6

Charles V, emperor (1516–66), 36, 52, 70, 117, 119, 138

Charles IX, king of France (1824–30), 85

Châtaigneraie, 247

Chaunu, Pierre, 116, 120

chestnut trees, 223, 244, 246, 247, 250, 290; bread from nuts, 247; flour from nuts, 356n29

Chiana, 231

China, 3, 82, 144, 264

Chioggia, war of, 34, 44, 48, 183

Chios (Sakız), 52, 160; Genoese, 95, 100, 112, 152, 194

*çiftlik*s (large-estates), 14, 21, 172–73, 214, 226–27, 233, 240; agricultural hands, 233, 285, 291, 304; and livestock husbandry, 132, 170, 205, 228

Cilicia, 17, 94, 124, 129, 171, 220, 221, 268; Armenian kingdom of, 94; resettlement of, 238, 285, 293, 296; sparseness of settlements in, 198, 205, 218

Cistercians, 163, 236

citrus, 245, 246, 306

Civil War (American), 159, 250

Coeur, Jacques, 35, 52

coffee, 262, 265; trade, 58, 176

Cognac, League of, 52

Coimbra, 99

Colbert, 166

coltura mistra (promiscua), 132, 163, 254, 266, 270

Columbian exchange, 15, 28, 268, 282, 299, 305, 306; dissemination of the crops of, 251, 252, 257, 258–59, 261

commission trade, 177, 179

Comnenian dynasty, 39, 40

Compagnie d'Afrique (France), 224

Constantine, 224

Constantinople (Istanbul), 38, 57, 134, 137, 165, 263, 265, 296; fall of, 43, 52–54, 180; and Florentine merchants, 35; and Genoese merchants, 40, 42, 93, 100; grain prices, 123, 213; northern presence in, 174; and *pax Mongolica*, 43, 93; provisioning of, 114–15, 224–25, 243; trade, 62, 92, 95, 178–79; and Venetian merchants, 93, 97, 99

Constanziaca, 195

contadi, 49, 85, 88, 115–16, 219; relocation of manufacturing into, 13, 138, 146, 154

Continental, 228

Continental Blockade, 168

copper, 94, 144; Tyrolean, 152

Córdoba, 77, 140, 153, 217

Corncro dynasty, 98

Corn Exchange (Amsterdam), 85, 211

Corn Laws (British), 21, 80, 168, 204, 213, 214

Coron, 95, 167

Corsi (Calabria), 223

Corsica, 17, 63, 217, 247

Cortes (Spanish), 126

Çorumlu, 130

corvée labor, 97–98

Cosimo, grand duke of Tuscany, 189

cottage industries, 26, 61, 69, 71, 73, 147, 155, 287; north of the Alps, 70, 77; peasants' second providence, 141–43, 138, 158

cotton, 51, 78, 109, 119, 176, 182, 250, 274; as cash crop, 26, 77, 96; demand for, 21, 57–58, 95, 102, 157, 206, 238, 297; dispersal of textile production, 145–53, 155–56, 159–61; exports from the Levant of, 46, 69, 102, 103, 104, 107, 114, 184; as plantation crop, 26, 94–96, 256, 261, 284–85; relocation of cultivation of, 56, 65–67, 90, 101, 199, 203; textiles, 62, 76, 102–3, 113, 118, 137–39

country trade, 59, 176, 181, 212, 319n138
Cracow, 12, 27, 72, 81, 98, 102, 164, 181
Cremona, 166
Crete, 17, 66, 95, 98, 111, 126, 301, 302; cash
  crops, 26, 95–96, 100, 152; Cretan crisis, 6, 71;
  Cretan war, 86; Venetian, 54, 65, 95
Crimean War, 250
crops: aquatic, 29, 193, 199, 225, 284; failure of,
  87, 126, 127, 131, 224; industrial, 144–73; irrigated, 105, 283; Mediterranean, 105–17; Oriental, 92–105; rotation of, 172, 260, 303, 304
Crosby, Alfred, 261
Cruquius, Nicolas, hydrologist, 234
crusades, 38, 92–94, 95–97, 222, 273, 275
Cuenca, 140, 142, 164
Çukurova, 229, 294
cypress, 299
Cyprus, 98, 99, 100, 148, 217, 221, 224, 229; cotton cultivation, 69, 100, 102, 152, 301–2; Ottoman conquest of, 53; sugarcane cultivation,
  66, 74, 78, 96, 98, 101, 301–2; timber trade,
  289; Venetian, 50, 54, 61, 65, 71, 95

Dalmatia, 93, 152
Damad Ibrahim, grand vizier, 138
Damascus, 124, 132, 167, 176, 218, 231, 280, 289;
  grain cultivation, 127–28, 132; rich trades, 94,
  179; textile industry, 140, 147, 160; tobacco
  cultivation, 265
Danube River, 93, 107, 123, 179, 229, 233
Danzig (Gdansk), 18, 177, 181, 189
deforestation, 17, 20, 27, 204, 274, 288, 289; dur-
  ing the Medieval Optimum, 190–92, 195–97
Denizli (Lazikiyye), 192, 193
Diu, 82
Djerba, 112
Dogano, The (kingdom of Naples), 14, 26, 192,
  230
Domesday Book, 163
Doria family, 5
drainage, 23, 197, 200, 215, 232, 253, 290, 297;
  in the Balkans, 254; channels, 286; companies, 205, 206; in Lower Languedoc, 220; in
  Lower Provence, 231, 235; in Lower Rhône
  region, 236, 237; lowlands, 21, 27, 88, 205,
  206, 220, 238; projects, 284, 297, 306; in
  Ribera Alta, 240; in southeastern Provence,
  236; swamps, 22, 198, 208, 219, 232, 240–41,
  283, 292, 350n169; in Venice, 234

drinks, 164, 268; distilled, 164
drought, 190, 193, 197, 201, 204, 208–9, 245, 302
Dubrovnik (Ragusa), 5, 51, 55, 118, 119, 180
Ducci, Gasparo, 84
Dulgadir principality, 198
Duoro valley, 11
Durance, 190
Durazzo plains, 210
Dutch. *See* Holland

East India Company (British), 6
Ebro River and valley, 17, 19, 166, 195, 237, 238;
  draining of, 241; marshes of, 221
Edirne. *See* Adrianople
Edward I, king of England (1307–27), 110
Egypt, 46, 108, 182, 261, 303; cotton and sugar-
  cane cultivation, 78, 96, 101, 182, 261; as granary, 48, 64, 132, 225; migrants from, 232,
  285
Elbe River, 73; east of, 63, 75, 123, 212, 228, 295;
  manorial estates and agrarian production, 14,
  18, 20, 26, 89, 143, 227
embroidery, 335n42
emigrants/emigration, 264, 290, 293
Emilia, 241, 286
England. *See* Britain
English Channel, 13, 28, 153, 172, 174, 207–8,
  261, 288; Baltic grain, 84; cotton industry, 101,
  139, 146
Ephesus. *See* Aya Solug
Epirus (Yanina), 233, 264
Erciyes, Mount, 229
erosion: cultivation of hillsides and, 287;
  deforestation and, 192, 196; increased wetness
  and, 215; minimize the impact of, 208, 266–
  67; and silting, 27, 190, 269; soil, 20, 190, 201,
  217, 230, 237, 287–89, 302; thinning tree
  cover and, 192, 194, 282; by the waters of
  Brenta, 195–96
Erzurum, 288
Esdraelon, plain of, 229, 294
Esparaguera, 156
Eu-Geneva line, 254
Evliyâ Çelebi, 101, 191, 198, 220, 221

fairs, 6, 12, 27, 72, 81, 142, 179, 181–83; Brie, 43;
  Burgos, 125; Champagne, 43, 45; Lyon, 4, 43,
  79, 175; Medina del Campo, 125. *See also specific fairs*

fallow, 119, 196, 207, 270, 282, 288; elimination of, 223, 239, 260, 266; fallowing effect, 196, 200

Famagusta (Cyprus), 94

famine, 15, 87, 89, 131, 224, 252, 288, 302; of 1590, 256; great, 196, 200, 207–8; and relief crops, 252, 257, 263, 305

fattoria, 14

Febvre, Lucien, 299

feudalism, 97–98

fires, 192

Flanders, 76, 112, 119, 150, 157, 255; Flemish merchants, 65, 67, 83, 163; and Florence manufacturing axis, 13, 68, 72; textiles, 78, 84, 103, 138, 145, 147, 150, 163

flax, 17, 148, 157, 218

fleets, 6, 59, 119, 120, 142, 273; Ayyubid, 33; Byzantine, 316n50; Genoese, 118; Ottoman, 51, 191, 192; Portuguese, 177–78; treasure, 55; Venetian, 34, 60, 144, 191

floods/flooding, 191, 195–96, 204–5, 208–9, 216–18, 236–37, 267, 289; on the Arno, 189–90; on the Ebro, 237; and hydraulics, 234, 236; on the Meander, 193; on the Po, 166; on the Rhône, 190, 196

Florence, 42, 52, 92, 102, 145–46, 237, 244, 314n13; and Ancona, 119; battle of Meloria, 41; companies, 47–49, 106, 147, 148, 181; and the eastern Mediterranean, 35, 47, 145, 244; and French throne, 25, 35; grain trade, 85, 88; inundation of Maremma, 189, 217; men of money and nobility, 35, 45, 82–83, 175, 180, 210, 273–74, 314n16; merchants, 37, 39, 47, 72, 82, 99, 113–14, 182; oleiculture, 165, 167; and Pisa, 116; textiles, 12, 13, 43, 70, 145, 149–50, 154–55, 178; wool imports, 43, 113. *See also* Tuscany

flour, 120, 126, 247, 255, 256, 327n129

fodder, 16, 209, 256, 257, 261, 263, 288

Foggia (Apulia), 63, 168, 198

food crops, 25, 254, 256–58, 261, 263, 265, 300

foothills, 220, 229, 237

forests, 170, 192–93, 215, 222, 273–74, 297, 302; assarting of, 20, 215, 228, 231, 234; Black Sea, 191; Cantabrian, 191; Castilian, 191; central Anatolian, 191; retreat of, 208, 231, 249, 288–90

France, 27, 36, 85, 93, 146, 296; agriculture, 85, 166, 231, 232, 236, 239, 244, 266; concentration of land, 64; Genoese capital in, 8, 12, 27, 184; industries, 13, 76, 156, 159, 250; merchants, 6, 58, 59, 154, 160, 176, 178, 179; northern, 76, 164; peasant revolts, 86; rural settlement patterns, 254, 276, 278, 280; sharecropping, 64, 307; southern, 23, 26, 73, 122, 150, 245; and wars of religion, 9, 27, 231, 235, 242

François I, king of France (1494–1547), 36

Frankfurt, 72, 156, 181

Frankfurt-am-Main, 183

Frederick II, Holy Roman Emperor (1212–50), 39

Fréjus, 220

Friuli, 152

Fugger, house of, 56, 66, 71, 123, 149, 151–52, 153; age of, 69–70, 87, 90, 135–36; bankers to the Spanish throne, 4, 70, 71, 117

fustian: manufacturing, 51, 67, 103, 149–52, 153, 155, 156; relocation by Venetian merchants of, 11, 68–69, 101–2, 146, 147

galere da mercato (Venetian), 118

Galicia, 125

galleons, 192

galleys: Genoese, 152; Ottoman, 192; Portuguese, 2; of Romania, 107

Gallipoli, 221

garden agriculture, 252, 256, 263, 267, 272, 274, 281; advance of, 141, 166, 168; crops, 254, 259, 262, 266, 268, 307; retreat of, 209. *See also* horticulture

Garonne River, 122, 170

Gascony, 11, 52, 111, 164, 263, 301

Gavi, 12

Gediz River, 193

Geneva, 7, 68, 71, 135, 153, 254; silk industry, 26, 70

Genoa, 3–5, 7, 35, 36, 41, 52, 59, 95, 99, 103, 106, 143, 146, 255, 286, 316n61; age of the Genoese, 61–74; banker-financiers, 58, 71, 73, 81, 116–17, 174–75, 246, 309n8; and the Black Sea, 41, 43–44, 50, 53, 92–93, 94; and the Byzantine empire, 42, 92–93, 95; galleys, 152; gold trade, 7, 50, 55, 64, 118; and the house of Habsburg, 25, 81, 116–17, 174, 175; and North Africa, 55–56, 63, 66, 113–14; *Officum victualium*, 48; retreat from commerce, and Sicily, 46, 49–50, 53, 55, 103, 109, 114; and slave trade, 44, 99, 104, 113; and the western Mediterranean, 35, 44–46, 50, 53, 56, 65, 74, 82. *See also* Liguria

Germany: fustian industry, 138, 149; mines and silver, 65, 84, 87, 89, 116; rural industry, 46, 71, 102–4, 151; southern, 11, 13, 26, 69, 78, 151–52, 181. See also *verlagssystem*

Ghaza, 97

Gibraltar, Straits of (Pillars of Hercules), 36, 66, 258, 274

glass making, 57, 105, 145, 151

Gnienzo, 181

Goa, 54

goats, 15, 209, 237, 249, 257, 288, 299

gold, 4, 50, 65, 90, 108–9, 118, 183; and Genoese bankers, 3, 7, 9, 36, 55, 64, 113; mines, 1, 65, 113; from Serbia and Hungary, 64, 113, 181; standard, 79, 199; West African, 112–13, 116; from the West Indies, 84, 86–87, 89, 120, 121–23, 168

grain: demand for, 18, 116, 119; "from the sea," 87, 88; *grano ciciliano*, 47, 106, 115, 117

grain trade, 87, 90, 106, 108–9, 114, 119–20, 213, 224; Baltic, 85, 116, 179, 197, 213; Bardi and Peruzzi, 228; Corn Exchange, 85, 211; Florentine companies, 47, 49, 106, 148; and Livorno, 80, 125; Pontic, 250; Sicilian, 117; and Toulouse, 261

Granada, 53, 95, 107, 113, 114

granum de Romania, 106, 115; Baltic, 84–85, 120–22, 126, 179, 189, 202, 213, 229; deficit in, 125, 128, 224, 255, 260; grain imperialism, 121; Morocco, 107, 109, 115, 121; municipal provisioning of, 48, 62, 224, 242–43; Pontic, 250, 297; porridge, 247; prices, 86–87, 122–23, 126, 157, 169, 219–20, 224; production per capita, 130–31; rotation, 252, 260, 284, 300; Sardinian, 22, 42–43, 48–50, 55, 63–64, 109, 211; scarcity of, 75, 88, 218; spring, 261; Venetian Grains Office, 256

Greece, 26, 263, 264, 280, 296; merchant marine, 59; merchants, 59, 73, 100, 166, 181–82, 212, 274

Grimaldi family, 5

Gruyères, 171

Guadalajara, 138, 158

Guadalquivir River, 63, 190, 206, 217

Guanches, 66, 113

guilds, 70, 157, 325n8, 338n130; bypassing of, 76, 102, 137, 139, 145

Gümüşgerdan family, 158

Habsburg: empire, 2, 8–9, 54, 64, 81, 120, 153, 178; and the Fuggers, 4, 70, 71; and the Genoese *nobili*, 3, 5–6, 12, 55, 59, 116–18, 183; house of, 34–36, 57, 118, 219; and Madrid, 180, 219, 245; revolts in, 242; sale of public lands, 64, 85; territorial expansion, 57, 63; wool trade, 119

Hafsid dynasty, 318n122

Hama, 96, 159, 160, 224, 293, 294

Hamburg, 63, 72, 181

hamlets, 29, 231, 253–54, 269, 277–79, 287, 289; turning into villages, 239, 297

Hanseatic League, 52

Hartz, 110

Havana, 191

hemp, 78, 147, 157

Henry II, king of France (1519–59), 36

Henry III, king of France (1574–89), 85

Henry IV, king of France (1598–1610), 235

Herzegovina, 264

hides, 72, 169, 181, 185, 226

hill settlements, 248, 253

Hoeffts, 23

Holland and Dutch (Netherlands, United Provinces), 6, 8, 154, 163, 175; merchants, 11, 12, 13, 36, 63, 65, 79, 174, 175, 178, 213; mother trade, 11, 18, 26, 213, 225; new draperies, 137, 156; *pax Neerlandica*, 10, 13, 18, 190, 199, 206, 211, 212, 213

Homs, 160, 293, 294

Hormuz, 54, 117, 118, 179

horses, 82, 172, 221, 257

horticulture, 17, 255, 256, 259, 262, 266, 283, 305; increasing significance of, 75, 133, 168, 199, 268, 306

Hüdavendigâr, 130

huerta, 245–46, 279

Hundred Years' War, 33, 196, 200

Hungary, 51, 129, 152, 164; nobility, 179; silver mining, 64

Ibn Battûta, 222

Ibrahim Pasha, of Egypt, 232, 285

Igualada, 156

import substitution, 57, 81, 105, 145, 151, 273

Indian Ocean, 34, 66, 82–83, 93–94, 153, 176,
 304; demand for silver, 3, 144; East India
 Company, 6; impact of the terrestrial central
 Asian route, 44, 92, 96; northern merchants
 in, 72, 146, 174; Ottoman presence in, 3, 53,
 54, 174; Portuguese presence in, 2, 61, 90, 135,
 147, 177–78, 264; and rich trades, 36, 39, 44,
 92, 117, 147, 174; twin cities and the crops of,
 65, 93, 162, 301; and westward migration of
 tropical crops, 162, 251, 256, 261, 265, 272,
 299, 303
irrigation, 48, 96, 234, 236, 251, 263, 301; canals,
 48, 234, 235, 236, 286; irrigated crops, 21, 29,
 100, 199, 225, 285
Istanbul. *See* Constantinople
Italy, 218, 263, 292, 296, 307; reclamation of
 swampy lands, 239, 241; unification of, 81,
 184. *See also specific cities and regions*
İzmir (Smyrna), 26, 148, 172, 183, 194, 225, 265,
 289; and northern merchants, 74, 174, 178,
 192, 211; rise of, 136–37, 180

Jabal Ansariyya, 160, 229, 250; and American
 crops, 262, 265; migration from, 293, 294
Jabal Nablus, 229
Jabbûl, 232
Jaén, 122
Jalal al-Din Pasha, governor-general of Aleppo,
 160
James I, king of Aragon (1213–76), 40
Jazira, 239
Jazzâr, Ahmad al-, governor of Acre (1775–
 1804), 160, 285
Jerez, 110, 122
Jerusalem, 96, 111, 259, 269
Jesuits, 235; and "Jesuit's bark," 22, 306
Jews: artisans, 134, 135, 136–37; merchants, 6,
 59, 78
Jordan, 294; valley, 96
Juan II, king of Aragón (1458–79), 53
Júcar River, 221, 245, 246
Jura, 171, 296

Karaburun, 194
Karaman, 191, 276
Karası principality, 45
Kars, 288, 294
Kastamonu, 159
Kayseri, 130

Kilia, 178
Kilis, 159, 239
Kisrawan, 243, 247, 280, 294
Kızılırmak River, 130
Konya, 131, 229, 231, 294, 346n52; colonization
 of the plain of, 129, 130, 206, 234
Kossovo-Metohija basin, 233
Kresses family, 149
Kuşadası, 195
Kutná Hora (Bohemia), 113

labor, 69, 70, 137, 140, 156, 233, 281, 285–87;
 coerced, 21, 63, 64, 97–98, 173, 205, 213, 285;
 migration from the mountains, 291–96;
 mobility, 89, 140, 233–34, 254; ratio, 170; regulations, 76; scarcity of, 99, 102, 110, 111, 147,
 151, 270; services, 124; slave, 62, 66, 75, 97–
 100, 113
lace making, 335n42
Lajazzo (Ayas), 94, 151, 222
La Mancha, 162
land: church, 85, 98, 106, 168, 170, 232–33, 240,
 247; codes, 211, 214, 240, 296; improvement
 and bonification of, 85, 166, 220, 230, 235,
 240–41, 245, 292; municipal, 168, 234, 240;
 periodic reallocation of, 233, 234; pious foundation, 121, 128, 166–67, 266; public/state,
 64, 85, 169, 211, 219, 227, 267, 291
landlords, 64, 75, 203, 223, 259, 261, 295; in
 Iberia, 126, 127; investment in land, 205, 206,
 210, 220
land reclamation, 167, 190, 195, 208, 231, 235,
 284; during the Medieval Optimum, 91, 120,
 128–29, 190–93, 197, 204, 217, 236; during
 the mid-Victorian boom, 206, 212, 238–39,
 240–41; of marshy land, 170, 241, 253, 273
Languedoc, 23, 78, 196, 208, 215, 223, 226, 229;
 Bas, 166, 217, 220, 297; chestnut zone, 247;
 cloth manufacturing, 76, 137, 138, 140, 148,
 158, 244; drainage, 23, 220; Haut, 263; livestock, 127; retreat of small holder, 219, 247;
 viticulture, 11, 127, 166, 297
La Rochelle, 166
Latakia, 160, 264, 293, 294
latifundia, 121, 205, 227, 281, 304
Lâzikiyye. *See* Denizli
leather trade, 72, 111, 169, 171, 176, 185, 226
legumes: American, 284, 304; widespread
 cultivation of, 18, 172, 252, 260, 263, 266, 268

Leiden, 71, 146

Leipzig fairs, 12, 27, 72, 156, 181, 183

Le Roy Ladurie, Emmanuel, 16

Levant, 52, 58, 71, 85, 112, 113, 148–51, 165, 172, 177, 179, 181, 191, 200, 244, 256, 259, 280, 302; cotton trade, 11, 46, 51, 68–69, 96, 102, 104; and the crusades, 38, 92, 93–94, 95; Genoese merchants, 5, 59; grain trade, 63–64, 86–87; northern merchants, 60; Potosí silver, 9, 37, 58, 90; rich trades, 19, 62, 91; slave trade, 97, 98; sugarcane cultivation, 98, 100–102; Venetian ducat in, 79; Venetian merchants, 5, 7, 50, 55–56, 81–82, 94–95

Levant Company (Britain), 156

Levante, 158, 176

Levant trade, 49, 119, 133, 143, 179, 180, 189, 201; eclipse of, 176, 178, 218, 232; hiatus in, 7, 67, 135; and Jewish merchants, 6, 135; Ottoman empire and, 34–35, 54, 174–75; profits accruing from, 67, 232; and Venetian merchants, 34–35, 66, 93, 103–5, 135, 150, 174–75, 178

Liège, 13, 76

Liguria, 48, 81, 111, 138, 246, 247, 268, 286; merchants, 43, 47, 53. *See also* Genoa

lions, 289

Lions, Gulf of, 47, 106

Lisbon, 7, 8, 115, 118, 122, 330n191; and Baltic grain, 84–85, 90

Little Ice Age, 16–19, 22–23, 199, 200–216, 217, 247, 290; and aquatic and irrigated crops, 78, 268; and big landholders, 173, 227; and climatic variability, 169, 257, 267, 288; close of, 209, 254, 297, 306; and ecological change, 16–18, 20–21, 195–97, 220, 229, 230; onset of, 16, 207–8, 237; resumption of, 27–29, 162, 164, 189–90, 270, 284, 302–3. *See also* Medieval Optimum

livestock: feed, 171, 172, 182, 185, 252, 255, 258, 261; husbandry's reign, 73, 75, 91, 120–21, 141, 167, 168–71, 274; sedentary sheep, 141, 170, 193, 243; small, 15, 24, 90, 132, 161, 168, 209, 305

Livorno (Leghorn), 47, 80, 125, 174, 339n174; and northern merchants, 6, 58, 74, 178, 211

lizards, 15, 209

locusts, 209

Lodi, Peace of (1454), 41, 116, 200

Loire River, 35, 110, 122, 190

Lombardy, 39, 116, 169, 197, 210, 263; merchants from, 35, 43; Milan and, 48, 116; rice cultivation, 105, 173, 286; textile manufacturing, 138, 145, 148, 149; tree crops, 105, 165, 167, 223. *See also* Milan

Lomelli family, 5

London, 71, 78, 146, 148, 206, 297

Lons-le-Saulnier, 4

looms, 137, 140, 152, 158

Louis XIV, king of France (1643–1715), 137

lowlands, 15–24, 29, 80, 90, 123–25, 139–61, 190, 203–10; malaria and, 25, 193, 198–99, 205, 211, 247, 286; reclamation of, 14, 239, 273, 284, 296; retreat from, 216–41; return to, 282–97

Lublin, 181

Lusignan dynasty, 96, 98, 99, 100, 102

luxury goods, 86, 156

Lwów, 72, 118, 178, 181

Lyon, 4–5, 8, 35, 79, 175, 180, 244; fairs, 43, 45, 135, 179; as a manufacturing center, 7, 68, 135, 137, 153, 156, 160

Maccarese, plain of, 241

Macedonia, 123, 164, 209, 213, 218, 281; grain trade, 64, 106; tobacco cultivation, 264–65

Madeira, 1, 11, 50, 106, 111, 114, 121, 302; sugarcane cultivation, 57, 99, 100

Madrid, 54, 67, 219, 243; as the new capital city, 180, 164–65, 245

Magistrato dell'Abbondanza (Florence), 48

maize, 225, 229, 249, 255–69, 284, 359n86; as animal feed, 171, 252, 255, 258; on higher elevations, 252, 254, 272, 304; and pellagra, 305; as substitute crop, 172, 286, 303, 304

Màlaga, 17, 110, 112

malaguetta, 101

malaria, 22, 194–95, 205, 209, 211, 237, 255, 291; and the crusades, 222; eradication of, 22, 238, 306; and periodic tilling of the lowlands, 217–18, 292; and quinine, 22, 235–36; and retreat from the lowlands, 25, 197–99, 216–30, 247, 258, 283, 285, 302–4; and rice cultivation, 240–41, 286; and swamps, 197, 208, 218, 235, 251, 253, 269, 285

Malta, 66, 96, 98, 159, 302

Mamluk empire, 42, 48, 66, 111, 128, 150, 167, 222; fall of, 54–55, 121; fleet, 33, 191; Genoese presence in, 44, 152; and the Mongols, 39, 45, 46, 93–96; rich trades, 35, 44, 54–55, 96–97, 108, 117, 144; and Venetian merchants, 44, 82, 118

Manastır, 136

Manchester, 206, 297

Manila galleons, 2

Manisa, 130, 131, 137, 192, 193, 198

Mansur, al-, emir of Morocco, 100

Manuel I Comnenus, Byzantine emperor, 40

manufacturing, 57, 76–78, 101–2, 134–37, 191;
city-states, 26, 51, 68; dispersal of, 13, 69–73,
134–42, 144–73; renaissance of, 67–71, 77–78

Mardin, 159

Maremma, plain of, 189, 198, 217, 223, 233, 287

Marinid dynasty, 101, 121, 318n122

Maritsa valley, 233

Maronites, 243, 294

Marqab, 160

marranos, 3, 61, 84

Marseille, 11, 35, 48, 53, 81, 224, 243, 244;
Levant trade, 52, 59, 152, 176–77

marshlands, 170, 190, 196, 198, 245, 271, 305,
352n233; and ascent of rural settlements,
216–30, 269, 271; expansion of, 17, 195, 208–
10, 244, 248, 253; improvement of, 22–23,
133, 230–41, 245, 253, 282–97

Massif Central, 220, 271, 296

mawât (waste) lands, 132, 211, 227, 230, 240,
348n120

mazra'a (adjunct settlements), 29, 128, 129, 276,
277, 278, 279, 281

Meander (Büyük Menderes) River and valley,
17, 129, 130, 190, 193, 265

meat: consumption, 91, 125–26, 164, 171, 209,
226, 272; production, 88; trade, 209

Mecca, 54, 225

Medieval (Little) Optimum, 197, 200, 201, 204,
209; expansion of vine growing, 162, 163, 165;
and land reclamation, 20, 197, 210, 230

Medina del Campo (Castile), 4, 122, 125

Meijer, Cornelius, 234

Meloria, battle of, 41

Mesetas, 46, 107, 119, 122, 125, 191, 245

Mesta, The (Castile), 14, 26, 71, 107, 111, 170, 171,
230; dissolution of, 22, 211; thinning of forest
cover and, 191, 192

metallurgy, 57, 272, 273

Michael VIII Palaelogus, Byzantine emperor,
40, 92

migration, 13, 70, 129, 134, 136, 138, 196, 238; to
cities, 168, 243; downstream, 29, 249, 250,
282–97; upward, 28, 247, 269–82

Milan, 3, 35, 70, 176; and Lombardy, 48, 116;
textile industry, 12, 70, 113, 146, 155, 178. *See
also* Lombardy

millet, 171, 172, 255, 257, 260, 361n135; humidity and, 267–68

Ming empire, 82, 144

Minorca, 43

mirror making, 57, 105, 145, 151

Mitidja, 19, 190, 206, 218, 241

Mocenigo, doge, 145

Modon, 95, 167

Moedernegotie (mother trade), 11, 18, 213, 225.
See also Holland

Moldavia, 164, 225

monemvasia vines, 111

Mongol empire, 42, 45, 49, 55, 93–94, 107, 200,
324n47; central Asian terrestrial trade route,
44, 82, 92; demise of, 43, 53, 99, 301; *pax
Mongolica*, 41, 43, 46, 49, 94; rise and expansion of, 39, 44, 92, 196

monoculture, 66, 132, 163, 164, 270, 289, 297,
305

Montenegro, 264, 296

Montluel, 4

Montpellier, 52, 95, 146, 166, 209, 271

Morea, 95, 96, 106, 112, 164

Moriscos, 121, 165, 217, 219, 245–46, 248

Morocco, 107, 109, 115, 121, 218, 263; sugarcane
cultivation in, 99, 100

mosquito, 22, 195, 306; and marshlands, 15, 216,
218, 220–21, 222, 232, 244, 304

Mosul, 159

mountain(s), 80, 142, 160, 167, 196–97, 205,
208–10, 221, 229; ascent of rural settlements, 239, 251, 254, 257, 269–82, 288, 295;
cloth, 159; cultivation, 164, 167, 205, 263,
270, 275, 281; denudation of, 190–91, 193,
201, 250, 282, 289; hamlets, 29, 231, 253–54,
269, 277–79, 287, 289, 297; migration from,
210, 250, 254, 274, 283–87, 291–96, 297;
and pastures, 28, 248, 257, 274, 283; renaissance of, 251, 273, 282; resources, 20, 142, 237,
249, 273, 289–90; villages, 158, 273, 280,
282, 285

Mount Lebanon, 17, 250, 254, 255, 258, 266,
294; maize in, 263, 265; and sericulture, 167,
182, 243–44, 250; timber trade, 290

Moya, 156

mulberry trees, 95, 245–46, 275; destruction of,

290; expanding dominion of, 105, 125, 157, 160, 167, 223–24, 244; and mixed cropping, 266, 267; plantations, 71

mules, 172

Murad II, Ottoman sultan (1421–51), 131

Murcia, 103, 125, 139, 140, 229, 231

Nablus, 129, 229, 263

Nantes, 235

Naples, kingdom of, 22, 53, 111, 122, 165, 182, 211, 242; city of, 115, 165; and Florence, 35, 47–48; Genoese merchants, 232; herding and the Dogana, 14, 171, 192, 272

Napoleonic era, 9, 18, 22, 160, 204, 240

Napoleonic Wars, 21, 58

Narbonne, 52, 220

Negroponte, 50, 95, 96, 98, 106, 108, 115

New Spain, 84, 87, 122, 165

Nice, 286

nobili vecchi (Genoa), 6, 71, 81, 116, 117, 174, 175

nomads, 11, 194, 198, 226, 240, 274, 275, 277; periodic tilling, 121, 218–19, 236, 291; riziculture, 172–73; settlement of, 120, 124, 128–29, 171, 192–93, 243, 295–96; transhumance, 124, 136; and wool, 136, 158

Normandy, 235, 236

North Africa (Maghreb), 53, 55, 111–12, 120, 138, 146, 152, 236; and Genoese merchants, 50, 56–57, 66, 107, 113; and gold trade, 108–9; grain trade, 48, 63; oleiculture, 26; oriental crops, 114, 148, 191–92, 221, 301; slave trade, 99; thinning of wood cover, 191–92, 231; vineyards, 165

North Sea, 3–4, 8, 24, 27, 85, 139, 162, 200; connection with the Mediterranean, 44, 52; destination of rich trades, 58, 85, 178; recentering of world-economic flows in, 27, 62, 72, 83, 173–74, 177–78, 180, 213

Nuremberg, 12, 70, 72, 149, 155, 156, 181, 183

oats, 16, 171, 172

Odessa, 213

Officum victualium (Genoa), 48

Öküz Mehmed Pasha, grand vizier, 194

Olesa, 156

Olivares, count-duke, 7, 175

olive oil, 112, 144, 162, 165, 167, 184, 339n172; oleiculture, 76, 86

olive trees, 111, 162, 164, 166–67, 224, 253, 258,

266, 275, 292, 299, 304n184, 341n199; and the Mediterranean triad, 91

Oran, 2, 53

orchards, 91, 162, 209, 246, 263, 266, 267, 281; conversion of the arable into, 88–89; expansion of, 26, 122, 125, 133, 141, 165–68, 223, 249

Oristano, Gulf of, 288

Orléans, 190

Osman, house of, 40, 42, 45, 54, 117, 121, 200; conquest of Constantinople, 52; and the house of Habsburg, 36; and the Serenissima, 6; Timur's advance, 33, 193

Ottoman empire: caravan trade, 176, 181; and Genoa, 44, 49, 53; grain trade, 87, 90, 114–15, 123, 131, 197, 224, 239; industry, 13, 134–46; land codes, 22, 129, 211, 214, 240, 267, 291, 296; land reclamation, 167, 230, 231; naval presence in Red Sea, 3, 54, 191; navy, 191; northern merchants, 174, 176, 178, 194; overland trade, 72, 118, 172, 178–83; and Persian Wars, 180; prebendal system, 35, 64, 124, 227; rice cultivation, 172, 226, 286; rich trades, 3, 5, 34–35, 54–55, 60; rural settlement pattern, 276–80; sheep tax, 171; shipping, 51; tax-farming, 64, 219, 227, 238, 280; tribal confederations in, 14, 26, 71, 272, 276, 285; and Venice, 6, 8, 49, 54, 60, 114–15, 118; wheat boom, 128

oxen, 226, 245, 266

Padua, 138

Palencia, 158

Paleologue family, 95, 200

Palestine, 17, 220, 261, 271, 279, 295; cotton cultivation in, 96, 285

Pamphylia, 129, 205, 218

panic, 268

paper making, 105, 145, 151

paprika, 262

Paris, 8, 35, 110, 175, 180, 296

pastures, 126, 170–71, 192, 245, 269, 283; conflict with bread lands, 120; conversion into arable, 22, 119, 238, 240, 272, 276; conversion of arable into, 88, 91, 127, 148, 169, 171, 207, 215; decline in the surface devoted to, 231; held in common, 233–34, 288; *kışlak*, 276; in lowlands, 229, 234; pastureland, 18, 124, 126, 215, 240, 301; summer, 91, 194, 237, 248, 274–75, pastures (*continued*)

283; *terres gastes*, 234; winter, 28, 124, 229, 248, 276, 283, 286, 296

Patras, 166, 182

Payas, 222

Paz family, 175

peasant/peasantry, 15, 16, 64, 78, 86, 87, 91, 110, 111, 123, 127, 139, 146, 154–56, 160, 165, 166, 208–10, 219, 221, 223–25, 228, 238, 239, 246, 253, 256, 261, 267, 290; and cash crops, 96, 125, 173, 203, 210, 240, 285, 297; crop diversification, 19, 87, 89, 126, 251; diet, 167, 170, 172, 252, 257, 262, 300, 305; dispossession of, 64, 133, 169, 211, 227, 280, 291; holdings, 132, 169, 173, 219, 223, 227, 266, 280; landless, 130, 249; loans and indebtedness, 43, 167, 219; migration, 193, 196, 210, 243, 247–52, 271–72, 279–80, 283–87; mobility of, 89, 140, 233, 254, 279–81, 283, 292, 294; revolts, 86; serfdom, 307; sharecropping, 63, 142, 213–14, 258, 268, 281, 285–87, 294–95

Pegolotti, F. Balducci, 49

pellagra, 258, 305

Peloponnese, 98, 111

Perez, Luis, 84

Perigord, 166

Perpignan, 122

Persian Gulf, 54, 92, 117, 191, 264

Peru, 22, 235, 265

Peruzzi, house of, 41, 47, 148, 228

Pescia, 162

Peter de Lusignan of Cyprus, 98

Petit-Poitou, 231

Philip II, king of Spain (1566–98), 2, 81, 85, 174, 191, 202; and Genoese bankers, 3–4, 71, 73, 117

Philip III, king of Spain (1598–1621), 175

Philip IV, king of Spain (1621–65), 5, 7

Philip le Bel, king of France (1268–1314), 39, 93

Phocaea (Foça), 53, 112, 192

phylloxera, 249, 90

Piacenza (Besançon) fairs, 4, 79, 175

Piedmont, 138, 286

pilgrimage, 176

Pinelli family, 5

piracy, 34, 181, 247, 251

Pisa, 39, 40, 41, 42, 92, 116, 150

Pistoia, 208, 236, 273

plantation: agriculture, 55, 56, 76, 245, 258,

284, 302; cinchona, 22; complex, 96–97, 300, 305; mulberry, 71; olive, 122, 302; oriental crops, 11, 14, 67, 74, 77, 94, 102, 152; tobacco, 303; tree crop, 185, 245

Poland, 81, 179, 181, 213, 228, 263; as breadbasket, 18, 51, 63, 73, 202, 213–14

polenta, 255, 256, 305

polyculture, 23, 205, 254, 263, 270, 281, 296, 297; as a strategy against erosion, 266, 268–69

Pontebba, 12

Pontine marshes, 211, 226, 241

Portugal, 50, 63, 109, 115, 121, 149, 264; in Africa, 66, 101; in the Atlantic, 65, 67–68, 83; bankers, 3, 36, 61, 175; in the Indian Ocean, 53, 117, 135, 147, 178, 179; merchants, 57, 85, 99, 117, 147; slave trade, 99

potatoes, 172, 255, 258, 260–62, 272, 284, 304; diffusion, 262–63; as a lesser cereal, 266, 303

Potosí silver, 4, 7, 9, 65, 87, 89, 168, 178

Poznan, 181

Provence, 46, 53, 111, 119, 120, 270, 288, 292; cereal culture, 108, 223, 243; investment in land, 169, 235, 206; marshes, 23, 221, 231, 234–35; rural settlement pattern, 276, 279; textile industry, 95, 148; viticulture, 127, 166, 182, 198, 288

Pruili, doge, 145

Puglia, 19, 63, 75, 119, 167, 223; exports of grain, 47, 48, 63, 106, 115; herding, 192, 205, 222; and Venetian merchants, 48

putting-out system, 60, 79, 137, 139, 158, 185, 285, 355n56; Venice and, 11, 13, 67, 70

Pyrenees, 14, 17, 245, 292, 296

quinine, 22, 218, 236, 306

Ragusa (Dubrovnik), 5, 118, 119, 180, 181; merchants, 51, 55

rainfall, 201, 208

raisins, 26, 166, 249

Reconquista (The Reconquest), 49, 51, 89, 103

Red Sea, 3, 36, 54, 92, 94, 191

reeds, 15, 195, 220, 234

Reformation, 164

Reims, 76

Rhineland, 51, 76, 110, 164

Rhine River, 164

Rhodes, 96, 98

Rhône River, 8, 17, 215, 223, 236, 247; and
    floods, 190, 196, 234, 237
Rialto (Venice), 34, 37
ribbon, 244
rice, 205, 257, 262, 268, 274, 284, 303; and
    marshes, 205; paddy fields, 105, 193, 217, 220,
    245, 246, 279, 286; riziculture, 172–73, 198,
    225–26, 229, 245, 285–87; as staple, 29, 286
rich trades, 5, 26, 49, 108–9, 173, 176–78, 203,
    303–4; central Asian nexus and the Black
    Sea, 58, 93–95, 147, 149; and Genoese mer-
    chants, 41, 44, 300; and the Mediterranean,
    52, 56, 91, 103, 133, 151, 183–84, 203; and
    northern merchants, 62, 78, 84, 90, 116; and
    Venetian merchants, 35, 48, 58, 144, 175, 300
rivers, 17, 72, 193, 201, 216, 269; beds, 17; chan-
    nels, 190, 195, 312n97; deltas, 190, 194, 221,
    241; valleys and banks, 123, 127, 193, 194, 209,
    237, 285. *See also specific rivers*
Roanne, 190
Robert the Wise, king of Naples, 40
Roman Campagna, 17
Romania, 93, 113, 125; galleys of, 107; *granum de
    Romania*, 18, 46, 48, 106, 114–15; Venetian
    and Genoese, 49–50, 63, 95, 109, 123
Rome, 167, 235
Rouen, 159
Roussillon, plain of, 19, 297
royal finance, 2, 40, 175, 219; Florentine bankers,
    175; the Fuggers, 117; Genoese bankers, 67,
    71, 74, 117, 174
Rumelia, 124, 182, 281, 287, 291, 293
Russia, southern plains, 73, 107, 213, 228
rye, 172, 225, 228, 257, 261, 267, 268, 303

Sabadell, 156
Safavid empire, 6, 9, 59, 82
Safed, 134, 157
Sahara Desert, 61, 92, 99, 109
Sahel (Tunisia), 164
sails, manufacturing of, 153
Saint Domingue. *See* Santa Domingo
Salonika, 19, 38, 53, 114–15, 117, 225, 229, 265;
    marshes, 241, 291; woolen industry, 73, 77,
    134–37, 138, 153, 154, 157–58
salt, 5, 49, 55, 94, 195, 271, 374n50
Samsun, 164
Sandomir, 181
San Pablo, 235

Santa Domingo (Haiti), 74–75
São Tomé, 50, 65, 99, 100, 101
Sardinia, 17, 40, 46, 55, 115, 159, 217, 241; as
    breadbasket of the Mediterranean, 18, 22, 48,
    63, 211; draining of Tirso, 288; Genoese mer-
    chants in, 42–44, 49–50, 55, 64, 109
Savoie, 171
Savona, 152
Scandinavia, 164
Scheldt River, 84, 89, 122
Schetz, Erasmus, 84
Schwerin, 171
Segovia, 73, 137, 153, 154–55, 158
Şehzade province, 137. *See also* Manisa
Seine River, 122, 164
Serbia, 64, 113, 181, 281
Serez (Serres), 136
serfdom, 63, 64, 98, 222, 307; second, 163, 213,
    227, 229, 295
sericulture, 68, 142, 160, 161, 167, 244–45
Serravalle, 12
Seville, 3–4, 7, 9–10, 35, 66, 74, 115, 164; and
    American treasure, 9, 87, 116, 175, 177, 183;
    and Atlantic trade, 54, 55, 59, 118, 122; cloth
    production, 140, 245; viniculture, 111
sharecropping, 14, 63, 142, 213, 223, 258, 268,
    307; and large estates, 281; and opening of
    new lands, 214, 233, 285–87, 294–95
sheep: epizootic diseases, 127; herding, 90–91,
    103, 109, 148, 170, 221, 257, 287; itinerant,
    248–49, 274; meat consumption, 209; Medi-
    terranean trinity, 15, 169, 185; and plow, 119,
    163, 227; sedentary, 141, 171; and tree cover,
    237, 288
sherry, 164
shipbuilding, 20, 62, 191, 192, 204, 273
Shuf mountains, 243, 247
Sicilies, Kingdom of Two, 48, 106, 200
Sicily, 34, 88, 125, 173, 200, 211, 223, 242; bread-
    basket of the Mediterranean, 22, 46, 63, 106,
    211; cotton cultivation, 148, 159; and Genoese
    merchants, 42–43, 44–45, 47, 48–51, 55, 64;
    grain trade, 46–49, 64, 85, 103, 108–9, 117,
    211, 224; and kingdom of Aragón, 40–41, 46,
    48, 53; silk, 77, 95, 113–14, 117, 148; sugar, 66,
    78, 99
Sidon, 96, 243
Siena, 198
Sierra de Espadán, 248–50

Sierra Espuña, 17

Sijilmasa, 116

Silesia, 18, 51, 63, 213

silk, 11, 65, 114, 118–19, 125, 156, 159, 162, 167, 176, 182, 250, 264; Aegean, 95; *arte della seta*, 70, 77; black taffeta, 244; Byzantine, 39; Calabrian, 95, 182; Caspian, 114, 146, 148, 159, 244; from Chios, 160; demand for, 156, 157, 246; fine wool replacing, 156; Genevan, 7, 26, 70; Granadian, 95, 113, 114; industry, 26, 47, 70, 95, 105, 113–14, 138–40, 146–49; Lombardian, 39, 105, 138, 233; Lyon, 79, 137, 156, 160, 244; Mount Lebanon, 167, 182, 243–44, 250; muscardine, 290; pebrine, 290; prices, 154, 156; ribbon, 244; Romanian, 95; Sicilian, 77, 95, 114, 117; taxes paid in, 167; textiles, 26, 39, 62, 77, 79, 105, 113, 136–40, 146–49, 152, 154, 156–57, 160–61, 175, 244, 246; twisting, 137; Valencia, 140, 160, 245–46. *See also* sericulture

silk trade, 76, 77, 95, 102–3, 113, 136, 146, 211; Bursa, 47, 77, 139; Genoese merchants and, 102–3, 184–85; Mount Lebanon, 243; Murcia, 139; and the Safavid empire, 2, 6, 9, 59; Sicily, 53, 95; Venetian merchants, 102–3

silver, 54, 57, 58–59, 79, 84, 116; American, 3–4, 8, 58, 85, 90, 120, 174, 175, 177, 242; German, 65, 84, 86, 87, 89, 116, 123; mines, 1, 64–65, 113

Simat de Valldigna, 246

slaves, 44, 66, 75, 97–101, 113, 119; trade of, 44, 97, 99, 104, 113, 116, 119, 325n67

Smyrna. *See* İzmir

snakes, 15, 209

snow trade, 17

soap, 165, 167; making, 26, 105, 111, 165; trade, 50

soil erosion, 27, 190, 192, 201, 217, 289

sorghum, 261

South China Sea, 92

Spain: agriculture, 168, 211, 231, 234, 240–41, 279, 284, 287; American crops, 259, 263; church lands, 168, 170, 211; crop shortages, 127; and the golden fleece, 107, 111, 119, 170–71, 191, 192, 211, 230; industry, 13, 156; land reclamation, 239, 240–41; and the Mesta, 22, 170, 171, 211; migration, 296; oleiculture, 165; rice cultivation, 205, 284, 286, 287; sale of public lands, 85; sericulture, 167; settlement pattern, 279; topography, 270; viticulture, 121, 126, 165, 287

Spanish contracts, *asientos*, 4, 5, 6, 7, 175

Spinola family, 5

Spoleto, 181

spring crops, 172, 260

squash, 259, 261, 262

St. Bernard Pass, 45

St. Chamon, 137

St. Étienne, 137

storks, 15

Styria, 68, 142

sugar, 50, 57, 94, 95, 99–101, 113–14, 122, 192

sugarcane, 109, 119, 192, 203, 218, 245, 274, 326n88; Atlantic-bound movement, 90, 95–102, 114, 302; egress from the Mediterranean, 78, 91, 102; slave labor, 113

Süleyman the Magnificent, Ottoman sultan (1520–66), 36, 81

Sursuq family, 240, 295

Swabia, 51, 67, 101, 146, 149, 151

swamps, 15, 27, 210, 232, 235, 271, 285; claims to, 240, 271, 283–84; and lowlands, 18, 21, 197, 205, 349n153; and malaria, 22, 218, 253; march of, 15, 22–23, 27, 194, 210–11, 219–20; reclamation of, 20, 22, 88, 170, 290

Sweden, 75, 225

Switzerland, 152, 164

Syria, 46, 111, 131, 148, 240, 262; colonization of new land in, 124–25, 129, 232, 239, 285, 293, 296; cotton cultivation, 51, 96, 150, 182, 232; rich trades, 35, 179; tobacco cultivation, 264–65

taffeta, 244

Takrūr, al-, 1, 64

Tana, 93, 97

Tarrasa, 156

Taurus Mountains, 17, 160, 239, 290, 293

Tavoliere of Fuggia (Italy), 19, 198, 223

taxation, 106, 123, 219, 259

tax-farming, 14, 219, 238, 280; lifetime, 64, 227

temperature changes, 82, 204, 208, 274, 295

Tenerife, 11

Terraferma (Venetian), 116, 118, 234, 256; cereal agriculture, 87, 115, 169, 219; manufacturing, 145, 154

Teutanic Knights (Tripoli), 96

textiles/cloth: cotton, 58, 76, 101–2, 113, 138,

145–51, 159; hemp, 78, 147, 157; industry, 70, 82, 105, 114, 137, 140, 144–61, 164; linen, 78, 79, 146, 147, 149, 153; woolen, 67–71, 76–77, 88, 113, 119, 134–40, 145–50, 153–59. *See also* fustian

Thessaly, 123–24, 129, 138, 140, 197, 238, 265; arable agriculture in, 64, 106, 114, 233

Thirty Years' War, 69, 155, 181

Thorun, 181

Thrace, 64, 123, 192, 233, 281

timar (prebendal) system, 35, 131, 227

timber, 268, 273, 290; demand for, 190, 204, 209, 237, 250, 273, 289, 344n13; scarcity of, 58; for shipbuilding, 191, 192; trade of, 289

Timurlane, 33, 58, 192

Tirso (Sardinia), 241

tobacco, 167, 198, 229, 252, 262–65, 272, 293; as cash crop, 250, 252, 258–59, 262, 284, 303

Tokat, 140, 159

Tokay region, 110

Toledo, 125, 140, 153, 158, 219, 245

tomato, 261, 262

Torcello, 195, 196

Toulouse, 261

transhumance, 91, 107, 124, 129, 148, 198, 218, 248; rise of long-distance, 168, 196, 237, 271–72, 274, 302

Transjordan, 239

tree crops, 162, 249, 253, 266, 272, 279; reemergence of, 161–73

tribal populations: in Ottoman lands, 14, 26, 71, 232, 285, 292; settlement of, 276, 295

Trieste, 58

Tripoli (Libya), 2, 116

Tripoli (Syria), 96, 124, 167, 179, 222, 325n57

Tunis, 34, 99, 111, 116, 165, 170

Tunisia, 164, 167, 182, 224

Turia River, 221

Tuscany, 37, 162, 190, 208, 209, 210, 237, 273; and Florence, 48, 116; flooding, 201, 217, 236; grain trade, 85, 189; merchants from, 35, 43; planting of trees, 166, 245; textile production, 148, 152. *See also* Florence

Twelve Years' Truce, 9

Tyre, 96

Tyrol, 149, 152

Ulm, 70, 151, 152

'Umar, Zâhir al-, governor of Acre (1730–71), 285

'Umarī, al-, 10

United Provinces. *See* Holland

Urfa, 159, 224, 239

Üsküp, 136

Valencia, 52, 108, 173, 208, 219, 231, 247, 248; Genoese merchants in, 103; reclamation of swamps, 240, 241; rural settlements, 198, 221–22; sericulture, 125, 245–46; textile industry, 140, 159, 160, 246

Valladoid, 85, 131, 180, 219

vegetables, 249, 252, 256, 262

Veneto (Venetia), 7, 88, 191, 195, 226, 286; and agricultural production, 252, 255, 256, 258; and textile industry, 26, 67, 68, 138; and Venice, 48. *See also* Venice

Venice: and Black Sea, 81, 93, 96–97, 99, 106, 114, 151; and Byzantine empire, 34–35, 39, 43, 45, 50, 82; *Camera frumenti*, 48; galleys, 107; and the Mamluk empire, 35, 82, 118; merchants, 39–41, 45–47, 51, 151–55, 174, 177, 178, 181–84; and Puglia, 48; rivalry with Genoa, 27, 40–41, 46, 55, 71, 81, 91, 112; and the Sublime Porte, 25, 54, 60, 81, 116–18; Terraferma, 87, 115, 118, 145, 154, 169, 219, 254; and Veneto, 48. *See also* Veneto

Ventoux, 17

verlagssystem, 70, 76

Verona, 68, 116, 138, 267

Via Egnatia, 73, 78, 182

Vienna, 63, 181

villages, 107, 124, 128, 170, 230, 240, 253; abandonment of, 75, 124, 196; adjunct, 29, 129, 219, 270, 275–79, 296; ascent of, 269–82; descent of, 15, 29, 282–308; double, 199, 253, 279–80; manufacturing in, 138, 140, 155–56, 159; mountain, 158, 273, 280, 282, 285; (re)establishment of, 96, 121, 124–25, 129–31, 200

vinegar, 338

vines, 111, 127, 162, 209, 253, 255, 275; growing popularity of, 90, 109, 110, 122, 125–27, 162, 244–46; and mixed cropping, 205, 251, 254–55, 266; monoculture, 297; and the triad of the Mediterranean, 14–15, 91, 161–68, 244–45, 283

vineyards, 90–91, 125, 193, 246, 249, 279, 281, 299; conversion of arable to, 89, 126–27, 133, 287; mixed cropping, 254, 266–67; and vineyards (*continued*)

    phylloxera, 290; retreat of, 110–11, 162, 209; spread of, 26, 84, 122, 141, 161–68, 198, 223

Vinland, 162

Vistula plains, 14

viticulture, 76, 110, 126–27, 131, 162–66, 279, 288, 301

Vivaldi brothers, 66

Vivarais, 137

Voltaggio, 12

Wallachia, 124

*waqf*s (mortmain), 160, 167, 223, 227, 236, 266, 294; and colonization of new lands, 121, 124, 131, 291, 307

wetlands, 15, 195, 196, 203, 217, 271

wheat, 65, 105–6, 108–9, 168, 261, 267, 284, 303; contraction of cultivation, 210–15, 216–41; demand for wheat after 1750, 21, 206–7, 238–39, 250; end of imperialism of, 15–16, 18–19, 257–58; imperialism of, 14, 28, 117–33; Mediterranean triad, 90–91, 163, 168–69, 171–72, 260; mixed farming, 254, 255, 267–68; trade of, 50, 55, 63, 207–8, 224–25; winter, 216, 248

White Sheep Turks, 107

wine, 120, 122, 162, 167, 246

wine trade, 11, 47, 108, 164; *monemvasia*, 111

wolves, 234, 289

woodlands, 110, 195, 207, 231, 233, 252, 269, 275; clearing of, 133, 166, 190–92, 201, 210, 288; growth of, 207, 249; resources of, 271, 273

Ximenez family, 85

Yeşilırmak River, 130

yields, grain, 87, 121, 258

Yozgad, 276

Zara, 95

Zurich, 152